# The Vasculitides

# The Vasculitides

## Science and practice

Edited by

## B.M. Ansell
*Consultant Rheumatologist and Former Head of Division of Rheumatology, Clinical Research Centre, Northwick Park Hospital, London, UK*

## P.A. Bacon
*Professor of Rheumatology and Head of the Rheumatology Department, University of Birmingham, Birmingham, UK*

## J.T. Lie
*Professor and Director, Division of Anatomic Pathology, University of California Davis School of Medicine and University of California Davis Medical Center, Sacramento, California, USA*

## and

## H. Yazıcı
*Professor and Chief, Division of Rheumatology, Department of Medicine, Cerrahpasa Medical Faculty, University of Istanbul, Turkey*

 Springer-Science+Business Media, B.V.

First edition 1996

© 1996 Springer Science+Business Media Dordrecht
Originally published by Chapman & Hall in 1996.
Softcover reprint of the hardcover 1st edition 1996

Typeset in 10/12 Palatino by J&L Composition Ltd, Filey, North Yorkshire

ISBN 978-0-412-64140-4      ISBN 978-1-4899-2889-4 (eBook)
DOI 10.1007/978-1-4899-2889-4

A catalogue record for this book is available from the British Library

Library of Congress Catalog Card Number: 95-78848
♾ Printed on acid free text paper, manufactured in accordance with ANSI/NISO Z39.48–1992 (Permanence of Paper).

# Contents

*Color plates appear between pages 232 and 233*

List of contributors     viii

Foreword     xxi
    *E.G.L. Bywaters*

Preface     xxiii

Part One
Classification and Histology     1

1 **Vasculitis – then and now**     3
    *P.A. Bacon*
2 **Classification and histopathologic specificity of systemic
    vasculitis**     21
    *J.T. Lie*

Part Two
Disease Mechanisms and Investigations     37

3 **Vasculitis: mechanisms of injury**     39
    *F.C. Breedveld and M.R. Daha*
4 **Antineutrophil cytoplasmic antibodies (ANCA): current
    perspectives**     48
    *C.G.M. Kallenberg*
5 **Antiendothelial cell antibodies, antiphospholipid antibodies
    and vascular disease**     65
    *D. D'Cruz and G. Hughes*

6 Adhesion molecules in disease: vascular cell adhesion
molecule-1 as a target for therapy    83
*J.M. Clements, R. Pigott, L.A. Needham and A.J.H. Gearing*

7 MRI and CT in vasculitis    99
*M. Heller, C. Muhle, M. Reuter and F. Schubert*

8 Angiography in vasculitis: its role in diagnosis and in the
management of complications    109
*J.E. Jackson*

Part Three
Primary and Systemic Vasculitides    119

9 Polyarteritis nodosa: clinical aspects    121
*L. Guillevin, F. Lhote and P. Casassus*

10 Microscopic polyarteritis (MPA)    135
*P.A. Bacon, C. Savage and D. Adu*

11 New concepts in Wegener's granulomatosis    145
*W.L. Gross*

12 Giant cell arteritis    171
*B.-Å. Bengtsson*

13 Takayasu arteritis: a current update    181
*J.T. Lie*

14 Behçet's syndrome with emphasis on clinical genetics and
mortality    199
*H. Yazıcı, B. Çiçek, G. Başaran et al.*

15 Endothelial cell dysfunction in Behçet's disease    207
*E. Kansu*

16 Pulmonary vasculitides    222
*L. Mouthon, F. Lhote and L. Guillevin*

17 Angiitis of the central nervous system    246
*J.T. Lie*

Part Four
Connective Tissue and Dermal Vasculitides    265

18 Rheumatoid vasculitis    267
*P.A. Bacon, R.A. Luqmani and D.G.I. Scott*

19 Vasculitis in primary Sjögren's syndrome    277
*A.G. Tzioufas, F.N. Skopouli, D. Boumba and H.M. Moutsopoulos*

20 Mixed cryoglobulinemia: the role of HCV and
organ-specific antibodies    283
*S. Bombardieri, A. Tavoni, M. Mosca et al.*

21 Cutaneous vasculitis    294
*J.J. Cream*

Part Five
Management                                                                 307

22 **Treatment of severe vasculitis: general principles and the
   use of corticosteroids**                                             309
   *J.S. Cameron*
23 **Therapy of vasculitis**                                            315
   *R.A. Luqmani, A. Pall, D. Adu et al.*
24 **Treatment of polyarteritis nodosa and Churg–Strauss
   syndrome**                                                           332
   *L. Guillevin and F. Lhote*
25 **Relapses of Wegener's granulomatosis: role of
   micro-organisms and prophylaxis with trimethoprim/
   sulfamethoxazole**                                                   356
   *C.G.M. Kallenberg*
26 **Alternative treatments for systemic vasculitis**                   362
   *C.M. Lockwood*

Part Six
Vasculitis in Childhood                                                   375

27 **Juvenile dermatomyositis**                                         377
   *B.M. Ansell*
28 **Kawasaki syndrome**                                                384
   *M.J. Dillon*
29 **Kawasaki syndrome: the Florentine experience**                     392
   *F. Falcini, M. Ermini, S. Trapani et al.*
30 **Vasculitis and its relation to the streptococcus**                 397
   *J. David*
31 **Schönlein–Henoch purpura nephritis**                               403
   *P. Niaudet*
32 **Vasculitis in familial Mediterranean fever**                       412
   *M. Pras, P. Langevitz, A. Livneh and D. Zemer*
33 **Behçet's syndrome in children**                                    417
   *H. Özdoğan*

**Index**                                                                 425

# Contributors

D. Adu
Birmingham Vasculitis Group
Departments of Rheumatology and Nephrology
Queen Elizabeth Medical Centre
University of Birmingham
Edgbaston
Birmingham
UK

B.M. Ansell
Stoke Poges
Berkshire
UK

P.A. Bacon
Department of Rheumatology
University of Birmingham
Edgbaston
Birmingham B15 2TT
UK

G. Bartolozzi
Department of Pediatrics
Rheumatology Unit
University of Florence
Florence
Italy

G. Başaran
Cerrahpasa Medical Faculty
Division of Rheumatology
University of Istanbul
Istanbul
Turkey

B.-Å. Bengtsson
Department of Medicine
Sahlgrenska University Hospital
Gothenburg
Sweden

S. Bombardieri
Clinical Immunology and Rheumatology Units
Istituto di Patologia Medica
University of Pisa
Via Roma 67
56126 Pisa
Italy

D. Boumba
Department of Internal Medicine
School of Medicine
University of Ioannina
Greece

F.C. Breedveld
Department of Rheumatology
Building 1, C4-R
PO Box 9600
2300 RC
Leiden
The Netherlands

E.G.L. Bywaters
Beaconsfield
Buckinghamshire
UK

N. Çakir
Cerrahpasa Medical Faculty
Division of Rheumatology
University of Istanbul
Istanbul
Turkey

J.S. Cameron
Clinical Science Laboratories
17th Floor Guy's Hospital
Guy's Campus UMDS
London SE1 9RT
UK

P. Casassus
Service de Médecine Interne
Hôpital Avicenne
125 rue de Stalingrad
F-/93009 Bobigny
France

B. Çiçek
Cerrahpasa Medical Faculty
Division of Rheumatology
University of Istanbul
Istanbul
Turkey

L. La Civita
Clinical Immunology and Rheumatology Units
Istituto di Patologia Medica
University of Pisa
Via Roma 67
56126 Pisa
Italy

J.M. Clements
British Biotech
Watlington Road
Cowley
Oxford OX4 5LY
UK

J.J. Cream
Charing Cross Hospital
Fulham Palace Road
London W6 8RF
UK

M.B. Daha
Department of Nephrology
Building 1, C3-P
PO Box 9600
2300 RC
Leiden
The Netherlands

J. David
Royal Berkshire & Battle Hospitals NHS Trust
Battle Hospital
Oxford Road
Reading RG3 1AG
UK

D. D'Cruz
Department of Rheumatology
St Bartholomew's Hospital
West Smithfield
London EC1A 7BE
UK

M.J. Dillon
The Hospital for Sick Children
Great Ormond Street
London
WC1N 3JH
UK

I. Dimitriatris
Department of Mathematics
Bogazici University
Turkey

H. Direskeneli
Division of Rheumatology
Department of Medicine
Marmara Medical Faculty
Marmara University
Istanbul
Turkey

M.P. Dolcher
Clinical Immunology and Rheumatology Units
Istituto di Patologia Medica
University of Pisa
Via Roma 67
56126 Pisa
Italy

P. Emery
Department of Rheumatology
University of Birmingham
Edgbaston
Birmingham B15 2TT
UK

M. Ermini
Department of Pediatrics
Rheumatology Unit
University of Florence
Florence
Italy

F. Falcini
Department of Pediatrics
Rheumatology Unit
University of Florence
Florence
Italy

A. Farsi
Institute of Medicine III
University of Florence
Florence
Italy

C. Ferri
Clinical Immunology and Rheumatology Units
Istituto di Patologia Medica
University of Pisa
Via Roma 67
56126 Pisa
Italy

A.J.H. Gearing
British Biotech
Watlington Road
Cowley
Oxford OX4 5LY
UK

W.L. Gross
Internal Medicine and Clinical Immunology
and Polyclinic for Rheumatology
Rheumaklinik Bad Bramstedt
Medizinische Universität Zu Lubeck
Oskar-Alexander Straße 26
24576 Bad Bramstedt
Germany

L. Guillevin
Service de Médecine Interne
Hôpital Avicenne
University of Paris
125 rue de Stalingrad
F-/93009 Bobigny
France

V. Hamuryudan
Cerrahpasa Medical Faculty
Division of Rheumatology
University of Istanbul
Istanbul
Turkey

M. Heller
Klinik für Radiologische Diagnostik
Christian-Albrechts-Universität zu Kiel
Arnold-Heller-Straße 9
D-24105 Kiel
Germany

A.J. Howie
Department of Rheumatology
University of Birmingham
Edgbaston
Birmingham B15 2TT
UK

G.R.V. Hughes
The Lupus Arthritis Research Unit
The Rayne Institute
St Thomas's Hospital
London SE1 7EH
UK

J.E. Jackson
Senior Lecturer and Honorary Consultant
Department of Diagnostic Radiology
Royal Postgraduate Medical School
Hammersmith Hospital
London
UK

C.G.M. Kallenberg
Department of Clinical Immunology
University Hospital
9713 EZ Groningen
The Netherlands

E. Kansu
Department of Internal Medicine
Faculty of Medicine
Hacettepe University
Ankara
Turkey

G. Keser
The Rayne Institute
Department of Immunology
St Thomas's Hospital
London
UK

M. Khamashta
The Rayne Institute
Department of Immunology
St Thomas's Hospital
London
UK

P. Langevitz
Heller Institute of Medical Research
Sheba Medical Center
Tel-Hashomer 52621
Israel

F. Lhote
Service de Médecine Interne
Hôpital Avicenne
University of Paris
125 rue de Stalingrad
F-/93009 Bobigny
France

J.T. Lie
Department of Pathology
University of California Davis Medical Center, PAT-1005
2315 Stockton Boulevard
Sacramento
CA 95817
USA

A. Livneh
Heller Institute of Medical Research
Sheba Medical Center
Tel-Hashomer 52621
Israel

C.M. Lockwood
School of Clinical Medicine
Addenbrooke's Hospital
Hills Road
Cambridge
CB2 2SP
UK

A. Lombardi
Institute of Clinical Medicine IV
University of Florence
Florence
Italy

F. Lombardini
Clinical Immunology and Rheumatology Units
Istituto di Patologia Medica
University of Pisa
Via Roma 67
56126 Pisa
Italy

R.A. Luqmani
Department of Rheumatology
Western General Hospital
Crewe Road South
Edinburgh EH4 2XU
UK

A. Manetti
Cardiology Unit
A. Aeyer Hospital
Florence
Italy

M.C. Mat
Department of Dermatology
Cerrahpasa Medical Faculty
University of Istanbul
Istanbul
Turkey

M. Matucci-Cerinic
Institute of Internal Medicine
University of Cagliari
Cagliari
Italy

J. Michael
Department of Rheumatology
University of Birmingham
Edgbaston
Birmingham B15 2TT
UK

P. Migliorini
Clinical Immunology and Rheumatology Units
Istituto di Patologia Medica
University of Pisa
Via Roma 67
56126 Pisa
Italy

R.J. Moots
Department of Rheumatology
University of Birmingham
Edgbaston
Birmingham B15 2TT
UK

M. Mosca
Clinical Immunology and Rheumatology Units
Istituto di Patologia Medica
University of Pisa
Via Roma 67
56126 Pisa
Italy

L. Mouthon
Service de Médecine Interne
Hôpital Avicenne
University of Paris
125 rue de Stalingrad
F-/93009 Bobigny
France

H.M. Moutsopoulos
Department of Pathophysiology
School of Medicine
National University of Athens
Athens
Greece

C. Muhle
Klinik für Radiologische Diagnostik
Christian-Albrechts-Universität zu Kiel
Arnold-Heller Straße 9
D-24105 Kiel
Germany

F. Moral
Cerrahpasa Medical Faculty
Division of Rheumatology
University of Istanbul
Istanbul
Turkey

L.A. Needham
British Biotech
Watlington Road
Cowley
Oxford OX4 5LY
UK

P. Niaudet
Service de Néphrologie Pédiatrique
Hôpital Necker-Enfants Malades
149 rue de Sèvres
75743 Paris Cedex 15
France

H. Özdoğan
Cerrahpasa Medical Faculty
Department of Internal Medicine
Rheumatology Unit
Istanbul 34303
Turkey

Y. Özyazgan
Cerrahpasa Medical Faculty
Division of Rheumatology
University of Istanbul
Istanbul
Turkey

A. Pall
Department of Rheumatology
University of Birmingham
Edgbaston
Birmingham B15 2TT
UK

R. Pigott
British Biotech
Watlington Road
Cowley
Oxford OX4 5LY
UK

M. Pras
Heller Institute of Medical Research
Sheba Medical Center
Tel-Hashomer 52621
Israel

M. Reuter
Klinik für Radiologische Diagnostik
Christian-Albrechts-Universität zu Kiel
Arnold-Heller Straße 9
D-24105 Kiel
Germany

N. Richards
Department of Rheumatology
University of Birmingham
Edgbaston
Birmingham B15 2TT
UK

C. Savage
Birmingham Vasculitis Group
Departments of Rheumatology and Nephrology
Queen Elizabeth Medical Centre
University of Birmingham
Edgbaston
Birmingham B15 2TT
UK

F. Schubert
Klinik für Radiologische Diagnostik
Christian-Albrechts-Universität zu Kiel
Arnold-Heller Straße 9
D-24105 Kiel
Germany

D.G.I. Scott
Department of Rheumatology
Norfolk and Norwich Hospital
Norwich
UK

L. De Simone
Cardiology Unit
A. Aeyer Hospital
Florence
Italy

F.N. Skopouli
Department of Internal Medicine
School of Medicine
University of Ioannina
Greece

G. Taccetti
Department of Pediatrics
Rheumatology Unit
University of Florence
Florence
Italy

A. Tavoni
Clinical Immunology and Rheumatology Units
Istituto di Patologia Medica
University of Pisa
Via Roma 67
56126 Pisa
Italy

S. Trapani
Department of Pediatrics
Rheumatology Unit
University of Florence
Florence
Italy

S. Turchini
Institute of Medicine III
University of Florence
Florence
Italy

Y. Tüzün
Cerrahpasa Medical Faculty
Division of Rheumatology
University of Istanbul
Istanbul
Turkey

A.G. Tzioufas
Department of Pathophysiology
School of Medicine
National University of Athens
Athens
Greece

H. Yazıcı
Cerrahpasa Medical Faculty
Division of Rheumatology
University of Istanbul
Istanbul
Turkey

S. Yurdakul
Cerrahpasa Medical Faculty
Division of Rheumatology
University of Istanbul
Istanbul
Turkey

D. Zemer
Heller Institute of Medical Research
Sheba Medical Center
Tel-Hashomer 52621
Israel

# Foreword

*Professor E.G.L. Bywaters* CBE, MO, FRCP, FACP

This book provides a summary – a distillation – of most of the presentation at the XIIIth Symposium of EULAR (The European League Against Rheumatism) on Vasculitis organized by Professor Hasan Yazici and Professor Paul Bacon. It was indeed a symposium, a festival of learning, science and medicine held in June 1994 in Istanbul, Turkey, in one of the newest and most lavishly fitted modern hotels of the twentieth century, with more extravagant fittings than any sultan at the height of his power was able to command and with a splendidly equipped conference center as well as a lovely view south-east over the Bosphorous with its busy water traffic. In contrast was the history by which we were surrounded, the early churches of Constantine, the mosques and palaces of the Ottoman sultans, a city embued with an illustrious past and which perhaps was at one time the most famous and the richest in the world.

The participants (numbering about 1000), like those early traders of the Golden Horn, united Eastern and Western worlds. They came from the Americas, from the Antipodes, from Asia and, of course, from Europe, bridging the world of science as did the ancient bridges and boats of the Bosphorous in their own time. Appropriately, the opening ceremony, beginning with Turkey's national anthem, was gloriously prefaced by an overview of the wonderful cultural history of Turkey performed and presented by an angel-voiced actress and a most distinguished historian in concert. They provided us with a most memorable, historical and cultural background to our meeting.

The scientific programme was well organized (as was the social programme with functions in palaces and memorable visits to mosques, museums, etc.). The presentations were of a high standard including also the poster sessions and drug firm participation.

The essence of the meeting can be judged by the contributions printed in this book, which sets out fully today's knowledge of vasculitis as seen by the top experts in the field and discussed by their peers.

Disease of the body's vessels can involve any organ or tissue and is, therefore, of general medical interest and concern even to the most specialist specialties. Up to now there have been very few advances on the subject – except for classifications galore – but the discovery of antibodies to endothelium and neutrophil cytoplasm has begun to transform the field: perhaps one of the prime reasons for initiating this fruitful meeting at this time; the meeting was made under EULAR auspices.

# Preface

The field of vasculitis is in a state of relative infancy. As such, nomenclature and its recognition are all important. Thus we open with history, classification and histology. Ensuing chapters on pathogenic mechanisms, laboratory tests and imaging are followed by sections devoted to either distinct clinical entities or discussions of organ systems where this group of diseases leave their greatest toll. Chapters on treatment intend to cover both state-of-the-art and investigational or even inspirational ways to control the vasculitides. One important asset of this book is the rather extensive cover of childhood vasculitides, about which precious little has previously seen the light of publication.

Vasculitis studies may be young but the field is in a healthy burgeoning state, although the relevant literature is divided across a number of areas. Where ideas are scattered facts never achieve the recognition they deserve. *The Vasculitides* is an attempt to put together in one volume what is established and not so established in facts and ideas within the broad topic of vasculitides. All our contributors are dedicated students of vasculitis in its many facets.

This book had its foundations in a busy three-day European symposium held in June 1994 in Istanbul, sponsored by EULAR. A well-attended international meeting devoted to a single topic suggests that the subject has come of age. Certainly, these diseases are no longer seen as mere curiosities, but as important diseases to all physicians, which are eminently treatable. This volume is not simply the proceedings of that symposium. It presents a clearly edited overview of the current state of knowledge across a broad spectrum of the vasculitides. We are grateful for the painstaking co-operation and understanding of all our contributors during the preparation of this volume within a short period. The same gratitude goes to Chapman & Hall, our publishers, for their prompt and expert help.

# Part One

## Classification and Histology

# 1

# Vasculitis – then and now

*P.A. Bacon*

## INTRODUCTION

A EULAR symposium devoted to vasculitis suggests a coming-of-age for the subject. It represents a suitable time to consider where the subject is, where it has come from, and where it is going to. Recorded medical history dates back at least two millennia, but remarkably little of that has involved the theme of vascular disease. Where is it mentioned in the writings of Hippocrates? Despite the initial rarity, the pace has quickened, particularly in the past century. In fact, the public image of vasculitis has changed immeasurably even since I was a student. My medical textbook of that time – *Price '56* – has only one mention of vasculitis in the index, where nodular vasculitis is described as a variant of erythema nodosum under skin diseases. There is more in the collagen disease section where polyarteritis and temporal arteritis are put together with systemic lupus erythematosus (SLE) and scleroderma. My interest in vasculitis developed during a fellowship year at UCLA in 1971. There, by contrast, 'life-threatening vasculitis' was diagnosed regularly and presented to the rheumatology rounds already on semi-lethal doses of steroids. Diagnostic procedures were few – this was clearly a case of Pellatan's dictum, *'l'occasion est urgente, le jugement difficile'* [1]. However, the value of adding immunosuppressive therapy was soon recognized [2]. Vasculitis research now has many fascinations, many achievements and a sense of a great future to come. It is now widely accepted that vasculitis is a set of important and not so rare diseases. We recognize that they are treatable – if not often curable. Indeed, aggressive therapy has converted systemic necrotizing vasculitides from an acute, life-threatening set of conditions to chronic relapsing

*The Vasculitides.* Edited by B.M. Ansell, P.A. Bacon, J.T. Lie and H. Yazici. Published in 1996 by Chapman & Hall, London. ISBN 978-0-412-64140-4.

diseases. In this respect, at least, they resemble multiple other rheumatological conditions and are a fitting subject for a rheumatology symposium. It may be instructive to look in more detail at the history to see how this remarkable change has come about.

## HISTORY

Historical descriptions of vasculitis are very thin on the ground. The one widely recognized as initiating the study of vasculitis is that of Kussmaul and Maier in 1866 [3]. They reported a 'so far not described peculiar disorder of the arteries, periarteritis nodosa, which presented together with Bright's disease and a rapidly progressive general muscular paralysis'. Their paper, a fairly low-key chatty description of disease presentation, evolution and post-mortem, gives a clear insight into the difficulties of practising medicine at that time. You had to be fit to get to the doctor – their patient, C.S., felt so weak he had to lie down after climbing 'the two high flights of stairs to the medical department'.

Their first comment was that 'this was one of those patients in whom one can tell the prognosis before the diagnosis'. He was cachectic and gave 'the impression of a lost person whose few days are counted'. This despite an acute illness of only eight days with diarrhea, fevers and sensory neuropathy. A 'flu going on to the chest' a few months earlier might also have been part of it. On examination he had a 'chlorotic marasmus' together with a most unusual and peculiar complex of symptoms which the doctors failed to fit to any specific diagnosis. The findings indicated widespread organ involvement with fever; persistent tachycardia; cough and wheeze; hematuria and proteinuria indicating Bright's disease; and intermittent diarrhea as well as marked weight loss. His neuropathy progressed from a sensory to a motor one, together with intensive fatigue and muscle tenderness, leading to a paralysis which proceeded relentlessly. He finally developed dementia and died within four weeks.

The most striking macroscopic finding at post-mortem was the 'unusual nodular thickening of the numerous arteries'. These changes involved the coronary arteries, the hepatic, the renal, the splenic and the mesenteric system. To a lesser extent, they were also found in fatty tissue, and subcutaneous tissue of thorax and abdomen as well. The highest degree of changes in the small arteries was encountered in the mesentery, leading to diffuse necrotizing enteritis. On microscopy this extensive pathological change was seen to be in arteries of the muscular type. They noted that there was no such change in veins and never in any capillaries.

Kussmaul and Maier thought that such remarkable thickenings in numerous branches of the arteries had never been described before.

Not that they were difficult to see, being present as 'localized nodular swellings found together in a string of pearls or distributed like apples on a tree'. They therefore thought this was a new disease – although it has subsequently become clear that there were some previous descriptions. Eppinger [4] reviewed a case previously described grossly by Rokitansky in 1852, again with small aneurysms present along the course of arteries throughout the body. Pelletan [1] reported briefly a case where he counted 63 small aneurysms. However, Lamb [5] credits Michaelis and Matani with the first gross description of the disease in 1755.

The reasons for this paucity of historical descriptions are probably multiple. On the one hand, it is possible that these are indeed new diseases, representing immune responses to new pathogens or to new drugs. The alternative view is that they have always existed but have not been recognized for what they are. The collagenoses have undoubtedly taken over from syphilis as the disease which can mimic all other diseases – and the vasculitides are no exception to this. Systemic or local diseases with a plethora of manifestations, they are frequently misdiagnosed even today. At initial presentation, the picture is often incomplete, so that the label 'suspected' or 'undifferentiated' vasculitis is more common than the single label. The classic picture of a primary systemic vasculitis is rarely present at the beginning so that time and disease development make an important contribution to the label. Even today, firm diagnosis rests on histopathological evidence of vascular inflammation, so it is perhaps not surprising that the classic descriptions of disease, such as those of Kussmaul and Maier or of Wegener, came from pathological study. Earlier descriptions of disease had come from anatomical studies, which would have a low chance of describing rarities. The blood vessels were examined, particularly the larger ones. For example, in 1757 William Hunter, a well-known physician and obstetrician in eighteenth century London who also ran an anatomy school, described syphilitic aneurysm of the aorta [6]. That was before atheroma had been described, which may be a comment on the frequency of the two diseases at that time, but more likely reflects the anatomist's interest in large structures rather than histology.

Voltaire's dictum that 'a man sees what he knows' is probably the major factor in explaining the slow development of interest in the vasculitides. Prior to accurate pathological descriptions, it was not possible to apply clear labels even when the disease was seen. All skin disease tended to be diagnosed as 'leprosy', even forms that looked nothing like leprosy as we recognize it today. For example, the Kariye museum in Istanbul contains a magnificent example of fourteenth century Byzantine art. One beautiful mosaic is entitled *Christ healing the leper,* but I would more likely diagnose Henoch–Schonlein purpura or some similar cutaneous vasculitis.

The most striking example of the difficulty in labeling an odd case of something not yet described is provided by Samuel Jones Gee from the teaching hospital, St Bartholomew's Hospital, London. The hospital dates back to 1123 but, unfortunately, only started publishing its series of St Bartholomew's Hospital Reports in 1865. They are a rich source of interest and might well have provided earlier descriptions of our diseases had the Reports been started earlier. Relevant to this discussion is the report by Henry Gee in 1871 of a boy of seven years, admitted with 'scarlatinal dropsy' who developed intercurrent pneumonia and died two days later [7]. Examination of the heart at post-mortem showed the coronary arteries dilated into aneurysms at three places, one near to the mouth of the coronary artery. In the post-mortem registry, Gee noted swelling of the neck glands, enlargement of the spleen as well as pleurisy and pneumonia. A scarlatinal dropsy would imply a rash with edematous extremities, suggesting that this was a description by Gee of the condition detailed many years later by Kawasaki in 1967 [8]. A more recent re-examination of the heart from the Bart's museum (by Dr A.G. Stansfield) showed a saccular aneurysm of the coronary artery entirely consistent with the diagnosis of mucocutaneous lymph node syndrome. However, Gee remains better known for his cough prescription – Gee's linctus.

The apparent rarity of vasculitis was reinforced by the confusion of incomplete forms, likened by Bywaters to the Cheshire cat effect. This is well illustrated by studies of Kawasaki's syndrome where more than 10 000 cases were described in the Japanese literature before the first English-language description. It was felt to be a Japanese disease – and, indeed, there are important geographical and ethnic effects which remain to be elucidated. However, after the initial English language descriptions it came to be widely recognized as occurring in the USA and Europe where it had not previously been recognized. About 10 cases a year are diagnosed in Birmingham's Children's Hospital, suggesting a minimal incidence of 5 per 100 000 children. Similarly, adult vasculitis, in the last decade in Birmingham, has gone from being 'so rare it is not worth bothering about' to being a major interest of rheumatologists and nephrologists, with at least 25 new cases seen over the last five years. A former colleague (D.G.I. Scott) has both discovered the cause of and documented an epidemic of vasculitis on moving to Norwich, in the east of England. His current figures show an incidence of systemic vasculitis of more than 4 per $10^5$, which is of the same order as SLE [9]. I suspect there are still unrecognized cases both in hospital clinics and in the general population.

Detailed epidemiological studies in many parts of the world are required to document both the total incidence and the geographical variations in types of vasculitis. The figures indicated above, and the interest in the EULAR Symposium, suggest that this would be a worthwhile exercise. Perhaps rheumatologists should start with systemic rheumatoid vasculitis, frequently thought to be a disappearing disease now, or at least steadily becoming a less serious problem. The Norwich figures, discussed elsewhere in this volume [10], show clearly that this is not so. Rheumatoid vasculitis remains a serious problem with important manifestations and major morbidity, although it responds to aggressive therapy. It is now more frequent than before, so presumably the cases in Norwich were largely unrecognized for their true selves before the arrival there of an interested physician.

This is not a new problem. Bannatyne, an energetic physician from the west of England, recorded his patients in detail, performed post-mortems when necessary, and wrote a popular textbook that ran to four editions within a few years [11]. In his book, he described 'small round cells around the blood vessels in the nerve sheath, together with thickening of the intima narrowing the lumen of the vessel', in a rheumatoid patient who had had clinical evidence of a peripheral neuropathy during life. He did not appear to appreciate the significance of this and nor did subsequent authors. Fifty years later, Ellman and Ball [12] ascribed the presence of arteritis in rheumatoid arthritis to the concurrence of polyarteritis nodosa (PAN) with rheumatoid arthritis (RA). It was not until the classic papers of Sokoloff and Bunim [13] and the detailed studies of Bywaters [14, 15], that the vascular disorder was recognized as a distinct complication of RA itself. Thus, the problems of recognition became overtaken by those of classification as arteritis was more widely recognized.

CLASSIFICATION

Another problem which has obscured the overall picture of vasculitis is that of classification. Historically, this springs from the expanding descriptions of disease which followed Kussmaul and Maier's original description of a classic case of PAN. The recognition that vascular inflammation could involve all organs at some time or another, in confusing combination, and can present at any age, completely distorted the original description and built up a multilayered vision of PAN. Indeed, for 100 years this term tended to be used to cover all types of vasculitis, often differing widely from the original description. The expansion started with the description of pulmonary involvement [16, 17] although early reports had frequently stated that the pulmonary arteries were rarely involved. Ophuls [17] in particular listed many features that differed from previous reports, particularly absence of aneurysms and presence of granulomatous lesions together with small vessel involvement and tissue infiltration with eosinophils. Asthma and eosinophilia was later documented in 18% of patients [18] before the term 'allergic granulomatous angiitis' was coined by Churg and Strauss in 1951 [19]. The presence of vasculitis in hypersensitivity states was documented both in animals [20, 21] and in humans [22, 23]. Vasculitis complicating other diseases, including rheumatic fever, has also been documented [24]. Although all called polyarteritis nodosa, a term introduced by Ferrari [25], it seemed unlikely that all this vasculitis represented the same disease.

This inspired a need for classification which tended to proceed by peeling off the layers of the vasculitic onion in piecemeal and confusing fashion. Zeek, in her critical review [26], noted that the term polyarteritis nodosa had gradually come to encompass a wide variety of vascular lesions, some of which had little in common with those of the original description. She made a brave attempt to deal with the increasing confusion of classification and the rapid increase in new cases by reviewing the literature to date and then proposing a new nomenclature. Her historical review covered three periods. The first, up to 1900, revealed reports of a rare but distinct disease entity, with macroscopic lesions widespread throughout the body, except in the lungs and brain. The lesion was an inflammatory one, commencing in the media of muscular arteries, with capillaries never involved and veins only occasionally (by spread from contiguous arterial inflammation). The unsolved etiology was thought to relate to an unidentified toxic or infectious agent, but the disease could be differentiated from syphilis. The second period, up to 1925, saw an extension of the range of reports of diseases in many different ways. Microscopic involvement as the sole feature was first noted in 1903 [27]. Lung disease was reported by Monckeberg [16] and again by Ophuls [17], who also described the presence of granuloma with eosinophilic infiltration of the tissues. The emphasis on infection was enhanced by the description of a case in an Axis deer. Initially one isolated case, the disease went on over the next decade to wipe out all but eight of the original herd in that zoological park, although the same species in other collections were not affected. Further descriptions followed in the pig [28], calves [29] and the dog [30] – all animals close to humans.

The third period, 1925 up to the review in 1952, was characterized by further expansion of the clinical profile together with a massive increase in documented cases. Only 70 cases were reported in the first six decades after the initial description of polyarteritis nodosa, yet by 1939, Boyd was able to review 395 reported cases [31]. This was largely due to better recognition by microscopy and increased ease of reporting cases. It also reflected a broader usage of the term which had now expanded to include further entities. Hypersensitivity was recognized as a major feature both in animal models [20, 21] and in humans [22, 23]. The animal models left many features unexplained, particularly the marked individual susceptibility to an identical induction regime [33]. In humans there was a major increase of reports in association with hypersensitivity in sulfonamide drugs. However, several authors emphasized that the necrotizing vasculitis seen in such hypersensitive lesions was not strictly comparable to a typical PAN and introduced various other terms, such as 'polyvasculitis'. Involvement of vasculitis in rheumatic fever was felt to fit into the hypersensitivity concept [24]. Involvement of the temporal arteries, first described by Horton, Magath and Brown in 1934, was clearly different in many ways [34]. Clinically, it was a non-fatal often spontaneously regressing and limited condition, established as a distinct entity by Kilbourne and Wolff [35]. Microscopically, it was characterized by the presence of prominent multinucleated giant cells in the arterial wall. Isolated vascular lesions limited to other locations were also described. For example, Plaut [36] reported 88 cases of isolated involvement of the appendix, where there was no correlation between the vascular lesions and other clinical pathology. Granuloma with giant cells and epithelioid cells was a conspicious feature in the necrotizing and granulomatous vasculitis described as 'allergic granuloma' by Churg and Strauss [19]. Again this had a longer disease duration, although it was still fatal in between three months and five years. Hypertension was first noted by Meyer [37] and its importance was stressed in the review of Boyd [38]. This period was thus characterized by a tendency to use the term polyarteritis nodosa for the lesions caused by hypersensitivity and even the attempt to find a new name for the condition described by Kussmaul and Maier.

Zeek noted the necessity to classify accurately before instituting therapy, which had become important with the use of corticosteroids. Subsequent to Zeek's review [26], many attempts have been made to define more closely the diseases included under the vasculitic umbrella, particularly those of the group known as polyarteritis nodosa. This was enlarged by Lie [39]. What is needed currently is a reconstruction of all these portions in either a logical complete form, or in a simple usable version serving the needs of today. The former would be more satisfying but is difficult at present since it requires detailed knowledge of the pathology and pathogenesis, together with an understanding of the details and extent of organ involvement including at the subclinical level.

## ATTEMPTS AT RATIONAL CLASSIFICATION

Many attempts have been made to classify vasculitis by etiological factors. The role of infection was stressed early and repeated on many occasions, although in the 1930s and 1940s hypersensitivity occupied center stage. In his series of cases with hypersensitivity vasculitis, Rich [23] pointed the finger at sulfonamides and horse serum. He was able to reproduce vasculitis experimentally with a single injection of horse serum, given together with or in the absence of sulfonamides [21]. However, in human disease he still speculated on the possible role of bacterial hypersensitivity. Arteritis could be induced in an animal model in relationship to hemolytic streptococci [40], which had also been implicated in human disease [41].

Others suggested that the etiology related to a filterable virus, a suggestion which received a major boost from the studies of hepatitis B-induced vasculitis [42] and the observations of circulating hepatitis B in immune complexes in spontaneous polyarteritis [43]. Unfortunately, the clinical and etiological features do not always overlap in an apparently logical way. For example, hepatitis B was originally described as a major etiological factor in PAN, but is clearly not the total answer to that disease for several reasons. The organism itself is not only associated with PAN, but also has been described in association with other vasculitic disorders, such as cryoglobulinemic vasculitis, polymyalgia/temporal arteritis and small-vessel vasculitis – as well as with liver disease which gave it its name. Secondly, PAN has also been associated with other viruses including hepatitis C, cytomegalovirus, parvovirus and HIV – the list is steadily increasing. Indeed, in many centers hepatitis B is an uncommon association with systemic necrotizing vasculitis and, in fact, the incidence appears to be decreasing. It may only be causative in areas of contaminated blood or high intravenous drug usage.

A similar picture is seen in relation to an important set of antibodies, the antineutrophil cytoplasmic antibodies (ANCA). ANCA has been claimed to be specific for Wegener's granulomatosis or to define a specific spectrum of ANCA-related diseases. In fact, the antibody was first described in glomerulonephritis [44] and is not always present even in classic cases of Wegener's. It is now clear that ANCA are a family of antibodies, similar to the antinuclear antibodies (ANA) family. Two of them – antiproteinase III and anti MPO antibodies – are clearly linked to vasculitis. However, others are seen in many diseases, ranging from cystic fibrosis to inflammatory bowel disease [45]. In these latter circumstances they do not appear to be related to vasculitis. This is perhaps not surprising in view of the ubiquitous role of polymorphs in acute inflammation in multiple organs and diseases. Indeed, it may be that the most important role of ANCA will be to have increased both the awareness and the understanding of vasculitis in centers where it has now become commonly used [46].

In the absence of clear current messages from etiologic studies, it is tempting to fall back on older concepts based on immunopathogenic mechanisms. However, that also presents problems. Immune complex-induced disease is the classic model for vascular damage – but activated T-cells can induce very similar lesions. The inflammatory lesions of spontaneous human disease often contain both polymorphs and lymphocytes – suggesting that both mechanisms may be involved in the total picture. Thus the intriguing findings of experimental lesions can do no more than suggest that such pathways are also worth looking for in human disease.

Pathology is still the sheet anchor for the diagnosis of vasculitis. As with antibodies, its diagnostic role is more important than its ability to define the precise type of vasculitis. In the latter, it has many problems for the unwary. A particular one is the time effect – since the picture of vascular inflammation in the acute, subacute and chronic stages can differ from a markedly inflammatory necrotizing vasculitis through to a chronic intimal proliferation with occlusion in the absence of inflammatory cells. Experiments with histamine skin tests have shown the spectrum of inflammatory cell changes which can occur even within 48 hours. The other problem with pathology is the spectrum of vessels involved, ranging from capillaries through to major arteries including the aorta. A single biopsy rarely encompasses more than one of these. Other methods, such as angiography, or newer techniques, such as a labeled leukocyte scan [47], are required to demonstrate the range and extent of vessels involved.

In view of all these problems, it is important for current clinical use to have a simple practical way of defining and describing these diseases. A meeting in Chapel Hill tried to provide this by looking for converging agreement and consensus rather than for absoute definitions. Importantly, the result takes full cognisance of the spectrum of vessel size involved, recognizing the importance of vessel size in previous attempts of classification [48] as well as the overlaps involved. A nomenclature is presented for primary systemic vasculitis only, since that was the focus of the meeting [49]. It does not exclude the recognition of either limited forms of these diseases or the systemic vasculitis secondary to connective tissue diseases, etc. Taken together these form a useful working nomenclature for the clinician. It thus represents a practical step towards standardization of practice, even if experts will disagree with the details [50].

## ASSESSMENT

In addition to classification, another aspect of vasculitis that is receiving increasing clinical attention is that of assessment of disease status. The improved survival that results from aggressive therapy provides its own problems of toxicity. This may pose the dilemma of whether to continue the therapy regime or to give less drug and risk relapse.

Accurate longitudinal measurement of disease evolution is essential in dealing with these problems. The spectrum of organ involvement between patients, as well as the differing degrees of disease extent at different times within patients followed serially, emphasizes the important need for standardized assessment in vasculitis. I suggest this is a good time to return to basics and place the emphasis on clinical assessment. This can then be used as a yardstick to determine the value of the increasing number of laboratory measures reported abnormal in vasculitis. Detailed assessment of patients has always been the strength of physicians. With sufficient rigor, clinical science can be as accurate, reproducible and sensitive to change as laboratory tests. It is no longer valid to rely on survival or mortality as measures to assess the efficacy of therapy. The initial results of aggressive treatment are good, but relapse is frequent [51] and the amount of time spent in disease-free survival is disappointingly low [52]. Thus detailed assessment of outcome must be obtained at a number of levels.

Disease activity is the most obvious measure to assess. This should reflect the current state of inflammation that relates specifically to vasculitis rather than to concomitant disease, such as infection or the effects of drugs. This is an important distinction, despite the fact that the three often occur together in real life, since the disease activity reflects the need for future therapy. Few validated indices of activity have been published. We have devised by consensus a disease activity index based on the degree of clinical involvement in nine separate organ-based systems [53]. The score for each system is weighted to reflect the clinical importance of the involvement of that system. Thus renal damage, the main contributor to mortality in systemic vasculitis, is weighted more heavily than cutaneous involvement. The weighting is recorded independently of the actual involvement and may need to be altered in the light of subsequent experience. This activity index allows both a total score to correlate with other tests and an appreciation of the number and extent of organs involved. In that respect, it is very similar to the well-validated BILAG instrument for assessing disease activity in SLE [54]. It is sensitive to change, in that the score is significantly higher during episodes of active disease than during disease inactivity when patients are followed serially. It correlates with two other indices of disease activity, is easy to use and available in a user friendly computerized version. Such an index of disease activity makes a useful addition to standardizing therapeutic studies in vasculitis. However, disease activity does not measure the entire outcome in vasculitis.

A damage index is also required, since damage is conceptually different from activity. It represents the accumulated scars of disease which theoretically will not heal even though functional improvement may occur. It is important to distinguish the two, since fixed damage does not usually require aggressive therapy but may indicate a neutral alternative approach, such as rehabilitation. However, the amount of damage and scars accumulating influence patients' vision of their progress considerably. The discordance often noted between assessment of disease activity and an overall score, such as the physician's or patient's global assessment, reflects the degree of damage. Recognition of this may limit the overenthusiastic use of drugs like cyclophosphamide, even when it is no longer appropriate. Experience in SLE has shown that a damage index can be useful as a serial measure and in assessing real outcome of any relapse [55]. The same principle has been applied to vasculitis with suitable adaptation to form a vasculitis damage index (VDI) which we have been applying to our patients recently.

The combination of activity and damage measure the current state and past progress of the disease, providing a severity index. They can also be used to assess the extent of disease, that is the number of organs involved. This may be particularly important in some vasculitides such as Wegener's granulomatosis, where a disease extension index has been recently devised building on the old ELK classification [56]. However, these are all measures of what the physician sees. It is important to add a functional index which reflects how the patient is coping with the burden of disease. It is common experience that the latter reflects other aspects, such as age and pre-existing physical and mental status, as well as social circumstances. Rheumatologists frequently devise functional indices specific to individual diseases. In these cost-conscious days, when there is a strong need to justify to government and other health agencies the importance of a therapy, there is much to be said for using a functional index widely used for many diseases. One such is the SF 36 originally devised in the USA, and validated in Europe [57]. It has been used for many conditions including immunodeficiency and neoplastic disease. We have been applying it to our patients in the vasculitis clinic. They have had little difficulty in filling it in and it adds a useful additional dimension to the overall assessment.

A package of activity, damage and functional indices together make a major advance in standardizing the approach to vasculitis, which will allow collaborative trials for uncommon disorders between many centers, as well as comparison between different studies. It will also be a useful measure to assess laboratory tests which might later be built in to the index itself if it can be shown to improve sensitivity and specificy. Immunological markers such as ANCA [58] and AECA [59] have been suggested as useful markers of disease activity, which need to be confirmed in prospective serial studies. A direct marker of endothelial cell damage would also be important. Von Willebrand factor (vWF) has been proposed for this role but the correlation with disease activity is disappointing. In serial studies, it appears that high levels are found in active disease but may persist after apparent clinical improvement [60]. *In vitro* studies demonstrate clearly that vWF is released from both damaged endothelial cells in culture and also from proliferating regenerating cells, which probably explains the clinical discrepancy [61]. Alternative assays of endothelial cell function need to be sought. Considerable interest has been aroused in the potential of soluble adhesion molecules, released from activated cells during inflammation. Some studies indicate that soluble vascular cell adhesion molecule (VCAM) may reflect disease activity in vasculitis, although it is not specific to this disease and similar levels are found in active SLE [62]. E-selectin, which is expressed only by activated endothelial cells, would appear an even more attractive candidate but in practice levels do not rise high nor do they reflect disease activity. The search will continue for better *ex vivo* markers of disease activity.

## THERAPY

The need for assessment is emphasized even more strongly by the changing perception of the effect of therapy in vasculitis. Cyclophosphamide was introduced two decades ago as an important therapy for vasculitis. Although it took time to recognize the marked superiority of this over steroid treatment alone in systemic necrotizing vasculitis, cyclophosphamide eventually became the paradigm for therapy [63]. However, recent reassessments have emphasized that this is not without cost. Review of the National Institutes of Health (NIH) series revealed the problems of continuous oral cyclophosphamide, where the dose is limited by toxicity [52]. We have always been impressed by the ability of pulse regimes to provide a higher immediate dose for lower toxicity [64]. Our current comparative study shows that pulse cyclophosphamide is as effective as oral cyclophosphamide in inducing remission, with a high survival rate – over 85% in both groups in the first year [65]. It has to be emphasized that our regime starts with pulses two weeks apart, which probably explains the different success rate we observed compared with that of the NIH group which used pulses at one month intervals. However, the longer term story is not so rosy, since long-term follow-up of a large group of patients shows a disturbingly high relapse rate [51]. This occurs in all forms of systemic vasculitis, does not appear to be related to the current therapy and can occur at any stage of disease. This suggests a grumbling level of inflammation controlled but not cured by cyclophosphamide. Such speculation is supported by continuing laboratory abnormalities, such as elevated vWF, circulating adhesion molecules or T cell alterations. Thus effective treatment of life-threatening illness has not provided the hoped-for cure. Rather it has translated these diseases into chronic relapsing conditions requiring life-

**Table 1.1** Therapeutic regimes for systemic vasculitis

| I. Induction of remission | Aggressive, brief:<br>• cyclophosphamide currently best<br>• adjuvant therapy may be required, e.g. IV Me Pred, plasma exchange, IVIg |
| --- | --- |
| II. Maintenance | Prolonged – non-toxic:<br>• new agents needed<br>• ? combination |
| III. Relapse/failed induction | New products of biotechnology:<br>• e.g. IVIg, monoclonal antibodies |

long attention and follow-up, very similar to many other rheumatological disorders.

FUTURE NEEDS

The effective control of vasculitis long term with current therapies, as well as the assessments of new therapies such as products of the biotechnology revolution, presents a challenge for the future. This would be most effectively addressed by collaboration between centers, which international symposia can do a great deal to foster. The ideal aim would be to have all patients diagnosed as having vasculitis involved in a controlled prospective therapy study and assessed according to a standardized database.

Therapy is the most obvious area requiring collaboration to the immediate benefit of patients. The success of cyclophosphamide in reducing the immediate mortality has given us time to think how to progress for the longer term. We require different regimes for different stages of the disease, each of which will be based upon combinations of drugs. For example, there is a clear need for induction regimes for acute vasculitis, based on the success of this approach in cancer. The induction regime requires to be aggressive but relatively brief to reduce toxicity. It could well be based on addition of adjuvant drugs unless improvement was seen within a set period. The combination of cyclophosphamide and steroid looks a good basis for induction, but is probably used for too long. Six months appears quite sufficient and the data from our prospective study [66] suggests that the major part of the clinical improvement has all occurred within three months. Several questions remain to be defined, particularly the place of additions, such as plasma exchange or pulse methylprednisolone, to the induction regime. These may well have a place, at least in those with severe renal involvement.

Maintenance regimes, clearly required over a long period to diminish the relapse rate, should preferably involve relatively non-toxic drugs to avoid the accumulative side-effects of long-term cytotoxics. Azathioprine and methotrexate have both been proposed as milder alternatives, but it may be questioned whether cytotoxic drugs are required at all at this stage. Trimethoprim sulfonamide has been proposed in Wegener's granulomatosis at this stage of the disease [67] and controlled trials are in progress. The mechanisms of action are unclear, but it is possible that such therapy depresses granuloma formation by affecting phagocytic cell function. Alternative drugs with similar actions should be sought. We have used both thalidomide and pentoxyphyline in a few cases. They have some effect, but do not appear to be the total answer. Some of the endothelial protective agents being devised in atheroma research may have applications in the healing stages of vasculitis. The third area of need is therapy for relapse. The newer products of the biotechnology revolution may be best applied here, but need to be compared with cyclophosphamide.

Progress in assessing new regimes such as these will require extensive collaboration between centers, with the aim of including every patient in a trial. A standardized database for assessment would provide the necessary tools for such collaboration. The essential requirement is for a core database which would allow any center to include its own local needs. The increasing use of computers in medicine, together with information highways serviced by the interuniversities' network, provides a practical basis for such a collaboration. Groups of interested super-specialists are already making proposals in this area, both under the aegis of the European Union (EU) and through a USA-led international co-operating group. It is important to extend this involvement into the working laboratories of ordinary rheumatologists who have any interest in the subject throughout Europe, in order to recruit sufficient patients for powerful studies which will provide real answers for the future. Collaborations with nephrologists and other specialists are also essential to cover the full spectrum of disease.

In parallel to this widespread clinical involvement, a focused effort to understand the pathogenesis of these diseases is progressing in a small number of academic centers. There have been enormous advances in understanding the molecules mediating the adhesion of circulating cells to the endothelium and to dissecting the mechanisms of migration into the tissues. This is an area of great complexity with tissue specific as well as general factors which have not been fully elucidated. The ability to study the trafficking of cells *in vivo* lags further behind but remains an important goal in seeking to understand immune inflammatory events throughout the body. However, great advances have been made, particularly in the dissection of autoantibodies, such as ANCA, which have enormous potential both as markers of disease and as direct mediators of endothelial cell damage. The latter is clearly a complex process which probably involves T cells as well as B cells. The future challenge is to fully dissect the basic pathological mechanisms, which will be essential both to provide rational therapeutic regimes for the future and improved markers specific for vascular inflammation that can be used to monitor such therapy.

## REFERENCES

1. Pelletan, Ph. J. (1810) Aneurismes particuliers, *Clinique Chirurgicale ou Memoires et Observations de Chirurgie Clinicale*, vol. 2, J.G. Dentu, Paris, p 1.
2. Leib, E.S., Restivo, C. and Paulus, H.E. (1979) Immunospressive and corticosteroid therapy of polyarteritis nodosa. *Am. J. Med.*, **67**.
3. Kussmaul, A. and Maier, R. (1866) Über eine bisher nicht beschreibene eigenthümliche Arterienerkrankung (periarteritis nodosa), die mit Morbus Brightii und rapid fortsch reitender allgemeiner Muskellähmung einhergeht. *Deutsches Arch. Klin. Med.*, **1**, 484–518.
4. Eppinger, H. (1887) Pathogenesis (Histogenesis und Aetiologie) der Aneurysmen einschliesslich des Aneurysma equi verminosum. *Arch. f. Klin. Cher.*, (Suppl) **1**, 553.
5. Lamb, A.R. (1914) Periarteritis nodosa – a clinical and pathological review. *Arch. Int. Med.*, **14**, 481–516.
6. Hunter, W. (1757) History of an Aneurysm. *Medical Observations & Inquiries*, **1**, 323.
7. Gee, S.J. (1871) Aneurysms of the coronary arteries in a boy. *St. Bartholomew's Hosp. Rep.*, **VII**.
8. Kawasaki, T. (1967) Acute febrile mucocutaneous syndrome with lymph node involvement with specific desquamation of the fingers and toes in children. *Arerugi*, **16**, 178–222.
9. Scott, D.G.I. and Watts, R. (personal communication).
10. Bacon, P.A., Scott, D.G.I. and Luqmani, R. (1996) Rheumatoid vasculitis, (see Chapter 18 of this volume).
11. Bannatyne, G.A. (1898) *Rheumatoid Arthritis*, 2nd edn, John Wright, Bristol.
12. Ellman, P. and Ball, R.E. (1948) 'Rheumatoid disease' with joint and pulmonary manifestations. *Br. Med. J.*, **ii**, 816–21.
13. Sokoloff, L. and Bunim, J.J. (1951) Arteritis of striated muscle in rheumatoid arthritis. *Am. J. Pathol.*, **27**, 157–73.
14. Bywaters, E.G.L. (1949) A variant of rheumatoid arthritis characterised by recurrent digital pad nodules and palmar fascitis, closely resembling palindromic rheumatism. *Ann. Rheum. Dis.*, **8**, 1–30.
15. Bywaters, E.G.L. (1957) Peripheral vascular obstruction in rheumatoid arthritis and its relationship to other vascular lesions. *Ann. Rheum. Dis.*, **16**, 84–103.
16. Monckeberg, J.G. (1905) Über Periarteriitis nodosa. *Beitr. path. Anat.*, **38**, 101–34.
17. Ophuls, W. (1923) Periarteritis acuta nodosa. *Arch. Int. Med.*, **32**, 870–98.
18. Wilson, K.S. and Alexander, H.L. (1945) The relation of periarteritis nodosa to bronchial asthma and other forms of human hypersensitiveness. *J. Lab. Clin. Med.*, **30**.

19. Churg, J. and Strauss, L. (1951) Allergic granulomatosis, allergic angiitis and periarteritis nodosa. *Am. J. Pathol.*, **27**, 277–302.
20. Klinge, F. (1930) Eisweiss uberempgindlichteit (Gewebsanaphylaxie) der Gelenke. Experimentelle pathologische – anatomische Studie zur Pathogenesedes Gelenkrheumatismus. *Beitr. path. Anat.*, **83**, 185–216.
21. Rich, A.R. and Gregory, J.E. (1943) The experimental demonstration that periarteritis nodosa is a manifestation of hypersensitivity. *Bull. Johns Hopkins Hosp.*, **72**, 65–88.
22. Clark, F. and Kaplan, B. (1937) Endocardial, arterial and other mesenchymal alterations associated with serum disease in man. *Arch. Path.*, **24**, 458–75.
23. Rich, A.R. (1942) The role of hypersensitivity in periarteritis nodosa. *Bull. Johns Hopkins Hosp.*, **71**, 123–40. (Additional evidence of the role of hypersensitivity in the etiology of periarteritis nodosa. *Ibid.*, **71**, 375–9.)
24. von Glahn, W.C. and Pappenheimer, A.M. (1926) Specific lesions of peripheral blood vessels in rheumatism. *Am. J. Path.*, **2**, 235–50.
25. Ferrari, E. (1903) Über Polyarteritis acute nodosa (sogenannte Periarteriitis nodosa) und ihre Beziehungen zur Polymyositis und Polyneuritis acuta. *Beitr. path. Anat.*, **34**, 1–25.
26. Zeek, P.M. (1952) Periarteritis nodosa: A critical review. *Am. J. Clin. Path.*, **22**, 777–90.
27. Veszpremi, D. and Jancso, M. (1903) Über einen Fall von Periarterits nodosa. *Beitr. path. Anat.*, **34**, 1–25.
28. Joest, E. and Harzer, J. (1921) Über Periarteritiitis nodosa beim Schwein. *Beitr. path. Anat.*, **69**, 85–102.
29. Guldner, E. (1915) Zwei neue Beobachtungen von Periarteriitis nodosa beim Menschen und beim haus Rinde. *Virchows Arch. f. path. Anat.*, **219**, 366–76.
30. Balo, J. (1924) Periarteriitis nodosa beim Hunde. *Virchows Arch. f. path. Anat.*, **248**, 337–44.
31. Boyd, L.J. (1938) The clinical aspects of periarteritis nodosa. *Bull. New York M. Coll.*, **1**, 219–25.
32. Metz, W. (1932) Die geweblichen Reaktionserscheinungen an der Gefasswand bei hyperergeichen Zustanden und deren Beziehungen zur Periarteritiitis nodosa. *Beitr. path. Anat.*, **88**, 17–36.
33. Smith, C.C. and Zeek, P.M. (1957) Studies on periarteritis nodosa II. The role of various factors in the etiology of periarteritis nodosa in experimental animals. *Am. J. Path.*, **23**, 148–57.
34. Horton, B.T., Magath, T.B. and Brown, G.E. (1934) Arteritis of the temporal vessels. *Arch. Int. Med.*, **53**, 400–9.
35. Kilbourne, E.D. and Wolff, H.G. (1946) Cranial arteritis. a critical evaluation of the syndrome of 'temporal arteritis' with report of a case. *Ann. Int. Med.*, **24**, 1–10.
36. Plaut, A. (1951) Asymptomatic focal arteritis of the appendix. Eighty-eight cases. *Am. J. Path.*, **27**, 247–64.
37. Meyer, P. (1878) Über Periarteritis nodosa oder multiple Aneurysmen der mittieren und kleineren Arterien. *Arch. path. Anat.*, **74**, 277–319.
38. Boyd, L.J. (1941) The renal and cardiac manifestations of periarteritis nodosa. *Bull. New York M. Coll.*, **4**.
39. Lie, J.T. (1996) Classification and histopathologic specificity of systemic vasculitis (see Chapter 2 of this volume).
40. Gruber, G.B. (1925) Zur Frage der Periarteritis nodosa, mit besonderer Berucksichtigung der Gallenblassen und Nieren Beteiligung. *Arch. path. Anat.*, **258**, 441–501.

41. Rose, G.A. and Spencer, H. (1957) Polyarteritis nodosa. *Q. J. Med.*, **26**.

42. Sergent, J.S., Lockshin, M.D., Christian, C.L. and Gocke, D.J. (1976) Vasculitis with hepatitis B antigenemia: long-term observations in nine patients. *Medicine (Baltimore)*, **55**, 1–18.

43. Trepo, C.G., Zuckerman, A.J., Bird, R.C. and Prince, A.M. (1974) The role of circulating hepatitis B antigen–antibody immune complexes in the pathogenesis of vascular and hepatic manifestations of polyarteritis nodosa. *J. Clin. Pathol.*, **27**, 863–8.

44. Davies, D.J., Moran, J.E., Niall, J.F. and Ryan, G.B. (1982) Segmental necrotizing glomerulonephritis with antineutrophil antibody. *Br. Med. J.*, **2**, 608–7.

45. Kallenberg, C.G.M. (1996) Anti-neutrophil cytoplasmic antibodies (ANCA): current perspectives. *Clin. Rheumatol.* (in press).

46. Feehaly, J. *et al.* (1990) Systemic vasculitis in the 1980s – is there an increasing incidence of Wegener's granulomatosis and microscopic polyarteritis? *J. Roy. Coll. Pys.*, **24**, 284–8.

47. Jonker, N.D., Peters, A.M., Gaskin, G. *et al.* (1992) A retrospective study of radiolabelled granulocyte kinetics in patients with systemic vasculitis. *J. Nucl. Med.*, **33**, 491–7.

48. Alarcon-Segovia, D. (1980) Classification of the necrotizing vasculitides in man. *Clin. Rheum. Dis.*, **6**, 223–32.

49. Jennette, J.C., Falk, R.J., Andrassy, K. *et al.* (1994) Nomenclature of systemic vasculitides – proposal of an international consensus conference. *Arthritis Rheum.*, **36(6)**, 185–95.

50. Lie, J.T. (1994) Nomenclature and classification of vasculitis: Plus Ça Change, Plus C'est La Même Chose. *Arthritis Rheum.*, **37(2)**, 181–6.

51. Gordon, M., Luqmani, R.A., Adu, D. *et al.* (1993) Relapses in patients with a systemic vasculitis. *Q. J. Med.*, **86**, 779–89.

52. Hoffman, G.S., Kerr, G.S., Leavitt, R.Y. *et al.* (1992) Wegener's granulomatosis: an analysis of 158 patients. *Ann. Intern. Med.*, **116**, 488–98.

53. Luqmani, R.A., Bacon, P.A., Moots, R.J. *et al.* (1996) Birmingham vasculitis activity score (BVAS) in systemic nectrotizing vasculitis. *Q. J. Med.* (in press).

54. Hay, E.M., Bacon, P.A., Gordon, C. *et al.* (1993) The BILAG index: a reliable and valid instrument for measuring clinical disease activity in systemic lupus erythematosus. *Q. J. Med.*, **86**, 447–58.

55. Gladman, D., Ginzler, E., Goldsmith, C. *et al.* (1996) The SLICC/ACR damage index for SLE. *Arthritis Rheum.* (in press).

56. Reinhold-Keller, E., Kekow, J., Schnabel, A. *et al.* (1994) Influence of disease manifestation and antineutrophil cytoplasmic antibody titer on the response to pulse cyclophosphamide therapy in patients with Wegener's granulomatosis. *Arthritis Rheum.*, **6**, 919–24.

57. Ware, J.E. and Sherbourne, C.D. (1992) The mos 36-item short-form health survey (SF-36) 1: conceptual framework and item selection. *Med. Care*, **30**, 473–83.

58. Cohen Taevert, J.W. *et al.* (1990) Prevention of relapses in Wegener's granulomatosis by treatment based on antineutrophil cytoplasmic antibody titre. *Lancet*, **ii**, 708-11.

59. DeCruz, D.P., Houssiau, F.A., Ramirez, G. *et al.* (1991) Antiendothelial cell antibodies in systemic lupus erythematosus: A potential marker for nephritis and vasculitis. *Clin. Exp. Immunol.*, **85**, 254–61.

60. Wolter, D., Exley, A.R., Moots, R.J. *et al.* (1994) Von Willebrand factor in the assessment of systemic vasculitis. *Clin. Rheumatol.*, **13**, Abst., 130.

61. Kitas, G.K., Saratzis, N., McBurney, A. *et al.* (1994) Free radical-mediated

ischaemia-reperfusion injury to endothelial cells in a human *in vivo* model. *Clin. Rheumatol.*, **13**, Abst., 102.

62. Janssen, B.A., Luqmani, R.A., Gordon, C. *et al.* (1994) Correlation of blood levels of soluble vascular cell adhesion molecule-1 (sVCAM-1) with disease activity in Systemic Lupus Erythematosus (SLE) and vasculitis. *BJR*, submitted.

63. Fauci, A.S., Haynes, B.F. and Katz, P. (1978) Cyclophosphamide induced remission in advanced polyarteritis nodosa. *Am. J. Med.*, **64**, 890–4.

64. Scott, D.G.I. and Bacon, P.A. (1984) Intravenous cyclophosphamide plus methylprednisolone treatment in systemic rheumatoid vasculitis. *Am. J. Med.*, **76**, 377–84.

65. Luqmani, R.A., Pall, A., Adu, D. *et al.* (1996) Therapy of vasculitis (see Chapter 23 of this volume).

67. Adu, D., Pall, A., Luqmani, R.A. *et al.* (1993) Controlled trial of treatment of vasculitis. *Clin. Exp. Rheumatol.*, **93**(84), Supp. 1.

67. De Remee, R.A., McDonald, and Weiland, C.H. (1985) Wegener's granulomatosis: observations on treatment with antimicrobial agents. *Mayo Clin. Proc.*, **60**, 27–32.

# 2

# Classification and histopathologic specificity of systemic vasculitis

*J.T. Lie*

Vasculitis is a disease of antiquity, predating the description of arteriosclerosis [1,2]. Vasculitis has a deceptively simple definition – inflammation, often with necrosis and occlusive changes of the blood vessels – but its clinical manifestations are diverse and complex. Vasculitis may be generalized or localized. It may occur *de novo* as an essential disorder of the blood vessels (primary vasculitis), or it may be associated with a variety of different underlying diseases (secondary vasculitis). The clinical spectrum of vasculitides therefore represents one of the most interesting and perplexing group of diseases in medicine. In the absence of pathognomonic clinical features and laboratory tests, the diagnosis of vasculitis still relies heavily on clinicopathologic correlation for the correct interpretation of biopsies because histologic changes may not be specific in any given case and they often overlap in different vasculitides [3–5].

The interpretation of histologic changes in a biopsy is also subject to such variables as the examining pathologist's interest and experience, tissue selection and sample size, chronologic age of the disease process and any drug treatment prior to the biopsy. A biopsy is adequate only when it provides a verifiable diagnosis. A positive biopsy is always helpful whereas a negative biopsy does not always exclude the disease under consideration because the vasculitis may be focal and segmental in distribution, and sampling errors are inherent to small biopsies in these situations [4].

*The Vasculitides*. Edited by B.M. Ansell, P.A. Bacon, J.T. Lie and H. Yazici.
Published in 1996 by Chapman & Hall, London. ISBN 978-0-412-64140-4.

CLASSIFICATION OF VASCULITIS

The prototype of systemic vasculitis, 'periarteritis nodosa' (or polyarteritis nodosa), was first described by Kussmaul and Maier in 1866 [6]. It was possible that the disease named periarteritis nodosa by Kussmaul and Maier had been recognized even earlier by Michaelis and Martani in 1755, by Pelletan in 1810, by von Rokitansky in 1852 and by Rudolf Virchow in 1863 [7]. But the 35-page-long article by Kussmaul and Maier [6] was a classic. They described the autopsy findings of a 27-year-old apprentice tailor who had a fulminant illness characterized by fever, productive cough, proteinuria, myalgia, peripheral neuritis and abdominal pain, with a fatal termination about one month later. The autopsy revealed widespread nodular inflammation of muscular arteries, especially at branching points, of the caliber of the hepatic and coronary arteries. The authors observed glomerular lesions and that the smallest branches of the coronary arteries were affected as well and, thus, Kussmaul and Maier had indeed described both the classic and microscopic forms of polyarteritis nodosa.

The descriptions of other necrotizing vasculitides followed, all of which may be linked to polyarteritis nodosa which is, as it were, the surrogate single parent of a family tree of necrotizing vasculitides (Figure 2.1). In 1926, von Glahn and Pappenheimer [8] observed vascular lesions in rheumatic disease but thought they were different from periarteritis nodosa. In 1933, Rössle [9], attempted to include rheumatic carditis and periarteritis in a circle of rheumatic collagen diseases. This circle consisted of diseases that present a combination of typical rheumatic lesions and necrotizing vascular lesions of the periarteritis nodosa type, in varying proportions and combinations. One of Rössle's patients (case IV) actually presented with granulomatous and necrotizing vascular lesions and involvement of the respiratory tract. This might have been an early example of the then unknown entity of Churg–Strauss syndrome or Wegener's granulomatosis. Only two years earlier, Klinger [10], in 1931, had described a patient with destructive sinusitis, nephritis and disseminated vasculitis. Klinger considered this to be a borderline case of periarteritis nodosa, but we now accept it as the first example of Wegener's granulomatosis. Wegener's own reports of 'generalized septic vascular disease' and 'rhinogenic granuloma' were published in 1936 and 1939 [11,12]. Godman and Churg [13] first published a review of Wegener's granulomatosis in 1954 and introduced the diagnostic triad of necrotizing granulomatous inflammation of the upper and lower respiratory tract, pulmonary and systemic vasculitis and glomerulonephritis.

The stage was then set for a classification of necrotizing systemic vasculitis to lessen the nosologic confusion. In 1952, Zeek [14] proposed

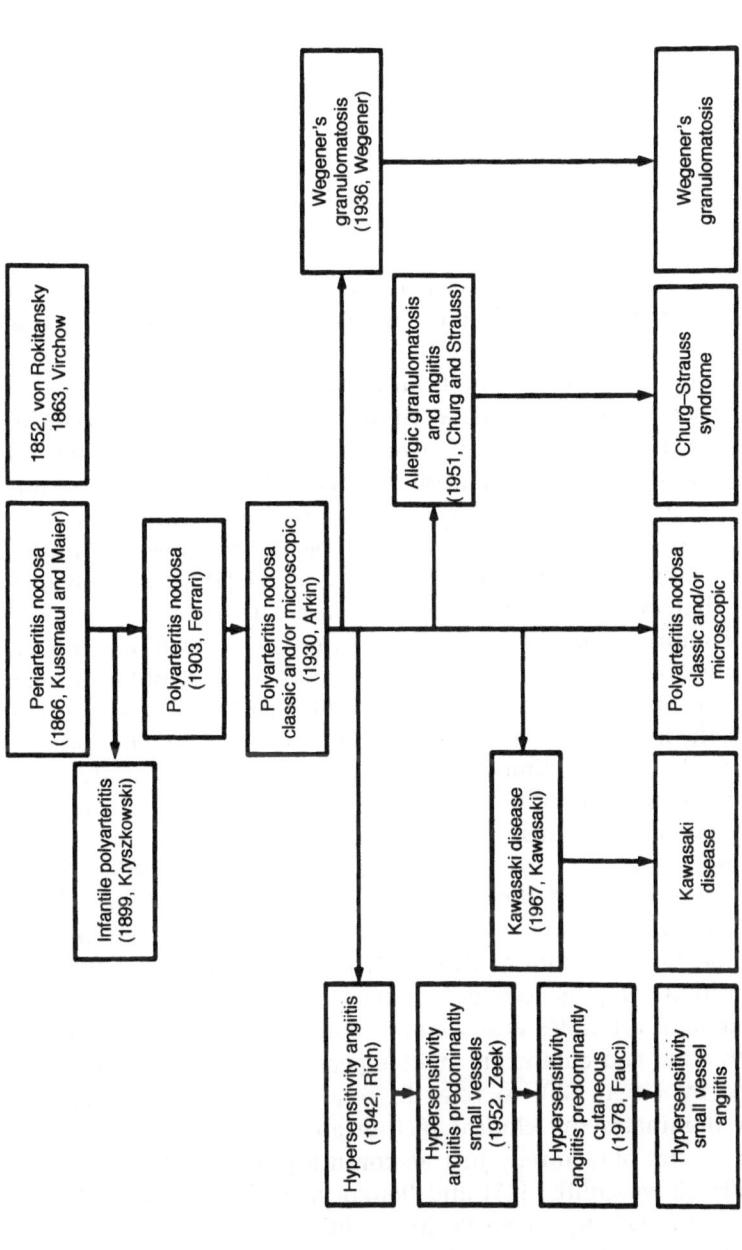

**Figure 2.1** A family tree of necrotizing vasculitis with periarteritis nodosa as the surrogate single parent. (Reproduced with permission from [2].)

the generic term 'necrotizing angiitis' to designate the vascular lesions, both arterial and venous, the fully developed stage of which consisted of fibrinoid necrosis and inflammatory reaction involving all three coats of vessel walls. Five types of necrotizing angiitis were recognized and designated as follows:

1. hypersensitivity angiitis
2. allergic granulomatous angiitis
3. rheumatic arteritis
4. periarteritis nodosa
5. temporal arteritis.

Zeek [14] was apparently either unacquainted with Wegener's granulomatosis or had grouped it together with allergic granulomatous angiitis in this classification.

Expanding on and modifying from the scheme devised by Zeek [14], there have been numerous other attempts to classify vasculitis according to the size and type of vessels affected, the presence or absence of lung involvement, the nature of cutaneous or renal lesions, and the presumptive etiologic factors and pathogenesis, none of which can be entirely satisfactory so long as the etiology and pathogenesis of most vasculitides are either unknown or incompletely understood [15]. The prevailing theory is that most vasculitis syndromes are caused by deposition of immune complexes or cell-mediated immune mechanism affecting the vessel walls [16–18]. The probable antigen has been identified in very few instances, notably in some patients with hepatitis B infection [19]. Moreover, the immune complex-mediated tissue injury does not produce a single or stereotypic clinical syndrome, but it tends to apply to the whole spectrum of vasculitides and to overlap with injuries caused by other immune mechanisms.

With the proliferation of competing schemes of classification, it becomes inevitable that the same name has been used to refer to different vasculitic syndromes and different names are adopted to describe the same condition. Zeek [14] had originally used the term 'hypersensitivity angiitis' to refer to patients with disseminated necrotizing vasculitis of arterioles and venules, both systemic and pulmonary, often with renal and cutaneous involvements. The same term has now been used by others [20] to refer to conditions in which cutaneous vasculitis dominates the clinical picture and visceral involvement is absent or occurs only rarely. Just as confusing is the adoption of the names 'allergic vasculitis' [21] and 'leukocytoclastic vasculitis' [22] as synonyms for Zeek's hypersensitivity angiitis.

In the past, an exact characterization of different types of vasculitis syndromes might have been principally a matter of academic interest since few effective therapeutic modalities existed. We now know that for many severe forms of vasculitis, such as Wegener's granulomatosis and systemic necrotizing vasculitis, the outlook may be vastly improved and long-term remissions made possible by treatment with corticosteroids and/or cyclophosphamide. Early and correct diagnosis of vasculitis becomes vital not only for prompt and aggressive immunosuppressive treatment of certain malignant forms of vasculitis to prevent irreversible tissue damage, but also for withholding cytotoxic agents in situations where their use is unwarranted and harmful.

A classification of primary and secondary vasculitis (Table 2.1), incorporating the predominant type and size of the blood vessel involvement, offers guidelines to the clinicians for when to suspect and what to expect in the histologic diagnosis of vasculitis [15]. The type of inflammatory cell infiltrate in vasculitis is independent of the size of blood vessels affected; and a mixed-cell infiltrate is the rule rather than an exception

**Table 2.1** A practical classification of vasculitis

---

*Primary vasculitis (according to different vessel size)*
Affecting large, medium-sized and small blood vessels:
- Takayasu arteritis;
- giant cell (temporal) arteritis;
- granulomatous angiitis of the central nervous system.

Affecting predominantly medium-sized and small blood vessels:
- polyarteritis nodosa;
- Churg–Strauss syndrome;
- Wegener's granulomatosis.

Affecting predominantly small blood vessels:
- microscopic polyangiitis;
- Schönlein–Henoch syndrome;
- cutaneous leukocytoclastic angiitis.

Miscellaneous conditions:
- Buerger's disease;
- Behçet's disease;
- Kawasaki disease.

Secondary vasculitis:
- infection-related vasculitis;
- serum sickness or drug hypersensitivity-related vasculitis;
- hypocomplementemic urticarial vasculitis;
- vasculitis associated with rheumatic connective tissue diseases;
- vasculitis associated with other systemic diseases;
- malignancy-related vasculitis;
- post-transplant vasculitis;
- pseudovasculitic syndromes (myxoma, endocarditis, Sneddon syndrome).

---

Modified from Lie [15].

for most vasculitides. Several histologic types of vasculitis are recognized: granulomatous vasculitis, with or without giant cells; large-vessel lymphoplasmacytic vasculitis; small-vessel lymphocytic vasculitis; eosinophilic vasculitis; necrotizing vasculitis, with or without fibrinoid necrosis; and leukocytoclastic vasculitis. Some vasculitides are focal and segmental in their anatomic distribution and, therefore, the examination of serial sections of a biopsy is often necessary to elicit a specific histologic marker of the vasculitis, e.g. giant cells in giant cell vasculitis. Involvement of veins can be an important discriminator, as in thromboangiitis obliterans (Buerger's disease).

## HISTOPATHOLOGIC SPECIFICITY OF VASCULITIS

Although a classification of vasculitis may be based on the predominant size of the blood vessel involvement, overlap frequently occurs among the major vasculitic syndromes (Figure 2.2). Overlapping in the size of blood vessel involvement also occurs within an individual disease entity. For example, vasculitis associated with rheumatoid arthritis and systemic lupus erythematosus (SLE) may affect the largest blood vessel in the body, the aorta (granulomatous aortitis), as well as the smallest cutaneous blood vessels (leukocytoclastic vasculitis), though not necessarily in the same person.

The type of inflammatory cell infiltrate in vasculitis is independent of the size of blood vessels affected, and a mixed-cell infiltrate is the rule rather than an exception for most vasculitides (Figure 2.3). Several distinct histologic types of vasculitis are recognized, namely granulomatous (giant cell) angiitis, thromboangiitis obliterans, necrotizing angiitis, eosinophilic (allergic) angiitis, leukocytoclastic vasculitis and lymphocytic vasculitis. These different histologic types of vasculitis usually have a predilection for affecting predominantly blood vessels of different calibers (Table 2.2). Involvement of veins in the lower or upper limbs can be an important diagnostic discriminator, as in thromboangiitis obliterans (Buerger's disease) [23]. In small vessel cutaneous angiitis, a distinction is often made between leukocytoclastic vasculitis (with neutrophilic infiltrate) and lymphocytic vasculitis (with lymphocytic infiltrate). However, a lymphocytic vasculitis may represent the resolving phase of an immune complex mediated neutrophilic vasculitis, as in urticarial vasculitis. Indeed, sequential biopsies have demonstrated that in cutaneous leukocytoclastic vasculitis, the character of the infiltrate progressively changed from a neutrophilic-predominant to a lymphocytic-predominant type over a 120-hour period [24].

The 1990 American College of Rheumatology's classification criteria of vasculitis [25] selected the following seven types as major categories of systemic vasculitis:

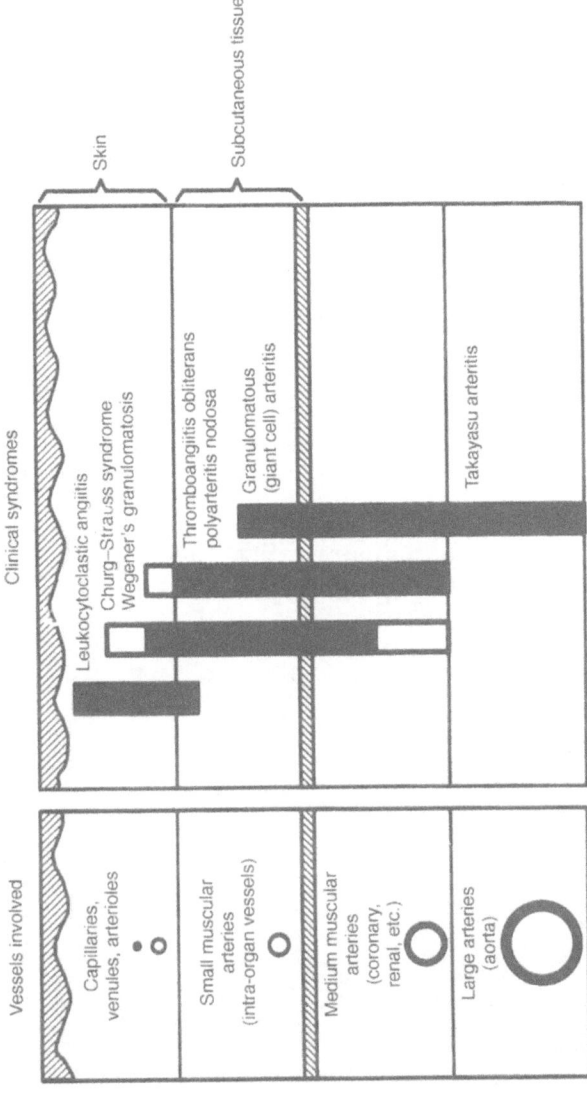

**Figure 2.2** A scheme depicting the overlapping nature of the level and size of blood vessels involved in major vasculitis syndromes. (Reproduced with permission from [2].)

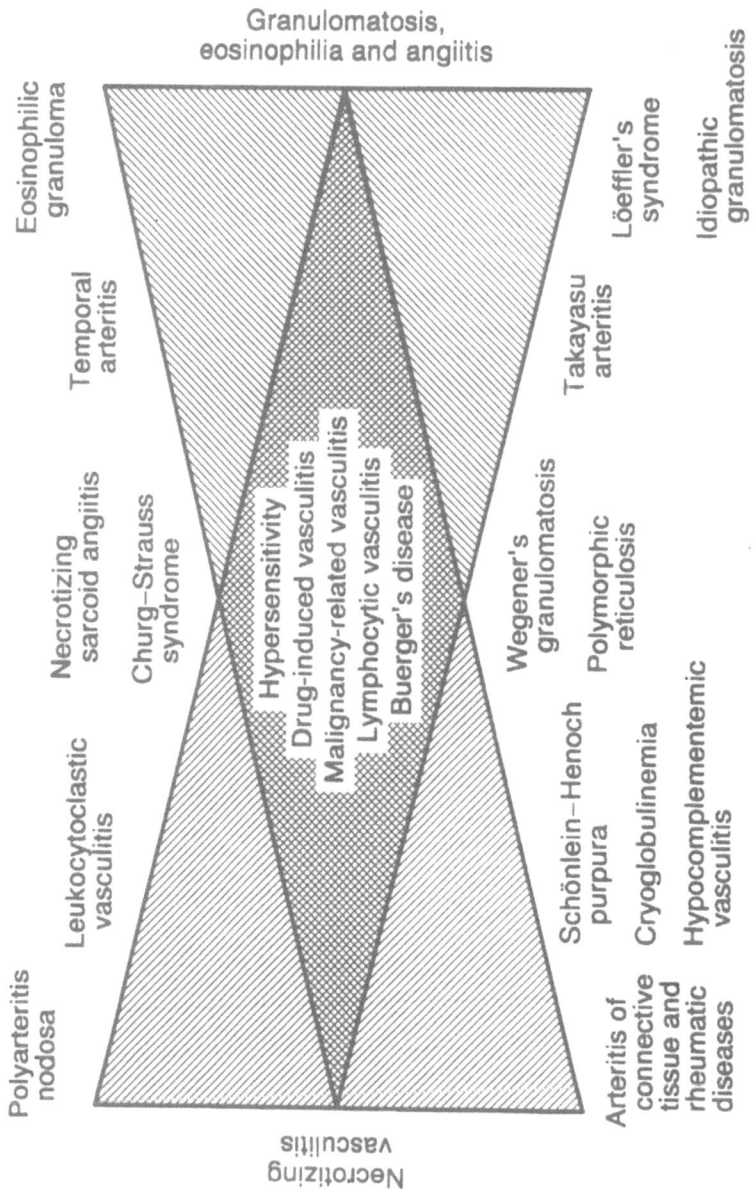

**Figure 2.3** Graphic representation of the overlapping spectra of various entities of granulomatous disease, granulomatous allergic angiitis and necrotizing angiitis. (Reproduced with permission from [2].)

**Table 2.2** Histopathology of vasculitides: morphology type versus vessel size

| Histopathology | Prototype example | Large arteries | Medium-sized arteries | Small arteries and arterioles | Veins and venules |
|---|---|---|---|---|---|
| Granulomatous (giant cell) angiitis | Temporal arteritis Takayasu arteritis | +++ | +++ | + | + |
| Thromboangiitis obliterans | Buerger's disease | + | +++ | +++ | +++ |
| Necrotizing vasculitis | Polyarteritis nodosa | 0 | +++ | +++ | + |
| Eosinophilic angiitis | Churg–Strauss syndrome | 0 | ++ | +++ | ++ |
| Leukocytoclastic vasculitis | Urticarial vasculitis | 0 | 0 | +++ | +++ |
| Lymphocytic vasculitis | Erythema nodosum | 0 | ++ | +++ | ++ |

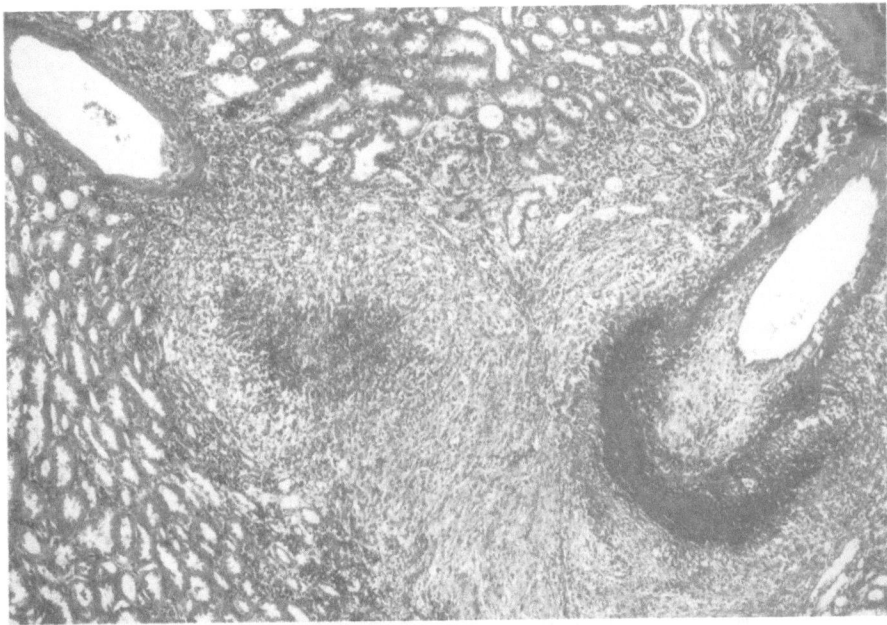

**Figure 2.4** Characteristic histopathology of classic polyarteritis nodosa in the kidney (from left to right): side-by-side coexistence of normal artery, artery with circumferential fibrinoid necrosis and artery with segmental necrosis and micro-aneurysm formation. (Hematoxylin-eosin, × 64.)

**Table 2.3** Comparison of the pathologic characteristics of selected vasculitis syndromes

| | Polyarteritis nodosa | Churg–Strauss syndrome | Wegener's granulomatosis | Hypersensitivity vasculitis | Henoch–Schönlein purpura | Giant cell (temporal) arteritis | Takayasu arteritis |
|---|---|---|---|---|---|---|---|
| Type of vessels involved | Medium and small muscular arteries, and sometimes arterioles | Small arteries and veins, often arterioles and venules | Usually small arteries and veins, sometimes larger vessels | Arterioles and venules, often small arteries and veins | Arterioles and venules, often small arteries and veins | Vessels of all sizes | Elastic arteries and selected muscular arteries |
| Distribution and localization | Visceral, cutaneous and, infrequently, cerebral vessels and lung | Upper and lower respiratory tract, viscera, heart and skin | Upper and lower respiratory tract, often kidney and, infrequently, skin, heart, viscera and brain | Predominantly skin and less commonly viscera, heart and synovium | Predominantly skin, gastro-intestinal, kidney and synovium | Predominantly temporal arteries, and less often any other large, medium and small vessels | Aorta, arch vessels and other major branches (coronary, renal, visceral) and pulmonary arteries |
| Type of vasculitis and inflammatory cell infiltrate | Necrotizing, with mixed cells and few eosinophils, rarely granulomatous | Necrotizing or granulomatous, with mixed cells and prominent eosinophils | Necrotizing or granulomatous, with mixed cells and occasional eosinophils | Leukocytoclastic or lymphocytic, with variable number of eosinophils, occasionally granulomatous | Leukocytoclastic, mixed-cell or lymphocytic, with variable number of eosinophils | Granulomatous, with variable number of giant cells, sometimes only lymphoplasmacytic | Granulomatous, with few giant cells in active phase, and sclerosing fibrosis in chronic stage, with scanty infiltrate |

| | | | | | | |
|---|---|---|---|---|---|---|
| Special features | Focal segmental involvement of vessels; coexisting acute and healed vascular lesions, or normal and affected vessels; microaneurysms | Extravascular necrotizing granulomas with prominent eosinophils; may manifest as 'limited form' | Geographic pattern of tissue necrosis and positive antineutrophil cytoplasmic antibodies; may manifest as 'limited form' | May be associated with myocarditis, interstitial nephritis or hepatitis | IgA immune deposits in affected tissue | Affected extracranial large vessels indistinguishable from Takayasu arteritis; may form aneurysm or cause dissection | Aneurysmal in 20%; may be segmental, and cause rupture or dissection |
| Demographic and environmental predisposition | Vascular lesion of infantile polyarteritis is indistinguishable from fatal cases of Kawasaki disease | Most patients have asthma or history of allergy | Occurs in all ages, with a slight male preponderance; associated with HLA-DR2; may respond to antimicrobial agents | Patients may have history of drug or chemical allergy, vaccination or occult malignancy | Predominantly children and young adults | Virtually all patients with temporal arteritis are over age 50; may be clinically asymptomatic | Most commonly in women of childbearing age; more prevalent in the Orient; an important cause of renovascular hypertension in adolescents |

Reproduced with permission from Lie, J.T. (1990) *Arthritis Rheum.*, 33, 1074–87.

1. polyarteritis nodosa
2. Churg–Strauss syndrome
3. Wegener's granulomatosis
4. hypersensitivity vasculitis
5. Henoch–Schönlein purpura (also known as Schönlein–Henoch purpura)
6. giant cell arteritis
7. Takayasu arteritis.

Each of these major vasculitic syndromes has its own unique histopathologic features, but overlap still occurs and often confuses the unwary or casual observer (Table 2.3).

Although few of these major vasculitic syndromes have absolutely diagnostic features, each and every one of them possesses certain characteristic histopathologic markers that, when present, lend specificity to the diagnosis. For classic polyarteritis nodosa, the characteristic markers are the side-by-side coexistence of normal arterial segments and segments with necrotizing vasculitis (Figure 2.4) as well as the coexistence of acute and healing vascular lesions (Figure 2.5); for Wegener's granulomatosis, vasculitis occurs in fewer than 50% of the cases and the

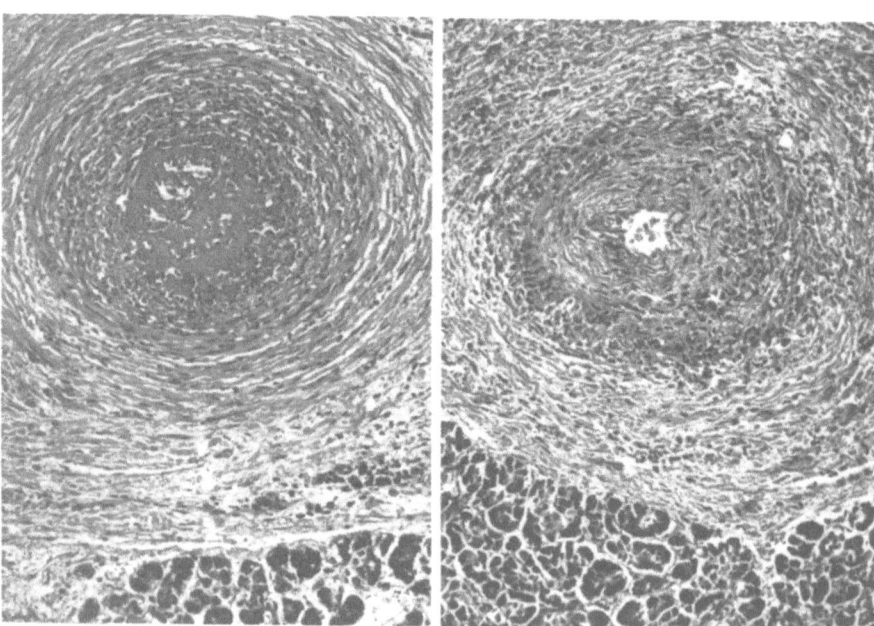

**Figure 2.5** Characteristic histopathology of classic polyarteritis nodosa in the pancreas: coexisting acute (left) and healing (right) necrotizing vasculitis. (Hematoxylin-eosin, × 160.)

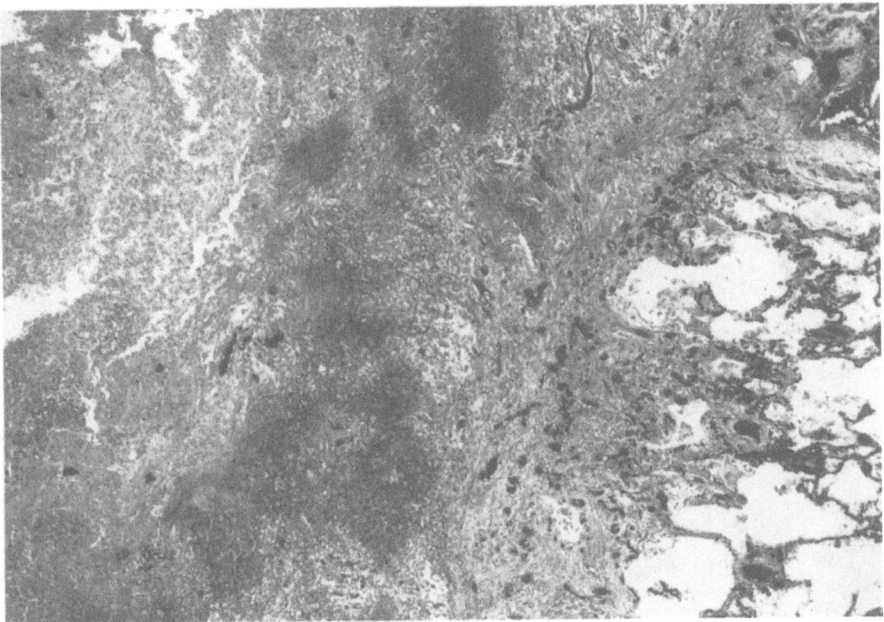

**Figure 2.6** Diagnostic histopathology of Wegener's granulomatosis in open lung biopsy: geographic pattern granulomatous necrosis. (Hematoxylin-eosin, × 64.)

geographic pattern necrosis (Figure 2.6) is diagnostic; for Churg–Strauss syndrome, it is the extravascular granuloma (Figure 2.7) more than the eosinophilic angiitis that is diagnostic; and for small vessel leukocytoclastic vasculitis, it is the involvement of venules and arterioles with karyorrhexis of the granulocytic infiltrate and extravasation of erythrocytes (Figure 2.8) that provides specificity, irrespective of the etiology and the underlying disease.

## CONCLUSION

The biopsy evaluation of vasculitis is a vital component in the clinical management of a patient, and a correct diagnosis usually cannot be made without correlation with the patient's clinical history, physical examination and angiographic findings; preferably all three. The interpretation of all biopsies for vasculitis must, in the final analysis, remain more an art than mere science.

## REFERENCES

1. Koelbing, H.M. (1975) Some remarks on the history of arterial pathology. *Pathol. Microbiol.*, **45**, 85–92.

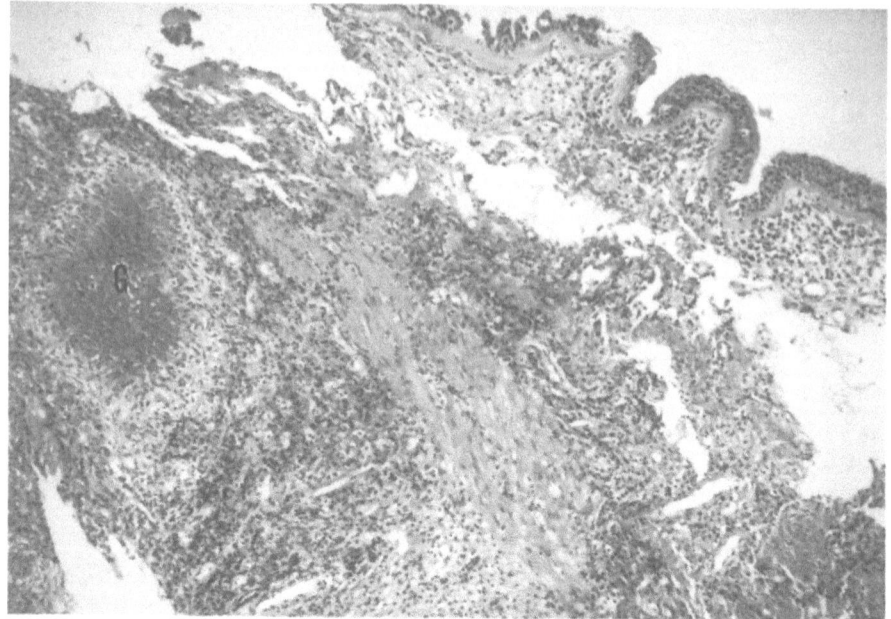

**Figure 2.7** Diagnostic histopathology of Churg–Strauss syndrome in lung biopsy: extravascular eosinophilic granuloma (*G*) in bronchial submucosa. (Hematoxylin-eosin, × 160.)

2. Lie, J.T. (1992) Vasculitis, 1815–1991: Classification and diagnostic specificity. (Dunlop-Dottridge Lecture) **19**, 83–89.
3. Lie, J.T. (1987) The classification and diagnosis of vasculitis in large and medium-sized blood vessels. *Pathol. Annu.*, **22**(part 1), 125–62.
4. Lie, J.T. (1989) Systemic and isolated vasculitis: A rational approach to classification and pathologic diagnosis. *Pathol. Annu.*, **24**(part 1), 25–114.
5. Lie, J.T. (1989) Classification of pulmonary angiitis and granulomatosis: Histopathologic perspectives. *Semin. Respir. Med.*, **10**, 111–21.
6. Kussmaul, A. and Maier, R. (1866) Über eine bisher nicht beschreibene eigenthümliche Arterienerkrankung (Periarteritis nodosa), die mit Morbus Brightii und rapid fortschreitender allgemeiner Muskellähmung einhergeht. *Dtsch. Arch. Klin. Med.*, **1**, 484–518.
7. Lamb, A.R. (1914) Periarteritis nodosa: A clinical and pathological review. *Arch. Intern. Med.*, **14**, 481–516.
8. von Glahn, W.C. and Pappenheimer, A.M. (1926) Specific lesions of peripheral blood vessels in rheumatism. *Am. J. Pathol.*, **2**, 235–50.
9. Rössle, R. (1933) Zum Formenkreis der rheumatischen Gewebsveränderungen mit besonderer Berücksichtigung der rheumatischen Gefässentzuendungen. *Virchows Arch. path. Anat.*, **288**, 780–832.
10. Klinger, H. (1931) Grenzformen der periarteritis nodosa. *Frankfurt Ztschr. Path.*, **42**, 455–80.
11. Wegener, F. (1936) Über generalisierte, septische Gefässerkrankugen. *Verhandl. deutsch. path. Gessellsch.*, **29**, 202–10.
12. Wegener, F. (1939) Über eine eigenartige rhinogene Granulomatose mit

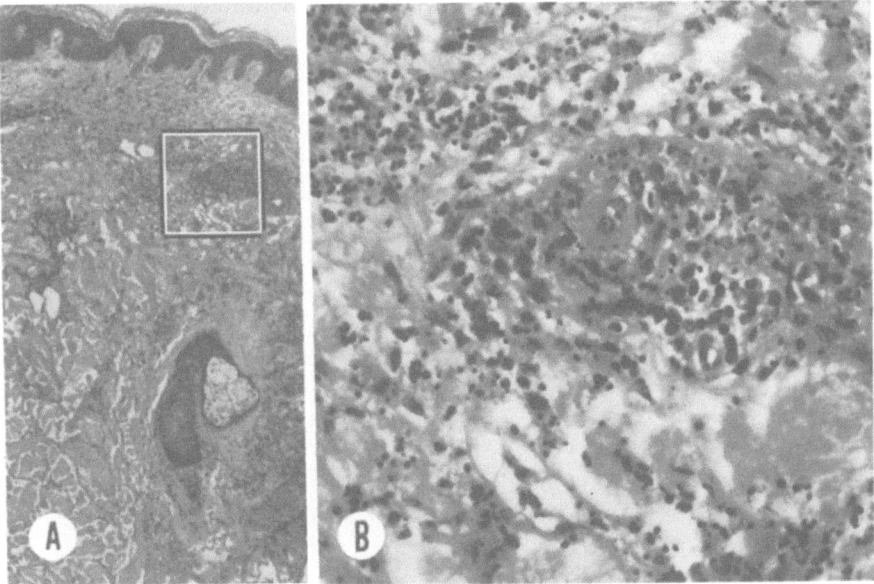

**Figure 2.8** A. Characteristic histopathology of cutaneous leukocytoclastic vasculitis. B. Close-up view of granulocyte karyorrhexis and small-vessel necrosis in the boxed area of 'A'. (Hematoxylin-eosin, A × 64, B× 400.)

besonderer Beteiligung des Arteriensystems und der Nieren. *Beitr. path. Anat.*, **102**, 36–68.

13. Godman, G. and Churg, J. (1954) Wegener's granulomatosis: Pathology and review of the literature. *Arch. Pathol.*, **58**, 533–53.
14. Zeek, P.M. (1952) Periarteritis nodosa: A critical review. *Am. J. Clin. Pathol.*, **22**, 777–90.
15. Lie, J.T. (1994) Nomenclature and classification of vasculitis: Plus ça change, plus c'est la même chose. *Arthritis Rheum.*, **37**, 181–6.
16. Mannik, M. (1987) Experimental models for immune complex-mediated vascular inflammation. *Acta Med. Scand.*, (Suppl.) **715**, 145–55.
17. Moore, P.M. (1989) Immune mechanisms in the primary and secondary vasculitides. *N. Neurol. Sci.*, **93**, 129–45.
18. Savage, C.O.S. (1991) Pathogenesis of systemic vasculitis, in *Systemic Vasculitides* (eds A. Churg and J. Churg), Igaku-Shoin, New York, pp. 7–30.
19. Sergent, J.S., Lickshin, M.D., Christian, C.L. and Gocke, D.J. (1976) Vasculitis with hepatitis B antigenemia: Long term observations in nine patients. *Medicine (Baltimore)*, **55** 1–18.
20. Cupps, T.R. and Fauci, A.S. (1965) *The vasculitides*, WB Saunders, Philadelphia, pp. 50–71.
21. McCombs, R.P. (1965) Systemic allergic vasculitis. *JAMA*, **194**, 1059–64.
22. Sams, W.M. Jr., Thorne, E.R., Small, P. *et al.* (1976) Leukocytoclastic vasculitis. *Arch. Dermatol.*, **112**, 219–26.
23. Lie, J.T. (1988) Thromboangiitis obliterans (Buerger's disease) revisited. *Pathol. Annu.*, **23**(part 2), 257–91.
24. Zax, R.H., Hodge, S.J. and Callen, J.P. (1990) Cutaneous leukocytoclastic

vasculitis: Serial histopathologic evaluation demonstrates the dynamic nature of the infiltrate. *Arch. Dermatol.*, **126**, 69–72.
25. The American College of Rheumatology (1990) Criteria for the Classification of Vasculitis. *Arthritis Rheum.*, **33**, 1065–144.

# Part Two

## Disease Mechanisms and Investigations

# 3

# Vasculitis: mechanisms of injury

*F.C. Breedveld and M.R. Daha*

In this review the possible pathogenetic mechanisms that lead to the inflammation and destruction of the vessel wall will be discussed. Both clinical and animal studies have shown that immunopathogenic mechanisms prevail in vasculitis although it is frequently not known how they are initiated and perpetuated. The following mechanisms presumably underlie most clinical types of vasculitis (Table 3.1). The most widely accepted mechanism is that circulating immune complexes deposited in the blood vessel walls instigate the pathological processes of several of the vasculitides. Endothelial cells, by forming the interphase between the bloodstream and the vessel wall, play a pivotal role in the prevention of lesions that occur in vasculitic syndromes. Therefore, many studies have focused on potential mechanisms of endothelial cell damage in vasculitis.

In the first place, endothelial cells may suffer from the binding of antibodies binding to antigenic determinants situated on the surface of endothelial cells or the binding of antibodies to antigens planted in the vessel wall. Furthermore, endothelial cells may suffer from the interac-

**Table 3.1** Mechanisms of immune vascular damage

- Deposition of immune complexes
- Antibodies against endothelial cells
- Antibodies against planted antigens
- Endothelial cell damage by phagocytes
- T cell reactivity against the vessel wall
- Deregulation of anti-inflammatory mechanisms at the endothelial level

*The Vasculitides*. Edited by B.M. Ansell, P.A. Bacon, J.T. Lie and H. Yazici.
Published in 1996 by Chapman & Hall, London. ISBN 978-0-412-64140-4.

tion with phagocytes or T cells, and finally deregulation of the anti-inflammatory activity of endothelial cells may be a factor underlying vasculitis lesions.

Already 50 years ago vasculitis was induced in rabbits by the injection of horse serum [1]. The intravenous injection of foreign proteins induces the production of antibodies that form immune complexes. Deposition of these circulating immune complexes in the vessel wall results in a necrotizing vasculitis and glomerulonephritis. These abnormalities predominantly occurred in the days when immune complexes are formed in the presence of antigen excess.

Once deposited in the vessel wall, immune complexes activate the complement cascade leading to the release of chemotactic factors and the recruitment of neutrophils. The release of proteolytic enzymes and toxic hydroxyl radicals then promotes vessel wall destruction. Thrombosis, occlusion and hemorrhage may occur, followed by ischemic changes in the tissue. These mechanisms probably occur in patients with allergic or leukocytoclastic vasculitides, which is seen in patients with viral infections, cryoglobulinemia, Henoch–Schönlein purpura, systemic lupus erythematosus (SLE) or rheumatoid arthritis (RA). Support for this view comes from the similarities in histology between experimental serum sickness and the histopathology found in these patients, as well as the presence of perivascular immune complex deposits, together with high levels of circulating immune complexes and hypocomplementemia in these diseases.

Westedt *et al.* studied the normal skin of 64 patients with rheumatoid arthritis and scored the biopsies for perivascular cell infiltrate (0, 1, 2, 3) and deposits of immunoglobulins or complement [2]. Perivascular cells were mainly found in seropositive patients, rarely in seronegative patients and not in healthy controls. Vascular deposits of IgM and C3 were seen in seropositive patients only (not in seronegative or healthy controls) and correlated to some extent with the score of cellular infiltrate. Perivascular deposits of IgG and IgA were only found in skin areas with leukocytoclastic vasculitis. Furthermore, rheumatoid arthritis (RA) patients with perivascular IgM deposits have significantly higher levels of circulating immune complexes as measured by the C1qBA. These correlations suggest that circulating immune complexes precipitate subendothelially and induce a cellular infiltrate in normal skin. This process is clinically relevant since longitudinal studies of this patient cohort showed that patients with the highest scores for cellular infiltrates developed the highest number of extra-articular features including vasculitis.

Obviously, this simple explanation brings forward many questions. Why are immune complexes deposited in some blood vessels and not in others? Why do immune complexes initiate inflammation at some places and not at others? Why are immune complexes not removed by the mononuclear phagocyte system?

Experimental work has produced some answers. A factor that influences the deposition of immune complexes is the permeability of the endothelial cell layer. In addition, injection of histamine in the normal skin of patients with allergic vasculitis induces local lesions [3]. The size of immune complexes also determines deposition. Vasculitis lesions predominantly occur in the first phase of serum sickness when immune complexes are formed in excess of antigen, which leads to the deposition of small complexes [4]. Physicochemical properties other than size, such as isotype and charge, are also important. Furthermore, hemodynamic factors, such as flow or hydrostatic pressure, were shown to be of influence, which explains the frequent localization of cutaneous vasculitis in the lower limbs.

The effectivity of the deposited immune complexes to activate the complement system may be the variable that determines whether deposited immune complexes initiate vasculitis or not. Immune complexes activate the classical pathway of the immune system, which results in the assembly of the so-called membrane attack complex (MAC), which has a cytolytic potential. Boom *et al.* studied the lesional and non-lesional skin of 15 patients with leukocytoclastic vasculitis with antibodies against the MAC [5]. Perivascular deposits of the MAC were found in all but two of the lesions and in only two biopsies of the uninvolved skin, which suggests that the activation of the terminal components of the complement system may play a role in the formation of lesions in leukocytoclastic vasculitis.

Another factor that may contribute to immune complex mediated vasculitis is a decreased clearance of circulating immune complexes. Circulating immune complexes are predominantly cleared by the liver and are transported to the liver by erythrocytes. Immune complexes bind complement and thereby acquire C3b on their surface, subsequently binding to the CR1 receptor, which in the circulation is mainly present on erythrocytes. The complexes are then transported to the liver, removed from erythrocytes and degraded. In experimental situations where the erythrocyte transport system was inactivated, the intravenous injection of immune complexes resulted in an accelerated disappearance of immune complexes from the blood, not only into the phagocytic system of liver and spleen but also because of systemic vascular deposition [6]. These phenomena can be observed in SLE. Radiolabeled immune complexes bind less efficiently to erythrocytes of SLE patients compared with those of RA or healthy controls, which could be attributed to a decreased expression of CR1 receptors. Halma *et al.* found that injecting radiolabeled immune complexes into 22 SLE patients and 12 healthy controls, produced a biexponential elimination curve with an initial phase of rapid elimination [7]. In this phase, scintigraphy revealed a decreased uptake of immune complexes by the liver and especially by the spleen, with an increased deposition of immune complexes mainly in the lungs and kidneys. An effect of genetic factors on the decreased expression of CR1 receptors in SLE has been reported, but there are now several reports that suggest that the decreased numbers of complement receptors on erythrocytes of SLE patients may be acquired and is therefore an epiphenomenon of immune complex formation [8, 9]. In both situations, however, this decrease, by enhancing the deposition of immune complexes outside the liver, can contribute to vasculitis.

Antibodies directed to antigenic components situated on the surface of endothelial cells have been strongly implicated in vasculitis. A breakthrough in the potential significance of antiendothelial antibodies was made in renal allograft rejection, where rejection was found to be correlated with antiendothelial antibodies [10]. Leung *et al.* have reported that > 60% of the patients with Kawasaki disease, a form of vasculitis in young children, have circulating IgM antibodies directed against endothelial cells [11]. These antibodies produce complement-dependent lysis of cultured endothelial cells only if these cells were previously stimulated with IL-1 or TNF. Both cytokines induce the expression of endothelial cell antigens. Cines *et al.* reported antiendothelial antibodies in SLE [12], and Heurkens *et al.* demonstrated these antibodies in SLE as well as in rheumatoid vasculitis [13]. Antibodies in these patients had no direct cytolytic potential. A minority of the serum samples induced antibody-dependent, cell-mediated cytotoxicity. Preincubation of endothelial cells with IL-1 and to a lesser extent with TNF or gamma–interferon, increased binding of the antibodies but rarely induced cytotoxicity. The heterogeneity of these antibodies was demonstrated by immunoblots prepared with membrane fragments of endothelial cells [14]. The nature of these antigens has not yet been defined. Initial investigations suggest that there is some cross-recognition of ANA, ANCA and antiphospholipid antibodies. The antibodies against endothelial cells are directed against a variety of antigens, which are not endothelial cell-specific since the antibody binding activity could be absorbed from serum both with endothelial cells and fibroblasts [13].

Despite the absence of specificity for endothelial cells or toxicity, these antibodies may initiate pathological changes by altering the biochemical behavior and function of endothelial cells [15]. *In vitro* experiments showed that incubation of endothelial cells with these antibodies leads to increased PMN+ platelet adhesion and increased procoagulant properties of these cells [12]. A pathogenetic role for these antibodies is suggested by their presence only in RA patients suffering from vasculitis, and not in patients with uncomplicated RA. Furthermore, longitudinal studies of patients with rheumatoid vasculitis showed that high levels of antiendothelial antibodies are present during active disease and low levels after successful treatment. The question of whether these antibodies arise from vasculitis or initiate and perpetuate the disease remains to be answered.

Antibodies relevant for vascular damage do not necessarily have to bind to the surface of endothelial cells. Binding to antigens planted in the vessel wall could lead to immune complex formation and complement activation. Antibodies binding to immune complexes, such as rheumatoid factor and anti-Clq antibodies, and antibodies binding to cell constituents, such as DNA or ANCA, were found to bind to antigens already planted in the blood vessels and the titers of these antibodies were found to be related to clinical activity of the vasculitic process [16, 17].

The vascular wall may suffer indirect damage due to the activation of neutrophils. This was demonstrated in experimental serum sickness where removal of circulating neutrophils prevented vasculitis. Neutrophils possess an extensive armory of lysosomal enzymes and reactive oxygen metabolites necessary for bacterial killing and the destruction of tissue. Working with endothelial cells cultured in monolayers, several investigators showed that neutrophils after activation can produce injury to the monolayers both through the generation of oxygen radicals and the release of granule proteases [18, 19]. Because of the presence of inhibitors in serum, intimate contact between neutrophils and endothelial cells is necessary to obtain this effect.

Knowledge of the signals that induce activation of neutrophils and adhesiveness to the vessel wall, as well as the role of endothelial cells in this adhesion, is growing rapidly. Binding of neutrophils to the blood vessel wall is mediated by membrane associated adhesion molecules that recognize complementary adhesion molecules on endothelial cells. The adhesion molecules on the surface of leukocytes and endothelial cells are summarized in Table 3.2. The expression of these molecules is augmented by the activity of inflammatory factors. When neutrophils and endothelial cells are activated by cytokines, such as TNF or IL-1, there is a rapid increase in the synthesis and expression of ILAM, as well as in the expression of CD11 adhesion molecules with a conformational

**Table 3.2** Adhesion molecules on the surface of endothelial cells and leukocytes

| *Adhesion molecule on endothelial cells* | *Counter-receptors on leukocytes* |
| --- | --- |
| ICAM-1 | LFA-1 / MAC-1 |
| ELAM-1 (E-selectin) | Lewis α-antigen |
| VCAM | VLA-4 |
| GMP-140 (P-selectin) | ? |

change in expression [20, 21]. Cytokines also induce the expression of selectins and VCAM-1. There is little doubt that cytokine-induced adhesion of neutrophils plays a role in vasculitis. The local administration of TNF produces microvascular injury, and its systemic administration causes transient neutropenia, whereas increased expression of adhesion molecules has been demonstrated in vasculitic lesions.

Lobatto *et al.* injected immune complexes into chimpanzees and found a transient neutropenia followed by neutrophilia which was due to complement activation [22]. Components of the complement cascade such as C3B may be deposited on the endothelial surface and contribute to neutrophil adherence since neutrophils possess C3B receptors. Cytokines and other inflammation mediators that induce adhesiveness were also shown to be effective activators of neutrophils. TNF stimulates ADCC activity of neutrophils, and both TNF and IL-1 augment the respiratory burst and degranulation of neutrophils that can destruct the vessel wall. TNF also enhances the response to a range of soluble or particulate stimuli. Several investigators have shown that groups of antibodies that bind to cytoplasmic antigens (p-ANCA and c-ANCA) can induce endothelial cell lysis by primed neutrophils. The cytoplasmic antigens, such as myeloperoxidase or proteinase-3, can be bound to the surface of neutrophils and endothelial cells, or can be increasingly expressed on the surface of these cells upon activation [23]. ANCA may bind to such surfaces and stimulate the adhesion of neutrophils to the endothelial cell surface, activate neutrophils and stimulate neutrophil-mediated damage to endothelial cells. Indirect evidence for the pathogenetic role of ANCA antibodies includes the correlation between ANCA positivity and vasculitis activity, and the low frequency of c-ANCA in non-vasculitic diseases.

Although definitive evidence for T cell mediated immune mechanisms in autoimmune vasculitis is lacking, the role of blood vessel wall components in the induction of T cell responses has begun to be investigated experimentally. In transplant rejection, the recognition of MHC molecules on the endothelial cells of the transplant by cytolytic T cells leads to a type of vasculitis. Human vascular endothelial cells can express MHC molecules and can act as antigen-presenting cells. In this respect, it is interesting that circulating T lymphocytes in patients with Wegener's granulomatosis, but not in control individuals, proliferate in response to proteinase-3 and other cytoplasmic proteins, whereas serum levels of soluble IL-2 receptor, a marker of T cell activation, were shown to correlate with disease activity [24, 25]. However, it has not yet been elucidated how T cells are involved in the production of ANCAs or the formation of destructive granulomas.

Another mechanism concerns the deregulation of the anti-inflammatory and the anticoagulant activity of endothelial cells. There is a need to explain why endothelium facilitates the development of inflammation in normal host defense and is a target for injury in vasculitis syndromes. Endothelial cells exhibit a number of anticoagulant, antiplatelet and fibrinolytic properties that maintain the integrity of blood cells as well as anti-inflammatory properties, such as the production of inhibitors of neutrophil adherence and chemotaxis. Among these anti-inflammatory activities is the expression of membrane-bound regulators of complement activation. It was discussed that complement-mediated damage occurs when the activation products of the terminal components are assembled on the surface in the form of a membrane attack cell complex. This assembly is controlled by endothelial cells by the expression of the homologous restriction factor, the membrane cofactor protein CD59, the C1q receptor and primarily by decay accelerating factor (DAF) that interferes with C3 and C5 activation. Boom *et al.* have investigated the expression of DAF in four patients with histologically proven cutaneous vasculitis by means of immune electromicroscopy [25]. The normal skin of a healthy volunteer shows that endothelial cells stain abundantly for DAF. However, studies of lesions show that DAF staining is normal on erythrocytes, but absent on most endothelial cells. The mechanism responsible for the depletion of DAF has not been studied in detail but it may be hypothesized that decreased DAF expression facilitates complement mediated cell destruction. In vasculitis, it may be that homeostatic mechanisms are compromised by an insult that modifies endothelial cell function, which may lead to further damage of the vessel wall.

CONCLUSION

In summary, the vessel wall can be a non-specific target for injury (being damaged by virtue of the anatomical position and physiological function), as well as a specific target for injury by the immune system. Obviously, there is no consensus about which of the destructive mechanisms is most relevant for the different clinical forms of vasculitis and many pieces of the puzzle are missing. New cytokines, lytic factors and adhesion molecules will certainly be found. At present the available information seems to point at a pivotal role for the vascular endothelium and its interaction with humoral and cellular components of the immune system. Whether this insight has resulted in a greater clinical distinction among various disorders or the availability of rational therapies will be made clear in the near future.

REFERENCES

1. Cochrane, C.G. and Dixon, F.J. (1976) Antigen-antibody complex induced disease, in *Textbook of immunopathology*, 2nd edn (eds P.A. Miescher and H.J. Müller-Eberhard), Grune & Stratton, New York, pp. 137–56.

2. Westedt, M.L., Meijer, C.J.L.M., Vermeer, B.J. *et al.* (1984) Rheumatoid arthritis: the clinical significance of histo- and immunopathological abnormalities in normal skin. *J. Rheumatol.*, **11**, 448–53.

3. Braverman, I.M. and Yen, A. (1975) Demonstration of immune complexes in spontaneous and histamine-induced lesions and in normal skin of patients with leukocytoclastic angiitis. *J. Invest. Dermatol.*, **64**, 105–12.

4. Mannik, M. (1982) Pathophysiology of circulating immune complexes. *Arthritis Rheum.*, **25**, 783–7.

5. Boom, B.W., Out-Luiting, C.J., Baldwin, W.M. *et al.* (1987) Membrane attack complex of complement in leukocytoclastic vasculitis of the skin. Presence and possible pathogenetic role. *Arch. Dermatol.*, **123**, 1192–5.

6. Waxman, F.J., Herbert, L.A. and Cosio, F.G. (1986) Differential binding of immunoglobulin A and immunoglobulin G immune complexes to primate erythrocytes *in vivo*. *J. Clin. Invest.*, **77**, 82–9.

7. Halma, C., Breedveld, F.C., Daha, M.R. *et al.* (1991) Elimination of soluble [123]I-labeled aggregates of IgG in patients with systemic lupus erythematosus. *Arthritis Rheum.*, **34**, 442–52.

8. Wilson, J.G., Wong, W.W., Schur, P.H. and Fearon, D.T. (1982) Mode of inheritance of decreased C3b receptors on erythrocytes of patients with systemic lupus erythematosus. *N. Engl. J. Med.*, **307**, 981–6.

9. Walport, M.J., Ross, G.D., Mackworth-Young, C. *et al.* (1985) Family studies of erythrocyte complement receptor type 1 levels: reduced levels in patients with SLE are acquired, not inherited. *Clin. Exp. Immunol.*, **59**, 547–54.

10. Lüscher, T.F., Richard, V., Tschudi, M. *et al.* (1990) Endothelial control of vascular tone in large and small arteries. *J. Am. Coll. Cardiol.*, **15**, 519.

11. Leung, D.Y.M., Geha, R.S., Newburger, J.W. *et al.* (1986) Two monokines, interleukin I and tumor necrosis factor, render cultured vascular endothelial cells susceptible to lysis by antibodies circulating during Kawasaki disease. *J. Exp. Med.*, **164**, 1958.

12. Cines, D.B., Lyss, A.P., Reeber, M. *et al.* (1984) Presence of complement-fixing anti-endothelial cell antibodies in systemic lupus erythematosus. *J. Clin. Invest.*, **73**, 611.

13. Heurkens, A.H.M., Hiemstra, P.S., Lafeber, G.J.M. *et al.* (1989) Anti-endothelial cell antibodies in patients with rheumatoid arthritis complicated by vasculitis. *Clin. Exp. Immunol.*, **78**, 7.

14. van der Zee, J.M., Heurkens, A.H.M., van der Voort, E.A.M. *et al.* (1991) Characterization of anti-endothelial antibodies in patients with rheumatoid arthritis complicated by vasculitis. *Clin. Exp. Rheumatol.*, **9**, 589–94.

15. Breedveld, F.C., Heurkens, A.H.M., Lafeber, G.J.M. *et al.* (1988) Immune complexes in sera from patients with rheumatoid vasculitis induce polymorphonuclear cell-mediated injury to endothelial cells. *Clin. Immunol. Immunopathol.*, **48**, 202.

16. Siegert, C.E.H., Daha, M.R., Halma, C. *et al.* (1992) IgG and IgA autoantibodies against C1q in systemic and renal diseases. *Clin. Exp. Rheumatol.*, **10**, 19–23.

17. van der Woude, F.J., Rasmussen, N., Lobatto, S. *et al.* (1985) The TH. Autoantibodies against neutrophils and monocytes: tool for diagnosis and marker of disease activity in Wegener's granulomatosis. *Lancet*, **1**, 425.

18. Harlan, J.M., Killen, P.D., Harker, L.A. and Striker, G.E. (1981) Neutrophil-mediated endothelial injury *in vitro*. *J. Clin. Invest.*, **68**, 1394.

19. Weiss, S.J., Curnutte, J.T. and Regiani, S. (1985) Neutrophil-mediated solubilization of the subendothelial matrix: oxidative and nonoxidative mechan-

isms of proteolysis used by normal and chronic granulomatous disease phagocytes. *J. Immunol.*, **136**, 636.

20. Carlos, T.M. and Harlan, J.M. (1990) Membrane proteins involved in phagocyte adherence to endothelium. *Immunol. Rev.*, **114**, 5.

21. Brown, K.A. (1994) Role of endothelial cells in the pathogenesis of vascular damage, in *Antibodies to endothelial cells and vascular damage* (eds R. Cevera, M.A. Khamashta and G.R.V. Hughes), CRC Press, Boca Raton, pp. 27–46.

22. Lobatto, S., Daha, M.R., Voetman, A.A. *et al.* (1987) Clearance of soluble aggregates of human immunoglobulin G in healthy volunteers and chimpanzees. *Clin. Exp. Immunol.*, **69**, 133–41.

23. Falk, R.J., Terrell, R.S., Charles, L.A. and Jennette, C. (1990) Anti-neutrophil cytoplasmic autoantibodies induce neutrophils to degranulate and produce oxygen radicals *in vitro*. *Proc. Nat. Acad. Sci. USA*, **87**, 4115.

24. Hagen, C., Ballieux, B.E.P.B., van Es, L.A. *et al.* (1993) Antineutrophil cytoplasmic autoantibodies: a review of the antigens involved, the assays, and the clinical and possible pathogenetic consequences. *Blood*, **81**, 1996–2002.

25. Boom, B.W., Mommaas, A.M., Daha, M.R. and Vermeer, B.J. (1991) Decreased expression of declay-accelerating factor on endothelial cells of immune complex-mediated vasculitic skin lesions. *J. Dermatol. Sci.*, **2**, 308–15.

# 4

# Antineutrophil cytoplasmic antibodies (ANCA): current perspectives

*C.G.M. Kallenberg*

## INTRODUCTION

The idiopathic systemic vasculitides constitute a group of disorders characterized by more or less widespread inflammation of vessel walls generally without an underlying cause or an associated condition. The idiopathic vasculitides are traditionally grouped according to the size of the vessels involved and the histopathology of the inflammatory process. Such a classification is shown in Table 4.1. A recent modification of this scheme based on a more strict application of histopathological criteria with respect to the size of the vessels involved [1], is given in Table 4.2. The definitions and classification of the vasculitides demonstrate that histopathological findings are essential for making a diagnosis of a distinct vasculitic disorder. In clinical practice, however, pathognomonic histopathological findings are hard to obtain in most patients suspected of vasculitis, even in cases where multiple biopsies have been performed [2]. For this and other reasons, classification criteria have been designed by the American College of Rheumatologists (ACR) that are mainly based on clinical signs and symptoms [3]. These criteria can be used for the classification of patients who have already have been proved to have vasculitis. As the ACR criteria for the several vasculitides are, however, not mutually exclusive, one particular patient may be classified as having two or more disease entities according to

*The Vasculitides.* Edited by B.M. Ansell, P.A. Bacon, J.T. Lie and H. Yazici.
Published in 1996 by Chapman & Hall, London. ISBN 978-0-412-64140-4.

**Table 4.1** Traditional classification of systemic idiopathic vasculitides

I. Affecting predominantly large- and medium-sized blood vessels
   1. Takayasu arteritis
   2. giant cell arteritis/temporal arteritis

II. Affecting predominantly medium- and small-sized blood vessels
   1. classic polyarteritis nodosa
   2. Churg–Strauss syndrome
   3. Wegener's granulomatosis
   4. polyangiitis overlap syndrome

III. Affecting predominantly small blood vessels
   1. microscopic polyarteritis
   2. Henoch–Schönlein purpura

**Table 4.2** Classification of the idiopathic vasculitides as proposed by an international study group at the Chapel Hill Consensus Conference on the Nomenclature of Systemic Vasculitis (adapted from [1])

I. Large-vessel vasculitis
   1. giant cell (temporal) arteritis
   2. Takayasu arteritis

II. Medium-sized vessel vasculitis
   1. polyarteritis nodosa
   2. Kawasaki disease

III. Small-vessel vasculitis
   1. Wegener's granulomatosis
   2. Churg–Strauss syndrome
   3. microscopic polyangiitis
   4. Henoch–Schönlein purpura
   5. essential cryoglobulinemic vasculitis
   6. cutaneous leukocytoclastic angiitis

those criteria. Thus, the ACR criteria not only have their limitations for the classification of a patient with vasculitis, but also require the *a priori* presence of vasculitis, as the criteria are meant for classification and not for diagnosis of vasculitis. The latter requirement, however, cannot always be achieved, as mentioned above. As such, there is a need for laboratory tests with high degrees of sensitivity and specificity for the various vasculitides. In this review I will discuss whether antineutrophil cytoplasmic antibodies (ANCA) fulfil those requirements.

ANCA were first described in 1982 by Davies *et al.* in a few patients with segmental necrotizing glomerulonephritis [4]. Only in 1985 did it become apparent that ANCA are a sensitive and specific marker for

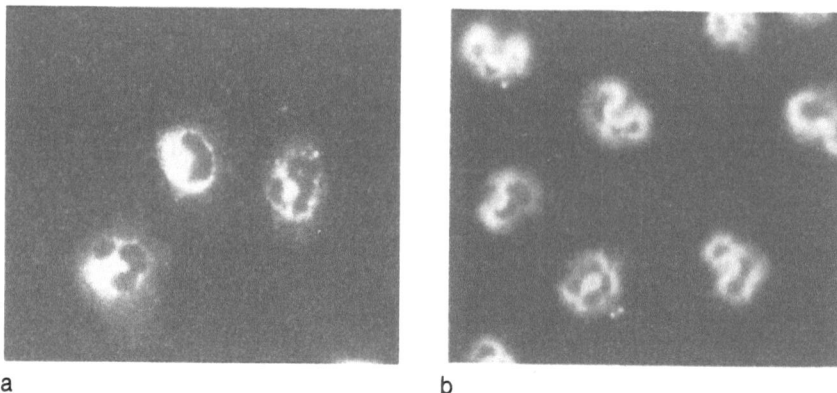

a                                              b

**Figure 4.1** (a) c-ANCA and (b) p-ANCA.

Wegener's granulomatosis (WG) [5]. ANCA in this disease were
described as autoantibodies that produce a characteristic cytoplasmic
fluorescent pattern by indirect immunofluorescence (IIF) on ethanol-
fixed neutrophils [6], which nowadays is called a c-ANCA pattern
(Figure 4.1). This pattern, which shows accentuation of the fluorescence
intensity in the area within the nuclear lobes, has been recognized as a
characteristic staining pattern produced by the sera of most patients
with WG [5,7], but also of some 50% of patients with microscopic
polyangiitis/arteritis and a minority of patients with other necrotizing
vasculitides or idiopathic necrotizing crescentic glomerulonephritis
(NCGN) [8].

Shortly after the introduction of ANCA testing, it became apparent
that a number of sera from patients suspected of vasculitis produced a
perinuclear fluorescence pattern on ethanol-fixed neutrophils. Those so-
called p-ANCA (Figure 4.1) were originally described in patients with
idiopathic and/or vasculitis-associated NCGN [9]. Further studies
showed that the p-ANCA pattern can be produced also by sera from
patients with vasculitis without renal involvement, and by sera from
patients with inflammatory bowel diseases, autoimmune liver diseases,
infectious diseases such as HIV infection, and rheumatic disorders such
as systemic lupus erythematosus (SLE) and rheumatoid arthritis [8,10].
An atypical cytoplasmic staining on ethanol-fixed neutrophils can be
produced by sera from patients with an even wider range of diseases,
including infectious diseases [10]. Thus, although a classical c-ANCA
pattern still suggests a diagnosis of WG, the mere presence of ANCA

as detected by indirect immunofluorescence has limited diagnostic value.

Recently it has become apparent that the antibodies producing a typical c-ANCA pattern in WG are directed against proteinase 3, a third serine protease from the azurophilic granules of neutrophils, different from elastase and cathepsin G, the other two serine proteases from these granules [11–13]. In addition, it proved that a considerable number of sera producing a p-ANCA pattern by IIF contained antibodies to myeloperoxidase (MPO) [9, 14]. A minority of p-ANCA positive sera were demonstrated to contain antibodies to elastase [11, 14, 15]. MPO and elastase are both cationic proteins that move during ethanol fixation to the negatively charged nuclear membrane [16], which explains the perinuclear fluorescence pattern. Following these studies, autoantibodies were described against other neutrophil granule constituents such as lactoferrin [17–23], cathepsin G [24] and β-glucuronidase [25] (Table 4.3). Thus, ANCA constitute a class of autoantibodies directed at various constituents of myeloid cells. As such, ANCA testing at present, comparable to ANA testing, should include antigen-specific assays whenever a positive ANCA result is obtained by IIF. Sensitivity and specificity of

**Table 4.3** Neutrophil granule constituents

| Class of constituent | Azurophilic granules | Specific granules |
|---|---|---|
| Microbial enzymes | Myeloperoxidase<br>Lysozyme | Lysozyme |
| Serine proteinases | Elastase<br>Cathepsin G<br>Proteinase 3<br>Azurocidin | |
| Metalloproteinases | | Collagenase<br>Gelatinase |
| Acid hydrolases | N-acetyl-β-<br>    glucosaminidase<br>Cathepsin B<br>Cathepsin D<br>β-glucoronidase<br>β-glycerophosphatase<br>α-mannosidase | |
| Other | Defensins | Lactoferrin<br>Vitamin B12-binding<br>protein |

ANCA for particular disorders should be based on antigen-specific assays as discussed below.

## ANCA, A CLASS OF AUTOANTIBODIES TO MYELOID GRANULAR PROTEINS

### Autoantibodies to proteinase 3

Following the detection of c-ANCA as a sensitive and specific marker for active WG [5], it was proved that the antibodies are directed to proteinase 3 [11–13]. Antibodies to proteinase 3 (PR3) occur in at least 90% of patients with extended WG characterized by the triad of granulomatous inflammation of the respiratory tract, systemic vasculitis, and necrotizing crescentic glomerulonephritis [7, 26, 27]. The sensitivity of anti-PR3 for limited WG, that is disease manifestations without obvious renal involvement, amounted to 75% [8]. Thus, the association between anti-PR3 and WG is not absolute. Although some patients with active WG are ANCA-negative, most patients who are negative for anti-PR3 have antibodies to MPO or leukocyte elastase [9, 15, 28, 29]. The specificity of anti-PR3 for active WG seems as high as 98% when sera are considered from patients with a wide variety of renal, autoimmune, vasculitis, or lymphoproliferative disorders [7, 26, 29]. Anti-PR3 are, however, also detected in patients with overlapping symptoms of WG and other forms of vasculitis, and in some patients with idiopathic necrotizing glomerulonephritis without systemic involvement [2, 26, 28, 30]. Microscopic polyarteritis or -angiitis (MPA) is such a syndrome. It is characterized by (pauci-immune) necrotizing vasculitis involving small-sized vessels without granuloma formation [31]. Clinically, it often manifests as a pulmonary-renal syndrome consisting of NCGN in combination with pulmonary capillaritis. Some 45% of patients with this disorder are positive for anti-PR3 (Table 4.4). It should be realized, in this context, that, in clinical practice, the distinction between WG, MPA, and idiopathic NCGN is far from absolute. Some patients who present with incomplete WG will evolve into definite (extended) WG. Thus, anti-PR3 characterize a group of patients within the spectrum of WG with idiopathic NCGN on the one side and extended WG on the other. Very recently, anti-PR3 antibodies have also been described in patients with invasive amebiasis [32] and in patients with thyroid disease treated with propylthiouracil [33]. These latter data have to be confirmed by other studies.

Besides their diagnostic value, the antibodies are relevant for the follow-up of patients with WG. Various longitudinal studies have shown that titers of c-ANCA (or anti-PR3) rise prior to a relapse of

**Table 4.4** Disease associations of antiproteinase 3 antibodies and antimyeloperoxidase antibodies*

| Disease entity | Sensitivity | |
|---|---|---|
| | antiproteinase 3 (%) | antimyeloperoxidase (%) |
| Wegener's granulomatosus | 85 | 10 |
| Microscopic polyangiitis | 45 | 45 |
| Idiopathic crescentic glomerulonephritis | 25 | 65 |
| Churg–Strauss syndrome | 10 | 60 |
| Polyarteritis nodosa | 5 | 15 |

*Data derived from the references cited in the text.

WG [7, 34]. The rise proved very sensitive for an ensuing relapse and was detectable a mean of 49 days before the moment of clinical relapse [7]. Based on these findings, a study was undertaken to test whether treatment based on changes in c-ANCA titer in WG could prevent the occurrence of relapses [35]. The study proved that treatment based on rising titers of c-ANCA indeed prevented relapses. In addition, this prophylactic way of treatment required less immunosuppressives and corticosteroids than conventional treatment. These data have been challenged by a recent report from the NIH group [36] demonstrating a rather poor correlation between changes in ANCA-titer and disease activity of WG. However, relapses occurring in patients who were previously in remission or had stable disease, were also preceded or paralleled by an increase in c-ANCA titer in the NIH study [37]. Otherwise, persistently or intermittently positive tests for ANCA in patients coming into remission are a considerable risk factor for an ensuing relapse [38].

Thus, the presence of anti-PR3 strongly suggests a diagnosis within the spectrum of WG, although the association is not absolute. Increasing titers should alert the clinician to the possibility of an ensuing relapse.

### Autoantibodies to myeloperoxidase (MPO)

In contrast to anti-PR3, anti-MPO are not specific for one disease entity within the spectrum of vasculitides, but occur in a variety of vasculitic disorders (Table 4.4). Their specificity for the group of idiopathic vasculitides including idiopathic NCGN is rather high [28]. Anti-MPO occur in the majority of patients with WG, microscopic polyangiitis or idiopathic NCGN, who are negative for anti-PR3. In addition, the antibodies occur in some 75% of patients with Churg–Strauss syndrome and in a minority of patients with classic polyarteritis nodosa [2, 39]. Most patients with classic PAN are ANCA-negative.

Anti-MPO have also been detected in 30–40% of patients with anti-GBM disease [40–42]. Their presence is in some cases associated with signs of systemic vasculitis, and among these patients recovery of renal function has been reported despite their initial need for hemodialysis [41]. Anti-MPO have additionally been described in patients with SLE [43, 44], but their prevalence in this disease differs between several series. Interestingly, anti-MPO have been detected in the sera of all six patients tested with hydralazine-induced lupus, in five of them in combination with anti-elastase antibodies [45].

The significance of following titers of anti-MPO in patients with anti-MPO associated vasculitides has not been studied in detail. Preliminary data suggest that titers of the antibodies parallel disease activity [28].

## Autoantibodies to leukocyte elastase

Anti-elastase antibodies occur very infrequently. During a period of four years we detected these antibodies (routinely tested in a large diagnostic laboratory) in seven patients only. Those patients generally had signs and symptoms suggestive of WG or a related disorder [15]. As mentioned earlier, anti-elastase antibodies have also been detected in patients with hydralazine-induced lupus, in conjunction with anti-MPO [45], and in patients who developed signs of vasculitis during treatment with propylthiouracil, in the latter study in conjunction with anti-PR3 and anti-MPO [33].

## Autoantibodies to lactoferrin

Antibodies to lactoferrin have been described particularly in patients with rheumatoid arthritis (RA). They have been specifically associated with vasculitis in RA [18] and with Felty's syndrome [20]. The association of anti-lactoferrin with vasculitis in RA has, however, not been confirmed by other studies [10, 17]. The latter studies actually suggest that ANCA in RA are a secondary phenomenon related to disease duration.

Anti-lactoferrin have also been detected in a minority of patients with ulcerative colitis [19], primary sclerosing cholangitis [46], and SLE [21].

## Autoantibodies to other neutrophil constituents

Antibodies to either cathepsin G [24] or β-glucuronidase [25] have been detected in patients with ulcerative colitis. These data have not been confirmed in other studies.

ROLE OF ANCA TESTING IN THE DIAGNOSTIC APPROACH OF THE PATIENT SUSPECTED OF VASCULITIS

ANCA testing should be included in every work-up of patients suspected of vasculitis. Screening is generally performed by IIF on ethanol-fixed leukocytes. Using the buffy coat as a substrate allows the distinction between real p-ANCA and ANA as real ANCA do not stain nuclei or nuclear rims of lymphocytes (when ANCA and ANA are, however, simultaneously present in a serum sample, this distinction cannot be made). The presence of a c-ANCA pattern by IIF may be considered as suggesting anti-proteinase 3 antibodies in experienced hands, but should ideally be followed by antigen-specific assays. The presence of a p-ANCA pattern by IIF should in any case be followed by antigen-specific assays. With respect to vasculitis those assays should include tests for the presence of anti-PR3, anti-MPO and, possibly, anti-elastase. In case of strong suspicion of idiopathic necrotizing vasculitis, tests for anti-PR3 and anti-MPO might be performed, even in the absence of a positive ANCA test by IIF, as cases have been described that are ANCA negative by IIF but positive by antigen-specific assays [47].

A positive test for anti-PR3 should alert the physician to the possible presence of WG or related conditions. A thorough clinical investigation should be performed, directed at signs and symptoms of WG. This should include ENT examination by an experienced ENT physician, and tests for renal function as well as microscopic examination of the urinary sediment. The same applies to the presence of anti-elastase antibodies. When anti-MPO antibodies are detected, the patient should be scrutinized for signs and symptoms of NCGN (renal function, urinary sediment, renal biopsy), Churg–Strauss syndrome (history of asthma, nasal polyposis, hypereosinophilia, mononeuritis multiplex, etc.), and polyarteritis nodosa (PAN) with visceral involvement. As muscle weakness and/or peripheral neuropathy in the context of general malaise and a raised erythrocyte sedimentation rate (ESR) frequently are presenting signs of systemic vasculitis for the rheumatologist, algorithms have been proposed (Figures 4.2–4.4) for patients suspected of vasculitis, based on the aforementioned findings and being positive for anti-PR3, positive for anti-MPO, or ANCA negative [48].

PATHOPHYSIOLOGIC ROLE OF ANCA IN SYSTEMIC VASCULITIS

A pathophysiologic role for ANCA has been suggested by clinical data showing a close association between the autoantibodies and idiopathic necrotizing vasculitis as well as by experimental data [49]. *In vitro* studies have shown that ANCA can activate neutrophils that are pretreated ('primed') with low dosage TNFα to the production of reactive

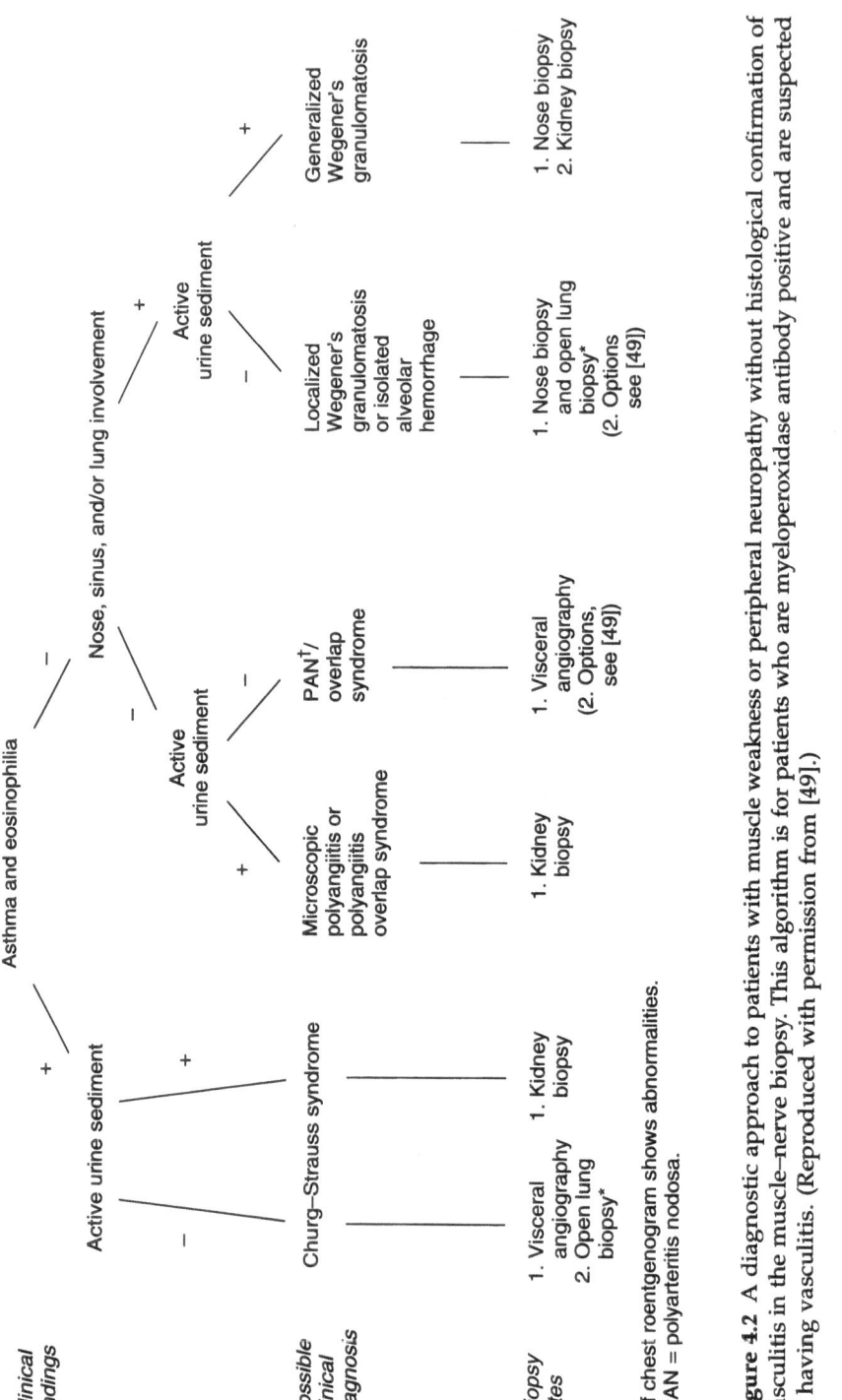

**Figure 4.2** A diagnostic approach to patients with muscle weakness or peripheral neuropathy without histological confirmation of vasculitis in the muscle–nerve biopsy. This algorithm is for patients who are myeloperoxidase antibody positive and are suspected of having vasculitis. (Reproduced with permission from [49].)

The figure contains the following labels and text:

**Clinical findings**

Asthma and eosinophilia

Nose, sinus, and/or lung involvement

Active urine sediment

Active urine sediment

Active urine sediment

**Possible clinical diagnosis**

Churg–Strauss syndrome

Microscopic polyangiitis or polyangiitis overlap syndrome

PAN†/ overlap syndrome

Localized Wegener's granulomatosis or isolated alveolar hemorrhage

Generalized Wegener's granulomatosis

**Biopsy sites**

1. Visceral angiography
2. Open lung biopsy*

1. Kidney biopsy

1. Kidney biopsy

1. Visceral angiography (2. Options, see [49])

1. Nose biopsy and open lung biopsy* (2. Options see [49])

1. Nose biopsy
2. Kidney biopsy

* If chest roentgenogram shows abnormalities.
† PAN = polyarteritis nodosa.

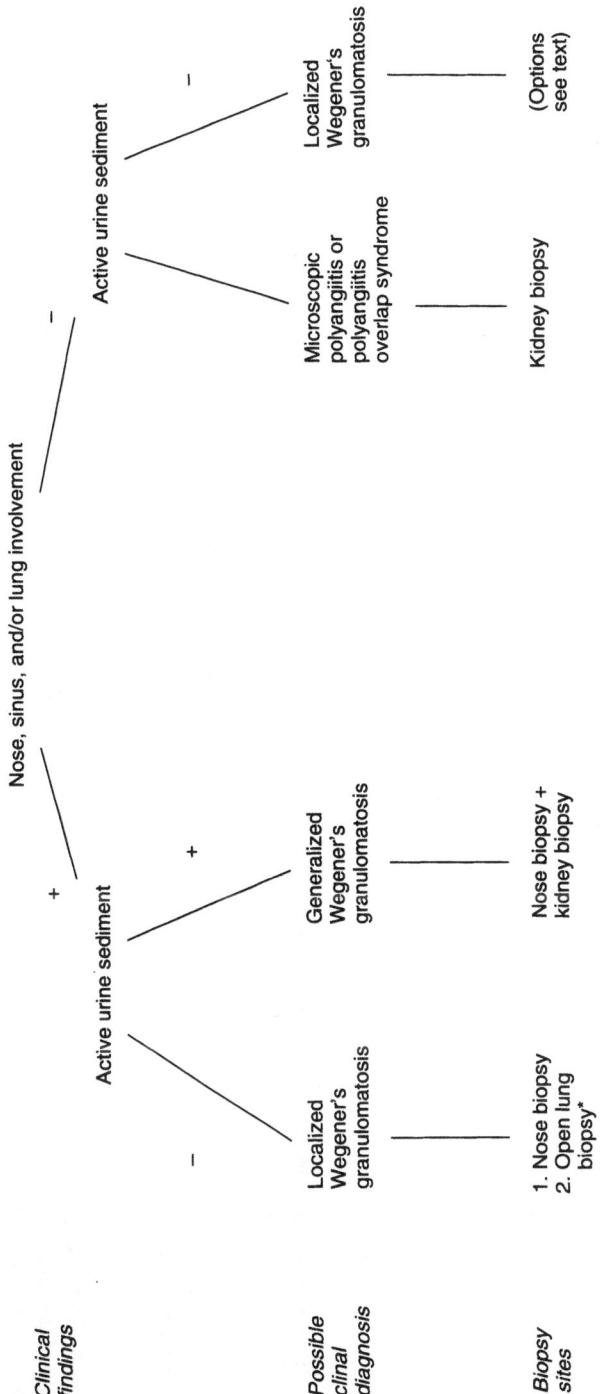

**Figure 4.3** A diagnostic approach for patients with muscle weakness or peripheral neuropathy who are proteinase 3 antibody positive and are suspected of having vasculitis but for whom the muscle–nerve biopsy results show no histologic confirmation. (Reproduced with permission from [49].)

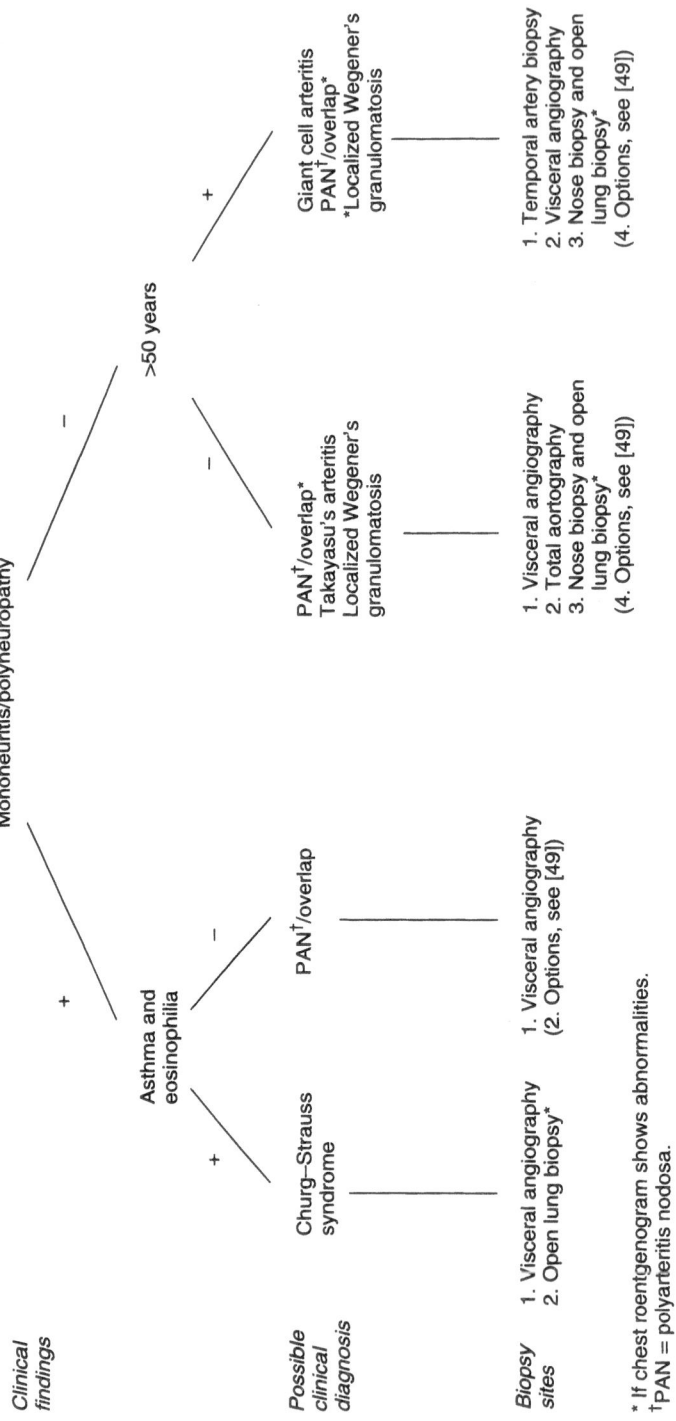

**Figure 4.4** A diagnostic approach for patients with muscle weakness or peripheral neuropathy who are proteinase 3 antibody and myeloperoxidase antibody negative and are suspected of having vasculitis but for whom the muscle–nerve biopsy results show no histologic confirmation. (Reproduced with permission from [49].)

oxygen species and the release of lysosomal enzymes [50]. Priming of neutrophils by TNFα, possibly occurring *in vivo* in the context of infection, results in the expression of lysosomal enzymes as PR3 and MPO at the cell surface [51]. Binding of ANCA to their target antigens at the cell surface results in further activation of the neutrophil, a process which is dependent also on the second Fc-γ receptor [52]. Another mechanism that may be relevant concerns the inhibition by anti-PR3 of the irreversible inactivation of proteinase 3 by $\alpha_1$-antitrypsin, its natural inhibitor [53]. In a longitudinal study in patients with WG it has been shown that disease activity correlates with the amount of this inhibitory activity of the serum, rather than with the titer of the anti-PR3 antibodies [54]. This suggests that escape of PR3 from its inactivation by $\alpha_1$-antitrypsin may contribute to the inflammatory process.

The *in vivo* role of anti-MPO with respect to the pathophysiology of pauci-immune NCGN has been studied by Brouwer *et al.* [55]. They immunized Brown–Norway rats with MPO and perfused, five weeks after immunization, the left kidney with products of activated neutrophils, that is lytic enzymes particularly PR3 and elastase, MPO and its substrate $H_2O_2$. The rats developed NCGN with interstitial infiltrate and vasculitis, whereas the lesions generally lacked immune deposits. These findings suggest that the initial step in the development of NCGN is (focal) immune complex formation. The presence of lytic enzymes, which were concomitantly perfused but may also have been released from neutrophils activated by ANCA, is probably responsible for the degradation of the immune deposits that were detected in the very early stage of the lesions.

A schematic representation of the possible pathophysiology of ANCA-associated necrotizing vasculitis/glomerulonephritis, which summarizes the items discussed above, is shown in Figure 4.5.

CONCLUSION

As stated in the introduction, there is a need for laboratory tests with high degrees of sensitivity and specificity for the various vasculitides. Do ANCA fulfil these requirements?

Anti-PR3 are highly sensitive and specific for active WG. In addition, changes in their titers reflect or even precede disease activity, and may therefore constitute guidelines for treatment. Nevertheless, as the association between anti-PR3 and WG is not absolute, the diagnosis of WG should, if possible, be based on histopathological findings. However, in case of suspicion of vasculitis in the presence of anti-PR3, treatment should not be postponed until a final histopathological diagnosis is reached.

Anti-MPO (and anti-elastase) are reasonably specific for one of the

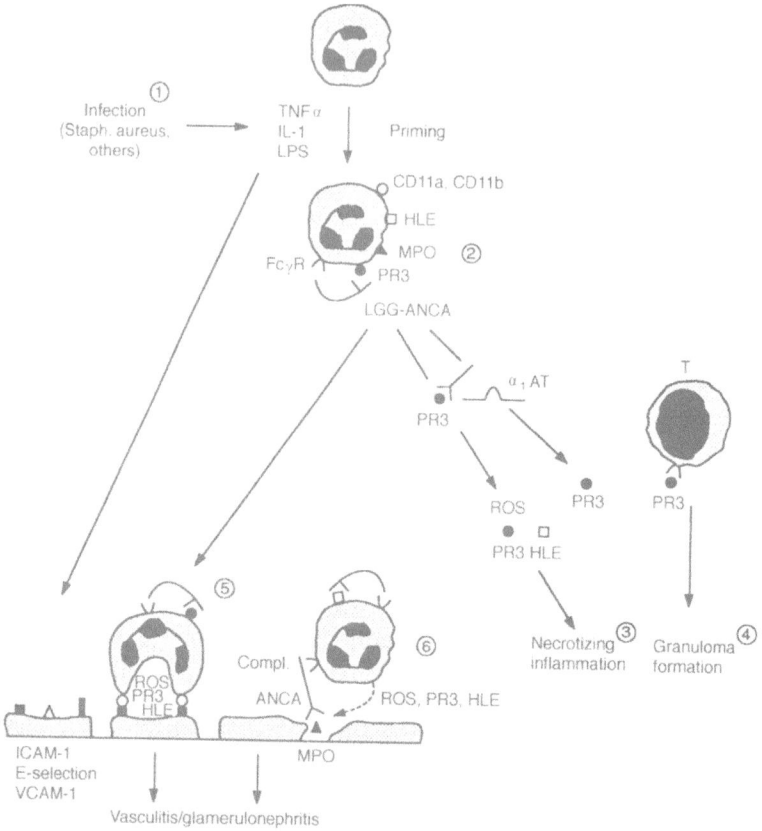

**Figure 4.5** Schematic representation of the possible pathophysiology of ANCA-associated necrotizing vasculitis/glomerulonephritis.

1. Neutrophils (PMN) are primed by low concentrations of cytokines resulting from (local) infection.
2. Primed PMN are further activated by IgG-ANCA to the production of reactive oxygen species (ROS) and the release of lytic enzymes such as proteinase 3 (PR3) and elastase (HLE).
3. Products of activated PMN induce necrotizing inflammation which might be potentiated by the inhibition by ANCA of the $\alpha_1$-antitrypsin ($\alpha_1$AT) induced inactivation of PR3.
4. CD4-positive T cells, possibly autoreactive to PR3, may contribute to granuloma formation.
5. Activated PMN may damage endothelial cells to which they adhere as a result of (focal) upregulation of adhesion molecules.
6. Binding of cationic MPO to polyanionic (basement) membranes is followed by *in situ* immune complex formation; attracted PMN additionally activated by ANCA evoke the initial step of vasculitis/glomerulonephritis and degrade immune complexes.

(Reproduced with permission from [49].)

necrotizing vasculitides, either Churg–Strauss syndrome, NCGN, (some cases of) WG, PAN with visceral involvement, or overlap syndromes. They are not highly sensitive for one specific disorder within the spectrum of vasculitides. Their presence should raise suspicion of vasculitis which should be ascertained by thorough clinical investigation. Here also, treatment should not be postponed in case of life-threatening disease.

The presence of ANCA of other, or undefined, specificities should at present not be considered as 'proof' of systemic vasculitis. As such, treatment should never be started on the basis of a positive ANCA test alone.

## REFERENCES

1. Jennette, J.C., Falk, R.J., Andrassy, K. *et al.* (1994) Nomenclature of systemic vasculitides. Proposal of an International Consensus Conference. *Arthritis Rheum.*, **37**, 187–92.
2. Cohen Tervaert, J.W., Limburg, P.C., Elema, J.D. *et al.* (1991) Detection of autoantibodies against myeloid lysosomal enzymes: a useful adjunct to classification of patients with biopsy-proven necrotizing arteritis. *Am. J. Med.*, **91**, 59–66.
3. Leavitt, R.Y., Fauci, A.S., Bloch, D.A. *et al.* (1990) The American College of Rheumatology 1990 criteria for the classification of Wegener's Granulomatosis. *Arthritis Rheum.*, **33**, 1101–7.
4. Davies, D.J., Moran, J.E., Niall, J.F. and Ryan, G.B. (1982) Segmental necrotizing glomerulonephritis with antineutrophil antibody: possible arbovirus aetiology? *Br. Med. J.*, **285**, 606.
5. Van der Woude, F.J., Rasmussen, N., Lobatto, S. *et al.* (1985) Autoantibodies to neutrophils and monocytes: a new tool for diagnosis and a marker of disease activity in Wegener's Granulomatosis. *Lancet*, **ii**, 425–9.
6. Wiik, A. (1989) Delineation for a standard procedure for indirect immunofluorescence detection of ANCA. *APMIS*, **97** (Suppl. 6), S12–S13.
7. Cohen Tervaert, J.W., van der Woude, F.J., Fauci, A.S. *et al* (1989) Association between active Wegener's Granulomatosis and anticytoplasmic antibodies. *Arch. Intern. Med.*, **149**, 2461–5.
8. Kallenberg, C.G.M., Mulder, A.H.L. and Cohen Tervaert, J.W. (1992) Antineutrophil cytoplasmic antibodies: a still growing class of autoantibodies in inflammatory disorders. *Am. J. Med.*, **93**, 675–82.
9. Falk, R.J. and Jennette, J.C. (1988) Anti-neutrophil cytoplasmic autoantibodies with specificity for myeloperoxidase in patients with systemic vasculitis and idiopathic necrotizing and crescentic glomerulonephritis. *N. Engl. J. Med.*, **318**, 1651–7.
10. Peter, H.H., Metzger, D., Rump. A. and Röther, E. (1993) ANCA in diseases other than systemic vasculitis. *Clin. Exp. Immunol.*, **91** (Suppl. 1), S12–S14.
11. Goldschmeding, R., Van der Schoot, C.E., Ten Bokkel Huinink, D. *et al.* (1989) Wegener's Granulomatosis autoantibodies identify a novel diisopropylfluorophosphate-binding protein in the lysosomes of normal human neutrophils. *J. Clin. Invest.*, **84**, 1577–87.
12. Niles, J.L., McCluskey, R.T., Ahmad, M.F. and Arnaout, M.A. (1989) Wegener's Granulomatosis autoantigen is a novel serine proteinase. *Blood*, **74**, 1888–93.

13. Lüdemann, J., Utecht, B. and Gross, W.I. (1990) Anti-neutrophil cytoplasm antibodies in Wegener's Granulomatosis recognize an elastinolytic enzyme. *J. Exp. Med.*, **171**, 357–62.

14. Goldschmeding, R., Cohen Tervaert, J.W., van der Schoot, C.E. *et al.* (1989) ANCA, anti-myeloperoxidase, and anti-elastase: three members of a novel class of autoantibodies against myeloid lysosomal enzymes. *APMIS*, **97**(S6), 48–9.

15. Cohen Tervaert, J.W., Mulder, A.H.L., Stegeman, C.A. *et al.* (1993) The occurrence of autoantibodies to human leukocyte elastase in Wegener's Granulomatosis and other inflammatory disorders. *Ann. Rheum. Dis.*, **52**, 115–20.

16. Charles, L.A., Falk, R.J. and Jennette, J.C. (1989) Reactivity of anti-neutrophil cytoplasmic autoantibodies with HL-60 cells. *Clin. Immunol. Immunopathol.*, **53**, 243–53.

17. Mulder, A.H.L., Horst, G., van Leeuwen, M.A. *et al.* (1993) Anti-neutrophil cytoplasmic antibodies in rheumatoid arthritis: characterization and clinical correlations. *Arthritis Rheum.*, **36**, 1054–60.

18. Coremans, E.M., Hagen, E.C., Daha, M.R. *et al.* (1992) Antilactoferrin antibodies in patients with rheumatoid arthritis are associated with vasculitis. *Arthritis Rheum.*, **35**, 1466–75.

19. Mulder, A.H.L., Broekroelofs, J., Horst, G. *et al.* (1994) Antineutrophil cytoplasmic antibodies (ANCA) in inflammatory bowel disease: characterization and clinical correlates. *Clin. Exp. Immunol.*, **95**, 490–7.

20. Coremans, I.E.M., Hagen, E.C., Daha, M.R. *et al.* (1993) Anti-lactoferrin antibodies in patients with Felty's syndrome. *Clin. Exp. Immunol.*, **93** (Suppl. 1), S33.

21. Lee, S.S., Lawton J.W.M. and Chan, C.E. (1992) Antilactoferrin antibody in systemic lupus erythematosus. *Brit. J. Rheumatol.*, **31**, 669–73.

22. Peen, E., Almer, S., Bodemar, G. *et al* (1993) Anti-lactoferrin antibodies and other types of ANCA in ulcerative colitis, primary sclerosing cholangitis and Crohn's disease. *Gut*, **34**, 56–62.

23. Sinico, R.A., Pozzi, C., Radice, A. *et al.* (1993) Clinical significance of anti-neutrophil cytoplasmic autoantibodies with specificity for lactoferrin in renal diseases. *Am. J. Kidney Dis.*, **22**, 253–60.

24. Halbwachs-Mecarelli, L.H., Nusbaum, L.H., Reumaux, D. *et al.* (1992) Anti-neutrophil cytoplasmic antibodies (ANCA) directed against Cathepsin G in ulcerative colitis, Crohn's disease and primary sclerosing cholangitis. *Clin. Exp. Immunol.*, **90**, 79–84.

25. Kaneko, K., Suzuki, Y., Yamashiro, Y. and Yabuta, S. (1993) Is p-ANCA in ulcerative colitis directed against β glucuronidase? *Lancet*, **341**, 320.

26. Nölle, B., Specks, U., Lüdemann, J. *et al.* (1989) Anticytoplasmic antibodies: their immunodiagnostic value in Wegener's Granulomatosis. *Ann. Intern. Med.*, **111**, 28–40.

27. Weber, M.F.A., Andrassy, K., Pullig, O. *et al.* (1992) Antineutrophil cytoplasmic antibodies and antiglomerular basement membrane antibodies in Goodpasture's syndrome and in Wegener's granulomatosis. *J. Am. Soc. Nephrol.*, **2**, 1227–34.

28. Cohen Tervaert, J.W., Goldschmeding, R., Elema, J.D. *et al.* (1990) Association of autoantibodies to myeloperoxidase with different forms of vasculitis. *Arthritis Rheum.*, **33**, 1264–72.

29. Kallenberg, C.G.M. and Cohen Tervaert, J.W. (1992) Anti-neutrophilic cytoplasmic antibodies: new tools in the diagnosis and follow-up of necrotizing

glomerulonephritis, in *International Yearbook of Nephrology* (eds V.E. Andreucci and L.G. Fine), pp. 313–36.

30. Cohen Tervaert, J.W., Goldschmeding, R., Elema, J.D. *et al* (1990) Autoantibodies against myeloid lysosomal enzymes in crescentic glomerulonephritis. *Kidney Int.*, **37**, 799–806.

31. Savage, C.O.S., Winearls, C.G., Evans, D.J. *et al.* (1985) Microscopic polyarteritis: presentation, pathology and prognosis. *Q. J. Med.*, **56**, 467–83.

32. Pudifin, D.J., Duursma, J., Gathiram, V. and Jackson, T.F.G.H. (1993) Serum from patients with invasive amoebiasis has anti-neutrophil cytoplasmic antibody activity. *Clin. Exp. Immunol.*, **93** (Suppl. 1), S33.

33. Dolman, K.M., Gans, R.O.B., Vervaart, T.H.J. *et al.* (1993) Vasculitic disorders and anti-neutrophil cytoplasmic autoantibodies associated with propylthiouracil therapy. *Lancet*, **342**, 651–2.

34. Egner, W. and Chapel, H.M. (1990) Titration of antibodies against neutrophil cytoplasmic antigens is useful in monitoring disease activity in systemic vasculitides. *Clin. Exp. Immunol.*, **82**, 244–9.

35. Cohen Tervaert, J.W., Huitema, M.G., Hené, R.J. *et al* (1990) Prevention of relapses in Wegener's granulomatosis by treatment based on antineutrophil cytoplasmic antibody titre. *Lancet*, **336**, 709–11.

36. Kerr, G.R., Fleischer, T.H.A., Hallahan, C.W. *et al.* (1993) Limited prognostic value of changes in antineutrophil cytoplasmic antibody titer in patients with Wegener's granulomatosis. *Arthritis Rheum.*, **36**, 365–71.

37. Cohen Tervaert, J.W., Stegeman, C.A., Huitema, M.G. and Kallenberg, C.G.M. (1994) Comments on article by Kerr *et al. Arthritis Rheum.*, **37**, 596–7.

38. Stegeman, C.A., Cohen Tervaert, J.W., Sluiter, W.J. *et al.* (1994) Association of chronic nasal carriage of *Staphylococcus aureus* and higher relapse rates in Wegener's Granulomatosis. *Ann. Intern. Med.*, **120**, 12–7.

39. Cohen Tervaert, J.W., Goldschmeding, R., Von dem Borne, A.E.G.K.R. and Kallenberg, C.G.M. (1991) Anti-myeloperoxidase antibodies in the Churg–Strauss syndrome. *Thorax*, **46**, 70–1.

40. Jayne, D.R.W., Marshall, P.D., Jones, S.J. and Lockwood, C.M. (1990) Autoantibodies to GBM and neutrophil cytoplasm in rapidly progressive glomerulonephritis. *Kidney Int.*, **37**, 965–70.

41. Bosch, X., Mirapeix, E., Font, J. *et al.* (1991) Prognostic implication of antineutrophil cytoplasmic autoantibodies with myeloperoxidase specificity in anti-glomerular basement membrane disease. *Clin. Nephrol.*, **36**, 107–13.

42. Niles, J.L., Pan, G., Collins, A.B. *et al.* (1991) Antigen-specific radio-immunoassay for anti-neutrophil cytoplasmic antibodies in the diagnosis of rapidly progressive glomerulonephritis. *J. Am. Soc. Nephrol.*, **2**, 27–36.

43. Gueirard, P., Delpech, A., Gilbert, D. *et al.* (1991) Anti-myeloperoxidase antibodies: immunological characteristics and clinical associations. *J. Autoimm.*, **4**, 517–27.

44. Gallichio, M.C. and Savige, J.A. (1991) Detection of anti-myeloperoxidase and anti-elastase antibodies in vasculitis and infections. *Clin. Exp. Immunol.*, **84**, 232–7.

45. Nässberger, L., Sjöholm, A.G., Jonsson, H. *et al.* (1990) Autoantibodies against neutrophil cytoplasm components in systemic lupus erythematosus and in hydralazine-induced lupus. *Clin. Exp. Immunol.*, **81**, 380–3.

46. Mulder, A.H.L., Horst, G., Haagsma, E.B. *et al.* (1993) Prevalence and characterization of anti-neutrophil cytoplasmic antibodies in autoimmune liver diseases. *Hepatology*, **17**, 411–17.

47. Kallenberg, C.G.M., Cohen Tervaert, J.W. and Limburg, P.C. (1991) ELISA for the detection of antibodies against neutrophil cytoplasm antigens, in *Techniques in diagnostic pathology*, Vol. 2 (eds G.R. Bullock, D. van Velzen, M.J. Warhol and P. Herbrink), Academic Press, London, pp. 43–60.
48. Cohen Tervaert, J.W. and Kallenberg, C.G.M. (1993) Neurologic manifestations of systemic vasculitides, in *Neurologic aspects of rheumatic diseases. Rheumatic Disease Clinics of North America* (ed. K.B. Elkon), Vol. 19, WB Saunders, Philadelphia, pp. 913–40.
49. Kallenberg, C.G.M., Brouwer, E., Weening, J.J. and Cohen Tervaert, J.W. (1994) Anti-neutrophil cytoplasmic antibodies: current diagnostic and pathophysiological potential. *Kidney Int.*, **46**, 1–15.
50. Falk, R.J., Terrell, R.S., Charles, L.A. and Jennette, J.C. (1990) Anti-neutrophil cytoplasmic autoantibodies induce neutrophils to degranulate and produce oxygen radicals *in vitro*. *Proc. Nat. Acad. Sci. USA*, **87**, 4115–9.
51. Charles, L.A., Caldas, M.L.R., Falk, R.J. *et al.* (1991) Antibodies against granule proteins activate neutrophils *in vitro*. *J. Leuk. Biol.*, **50**, 539–46.
52. Mulder, A.H.L., Horst, G., Limburg, P.C. and Kallenberg, C.G.M. (1993) Activation of neutrophils by anti-neutrophil cytoplasmic antibodies is FcR-dependent. *Clin. Exp. Immunol.*, **93** (Suppl. 1), S16.
53. Van de Wiel, A., Dolman, K.M., van der Meer-Gerritsen, C.H. *et al.* (1992) Interference of Wegener's Granulomatosis autoantibodies with neutrophil proteinase 3 activity. *Clin. Exp. Immunol.*, **90**, 409–14.
54. Dolman, K.M., Stegeman, C.A., van de Wiel, B.A. *et al.* (1993) Relevance of classic anti-neutrophil cytoplasmic autoantibody (c-ANCA)-mediated inhibition of proteinase 3-$\alpha_1$-antitrypsin complexation to disease activity in Wegener's Granulomatosis. *Clin. Exp. Immunol.*, **93**, 405–10.
55. Brouwer, E., Huitema, M.G., Klok, P.A. *et al*, (1993) Anti-myeloperoxidase associated proliferative glomerulonephritis: an animal model. *J. Exp. Med.*, **177**, 905–14.

# 5

# Antiendothelial cell antibodies, antiphospholipid antibodies and vascular disease

*D. D'Cruz and G. Hughes*

## INTRODUCTION

Over the last two decades it has become increasingly clear that endothelial cells play a central role in the pathogenesis of a variety of vascular diseases including atherosclerosis, diabetic vasculopathy, graft rejection, tumor or synovial angiogenesis, vascular inflammation and the connective tissue diseases. This understanding of endothelial cell pathophysiology has advanced considerably following the development of techniques to isolate and culture pure endothelial cells *in vitro* [1,2].

Endothelial damage, either by complement fixing antibodies or by cell-mediated immunity, offers an attractive explanation for some of the clinical manifestations of the connective tissue diseases, including systemic lupus erythematosus (SLE) and the vasculitides. The discovery of antibodies directed against endothelial cells has resulted in widespread interest in the prevalence of these antibodies in a variety of connective tissue and other diseases. The aim of this chapter is to discuss how antiendothelial cell antibodies may be measured, their prevalence among diseases associated with vascular damage and the correlation with vasculitis, nephritis and thrombosis in SLE. Their binding characteristics, antigenic specificity and potential cytotoxicity *in vitro* will also be discussed.

*The Vasculitides.* Edited by B.M. Ansell, P.A. Bacon, J.T. Lie and H. Yazici.
Published in 1996 by Chapman & Hall, London. ISBN 978-0-412-64140-4.

**Table 5.1** Rheumatic and other diseases associated with antiendothelial cell antibodies

| | |
|---|---|
| Systemic lupus erythematosus | Thrombotic thrombocytopenic purpura |
| Rheumatoid arthritis | Hemolytic uremic syndrome |
| Mixed connective tissue disease | Multiple sclerosis |
| Scleroderma | HLA matched graft rejection |
| Polymyositis/dermatomyositis | Immune mediated hypoparathyroidism |
| Kawasaki disease | Episodic angioedema/hypereosinophilia |
| Systemic vasculitides | Acute pre-eclampsia |
| Inflammatory bowel disease | Acute asthma |
| Diabetes mellitus | Thyroid disease |
| Endometriosis | Heparin associated thrombocytopenia |

## Antiendothelial cell antibodies (aECA)

aECA were first described in 1971–2 using frozen mouse kidney sections in indirect immunofluorescence studies [3,4]. Sera from patients with SLE (15%), inflammatory myopathies (46%), scleroderma (55%) and rheumatoid arthritis (32%) reacted with mouse glomerular endothelium and peritubular capillaries producing a diffuse cytoplasmic staining pattern. Sera reacted with both membrane and cytoplasmic antigens and there was no correlation with blood group antigens or with antinuclear antibodies. Both papers described evidence of complement factor C3 fixation, supporting an antibody–antigen reaction. Over the two decades since these papers were published, aECA have been demonstrated in a wide variety of rheumatic and other diseases (Table 5.1) [3–27].

## Detection and binding characteristics aECA

A number of methods, all of which have some limitations, have been used for the detection of aECA, including immunofluorescence, radioimmunoassay and microcytotoxicity. For example, Cines *et al.* [24] used radioimmunoassay and flow cytometry techniques to study the binding characteristics of aECA in SLE in detail. IgG aECA and heat aggregated IgG bound to cultured endothelial cells, initiated complement activation and resulted in the deposition of the third component of complement (C3) on to the endothelium. Preincubation of IgG complexes with a serum source of complement augmented the binding of IgG and C3 and this was compatible with the previously demonstrated presence of C1q and C3b receptors on the endothelial surface. In some experiments disruption of the endothelial monolayer was seen in the presence of aECA. The majority of the aECA binding was shown to be via the

F(ab')$_2$ portion of monomeric IgG rather than via Fc receptors. In addition, binding of IgG to aECA and heat aggregated IgG resulted in the increased secretion of prostacyclin and increased adhesion of platelets to the endothelium.

Since a highly reproducible, sensitive and quantitative ELISA for detecting aECA was developed [25] this method has been widely adopted. Nevertheless, no standard assay or controls have been agreed and different groups have their own variations of the technique. This has led to considerable difficulties in making direct comparisons of results between laboratories.

## aECA AND CLINICAL ASPECTS OF SLE

The production of autoantibodies against tissue and cell specific antigens is one of the hallmarks of SLE, indeed aECA were first described in this condition. A number of studies have assessed the prevalence of aECA in lupus patients. Table 5.2 shows that the figures vary from 15% to 89% depending on the different populations studied, the methods used for the detection of aECA and the different disease characteristics of the

**Table 5.2** Prevalence of aECA in SLE determined by a variety of methods in different lupus populations

| Author | Year | Method | Patients studied | Prevalence (%) of aECA | |
|--------|------|--------|------------------|------|------|
| | | | | IgG | IgM |
| Lindqvist | 1971 | IIF | Not known | 17 | ND |
| Tan | 1972 | IIF | 40 | 15 | ND |
| Shingu | 1981 | IIF | 18 | 50 | 0 |
| Cines | 1984 | RIA | 27 | 89 | ND |
| Le Roux | 1986 | RIA | 20 | 65 | ND |
| Hashemi | 1987 | ELISA | 28 | 43 | 21 |
| Baguley | 1987 | ELISA | 82 | 22* | ND |
| Rosenbaum | 1988 | ELISA | 43 | 74 | ND |
| Vismara | 1988 | RIA | 51 | 39 | 45 |
| Quadros | 1990 | ELISA | 30 | 50 | 13 |
| D'Cruz | 1991 | ELISA | 107 | 81 | 58 |
| Cervera | 1991 | ELISA | 30 | 53 | 3 |
| Sebastiani | 1991 | ELISA | 42 | 69 | ND |
| van der Zee | 1991 | ELISA | 64 | 88 | ND |
| Perry | 1992 | ELISA | 22 | 80 | ND |

ELISA = enzyme linked immunosorbent assay.
RIA = radioimmunoassay.
ND = Not done.
IIF = indirect immunofluorescence.
* = polyclonal conjugate used.

patients. Many of these investigators showed a correlation between aECA levels and disease activity.

## Vasculitis

Endothelial cells clearly have a role in the pathogenesis of vasculitis. They are able to express complement receptors, in particular C1q receptors, as well as Fc receptors and so have the capacity to bind immune complexes [28]. Subsequent complement activation with the formation of membrane attack complexes could then result in vascular damage. This explanation may be somewhat simplistic since cellular immunity involving activation and adherence of lymphocytes to the vascular endothelium may also be part of the process that adds to the vascular damage.

aECA provide a possible mechanism for the vasculitis seen in lupus. In the original papers, aECA were associated with the vascular lesions seen in lupus and we and others have confirmed this correlation of aECA with lupus vasculitis using cultured human endothelial cells [5,17]. In a cohort of 107 lupus patients, 25 had active cutaneous or digital vasculitis

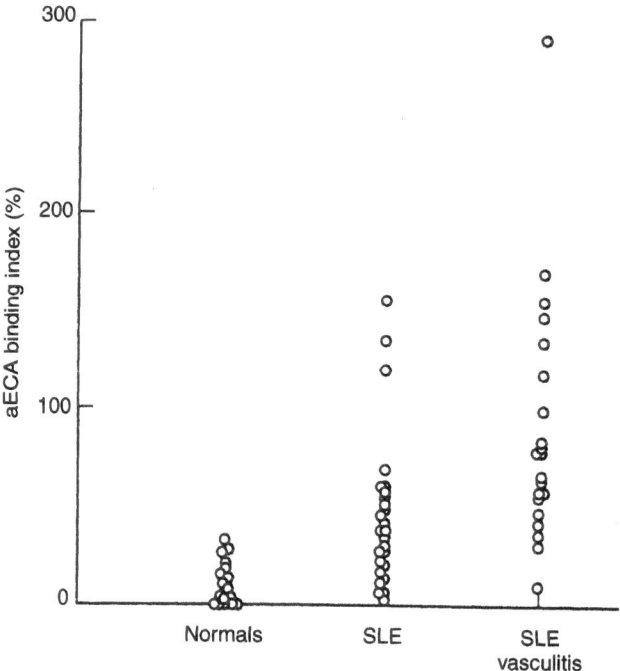

**Figure 5.1** aECA levels are significantly higher in lupus patients with vasculitis.

and these patients had significantly higher levels of aECA when compared with either normal controls or lupus patients without vasculitis (Figure 5.1). In keeping with other investigators, we were unable to show that aECA were cytotoxic to unstimulated endothelium even in the presence of fresh complement. Although aECA are clearly associated with clinical episodes of vasculitis in SLE, a possible pathogenetic role has yet to be clarified and the possibility that they may arise as an epiphenomenon in response to vascular damage cannot be discounted.

## Lupus nephritis

The pathogenesis of lupus nephritis is extremely complex and continues to defy efforts to elucidate its etiology. In mice, anti-DNA antibodies are associated with lupus nephritis. For example, injected monoclonal anti-DNA antibodies bind directly to glomerular antigens to form immune deposits and result in a proliferative glomerulonephritis [29]. In humans, anti-DNA antibodies may bind DNA with attachment of this complex to the heparan sulfate component of the glomerular basement membrane occuring via histones in the DNA double helix [30].

The first indications that aECA may be associated with nephritis came from animal experiments where labeled aECA were injected into rabbits [31]. Renal histology demonstrated that the labeled antibodies were bound to the surface of the glomerular endothelial cells and a mild, transient but consistently reproducible glomerulonephritis was documented. Furthermore, guinea pigs immunized with plasma membrane products derived from cultured rat brain and human endothelium developed aECA, circulating immune complexes and clinically and histogically proven nephritis [32]. It is possible that aECA reacted with a surface antigen on glomerular endothelium and resulted in the shedding of immune complexes from the cell surface into the subendothelial region as electron dense deposits. In humans, IgA aECA were found in 32% of patients with IgA nephropathy compared with 9% of other patients with primary glomerulonephritis [33].

Our own studies in SLE [17] showed that 81% of lupus nephritis patients had high levels of aECA compared with 44% of lupus patients who had normal renal function and 1% of normal controls (Figure 5.2). There was a correlation with renal histology in that the highest values were seen in patients with diffuse proliferative lupus nephritis (WHO Class IV). High aECA levels were associated with biopsies with high activity indices and also with patients with nephrotic syndrome at the time of biopsy. Serologically, aECA correlated with reduced complement levels but not with anti-DNA antibodies, antibodies against extractable nuclear antigens or serum immunoglobulin levels. aECA levels fell significantly following treatment and Figure 5.3 shows a representative

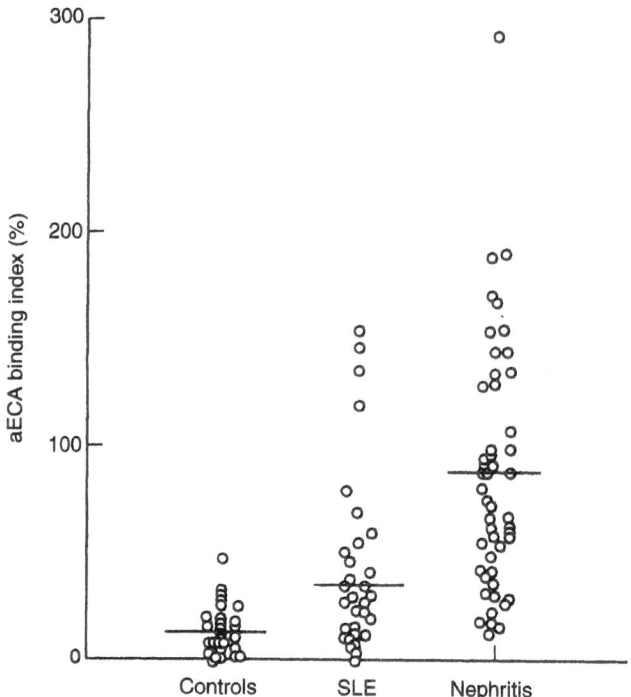

**Figure 5.2** aECA levels in lupus patients with nephritis. (Reprinted with permission.)

patient's profile. Although Cines *et al.* [24] showed that heat aggregated immune complexes could bind to endothelial cells, we and others [25, 27, 33] showed that removal of circulating immune complexes from the sera of these patients did not affect aECA binding.

Other authors have since confirmed that aECA is associated with lupus nephritis [34]. They were unable to show associations with renal histology or clinical features, such as nephrotic syndrome, though the number of patients they studied was small. They also found that some aECA reactivity could be partially absorbed by immobilized DNA and that purified polyclonal anti-DNA antibodies bound to endothelial cells, especially in the presence of histone.

Other markers of endothelial damage, such as von Willebrand factor antigen, are significantly raised in patients with lupus nephritis compared with lupus controls without nephritis [35]. Similarly, laminin is a basement membrane component that is released into the circulation when damage occurs and levels of the P1 fragment were significantly elevated, further supporting the evidence for endothelial and basement membrane damage in lupus nephritis [36].

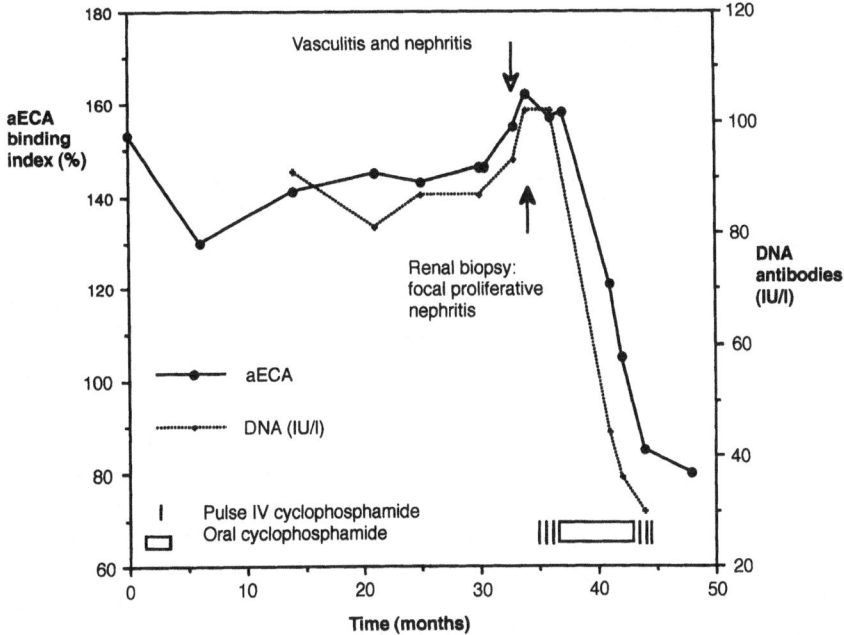

**Figure 5.3** Serological profile of a patient with lupus vasculitis and nephritis showing the change in aECA and DNA binding levels following treatment. (Reprinted with permission.)

Clinically, therefore, aECA may have a role as a serological marker for the development of lupus nephritis. Whether these antibodies have a role similar to that of the experimental nephritis models, or whether they are merely a response to glomerular damage caused by another process remains to be defined. Another consideration is that glomerular endothelial cells differ significantly in morphology from umbilical vein endothelial cells and may also differ in their surface glycoprotein characteristics. This raises the possibility that antibodies in lupus sera to umbilical vein endothelial cells may bind to different epitopes on glomerular endothelial cells with differing consequences.

## Thrombosis

The antiphospholipid syndrome, with venous and arterial thrombosis, strokes, migraine, recurrent miscarriages, livedo reticularis and thrombocytopenia, is now recognized as a distinct clinical entity [37]. It is frequently associated with diseases such as SLE, but may occur in isolation.

While the clinical syndrome has been well described, the reasons for the clotting tendency and associated vascular occlusions remain elusive and the evidence that antiphospholipid antibodies (aPL) actually cause thrombosis remains circumstantial. In animal studies, for example, injection of mouse monoclonal or human polyclonal aPL is associated with thrombocytopenia, impaired placental development and fetal loss [38]. A cofactor, $\beta$ 2 glycoprotein 1, is an absolute requirement for aPL binding [39], suggesting that the epitopes recognized by aPL are phospholipid–protein complexes. Interestingly, injection of this cofactor into animals results in the development of circulating aPL [40].

Vascular endothelium regulates the balance of coagulation *in vivo* and endothelial damage or dysfunction may be associated with thrombosis. Nitric oxide and prostacyclin are both synthesized by endothelial cells and acting together are powerful vasodilators and inhibitors of platelet function and adhesion. Inhibition of prostacyclin production by aPL or aECA could therefore be a mechanism for thrombosis. Carreras and Vermylen [41], using fresh rings of rat aorta, first suggested that the action of lupus anticoagulants may be mediated through the inhibition of prostacyclin production. This work stimulated a number of further studies with conflicting results [42, 43]. The variation in the conclusions appeared to be related to the source of the endothelium used – those studies that used rat or bovine endothelium supported Carreras and Vemylen's data, but those using human endothelium did not.

Several other mechanisms involving the clotting cascade and endothelium have been proposed to explain the thrombotic tendency associated with aPL. These include impaired fibrinolytic activity [44], enhanced platelet adhesion to the subendothelium [45], impairment of antithrombin III [46] and inhibition of protein C activation [47]. The cytokines IL-1 and tumor necrosis factor (TNF) markedly increase tissue factor-like procoagulant activity in cultured endothelium. They also decrease endothelial surface thrombomodulin, thus inhibiting the anticoagulant effects of protein C and protein S and also increase the secretion of the inhibitor of tissue plasminogen activator [48]. The overall effects of these cytokines is thus towards a procoagulant effect with the deposition of fibrin on the endothelial surface, as well as intravascular coagulation.

This picture has been complicated by the possibility that aECA may also have a role in thrombosis [49, 50]. For example, IgG and IgG-containing immune complexes may be deposited on to endothelial cells in association with the lupus anticoagulant [51]. Binding of aECA may increase the adherence of gel-filtered platelets to endothelium and endothelial cells express cell-associated procoagulant activity when purified monomeric IgG, IgM and IgA fractions of SLE sera are added [24, 52]. Furthermore, IgG complexes may inhibit the fibrinolytic activity of endothelial cells [24].

The possibility that aPL may be part of this very heterogeneous population of aECA has been explored in several studies. Some of these studies support the idea of serologically distinct antibody populations, though others support the view that aPL correlate with aECA and/or thrombosis and may cross react with antigens on the endothelial surface [49, 50, 53]. The major problem with trying to ascertain whether these two antibody populations are related is the paucity of information on the precise antigenic specificities of aECA. A further problem is the lack of a standardized assay for the quantification of aECA, making direct comparison betwen groups difficult. Efforts are being made to address this problem.

Many of the studies cited above regarding prostacyclin release from endothelial cells did not specifically measure aECA. Walker *et al.* [54] found that 10 out of 19 lupus anticoagulant-containing plasmas also contained aECA, but there was no correlation between aECA or aPL and a past history of thrombosis. When they examined the effect of these plasmas on prostacyclin production from human umbilical vein endothelial cells, six in fact **stimulated** prostacyclin production, one was inhibitory and seven were equivocal, in marked contrast to previous studies. Another study [55] overall showed no evidence for prostacyclin release from endothelial cells exposed to IgG from patients with SLE and the antiphospholipid syndrome, although they did not use affinity purified aPL. Statistical associations suggested that increasing levels of both aPL and aECA were correlated with each other and with inhibition of the prostacyclin stimulation index. aECA but not aPL correlated with prostacyclin inhibition as well as thrombotic events.

There are several difficulties inherent in the measurement of prostacyclin production from human endothelial cells. These include marked batch-to-batch variation and this together with interassay variation in the aECA assay may make interpretation difficult. Nevertheless, future studies may clarify the potential relationship between aECA, aPL and thrombosis.

## Central nervous sytem (CNS) disease

To date there have been few reports of aECA specifically associated with CNS disease in lupus. Patients with CNS disease in one study had higher aECA levels than those without, but the differences did not reach statistical significance [56]. Future studies in this field may be complicated by the fact that patients with active CNS disease frequently have renal disease and/or cutaneous vasculitis. It would not, therefore, be surprising to find elevated levels of aECA in these patients, but this may merely reflect disease activity. Animal studies certainly provide some evidence that aECA may be associated with neurological disease. Guinea pigs immunized with endothelial plasma membrane products developed aECA and brain and retinal lesions that were similar to the lesions seen in multiple sclerosis and experimental allergic encephalomyelitis [32]. Interestingly, aECA have been described in association with multiple sclerosis [12], and it is tempting to speculate that these antibodies may also have a role in the CNS disease seen in lupus.

### Genetics of aECA in SLE

There have been two studies from the same group examining the role of HLA alleles in association with aECA in SLE. In their first study [57], they reported an association of aECA with HLA-DR7 and DQw2. In a second study examining HLA-DP alleles, they found a weak link with DPB1*1401 and a possible protective effect of DPB1*0401 [58]. They suggested that the association of these DP beta alleles was independent of linkage to the DR7 and DQw2 associated susceptibility antigens. This group also reported an association between anticardiolipin antibodies and the DPB1*1401 allele [59] and this is reminiscent of the possible association between aECA and anticardiolipin antibodies in SLE patients. Taken as a whole, these studies suggest a possible role for the MHC in controlling aECA production in these patients.

### aECA AND PATHOGENETIC MECHANISMS IN SYSTEMIC VASCULITIS

aECA may be present, often in high levels, in some patients with systemic vasculitis [15], Wegener's granulomatosis and microscopic polyangiitis [60–64]. In some patients, aECA levels fluctuate with clinical disease activity, suggesting that it might be possible to use these antibodies to monitor disease activity and response to treatment. aECA are probably separate from antineutrophil cytoplasmic antibodies (ANCA) and there is little cross reactivity between these two antibody populations [61, 62]. In contrast to ANCA which appear to be a good marker for Wegener's granulomatosis, aECA occur in a variety of the systemic vasculitides and are not disease specific.

Vascular endothelial cells have tissue specific antigens on their surface that may be induced by cytokines and could be potent immunogens in systemic vasculitis. For example, IL-1 and TNF induce the expression of adhesion molecules such as endothelial leukocyte adhesion molecule (ELAM-1) and intercellular adhesion molecule (ICAM-1). INF-γ also induces ICAM-1 expression as well as Class I and Class II MHC antigens. Thus, endothelial cells may be capable of presenting antigen in much the same way as classic antigen-presenting cells [63].

It is possible that an antigen presented on the endothelial cell surface is recognized by T cells which stimulate the production, via helper cells, of pathogenetic antibodies from B cells. These antibodies may then bind to the endothelial surface initiating damage in conjunction with complement activation and membrane attack complexes. This hypothesis depends on the premise that endothelial cells are capable of processing and presenting antigen via the MHC molecules on the endothelial surface to CD4+ T cells. There is good *in vitro* and persuasive *in vivo* evidence to support this idea. Class II expression on the endothelial surface is markedly upregulated by INF-γ and these inducible molecules are capable of presenting nominal protein antigens to T cells in an HLA-DR restricted manner [64]. Endothelial cells may also provide costimulatory signals to T cells by synthesis of IL-1 and IL-6, and may modulate T cell IL-2 synthesis which increases when T cells adhere to endothelial cells via the CD2:LFA 3 ligand pair [64]. These findings certainly suggest a close interaction between endothelial cells and T cells in the pathogenesis of vasculitis. How aECA and/or ANCA may be relevant to these cellular mechanisms remains unclear.

## aECA and other vasculitic diseases

aECA have been described in patients with both isolated retinal vasculitis and patients with systemic diseases with ocular involvement [65]. In Behçet's disease, aECA occur in 15–30% of patients and may correlate with venous and arterial thrombosis, but do not exhibit complement-mediated cytotoxicity [66]. No relationship with aPL was found – indeed aPL are only rarely found in this often thrombotic disease.

In the idiopathic inflammatory myopathies aECA correlate with the presence of interstitial pulmonary fibrosis independently of anti-Jo1 antibodies [67]. This correlation is interesting since dermatomyositis in particular is characterized by vascular endothelial damage.

## aECA and endothelial cytoxicity

One of the most interesting aspects of aECA is whether they may be pathogenic and many investigators have studied the potential of these antibodies to display cytotoxic effects *in vitro*. Although Cines initially showed that aECA from lupus patients may disrupt endothelial monolayers, a number of other groups have failed to show direct complement-dependent serum cytotoxicity except in a few selected patients with SLE [68].

Kawasaki disease is characterized by a panvasculitis, and aECA have been detected in the sera of patients in the acute phase of the disease which subsequently disappear in the convalescent phase. aECA in these sera displayed a complement-dependent endothelial cytotoxicity on cytokine-treated but not resting endothelial cells [9,69]. The target antigens induced by INF-γ seemed to be quite distinct from those induced by IL-1 or TNF as suggested by the different time course of antigen expression and by absorption experiments. Production of large amounts of IL-1, TNF and INF-γ may be central to the vascular damage seen in these children.

In the systemic vasculitides, approximately half of the aECA positive sera were shown to mediate endothelial cytotoxicity in the presence of normal peripheral blood mononuclear cells [70]. The phenomenon was specific and reproducible with purified IgG from aECA positive sera but required high effector/target ratios and was not increased by pretreatment of the endothelium with cytokines. It is possible that this cytotoxicity was mediated by FcR–γIII positive lymphocytes since these cells may lyse targets previously sensitized by IgG.

Thus the evidence for a cytotoxic effect of sera from connective tissue disease patients on endothelium, mediated either directly by aECA or through antibody dependent cellular cytotoxicity, is conflicting, though overall the data tend to support a cytotoxic effect in some patients with certain diseases. It is also possible, however, that some of the cytotoxic effects may be due to other factors such as oxidized lipoproteins, free radicals or storage artifact [71]. A clearer picture emerges in renal allograft rejection where 76–100% of sera from patients suffering HLA-matched renal graft rejection had cytotoxic aECA [72] where the antibodies are directed against an endothelial–monocyte antigen system [14].

### Antigenic specificity of aECA

Several studies suggest that aECA bind to endothelial membrane antigens by the F(ab')$_2$ and not the Fc portion. Adsorption of aECA positive sera by other cells such as fibroblasts and, to a lesser extent, by leukocytes and monocytes, reduces aECA binding. This suggests that the antigenic epitopes are not specific for endothelial cells and that aECA are probably a heterogeneous group of antibodies that either recognize a number of different antigens on different cells, or that similar epitopes are shared by a number of cells. There is a consensus that aECA reactivity is unrelated to the presence of antinuclear antibodies, antibodies to extractable nuclear antigens or rheumatoid factor.

In all the studies described above, very little data has been obtained about the precise antigens recognized by aECA in lupus sera. The information available suggests that aECA are unrelated to blood group antigens and binding is not increased by cytokines that upregulate HLA Class I or II expression. The antigens are present on endothelial cells derived from different donors and aECA activity can partially be absorbed out by peripheral blood mononuclear cells, leukocytes and fibroblasts [26, 56], cell lines such as K562 (erythroleukemia), HL 60 (acute promyelocytic leukemia) and Raji cells (Burkitt's lymphoma) [16], cardiolipin [26, 49] and DNA. This suggests that aECA are a very heterogeneous population of antibodies that recognize a wide variety of antigens on the endothelial surface.

Characterization of these antigens has thus proved extremely difficult, particularly in lupus sera where there may be a large variety of auto-antibodies present. van der Zee *et al.* [56] obtained a membrane preparation from human umbilical vein endothelial cells and used this in immunoblotting experiments. aECA reacted with a variety of antigens between 15 and 200 kD in size and lupus nephritis sera were particularly reactive against three antigens of 38, 41 and 150 kD. They thus confirmed the heterogeneity of these antibodies and suggested that certain antigens may be associated with the development of nephritis. Our own observations [73] suggest that lupus nephritis sera react with a larger number of antigens than control lupus sera, and we also noted that the 38, 41 and 150 kD antigens were more frequently recognized by the nephritis sera. Sera from patients with Wegener's granulomatosis recognized fewer antigens (180, 155, 125, 68 and 25 kD) than SLE sera [74]. Some endothelial antigens were precipitated only by Wegener's sera (125 kD) or by SLE sera (200 kD), suggesting different endothelial reactivity in different vasculitic processes. Further studies in vasculitis sera [75] demonstrated that the antigens recognized by aECA are expressed in the subendothelial matrix and that anticollagen antibodies may form part of the aECA population.

## FUTURE DIRECTIONS FOR RESEARCH

It is clear that aECA are associated with vascular damage, though the evidence that they may be directly pathogenic remains inconclusive. Future research could focus on the possibility that aECA may have functional effects on the endothelium that alters its physiological functions without killing the cells. Other intriguing possibilities include the recognition of proteinase 3 on the surface of stimulated endothelial cells by ANCA, raising the hypothesis that ANCA may directly bind to and damage endothelial cells.

Full characterization of the aECA antigens has proved to be a difficult enterprise. However, when more information becomes available, more specific ELISA techniques can be developed that would allow some standardization to be achieved and may also give further insights into the possible relevance of aECA in the connective tissue diseases.

REFERENCES

1. Jaffe, E.A., Nachmann, R.L., Becker, C.G. and Minick, C.R. (1973) Culture of human endothelial cells derived from umbilical veins: identification by morphologic and immunologic criteria. *J. Clin. Invest.*, **52**, 2745–56.
2. Ziff, M. (1989) Role of endothelium in chronic inflammation. *Springer Semin. Immunopathol.*, **11**, 199–214.
3. Lindqvist, K.J. and Osterland, C.K. (1971) Human antibodies to vascular endothelium. *Clin. Exp. Immunol.*, **9**, 753–60.
4. Tan, E.M. and Pearson, C.M. (1972) Rheumatic disease sera reactive with capillaries in the mouse kidney. *Arthritis Rheum.*, **15**, 23–8.
5. Shingu, M. and Hurd, E.R. (1981) Sera from patients with systemic lupus erythematosus reactive with human endothelial cells. *J. Rheumatol.*, **8**, 581–6.
6. Fattorossi, A., Aurbach, G.D., Sakaguchi, K. *et al.* (1988) Anti-endothelial cell antibodies: Detection and characterization in sera from patients with autoimmune hypoparathyroidism. *Proc. Nat. Acad. Sci. USA*, **85**, 4015–19.
7. Baguley, E. and Hughes, G.R.V. (1988) Lytic IgG antiendothelial cell antibodies in vasculitis. *Lancet*, **ii** (8616), **907**.
8. Cines, D.B. (1989) Disorders associated with antibodies to endothelial cells. *Rev. Infect. Dis.*, **11**, 705–11.
9. Leung, D.Y.M., Collins, T., Lapierre, L.A. *et al.* (1986) Immunoglobulin M antibodies present in the acute phase of Kawasaki syndrome lyse cultured vascular endothelial cells stimulated by gamma interferon. *J. Clin. Invest.*, **77**, 1428–35.
10. Leung, D.Y.M., Moake, J.L., Havens, P.L. *et al.* (1988) Lytic antiendothelial cell antibodies in haemolytic uraemic syndrome. *Lancet*, **2**, 183–6.
11. Burns, E.R. and Zucker-Franklin, D. (1982) Pathologic effects of plasma from patients with Thrombotic Thrombocytopenic Purpura on platelets and cultured vascular endothelial cells. *Blood*, **60**, 1030–7.
12. Tanaka, Y., Tsukada, N., Koh, Ch.-S. and Yanagisawa, N. (1987) Antiendothelial cell antibodies and circulating immune complexes in the sera of patients with multiple sclerosis. *J. Neuroimmunol.*, **17**, 49–59.
13. Rappaport, V.J., Hirata, G., Yap, H.K. and Jordan, S.C. (1990) Antivascular endothelial cell antibodies in severe pre-eclampsia. *Am. J. Obstet. Gynecol.*, **162**, 138–46.
14. Cerilli, J., Holliday, J.E., Fesperman, D.P. and Folger, M.A. (1977) Antivascular endothelial cell antibody – its role in transplantation. *Surgery*, **81**, 132–8.
15. Brasile, L., Kremer, J.M. Clarke, J.L. and Cerilli, J. (1989) Identification of an autoantibody to vascular endothelial cell-specific antigens in patients with systemic vasculitis. *Am. J. Med.*, **87**, 74–80.
16. Quadros, N.P., Roberts-Thompson, P.J. and Gallus, A.S. (1990) IgG and IgM anti-endothelial cell antibodies in patients with collagen-vascular disorders. *Rheum. Int.*, **10**, 113–19.
17. D'Cruz, D.P., Houssiau, F.A., Ramirez, G. *et al.* (1991) Antiendothelial cell antibodies in systemic lupus erythematosus: A potential marker for nephritis and vasculitis. *Clin. Exp. Immunol.*, **85**, 254–61.
18. Fernandez-Shaw, S., Hicks, B.R., Yudkin, P.L. *et al.* (1993) Anti-endometrial and anti-endothelial autoantibodies in women with endometriosis. *Hum. Reprod.*, **8**, 310–15.
19. Lassalle, P., Gosset, P., Gruart, V. *et al.* (1990) Presence of antibodies against endothelial cells in the sera of patients with episodic angioedema and hypereosinophilia. *Clin. Exp. Immunol.*, **82**, 38–43.

20. Lassalle, P., Delneste, Y., Gosset, P. *et al.* (1993) T and B cell immune response to a 55-kDa endothelial cell-derived antigen in severe asthma. *Eur. J. Immunol.*, **23**, 796–803.

21. Wangel, A.G., Kontianen, S., Scheinin, T. *et al.* (1992) Anti-endothelial cell antibodies in insulin-dependent diabetes mellitus. *Clin. Exp. Immunol.*, **88**, 410–13.

22. Wangel, A.G., Kontianen, S., Melamies, L. and Weber, T. (1993) Hypothyroidism and anti-endothelial cell antibodies. *APMIS*, **101**, 91–4.

23. Stevens, T.R., Harley, S.L., Groom, J.S. *et al.* (1993) Anti-endothelial cell antibodies in inflammatory bowel disease. *Dig. Dis. Sci.*, **38**, 426–32.

24. Cines, D.B., Lyss, A.P., Reeber, M. *et al.* (1984) Presence of complement fixing anti-endothelial cell antibodies in systemic lupus erythematosus. *J. Clin. Invest.*, **73**, 611–25.

25. Hashemi, S., Smith, C.D. and Izaguirre, C.A. (1987) Antiendothelial cell antibodies: Detection and characterisation using a cellular enzyme-linked immunosorbent assay. *J. Lab. Clin. Med.*, **109**, 434–40.

26. Rosenbaum, J., Pottinger, B.E., Woo, P. *et al.* (1988) Measurement and characterisation of circulating anti-endothelial cell IgG in connective tissue diseases. *Clin. Exp. Immunol.*, **72**, 450–6.

27. Heurkens, A.H.M., Hiemstra, P.S., Lafeber, G.J.M. *et al.* (1989) Antiendothelial cell antibodies in patients with rheumatoid arthritis complicated by vasculitis. *Clin. Exp. Immunol.*, **78**, 7–12.

28. Jaffe, E.A. (1987) Cell biology of endothelial cells. *Hum. Pathol.*, **18**, 234–9.

29. Raz, E., Brezis, M., Rosenmann, E. and Eilat, D. (1989) Anti-DNA antibodies bind directly to renal antigens and induce kidney dysfunction in the isolated perfused rat kidney. *J. Immunol.*, **142**, 3076–82.

30. Brinkman, K., Termaat, R., Berden, J.H.M. and Smeenk, R.J.T. (1990) Anti-DNA antibodies and lupus nephritis: the complexity of crossreactivity. *Immunol. Today*, **11**, 232–4.

31. Matsuo, S., Fukatsu, A., Taub, M.L. *et al.* (1987) Glomerulonephritis induced in the rabbit by antiendothelial cell antibodies. *J. Clin. Invest.*, **79**, 1798.

32. Matsuda, M. (1988) Experimental glomerular tissue injury induced by immunisation with cultured endothelial cell plasma membrane. *Acta Pathol. Japan*, **38**, 823.

33. Yap, H.K., Sakai, R.S., Bahn, L. *et al.* (1988) Antivascular endothelial cell antibodies in patients with IgA nephropathy: frequency and clinical significance. *Clin. Immunol. Immunopathol.*, **49**, 450.

34. Perry, G.J., Roedecker, T., Chan, T.M. (1992) Antiendothelial cell antibodies and their functional and histological associations in lupus nephritis. *Lupus*, **1**, 9 (Abstract).

35. D'Cruz, D., Jedryka-Goral, A., Khamashta, M.A. and Hughes, G.R.V. (1991) von Willebrand factor antigen in lupus nephritis. *Arthritis Rheum.*, **34** (supplement 9), 130.

36. D'Cruz, D., Schneider, M., Khamashta, M.A. and Hughes, G.R.V. (1992) Laminin P1 levels in lupus nephritis. *Arthritis Rheum.*, **35**(Suppl.), 109.

37. Hughes, G.R.V. (1993) The antiphospholipid syndrome – 10 years on. *Lancet*, **342**, 341–4.

38. Blank, M.J., Cohen, V., Toder, V and Schonfeld, Y. (1991) Induction of antiphospholipid syndrome in naive mice with mouse lupus monoclonal and human polyclonal anticardiolipin antibodies. *Proc. Nat. Acad. Sci. USA*, **88**, 3069–73.

39. McNeil, H.P., Simpson, R., Chesterman, C.N. and Krilis, S.A. (1990) Antiphos-

pholipid antibodies are directed against a complex antigen that includes a lipid-binding inhibitor of coagulation: β2-glycoprotein 1 (apolipoprotein H). *Proc. Nat. Acad. Sci. USA*, **87**, 4120–4.

40. Gharavi, A.E., Sammaritano, L.R., Wen, J. and Elkon, K.B. (1992) Induction of antiphospholipid antibodies by immunisation with a lipid binding protein β2-glycoprotein I. *Lupus*, **1**(Suppl. 1), 150.

41. Carreras, L.O. and Vermylen, J.G. (1982) 'Lupus' anticoagulant and thrombosis – possible role of inhibition of prostacyclin formation. *Thromb. Haemostas.*, **48**, 38–40.

42. Watson, K.V. and Schorer, A.E. (1991) Lupus anticoagulant inhibition of *in vitro* protacyclin release is associated with a thrombosis-prone subset of patients. *Am. J. Med.*, **90**, 47–53.

43. Hasselaar, P., Derksen, R.H.W.M., Blokzijl, L. and DeGroot, P.G. (1985) Thrombosis associated with antiphospholipid antibodies cannot be explained by effects on endothelial cell–platelet prostanoid synthesis. *Thromb. Haemostas.*, **59**, 80–5.

44. Tsakiris, D.A., Marbet, G.A., Makris, P.E. *et al.* (1989) Impaired fibrinolysis as an essential contribution to thrombosis in patients with lupus anticoagulant. *Thromb. Haemost.*, **69**, 175–77.

45. Escolar, G., Font, J., Reverter, J.C. *et al.* (1992) Plasma from systemic lupus erythematosus patients with antiphospholipid antibodies promotes platelet aggregation. Studies in a perfusion system. *Arterioscler. Thromb.*, **12**, 196–200.

46. Boey, M.L., Loizou, S., Colaco, B. *et al.* (1984) Antithrombin III in systemic lupus erythematosus. *Clini. Exp. Rheumatol.*, **2**, 53–6.

47. Cariou, R., Tobelem, G., Belluci, S. *et al.* (1988) Effect of lupus anticoagulant on antithrombogenic properties of endothelial cells – inhibition of thrombomodulin-dependent protein C activation. *Thromb. Haemost.*, **60**, 54–8.

48. Cotran, R.S. and Pober, J.S. (1989) Effects of cytokines on vascular endothelium: Their role in vascular and immune injury. *Kidney Int.*, **35** 969–75.

49. Cervera, R., Khamashta, M.A., Font, J. *et al.* (1991) Antiendothelial cell antibodies in patients with the antiphospholipid syndrome. *Autoimmunity*, **11**, 1–6.

50. McCrae, K.R., DeMichele, P., Samuels, P. *et al.* (1991) Detection of endothelial cell-reactive immunoglobulin in patients with anti-phospholipid antibodies. *Br. J. Haematol.*, **79**, 595–605.

51. LeRoux, G., Wautier, M.P., Guillevin, L. and Wautier, J.L. (1986) IgG binding to endothelial cells in systemic lupus erythematosus. *Thromb. Haemostas.*, **56**, 144.

52. Tannebaum, S.H., Finko, R. and Cines, D.B. (1986) Antibody and immune complexes induce tissue factor production by human endothelial cells. *J. Immunol.*, **137**, 1532–7.

53. Hasselaar, P., Derksen, R.H.W.M., Blokzijl, L. and DeGroot, P.G. (1990) Crossreactivity of antibodies directed against cardiolipin, DNA, endothelial cells and blood platelets. *Thromb. Haemostas.*, **63**, 169–73.

54. Walker, T.S., Triplett, D.A., Javed, N. and Musgrave, K. (1988) Evaluation of lupus anticoagulants: antiphospholipid antibodies, endothelium associated immunoglobuin. endothelial prostacyclin secretion and antigenic protein S levels. *Thrombosis Res.*, **51**, 267–81.

55. Lindsey, N.J., Henderson, F.I., Malia, R. *et al.* (1993) Inhibition of prostacyclin release by endothelial binding anticardiolipin antibodies in thrombosis prone patients with systemic lupus erythematosus and the antiphospholipid syndrome. *Br. J. Rheumatol.*, **33**, 20–6.

56. van der Zee, J.M., Siegert, C.E.H., De Vreede, T.A. *et al.* (1991) Characterisation of anti-endothelial cell antibodies in systemic lupus erythematosus. *Clin. Exp. Immunol.*, **84**, 238–44.

57. Sebastiani, G.D., Passiu, G., Lulli, A. *et al.* (1991) Association of HLA DR-7 and DQw2 with antiendothelial cell antibodies in systemic lupus erythematosus. *Abstracts of the XIIth European Congress of Rheumatology: Hungarian Rheumatology*, **32**, 208.

58. Sebastiani, G.D., Galeazzi, M., Passiu, G. *et al.* (1992) Genetics of antiendothelial cell antibodies in systemic lupus erythematosus: the role of HLA-DP alleles. *Contrib. Nephrol.*, **99**, 102–7.

59. Galeazzi, M., Sebastiani, G.D., Passiu, G. *et al.* (1992) HLA-DP genotyping in patients with systemic lupus erythematosus: correlations with autoantibody subsets. *J. Rheumatol.*, **19**, 42–6.

60. Ferraro, G., Meroni, P.L., Tincani, A. *et al.* (1990) Antiendothelial cell antibodies in patients with Wegener's Granulomatosis and micropolyarteritis. *Clin. Exp. Immunol.*, **79**, 47–53.

61. Frampton, G., Jayne, D.R.W., Perry, G.J. *et al.* (1990) Autoantibodies to endothelial cells and neutrophil cytoplasmic antigens in systemic vasculitis. *Clin. Exp. Immunol.*, **82**, 227–32.

62. Savage, C.O.S., Pottinger, B.E., Gaskin, G. *et al.* (1991) Vascular damage in Wegener's granulomatosis and microscopic polyarteritis: presence of antiendothelial cell antibodies and their relation to anti-neutrophil cytoplasm antibodies. *Clin. Exp. Immunol.*, **85**, 14–19.

63. Hirschberg, H, Bergh, O.J. and Thorsby, E. (1980) Antigen-presenting properties of human vascular endothelial cells. *J. Exp. Med.*, **152**, 249–55.

64. Hughes, C.C.W., Savage, C.O.S. and Pober, J.S. (1990) The endothelial cell as a regulator of T cell function. *Immunol. Reviews.*, **117**, 85–102.

65. Edelsten, C., D'Cruz, D., Hughes, G.R.V. and Graham, E.M. (1992) Antiendothelial cell antibodies in patients with idiopathic retinal vasculitis. *Curr. Eye Res.*, **11**, 203–8.

66. Aydintug, A.O., Tokgoz, G., D'Cruz, D. *et al.* (1993) Antiendothelial cell antibodies in patients with Behçet's disease. *Clin. Immunol. Immunopathol.*, **67**, 157–62.

67. Cervera, R., Ramirez, G., Fernandez-Sola, J. *et al.* (1991) Antibodies to endothelial cells in dermatomyositis: association with interstitial lung disease. *Br. Med. J.*, **302**, 880–1.

68. Penning, C.A., French, M.A.H., Rowell, N.R. and Hughes, P. (1985) Antibody-dependent cellular cytotoxicity of human vascular endothelium in systemic lupus erythematosus. *J. Clin. Lab. Immunol.*, **17**, 125–30.

69. Tizard, J., Baguley, E., Hughes, G.R.V. and Dillon, M. (1991) Antiendothelial cell antibodies detected by a cellular based ELISA in Kawasaki's disease. *Arch. Dis. Child.*, **66**, 189–92.

70. Del Papa, N., Meroni, P.L., Barcellini, W. *et al.* (1992) Antibodies to endothelial cells in primary vasculitides mediate *in vitro* endothelial cytotoxicity in the presence of normal peripheral blood mononuclear cells. *Clin. Immunol. Immunopathol.*, **63**, 267–74.

71. Blake, D.R., Winyard, P., Scott, D.G.I. *et al.* (1985) Endothelial cell cytotoxicity in inflammatory vascular diseases – the possible role of oxidised lipoproteins. *Ann. Rheum. Dis.*, **44**, 176–82.

72. Xiuming, W., Huiping, Z., Shaofeng, Z. and Jieping, C. (1989) Detection of antivascular endothelial cell antibodies with microcytotoxicity testing. *J. Clin. Lab. Immunol.*, **28**, 73–8.

73. D'Cruz, D.P., Khamashta, M.A. and Hughes, G.R.V. (1993) Characterisation of anti-endothelial cell antibody binding in lupus nephritis. *Br. J. Rheum.*, **32**(Suppl. 1), 23.
74. Del Papa, N., Conforti, G., Gambini, D. *et al.* (1994) Characterization of the endothelial surface proteins recognized by anti-endothelial antibodies in primary and secondary autoimmune vasculitis. *Clin. Immunol. Immunopathol.*, **70**, 211–16.
75. Direskeneli, H., D'Cruz, D., Khamashta, M.A. and Hughes, G.R.V. (1994) Autoantibodies against endothelial cells, extracellular matrix and human collagen type IV in patients with systemic vasculitis. *Clin. Immunol. Immunopathol.*, **70**, 206–10.

# 6

# Adhesion molecules in disease: vascular cell adhesion molecule-1 as a target for therapy

*J.M. Clements, R. Pigott, L.A. Needham and A.J.H. Gearing*

## INTRODUCTION

During the normal course of the inflammatory response, leukocytes leave the circulation by first adhering to the endothelium, before migrating to the site of infection or injury. This process is controlled, both temporally and spatially, and occurs at sites adjacent to inflammatory lesions, primarily from postcapillary venules. There is an early, rapid influx of neutrophils followed by monocytes and lymphocytes. This complex process is controlled in part by modulating both the expression and affinity of specific adhesion molecules expressed on leukocytes and endothelial cells. Adhesion molecules are involved in all stages of tissue infiltration, from the initial tethering of the leukocyte to the endothelial cell, through to migration across the endothelium.

Much progress has been made in understanding the molecular basis of these interactions. Cell adhesion molecules (CAMs) are large membrane bound proteins, which were originally defined by monoclonal antibodies (mAbs) which blocked adhesive interactions. The development of expression cloning techniques has led to the identification of cDNAs for many adhesion molecules [1]. It is possible to divide CAMs into families based on homologies in their amino acid sequence (Figure 6.1). Three major classes of adhesion molecules have been identified: the integrins,

*The Vasculitides*. Edited by B.M. Ansell, P.A. Bacon, J.T. Lie and H. Yazici. Published in 1996 by Chapman & Hall, London. ISBN 978-0-412-64140-4.

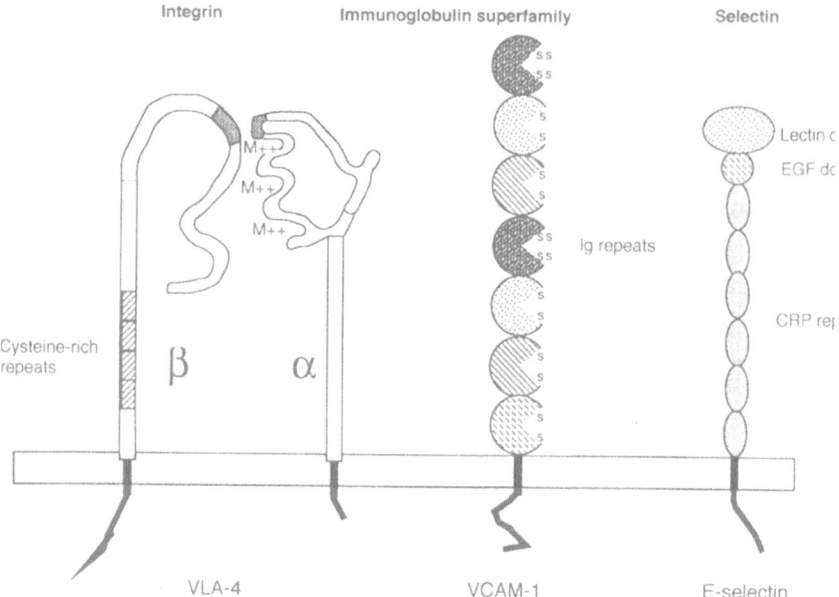

**Figure 6.1** The three main classes of cell adhesion molecules are shown in schematic form. The integrin family consists of α and β subunits which are non-covalently associated, the shared ligand-binding region is shown as a shaded area, metal-binding sites and the cysteine-rich repeats are indicated. The immunoglobulin superfamily is composed of immunoglobulin repeats of 90–100 amino acids, the disulfide bonds are indicated by the 's'. The selectins have an N-terminal domain of 120 amino acids, followed by a growth factor-like repeat, then short repeats of a motif found in complement regulatory proteins.

members of the immunoglobulin superfamily and the selectins. There is now good evidence that members of these families function as an integrated network to regulate leukocyte adhesion. An initial, weak and reversible adhesive interaction is mediated by members of the selectin family binding to carbohydrate ligands, followed by consolidation of adhesion involving interactions between members of the integrin and immunoglobulin families.

## CELL ADHESION MOLECULES

### Integrins

The integrins are the most versatile members of the adhesion molecule family. They are multifunctional receptors for a range of matrix components and for other CAMs. Each integrin consists of an α and β subunit associated in a heterodimeric complex. They can be divided into three subfamilies based on a common β chain; the very late antigen (VLA) proteins, the leukocyte integrins and the cytoadhesins. The name integrin derives from the fact that the cytoplasmic domains of integrins interact with components of the cytoskeleton, and hence integrate the cytoskeleton with the extracellular environment. Many integrins require activation to be fully functional, the best example being the platelet IIb/IIIa integrin, which can be activated by thrombin stimulation of platelets. Integrins can also act as signaling molecules to activate cellular functions. For example, certain monoclonal antibodies against VLA-4 (α4β1) have been shown to enhance binding to several ligands. Integrins can also bind multiple ligands, for example VLA-4 expressed on lymphocytes and monocytes functions as both a matrix receptor for fibronectin, and as a cell receptor for vascular cell adhesion molecule-1 (VCAM-1). At least 13 α and eight β chains have been identified and, as both α and β chains contribute to the binding specificity of the integrins, there is clearly an enormous potential for different binding specificites in this group of adhesion molecules.

**Immunoglobulin superfamily receptors**

More than 100 cell surface proteins contain extracellular immunoglobulin (IG) domains, and several of these proteins mediate intercellular adhesion. They are characterized by repeated domains of 90–100 amino acids, similar to those found in immunoglobulins, consisting of two sheets of antiparallel β-strands, stabilized by a disulfide bond. In general, they bind to integrin receptors found on leukocytes. The ICAM-1/LFA-1 adhesive pair was the first to be identified mediating leukocyte/leukocyte and leukocyte/endothelium binding. Subsequently, other IG superfamily adhesion molecule/β2 or β1 integrins pairs have been identified. The cell surface expression of adhesion molecules, such as intercellular cell adhesion molecule-1 (ICAM-1) and VCAM-1, are upregulated in immune responses and inflammation. ICAM-1 is expressed on endothelial, epithelial, fibroblast, leukocyte and many tumor cells, and VCAM-1 is also widely distributed on endothelial, epithelial, macrophage and dendritic cells. ICAM-1 and VCAM-1 on endothelial cells appear to be particularly important for the firm attachment and transendothelial migration of leukocytes.

**Selectins**

It had long been postulated that carbohydrates might mediate intercellular interactions; this has recently been vindicated with the identification of a new family of CAMs which bind to carbohydrates. The selectin family to date comprises three members, E-, P- and L- selectin which are involved in the initial interactions between leukocytes and vascular endothelium, platelet binding and lymphoycyte homing. They display a similar domain structure with a calcium-dependent lectin (carbohydrate binding) domain at their N-terminus. This is followed by an epidermal growth factor-like domain, a varying number of short repeats which share homology with complement regulatory proteins, a transmembrane domain and short cytoplasmic tail. E-, P- and L- selectin all appear to recognize sialyl-Lewis$^x$ and related oligosaccharides.

E-selectin is expressed specifically on endothelial cells in response to inflammatory stimuli, including IL-1, tumor necrosis factor (TNF) and bacterial lipopolysaccharide. It reaches a peak four hours after stimulation, declining rapidly thereafter. Although only expressed transiently *in vitro*, it can be expressed chronically in some disease states. P-selectin (CD62) is a receptor for neutrophils and monocytes that is stored for rapid release in the alpha granules of platelets or the Weibel-palade bodies of endothelial cells and is rapidly translocated to the plasma membrane of activated platelets and endothelial cells in response to stimulation by histamine and thrombin. Finally, L-selectin, previously known as the lymph node homing receptor LAM-1 or leu8, is expressed on lymphocytes, neutrophils and monocytes. It mediates the binding of lymphocytes to the high walled endothelium (HEV) in the postcapillary venules found in lymphoid organs and has a role in the recruitment of leukocytes to inflammatory sites. Recently, mucin-like glycoproteins CD34 and GlyCAM-1 on HEV have been identified as ligands for L-selectin [2, 3].

ROLE OF ENDOTHELIAL ADHESION MOLECULES

To be effective, leukocytes must both circulate as non-adherent cells and, in response to foreign antigen, become adherent cells able to leave the circulation at sites of infection (Figure 6.2). Recruitment of inflammatory cells to sites of infection is a multiple stage process which is regulated by a number of adhesion molecules. Initially, inflammatory mediators, such as IL-1, TNF-α histamine and lipopolysaccharide, induce changes in the endothelial cells that line the vessel wall. This activates the expression of a number of CAMs. P-selectin is expressed within minutes, whereas E-selectin and VCAM-1 appear over a two to six hour period. Leukocytes, then form a transient attachment to the endothelium and begin to 'roll' along the vessel. The interaction with the selectin slows the leukocytes sufficiently, despite the high shear forces existing in the vessel, to enable more stable integrin/Ig superfamily contacts to be made. The leukocytes disengage and migrate between adjacent endothelial cells, crossing the basement membrane underlying the endothelium, enter the tissue and

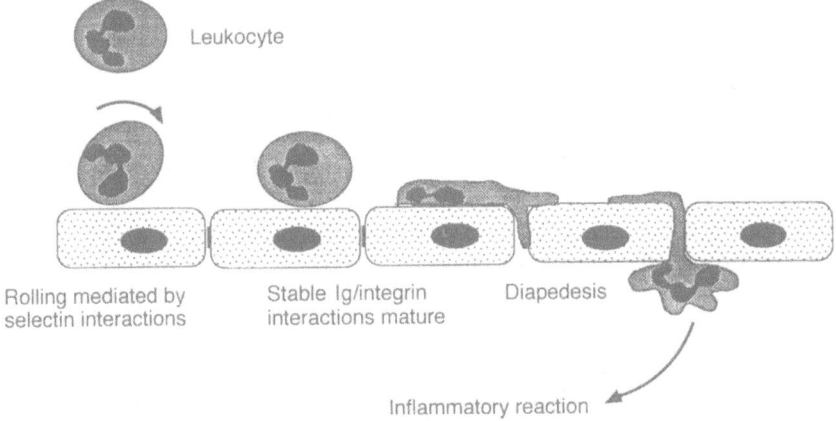

**Figure 6.2** Leukocyte extravasation. Leukocytes leave the bloodstream to sites of infection via a multiple stage process regulated by CAMs expressed on the endothelial cells and leukocyte.

migrate towards the source of infection using integrin mediated inter-actions.

Adhesion molecules also play a crucial role in the circulation of lymphocytes. It is known that lymphocytes follow a defined path through the body. It is possible to identify lymphocyte subpopulations that 'home' specifically to the peripheral lymph nodes, to the mucosal lymph nodes that serve the gut, or to the skin. These subsets of lympho-cytes may be specialized to deal with pathogens that are most prevalent in a particular tissue. The 'programming' of trafficking is known to involve CAMs displayed by both the lymphocytes and by specialized high endothelium in the postcapillary venules where the lymphocytes leave the bloodstream. For example, lymphocytes that home to the peripheral lymph nodes display L-selectin. L-selectin recognizes a mucin carbohydrate ligand displayed on the endothelium associated with the peripheral lymph nodes.

## CIRCULATING ADHESION MOLECULES IN DISEASE

Inflammation is normally a beneficial process; however, many patholo-gical conditions involve the inappropriate or excessive expression of inflammatory responses. Monitoring the expression of cell surface adhe-sion molecules can provide useful information on the activation state of tissues. Leukocyte adhesion molecules can be quantified by FACS ana-lysis of peripheral blood samples. Expression on other cells requires histological analysis of tissue biopsies. Immunohistology provides a detailed picture of adhesion molecule expression at a particular moment, but repeated biopsy is not a viable means of sampling patients to generate a kinetic picture of disease progression. In most diseases blood is the only tissue which can be regularly sampled. Work in our and other laboratories over the last three to four years has demonstrated that many adhesion molecules can be shed from the surface of endothelial, leukocytic and tumor cells, and that these soluble forms of the adhesion molecules can be detected in body fluids [4]. It has now been demonstrated that monitoring the levels of soluble adhesion molecules can provide useful information on the progress of experimental and human diseases.

### L-selectin

Soluble L-selectin circulates in at least two isoforms, one of 62 Kd and the other approximately 75–100 Kd [5]. These are thought to represent essentially the same fragment differentially glycosylated and derived from lymphocytes and neutrophils respectively. sL-selectin is 3–5 kD smaller than the corresponding cell surface form of the receptor. It can be released from the cell surface within minutes of activation of lymphocytes by phorbol esters, and from neutrophils by f-met-leu-phe, IL-8, bacterial lipopolysaccharide and GM-CSF. sL-selectin is also released by lymphocytes following activation with mitogens, but at a slower rate. Initial studies of patients with sepsis show a two- to three-fold increase in the sL-selectin levels compared to normals [6].

### P-selectin

P-selectin is found in the plasma of normal individuals [7], and it appears to be approximately 3 kD smaller than native P-selectin. Evidence is emerging that the majority of sP-selectin in blood is the alternatively spliced form lacking a transmembrane domain [M. Handa, personal communication] with the N-terminus intact. Since transcripts of P-selectin with or without the transmembrane form exist in both endothelial cells and megakaryocytes, it seems plausible that circulating levels may reflect some contribution by both cell types. Currently, there is little published information on the variations in P-selectin levels in disease states. Circulating P-selectin values in patients with hemolytic uremic syndrome and thrombotic thrombocytopenic purpura are elevated two- to three-fold over normal values [8].

### E-selectin

E-selectin is particularly interesting because it is found only on activated endothelium, in contrast to other adhesion molecules which have a wider tissue distribution. Endothelial cells have been shown to release

E-selectin following *in vitro* activation [9]. The demonstration of soluble E-selectin in blood would therefore be taken as conclusive evidence of endothelial activation. In comparison, thrombomodulin, endothelin and von Willebrand factor levels in blood have been used as circulating markers of vascular endothelial damage, but they are in fact, not endothelial specific. The mechanism of release of sE-selectin has not been established, but immunochemical evidence suggests a lost or defective cytoplasmic domain in the sE-selectin found in plasma [10]. Like L-selectin, E-selectin does not have an alternately spliced form lacking a transmembrane domain.

Elevated levels of sE-selectin have been found in diabetic patients independent of hypertension, nephropathy or renal failure, and whether or not they were insulin dependent [11; A. Rees, personal communication]. Similarly, sE-selectin was detected in patients with the vasculidities, polyarteritis nodosum, giant cell arteritis and scleroderma, with higher levels noted in lupus patients. Overall, however, there was no correlation with disease activity and only a weak correlation with the degree of organ involvement [12]. These studies suggest that the endothelium in these patients is activated, and that overt disease activity involves additional factors. Further longitudinal studies are needed to define possible associations.

sE-selectin levels have also shown marked elevations in sepsis, in one study a 20-fold increase over the normal range [10]. In this and another study, the levels of sE-selectin appear to correlate with disease severity and/or outcome [13]. Higher levels or persistent elevation were associated with greater mortality and this is not unexpected in view of the widespread expression of E-selectin in most vessels in primates challenged with a lethal dose of live *Escherichia coli* [14].

## Soluble immunoglobulin superfamily adhesins

### ICAM-1

The ready availability of antibodies to ICAM-1 led to the early development of immunological methods for the detection of circulating ICAM-1. A number of studies have now been published which give estimates of the mean level of circulating ICAM as between 102 and 450 ng/ml. ICAM levels have been reported to be elevated in inflammation, infection and cancers [4] (Table 6.1). The clinical utility of measuring levels of circulating ICAM remains to be established. ICAM levels are reported to correlate with liver metastasis in a variety of tumors, and with disease progression in melanoma [15]. It has been shown in nude mouse models that the melanoma cells themselves are a source of soluble ICAM [16].

**Table 6.1** Blood levels of circulating adhesion molecules

| Adhesion molecule | Cell source | Levels in healthy individuals* | Fold elevation in disease |
|---|---|---|---|
| VCAM-1 | Endothelium, epithelium, dendritic cells, macrophages, smooth muscle | 460 ng/ml | ×8 Septic shock<br>×4 Systemic lupus erythematosus<br>×2 Rheumatoid arthritis<br>×3 Vasculitis<br>×6 Renal allograft |
| ICAM-1 | Leukocytes, endothelium, epithelium, hepatocytes, smooth muscle | 240 ng/ml | ×2 Diabetes<br>×3 Septic shock<br>×2 Ulcerative colitis<br>×2 Crohn's disease |
| L-selectin | Leukocytes | 1700 ng/ml | ×2 Sepsis |
| P-selectin | Platelets, endothelium | 160 ng/ml | ×3 Thrombotic thrombocytopenic purpura |
| E-selectin | Endothelium | 28 ng/ml | ×2 Diabetes<br>×23 Septic shock<br>×4 Systemic lupus erythematosus<br>×2 Giant cell arteritis<br>×2 Polyarteritis nodosa |

*Levels are the mean of published values as adapted from Gearing and Newman [4].

There have been few longitudinal surveys of serum levels of ICAM-1. Shijubo *et al.* suggest that ICAM levels rise prior to death in idiopathic pulmonary fibrosis [17], and we have recently completed a survey of renal allograft patients and have seen rises in ICAM-1 associated with episodes of graft rejection [18]. ICAM-1 has also been detected in synovial fluids from patients with rheumatoid arthritis [19], and in the bronchioalveolar lavage fluid of patients with interstitial lung disease [15].

## VCAM-1

Soluble VCAM is released by activated endothelial cells in culture [9]; however, there is limited published information on levels of soluble VCAM in the blood. Estimates of the mean level of VCAM range from

431–504 ng/ml in normal individuals [4] (Table 6.1). VCAM in blood has been purified by affinity chromatography with bands of 80 and 50 kD, the 80 kD form was shown to be N-terminally identical to cell surface VCAM [I. Hemingway, unpublished results]. Haskard has reported a MW of 85–90 kD for plasma soluble VCAM-1, which was shown to be capable of supporting the adhesion of T cells [20]. VCAM levels are elevated in serum from patients with cancer and with inflammatory diseases [4] (Table 6.1). Longitudinal studies have shown increases in VCAM associated with renal allograft rejection episodes (Figure 6.2), and also with cytomegalovirus infection [4]. VCAM has recently been demonstrated in the synovial fluids of patients with rheumatoid arthritis [19]. VCAM levels appear to show an excellent correlation with disease activity in systemic lupus erythematosus (SLE) [21]. In rheumatoid arthritis patients, VCAM levels appeared to correlate with erythrocyte sedimentation rate and C-reactive protein levels, whereas ICAM did not [19].

In summary, the clinical utility of monitoring levels of soluble adhesion molecules has yet to be established; however, initial indications suggests that further well-designed studies are warranted. The availability of commercial assay kits should allow their widespread evaluation in a clinical setting.

## VCAM/VLA-4 INTERACTIONS AS A TARGET FOR THERAPY

Cell adhesion is an attractive target for therapy because it represents a very early stage in pathology. Thus it is better to block recruitment of inflammatory cells than to cope with the consequences of their activation in the tissue. We have chosen to focus on the VCAM/VLA-4 adhesion pathway as a target for therapy. VCAM-1 is expressed on endothelial cells only at sites of inflammation, and is important for the adhesion of lymphocytes, monocytes and eosinophils to the endothelium. Inhibition of this interaction therefore represents a novel therapeutic target for a range of inflammatory conditions characterized by the infiltration of these cell types, including rheumatoid arthritis, graft rejection, chronic inflammatory skin disease and asthma. In addition to the role of VCAM/VLA-4 in leukocyte adhesion, there is evidence that the VCAM/VLA-4 adhesion pathway may be subverted by tumor cells during metastasis, allowing them to leave the circulation and establish secondary tumors. Inhibition of tumor cell adhesion via blockade of VCAM/VLA-4 may therefore prevent metastatic spread from primary tumors.

Recent *in vivo* studies have shown that anti-VLA-4 monoclonal antibodies block emigration of lymphocytes, monocytes and eosinophils into tissue [22–25]. For example, in a model of the migration of lymphocytes, pretreatment with an anti-VLA-4 antibody inhibited the migration of a population of inflammation seeking lymphocytes to sites of cutaneous inflammation or peripheral lymph by up to 65% [25]. In addition, blockade of VLA-4 has been shown to be beneficial in rat models of experimental autoimmune encephalomyelitis [26] and vasculitis following cardiac allograft [27].

In rheumatoid arthritis, there is evidence that both the recruitment and the retention of lymphocytes within the synovium involve the interaction of VCAM-VLA-4. A number of studies have shown the presence of VCAM-1 both on endothelial cells within the synovium and on the synovial fibroblasts [28]. Complementary studies have shown that a high proportion of T cells within the synovium and synovial fluid express increased amounts of VLA-4 in comparison to peripheral blood T cells [29]. *In vitro* studies by van Dinther-Janssen *et al.* [28] have shown that isolated unstimulated lymphocytes are able to adhere to sections of inflamed synovium and that this interaction can be strongly inhibited by antibodies to either VCAM-1 or VLA-4. Studies of Postigo *et al.* [30] have produced similar results and shown T cells isolated from rheumatoid synovial fluid or synovial membrane with an activated or memory phenotype were found to have an enhanced capacity to bind to recombinant VCAM-1. In addition, T cell interaction with VCAM-1 was found to induce strong proliferative responses in the presence of submitogenic doses of an anti-CD3 antibody. In an adjuvant-induced arthritis model in rats, T cells from most of the recirculating pool (spleen, peritoneal, Peyer's patch and peripheral lymph node) were found to migrate into the inflamed synovium [22]. However, an antibody to VLA-4 was found to inhibit only the accumulation of a population of inflammation seeking lymphocytes (activated and memory cells) isolated from peritoneal exudates.

Although anti-VLA-4 monoclonal antibodies represent potential therapeutics, several problems can be associated with the use of such protein drugs. These include antigenicity, the need for delivery by injection and cost of production, all of which restrict the scope for their use. Thus in the longer term the development of an orally available low molecular weight adhesion blocker is the ultimate goal. This is an area where molecular biology and medicinal chemistry combine. The production of pure forms of adhesion molecules makes possible the development of assays for 'high throughput screens' for compound libraries and detailed structural-function studies on CAMs. The structures emerging from these screens, and knowledge of the structure of a CAM, will allow drug candidates to be developed. Our approach has been to understand the molecular basis of the VCAM/VLA-4 interaction as a first step in designing a low molecular weight antagonist of this interaction.

VCAM was originally identified at British Biotech by isolating mAbs

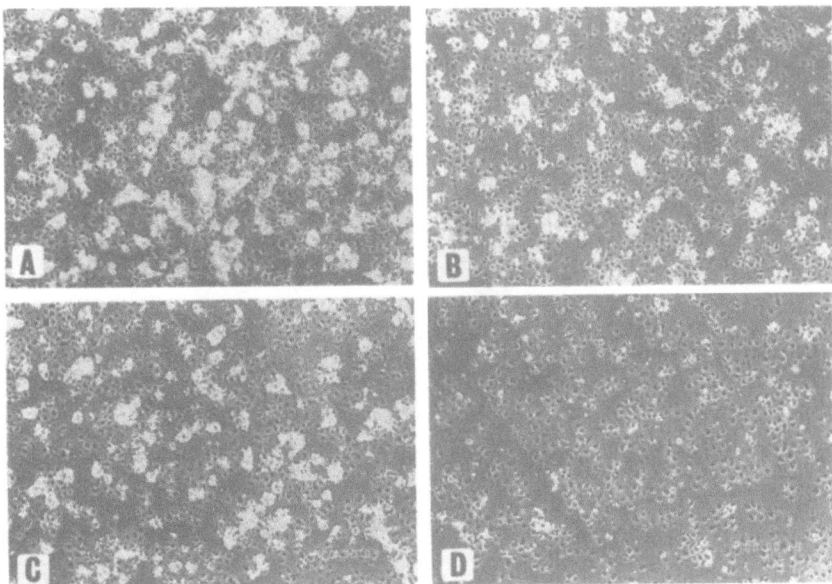

**Figure 6.3** U937 adhesion to VCAM-7d transiently expressed on Cos-7 cells: effect of temperature and anti-VCAM mAb 4B2. U937 adhesion was measured either at 37°C for 40 minutes (A and B) or at 4°C for 90 minutes in the presence of 0.1% sodium azide (C and D) in the presence (B and D) or absence (A and C) of anti-VCAM mAb 4B2. After fixation in 0.5% glutaraldehyde results were assessed by Nikon Diaphot TMD phase contrast microscopy (× 47).

which could inhibit U937 cell adhesion to cytokine activated HUVECs that was not mediated by E-selectin [1]. This assay identified two mAbs which could almost completely abrogate the remaining adhesion of U937 cells to HUVECS. The antigen recognized by these mAbs was cloned and found to encode the 7 domain form of VCAM-1. VCAM-1 is found as two alternately spliced forms, a 7 Ig domain form or a 6 domain form in which domain 4 is absent [31]. Alignment of the VCAM-1 domain reveals sequence identity of greater than 50% for the pairwise alignments of domain 1 and 4, 2 and 5, and 3 and 6, with domain 7 sharing little identity with the other six domains. The 7 domain form of VCAM-1 is believed to have arisen from tandem duplication of this 3 domain module. Both 6 and 7 domain forms are known to be present on cytokine-activated endothelial cells.

Using VCAM-1 deletion mutants, VCAM-transfected cell lines and monoclonal antibodies, we mapped two VLA-4 binding sites on VCAM-1 located in domains 1 and 4. The properties of these binding sites are different in that binding to domain 1 is temperature independent and insensitive to phorbol 12-myristate 13-acetate (PMA) activa-

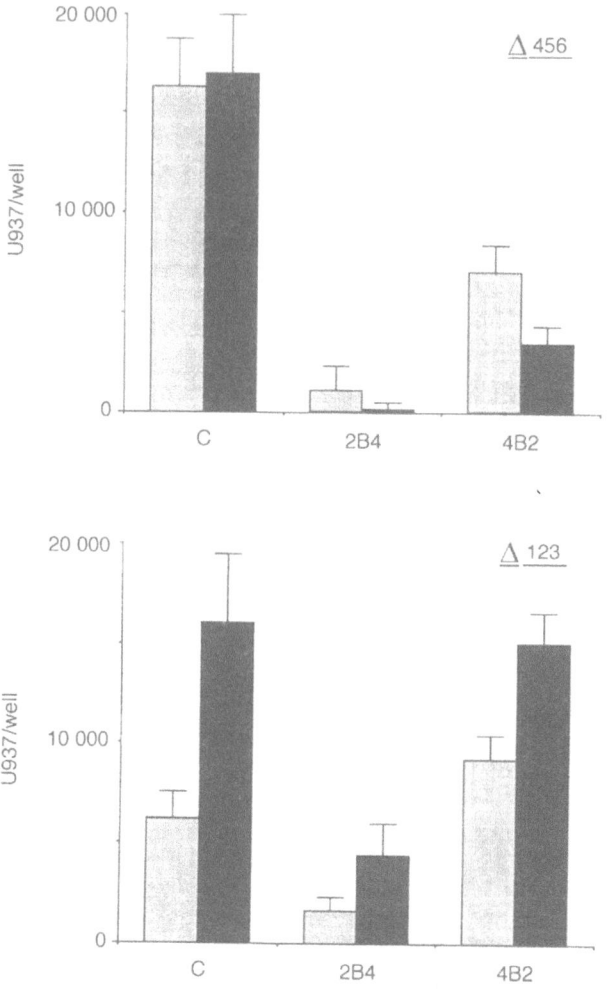

**Figure 6.4** Effect of PMA on U937 adhesion to Cos-7 cells transiently expressing either Δ123 or Δ456 VCAM domain deletion mutants. Inhibitory effects of co-incubation with the anti-VCAM mAb 4B2 or the anti-VLA$_4$ mAb 2B4 on adhesion of radiolabeled U937 cells to Cos-7 cells transfected with either Δ123 or Δ456 VCAM deletion mutants, was measured after 40 minutes at 37°C in the absence (shaded bar) or presence (filled bar) of PMA (20 ng/ml).

tion. In contrast, adhesion to domain 4 is temperature sensitive and augmented by PMA (Figure 6.4). Using CHO cell lines expressing VCAM-1, we were able to demonstrate temporal differences in the adhesion to the two binding sites and this lead us to suggest a two-stage ligand-receptor interaction that involves activation-induced changes in the avidity of VLA-4 for domain 4 of VCAM-1 [32].

## Identification of VLA-4 binding sites on VCAM-1 domains 1 and 4

Having identified the VLA-4 interactive sites on VCAM-1 as the homologous domains 1 and 4, we proceeded to map the critical residues on VCAM-1 domain 1 for VLA-4 binding. Clues to the VLA-4 binding sites in VCAM-1 can be derived from studies on fibronectin where VLA-4 has been shown to bind to the C-terminal heparin-binding domain type III connecting segment of fibronectin (HepII/IIICS). The dominant binding site in this fragment is a 25-mer peptide CS1 which contains the tripeptide Leu-Asp-Val (LDV) as its minimal active site. Recently, it has been shown that the CS1 peptide could compete for binding of VCAM-1 to VLA-4, indicating that fibronectin and VCAM-1 share either identically or spatially close binding sites on VLA-4 [33]. This raises the question of whether VCAM-1 possesses CS1-like sequences, or are the two binding sites different in nature.

Surface accessibility predictions in conjunction with optimal sequence alignment of the human, mouse and rat domains 1 and 4 suggested candidates for the site of the VLA-4 interaction. We then probed these potential receptor binding sites in VCAM-1 by rational site-directed mutagenesis and synthetic peptide studies [34]. By measuring the ability of VCAM-1 mutants to support adhesion of the U937 cells, and simultaneous binding of a panel of anti-VCM-1 mAbs, we identified epitopes involved in VCAM-1 adhesion. The most pronounced effect was obtained by mutation of aspartate 40 within the Ile-Asp-Ser-Pro (IDSP) motif. This mutation abolished domain 1 mediated adhesion (Figure 6.5). The equivalent mutation in domain 4 also abolished

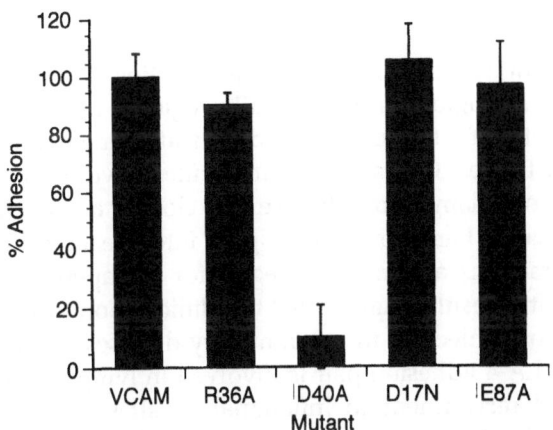

**Figure 6.5** Adhesion assays Cos-7 cells transiently transfected with the indicated VCAM mutant were seeded into 96 well plates as indicated. Adhesion of radiolabeled U937 cells at 37°C for 40 minutes was measured.

domain 4 mediated adhesion. Mutation of this site had no effect on the binding of two conformation-dependent anti-domain 1 monoclonal antibodies.

To further probe the functionality of the motif containing Asp 40, a peptide which spans this region (Thr-Gln-Ile-Asp-Ser-Pro-Leu-Asn) was synthesized and shown to inhibit the spreading of melanoma cells on soluble VCAM-1 with an IC50 of 2.2 mM [34]. A peptide based on the LDV sequence from the CS1 region of fibronectin inhibits this interaction with an IC50 of 0.6 mM. These IC50s compare well with those found for Arg-Gly-Asp (RGD) peptides blocking integrin binding.

Thus the IDSP motif is another sequence identified as a functional integrin-binding site, related to RGD. It is significant that all other integrin-binding immunoglobulin domains contain an aspartate or glutamate residue in their N-terminal Ig domain in a position analogous to the IDSP of VCAM-1. There is now considerable evidence that these residues contribute to function. Mutation of glutamate 34 in ICAM-1 domain 1 and glutamate 37 within ICAM-3 inhibits LFA-1 binding, and mutation of aspartate 41 within MAdCAM-1 inhibits α4β7 binding. Thus a consensus integrin binding motif on cell adhesion molecules becomes apparent: (1) aliphatic, (2) aspartate or glutamate, (3) serine or threonine, (4) proline or hydrophilic. In the future, it will be important to determine the effect of similar peptides on these integrin-immunoglobulin interactions, and to use them as the basis for the designing of novel classes of anti-adhesive agents.


SUMMARY

Three families of adhesion receptors, the selectins, integrins and members of the immunoglobulin superfamily regulate the extravasation of leukocytes to sites of inflammation, and the trafficking of lymphocytes to distinct target tissues. Cell adhesion molecules have been shown to be key mediators of inflammatory disease. The cloning and characterization of these molecules has provided insights into the molecular basis of adhesive interactions and new strategies for therapeutic intervention. In addition, initial results suggest that the clinical monitoring of levels of soluble adhesion molecules in inflammatory diseases will be of benefit. The VCAM/VLA-4 adhesion pair is involved in lymphocyte, monocyte and eosinophil recruitment to inflammatory sites, and in models of disease, blockade of VLA-4 has been beneficial. We have chosen to focus on mapping the molecular detail of the VCAM/VLA-4 pathway as a first step in the design of low molecular weight inhibitors which should be of clinical benefit.

REFERENCES

1. Simmons, D.L. and Needham, L.A. (1991) Cloning cell surface molecules using monoclonal antibodies, in *Vascular Endothelium: Interactions with Circulating Cells* (ed. J.L. Gordon), Elsevier Science Publishers B.V., Amsterdam, 3–29.
2. Lasky, L.A., Singer, M.S., Dowbenko, D. *et al.* (1992) An endothelial ligand for L-selectin is a novel mucin-like molecule. *Cell*, **69**, 927–38.
3. Baumheter, S., Singer, M.S., Henzel, W. *et al.* (1993) Binding of L-selectin to the vascular sialomucin CD34. *Science*, **262**, 436–8.
4. Gearing, A.J.H. and Newman, W. (1993) Circulating adhesion molecules in disease. *Immunol. Today.*, **14**, 506–12.
5. Schleiffenbaum, B., Spertini, O. and Tedder, T.F. (1992) Soluble L-selectin is present in human plasma at high levels and retains functional activity. *J. Cell Biol.* **119**, 229–38.
6. Spertini, O., Schleiffenbaum, B., White-Owen, C. *et al.* (1992) ELISA for quantitation of L-selectin shed from leucocytes *in vivo*. *J. Immunol. Meths*, **156**, 115–23.
7. Katayama, M., Handa, M., Hironobu, A. *et al.* (1992) A monoclonal antibody based ELISA for GMP-140/P-selectin. *J. Immunol. Meths*, **153**, 41–8.
8. Katayama, M., Hanada, M., Araki, Y. *et al.* (1993) Soluble P-selectin is present in normal circulation and its plasma level is elevated in patients with thrombotic thromocytopenic purpura and haemolytic uraemic syndrome. *Br. J. Hematol.*, **84**, 702–10.
9. Pigott, R., Dillon, L.P., Hemingway, I.K. and Gearing, A.J.H. (1992) Soluble forms of E-selectin, ICAM-1, and VCAM-1 are present in the supernatants of cytokine-activated endothelial cells. *Biochem. Biophys. Res. Comm.*, **187**, 584–9.
10. Newman, W., Beall, D.B., Carson, C.W. *et al.* (1993) Soluble E-selectin is found in supernatants of activated endothelial cells and is elevated in the serum of patients with septic shock. *J. Immunol.*, **150**, 633–54.
11. Gearing, A.J.H., Hemingway, I., Pigott, R. *et al.* (1992) Soluble forms of vascular adhesion molecules E-selectin, ICAM-1 and VCAM-1: pathological significance. *An. N.Y. Acad. Sci.*, **667**, 324–31.
12. Carson, C.W., Beall, L.D., Hunder, G.G. *et al.* (1993) Serum ELAM-1 is increased in vasculitis, scleroderma and systemic lupus erythematosus. *J. Rheumatol.*, **20**, 809–12.
13. Cowley, H.C., Heney, D., Gearing, A.J. *et al.* (1994) Increased circulating adhesion molecule concentrations in patients with systemic inflammatory response syndrome: a prospective cohort study. *Critical Care Med.*, **22**, 651–7.
14. Redl, H., Dinges, H.P., Buurman, W.A. *et al.* (1991) Expression of ELAM-1 in septic but not traumatic/hypovolemic shock in the baboon. *Am. J. Pathol.*, **139**, 461–6.
15. Banks, R.E. Gearing, A.J.H., Hemingway, I.K. *et al.* (1993) Circulating intercellular adhesion molecule-1, E-selectin and vascular cell adhesion molecule-1 in human malignancies. *Br. J. Cancer*, **68**, 122–4.
16. Giavazzi, R., Chirivi, R.G.S., Garofalo, A. *et al.* (1992) Soluble intercellular adhesion molecule 1 is released by human melanoma cells and is associated with tumor growth in nude mice. *Cancer Res.*, **52**, 2628–30.
17. Shijubo, N., Imai, N., Aoki, S. *et al.* (1992) Circulating ICAM-1 antigen in sera of patients with idiopathic pulmonary fibrosis. *Clin. Exp. Immunol.*, **89**, 58–62.
18. Gordon, J.L., Edwards, R.M., Cashman, S.J. *et al.* (1993) Vascular endothe-

lium: physiological basis of clinical problems II. *Proceedings from the NATO Advanced Study Institute.*

19. Mason, J.C., Kapahi, P. and Haskard, D.O. (1993) Detection of increased levels of circulating ICAM-1 in some patients with rheumatoid arthritis but not in patients with systemic lupus erythematosus. *Arth. Rheumatism*, **36**, 519–27.

20. Wellicome, S.M., Kapahi, P., Mason, J.C. *et al.* (1993) Detection of a circulating form of VCAM-1: raised levels in rheumatoid arthritis and systemic lupus erthematosus. *Clin. Exp. Immunol.*, **92**, 412–18.

21. Janssen, B.A., Luqmani, R.A., Gordon, C. *et al.* (1994) Correlation of blood levels of soluble vascular cell adhesion molecule-1 with disease activity in systemic lupus erythematosus and vasculitis. *Br. J. Rheum.*, **33**, 1112–16.

22. Issekutz, T.B. and Issekutz, A.C. (1991) T lymphocyte migration to arthritic joints and dermal inflammation in the rat: Differing migration patterns and the involvement of VLA-4. *Clin. Immunol. and Immunopath.*, **61**, 436–47.

23. Weg, V.B., Williams, T.J., Lobb, R.R. and Nourshargh, S. (1993) A monoclonal antibody recognizing VLA-4 inhibits eosinophil accumulation *in vivo*. *J. Exp. Med.*, **177**, 561–6.

24. Elices, M.J., Tamraz, S. and Vollger, T.L.W. (1993) The integrin VLA-4 mediates leukocyte recruitment to skin inflammatory sites *in vivo*. *Clin. Exp. Rheumatol.*, **11** (Suppl. 8), S77–S80.

25. Issekutz, T.B. (1991) Inhibition of *in vivo* lymphocyte migration to inflammation and homing to lymphoid tissues by the TA-2 monoclonal antibody. *J. Immunol.*, **147**, 4178.

26. Yednock, T.A., Cannon, C., Fritz, L.C. *et al.* (1992) Prevention of experimental autoimmune encephalomyelitis by antibodies against α4β1 integrin. *Nature*, **356**, 63–6.

27. Paul, L.C., Davidoff, A., Benediktsson, H. and Issekutz, T.B. (1993) The efficacy of LFA-1 and VLA-4 antibody treatment in rat vascularized cardiac allograft rejection. *Transplantation*, **55**, 1196–9.

28. van Dinther-Janssen, A.C.H.M., Horst, E., Koopman, G. *et al.* (1991) The VLA-4/VCAM-1 pathway is involved in lymphocyte adhesion to endothelium in rheumatoid synovium. *J. Immunol.*, **147**, 4207–10.

29. Laffon, A., Garcia-Vicuna, R., Humbria, A. *et al.* (1991) Upregulated expression of VLA-4 fibronectin receptors on human activated T-cells in rheumatoid arthritis. *J. Clin. Invest.*, **88**, 546–52.

30. Postigo, A.A., Garcia-Vicuna, R., Diaz-Gonzalez, F. *et al.* (1992) Increased binding of synovial T lymphocytes from rheumatoid arthritis to endothelial-leukocyte adhesion molecule-1 and vascular cell adhesion molecule-1. *J. Clin. Invest.*, **89**, 1445–52.

31. Polte, T., Newman, W., Raghunsthen, G. and Venkat-Gopal, T. (1991) Structural and functional studies of full length VCAM-1; internal duplication and homology to several adhesion proteins. *DNA and Cell. Biol.*, **10**, 349–57.

32. Needham, L.A., Van Dijk, S., Pigott, R. *et al.* (1994) Activation dependent and independent VLA-4 binding sites on VCAM-1. *Cell Ad. Comm.*, **2**, 87–99.

33. Makarem, R., Newham, P., Askari, J.A. *et al.* (1994) Competitive binding of vascular cell adhesion molecule-1 and the HepII/IIICS domain of fibronectin to the integrin α4β1. *J. Biol. Chem.*, **269**, 4005–11.

34. Clements, J.M., Newham, P., Shepherd, M. *et al.* (1994) Identification of a key integrin-binding sequence in VCAM-1 homologous to the LDV active site in fibronectin. *J. Cell Science*, **107**, 2127–35.

# 7

# MRI and CT in vasculitis

*M. Heller, C. Muhle, M. Reuter and F. Schubert*

## INTRODUCTION

Wegener's granulomatosis (WG) is diagnosed clinically by symptoms affecting paranasal sinuses, lungs and kidneys; histologically by typical findings of granulomatous vasculitis; and biochemically by the presence of classic antineutrophil cytoplasmic antibodies (cANCA) using indirect immunofluorescence [1]. Although virtually any organ system can be affected by WG, initial manifestation is most often found in the upper respiratory tract (paranasal sinuses) and/or the orbits and the middle ear [1].

In the last three years, about 350 examinations of patients with vasculitis were performed using magnetic resonance imaging (MRI), and about 150 examinations by means of computed tomography (CT).

The rationale for these examinations [2–6] was to find diagnostic criteria for granulomatous vasculitis, to establish examination techniques, to analyze the pathomorphology and to define the role of MRI of the skull and brain. In addition, the value of pulmonary CT with regard to diagnosis, staging and follow-up was clarified.

Until now we have systematically analyzed MRI of 50 patients with WG of the skull and brain [2–4], high resolution-CT of 35 patients [5] and quantitative pulmo-CT of 27 patients with WG of the lung [6].

## METHODS

MRI was generated on a 1.5 tesla whole body system (Gyroscan S 15, Philips) using a head coil for signal reception. Axial proton- and T2-

*The Vasculitides*. Edited by B.M. Ansell, P.A. Bacon, J.T. Lie and H. Yazici.
Published in 1996 by Chapman & Hall, London. ISBN 978-0-412-64140-4.

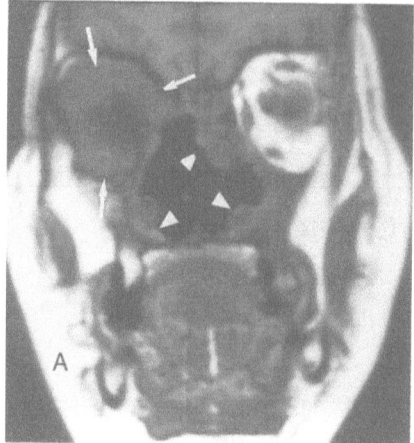 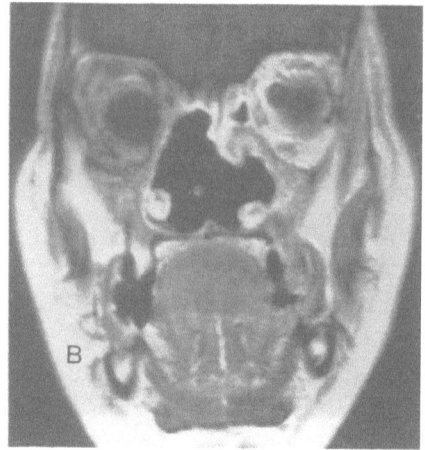

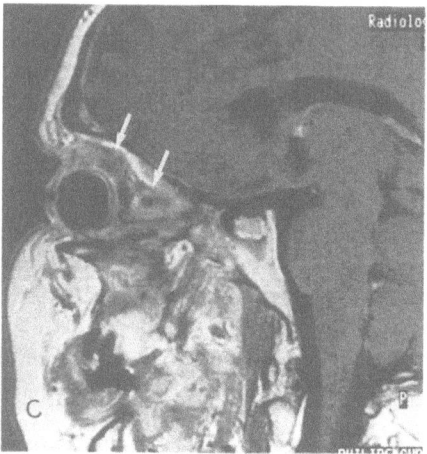

**Figure 7.1** Granulomatous involvement of the nasal cavity, the maxillary sinuses, the ethmoidal cells and the right orbit.

(A) Coronal non-enhanced T1-weighted image: low signal intensity masses outlining the destroyed nasal cavity and the paranasal sinuses (arrow heads). Obliteration of the right orbit, encasing the bulb and infiltrating the intraorbital muscles (arrows).

(B) Identical plane after contrast administration: inhomogeneous enhancement of the granulomatous mucosa of the nasal cavity, the maxillary sinuses, the ethmoid and the orbital granuloma.

(C) Sagittal contrast-enhanced image of the right orbit: encasement of the optic nerve, deformity of the bulb, inhomogeneous enhancement of the granulomas (arrows), proptosis.

weighted spin-echo images (TR 3000 ms; TE 20–90 ms) and coronal T1-weighted spin-echo sequences (TR 400–650 ms; TE 15–20 ms) were obtained (slice thickness: 5 mm; scan matrix: 256 × 256). Series were carried out native and after intravenous application of Gd-DTPA (Magnevist[R], Schering AG; 0.1 mmol/kg body weight).

CT examinations were performed on a Somatom Plus-S (Siemens) and a Tomoscan LX (Philips) scanner as HR-CT (slice thickness 1–1.5 mm; scan time 1–2 sec; high resolution algorithm) and additionally as Q-P-CT (spirometer directed gating of CT scanner at 50% of vital capacity; slice thickness 1 mm; table feed 5 cm with three sections at the carinal level; scan time 1 sec; isolation of pulmonary parenchyma using a fast contour tracing algorithm; assessment of whole lung density; calculation of individual age-matched normative values).

## RESULTS OF MRI STUDY

Forty-five cases had pathological MRI findings as incomplete or complete mucosal thickening of the nasal cavity and paranasal sinuses (Figures 7.1–7.3). Mucosal inflammatory alterations were demonstrated as high signal intensity structures on T2-weighted images and showed a low signal on T1-weighted images. Following the application of the contrast agent, the linings of the sinuses showed an increased signal,

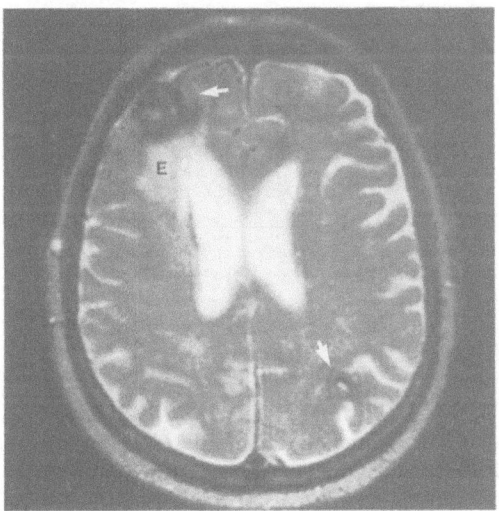

**Figure 7.2** Cerebral involvement. Axial T2-weighted image: right frontal and left occipital bleeding in active disease (arrows). Perifocal frontal white matter edema (E) of high signal intensity.

**Table 7.1** Distribution of pathological MRI findings of the skull and brain in 50 patients with Wegener's granulomatosis

| Feature | Patients (n) | % |
|---|---|---|
| *Mucosal thickening* | 44 | 88 |
| 1. Paranasal sinuses | 39 | 78 |
|     maxillary sinus | 35 | 70 |
|     ethmoidal cells | 22 | 44 |
|     sphenoid sinuses | 11 | 22 |
|     frontal sinus | 5 | 10 |
| 2. Mastoid/middle ear | 16 | 32 |
| *Granuloma* | 11 | 22 |
| 1. Paranasal sinuses | 4 | 8 |
| 2. Orbits | 5 | 10 |
| 3. Orbits and paranasal sinuses | 2 | 4 |
| *Bone destruction* | 4 | 8 |
| *White matter pesious* | 12 | 24 |

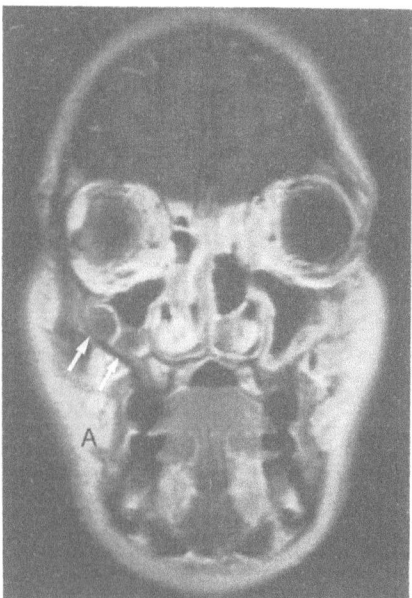 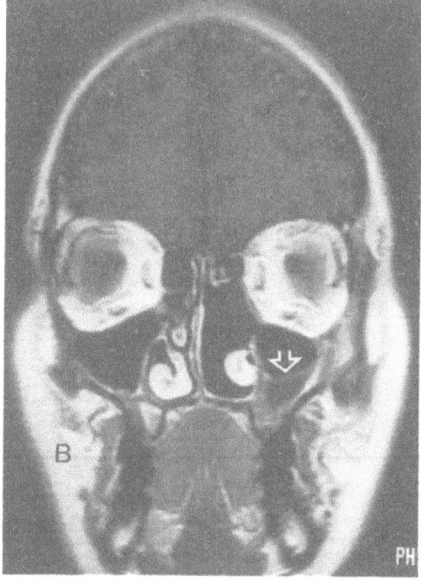

**Figure 7.3** Extended involvement of the nasal cavity, both maxillary sinuses, and the ethmoid. Coronal T1-weighted postcontrast image.

(A) Before treatment: bony destruction of the left nasal cavity. Mucosal thickening. Inhomogeneous enhancement. Two large granulomas at the lateral right maxillary sinus (arrows).

(B) Six months after treatment: only a slight mucosal thickening remains. Mucus retention at the left maxillary sinus (open arrow).

whereas the retained mucus or serous fluid was demonstrated as circumscribed areas of low signal intensity.

Mucosal thickening/fluid retention of the mastoid and middle ear was noted in 16 patients. ENT examinations revealed a conductive hearing loss in two of them.

Granulomas of the paranasal sinuses and orbits were detected in 11 patients. Compared to muscles of the face, granulomas had a characteristic appearance: low signal intensity in T1- and T2-weighted sequences, and inhomogeneous signal enhancement after intravenous contrast agent application (Figures 7.1 and 7.3). Orbital granulomas varied in size from $2 \times 1$ cm to an almost complete obstruction of the orbit. They were located either in the lateral intraconal space, encompassing the lateral rectus muscle, or in the retroconal space, encasing the optic nerve and extraocular muscles (Figure 7.1(c)), thus causing clinical complications as loss of vision (seven patients), ophthalmoplegia (four patients) and proptosis (six patients).

Bony defects corresponding to previous nasal septum and maxillary sinus surgery were exhibited by MRI. Also bony destructions due to contiguous granulomatous tissue infiltration from the right maxillary sinus into the orbit, accompanied by destruction of the nasal cavity and the medial walls of the maxillary sinuses, was detected. In general, these bony destructions and a pattern resembling a sclerosing osteomyelitis were best demonstrated by CT.

In 12 patients, unspecific white matter lesions of the brain were depicted with only one of them having clinical evidence of central nervous system (CNS) involvement.

The exact localization of a pathologic lesion/granuloma using MRI helped in directing biopsies.

RESULTS OF HR-CT STUDY

HR-CT of the lung and plain film X-ray were achieved to reveal pulmonary manifestation of WG. Pleural and parenchymal pathology was detected on chest X-rays in 57% in comparison to 86% using HR-CT. Pulmonary and pleural (Figure 7.5) granulomas with smooth or spiculated margins and with (Figure 7.4) or without cavitations (Figure 7.5) were deemed pathognomonic. Infiltrations, thickened interlobular septa and fibrotic scars of the parenchyma (Figures 7.4, 7.5, 7.7(b)) and pleura (Figure 7.4) were regarded as non-specific findings. Ground glass opacities (Figure 7.6), traction bronchiectases (Figure 7.4) and cyst-like lesions were only visible on HR-CT.

As expected, HR-CT proved to be more sensitive in detecting subtle lung alterations than plain film X-ray. The differentiation between

**Table 7.2** Pathological HR-CT findings of the lung in 35 patients with Wegener's granulomatosis

| Feature | Patients (n) | % |
|---|---|---|
| Granuloma | 3 | 9 |
| Granuloma with cavitation | 5 | 14 |
| Infiltration | 4 | 11 |
| Ground glass opacity | 9 | 26 |
| Thickened interlobular septa | 15 | 43 |
| Bronchiectasis | 3 | 9 |
| Cystic lesion | 2 | 6 |
| Pleural thickening | 13 | 37 |

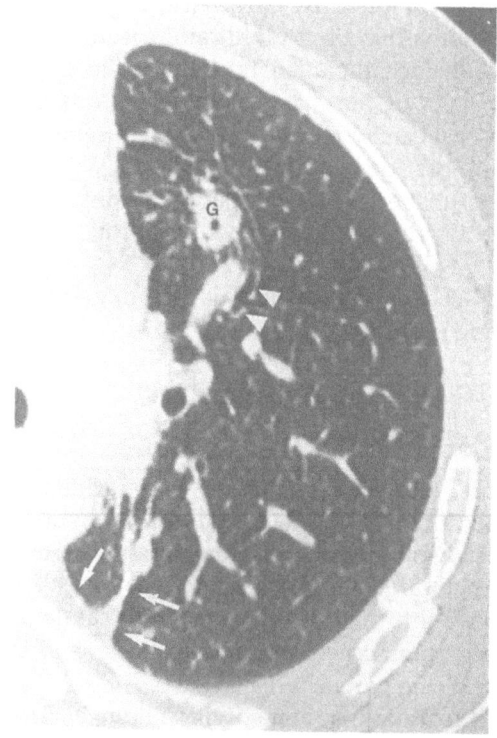

**Figure 7.4** HR-CT of the left lung: spiculated granuloma with central cavitation (G); bronchiectasis (arrow head); fibrotic scar formation with pleural thickening (arrows).

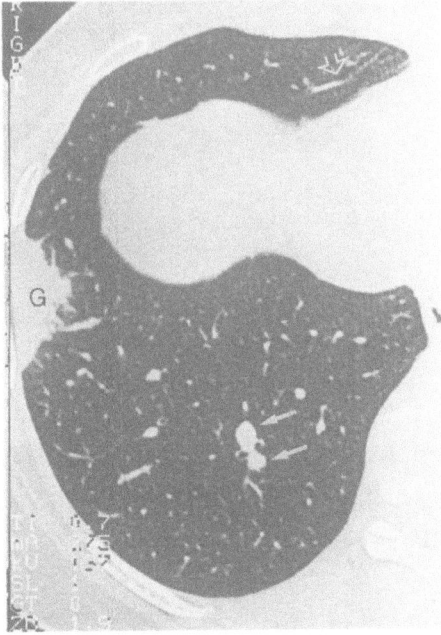

**Figure 7.5** HR-CT of the right lung: spiculated pleural granuloma (G); paren-chymal granulomas without cavitation (arrows); fibrotic scar (open arrow).

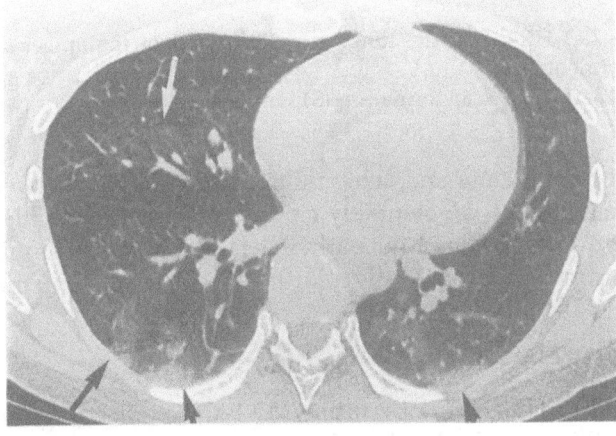

**Figure 7.6** HR-CT of the lung: ground glass opacities of the left lower lobe, the right middle and lower lobe (arrows).

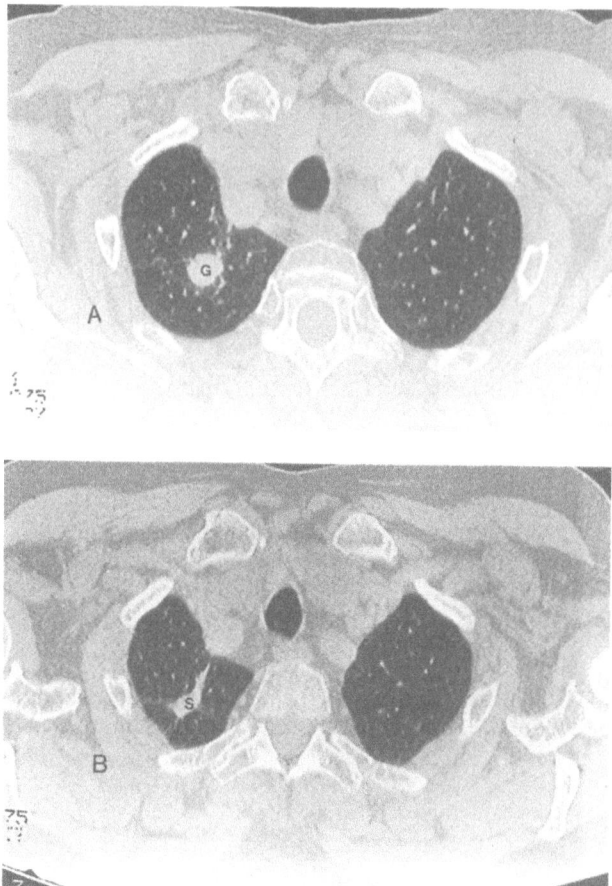

**Figure 7.7** (A) HR-CT of the lung: granuloma (G) of the apical segment just before treatment.
(B) After treatment a scar formation (S) remains.

pulmonary granuloma and acute inflammation, that is between chronic fibrotic changes and a potentially curable process, is possible.

The exact localization of the pathological lesion by performing HR-CT helped in directing bronchioalveolar lavage (BAL) or lung biopsy.

RESULTS OF Q-P-CT STUDY

In addition to HR-CT the performed Q-P-CT studies (Figure 7.8) indicated that this recently developed method may help to detect subclincal parenchymal lesions without corresponding HR-CT pathomorphology. Continued studies have to clarify if Q-P-CT will help to detect early

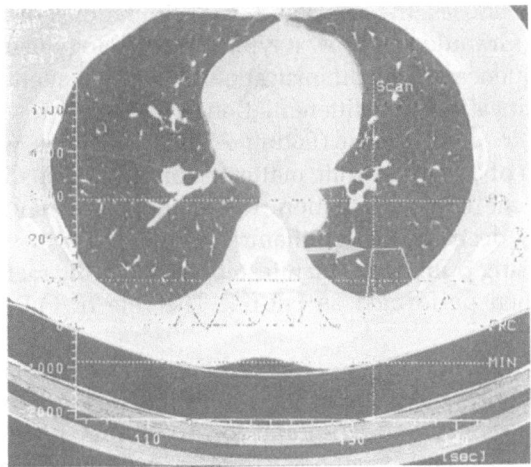

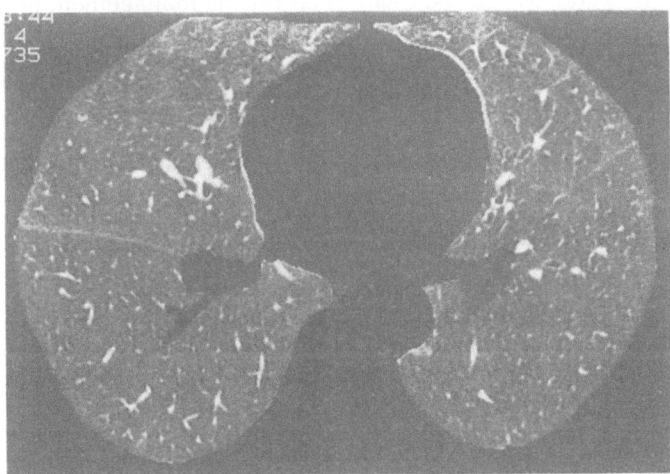

**Figure 7.8** (a) Q-P-CT: spirometer directed gating of CT at 50% (arrow) of vital capacity.
(b) Q-P-CT: semiautomatic isolation of the lung parenchyma for density measurement.

pulmonary manifestations and allow us to differentiate between active and chronic disease on the one hand, and between active inflammation and therapy induced alterations on the other hand.

## CONCLUSIONS

MRI is highly sensitive in detecting and distinguishing granulomas from unspecified mucosal inflammation or other masses of the nasal cavity, the paranasal sinuses, the orbits and, with limitations, of the middle ear and mastoid. Granulomas show a typical morphology using T1- and T2-weighted sequences. The administration of contrast agent is mandatory for better delineation and differentiation.

Typical brain involvement (Figure 7.2) is depictable with MRI. The interpretation of common white matter lesions in patients with vasculitis is related to the clinical situation. CT of the head may be helpful in defining bony destructions or inflammation.

For diagnosing pulmonary manifestations, CT is the method of choice, especially when performed as HR-CT. The role of Q-P-CT has to be defined.

In general, MRI and CT provide additional information for the diagnosing and staging of vasculitis as well as for directing invasive procedures such as biopsy and BAL, and for providing sufficient follow-up (Figures 7.3, 7.7). Both methods supplement clinical and laboratory findings and should thus be established for the management of patients suffering from vasculitis.

## REFERENCES

1. Gross, W.L. (1991) Neue Aspekte bei der Wegenerschen Granulomatose. *Dt. Ärztebl.*, **88**, 28–34.
2. Asmus, R., Muhle, C., Koltze, H. *et al.* (1992) Magnetresonanztomographische Kennzeichen der Wegenerschen Granulomatose im Kopfbereich. *Fortschr. Röntgenstr.*, **157**, 11–14.
3. Duncker, G., Gross, W.L., Nölle, B. *et al.* (1992) Orbitale Beteiligung bei Wegenerscher Granulomatose. *Klin. Mbl. Augenhelik.*, **201**, 309–16.
4. Muhle, C., Nölle, B., Brinkmann, G. *et al.* (1994) Magnetresonanztomographie und Computertomographie der Wegenerschen Graunulomatose der Orbita. *Akt. Radiol.*, **4**, 229–34.
5. Muhle, C., Koltze, H., Reinhold-Keller, E. *et al.* (in press) MRI of the paranasal sinuses, the orbits and the brain in Wegener's Granulomatosis.
6. Schubert, F., Muhle, C., Schnabel, A. *et al.* (1994) High-resolution CT (HRCT) der Lunge bei Wegenerscher Granulomatose. *Fortschr. Röntgenstr.*, **161**, 19–24.
7. Reuter, M., Schubert, F., Freund, M. and Reinhold-Keller, E. (1994) Die quantitative Computertomographie der Lunge bei systemischen Vaskuliti-den: erste Ergebnisse und Perspektiven. *Zbl. Rad.*, **150**, 19.

# 8

# Angiography in vasculitis: its role in diagnosis and in the management of complications

*J.E. Jackson*

## INTRODUCTION

The vasculitides are a heterogenous group of uncommon diseases characterized by inflammation of blood vessels. Diagnosis is usually based upon clinical and pathological findings, together with immunodiagnostic laboratory tests. Angiography may, however, provide important information in certain patients, both in terms of diagnosis and in the assessment of disease extent.

The role of angiography in vasculitis is best addressed by considering the various disorders in which it may occur under different headings according to the size of the vessel predominantly involved.

## SYSTEMIC NECROTIZING ARTERITIS

### Classical polyarteritis nodosa (PAN)

Classical PAN is a systemic necrotizing vasculitis involving small- and medium-sized arteries. The clinical presentation is dominated by organ infarction and hemorrhage, neuropathy and myalgia in the context of a systemic illness. Visceral arterial aneurysms were noted in the original description of the condition by Kussmaul and Maier in 1866 [1] and were first demonstrated angiographically in 1965 by Fleming and Stern [2].

*The Vasculitides.* Edited by B.M. Ansell, P.A. Bacon, J.T. Lie and H. Yazici. Published in 1996 by Chapman & Hall, London. ISBN 978-0-412-64140-4.

Since then several papers have documented the usefulness of angiography in the diagnosis of this disease.

The angiographic findings in classical PAN include aneurysms, irregular beading and occlusions of small- and medium-sized vessels either in isolation or in combination. These changes are often best appreciated within the renal and hepatic vascular territories (Figures 8.1 and 8.2), although other organs may be involved. Aneurysms may vary in size from 1–12 mm, but tend to be of similar dimensions in individual patients. Multiple aneurysms (more than 10) are often present.

Arterial abnormality is present in the majority of patients with PAN, although there appears to be little, if any, relationship between the presence of organ involvement clinically and the finding of arteritic changes within that organ.

On selective renal angiography, between 83 and 100% of patients will have evidence of an arteriopathy (arterial beading, stenoses, occlusions) and between 44 and 94% will show aneurysms. Within the liver, arteriopathic changes may be visible in 26–94% of patients, and aneurysms will be present in approximately 50%. The mesenteric circulation is much less

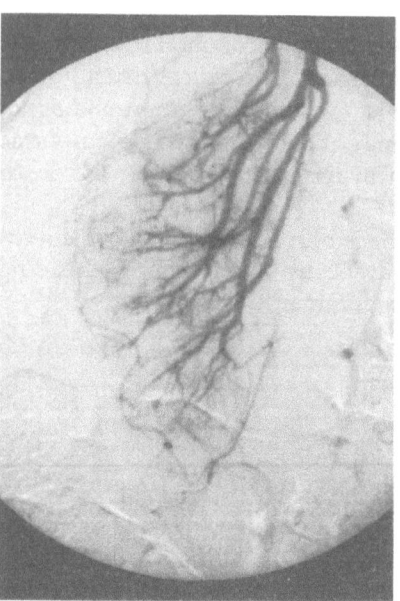 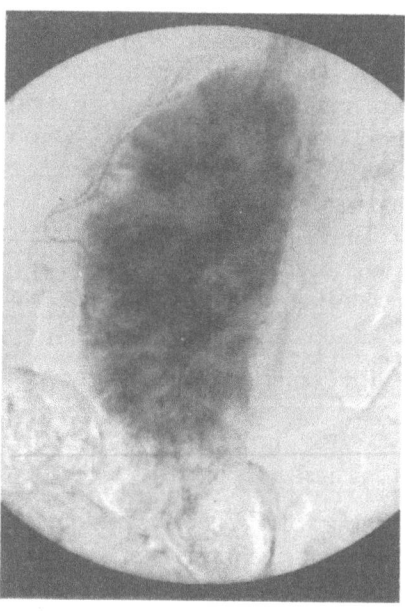

a                                        b

**Figure 8.1** Right renal arteriogram in a patient with PAN. (a) Severe arteritic changes are present with vessel irregularity, multiple peripheral arterial occlusions and several small (2–4 mm) aneurysms. Note the capsular vessels which have hypertrophied as a result of the peripheral renal arterial disease. (b) Typical 'moth-eaten' angiographic nephrogram due to peripheral organ infarction. Several small aneurysms remain visible, best seen in the lower renal pole.

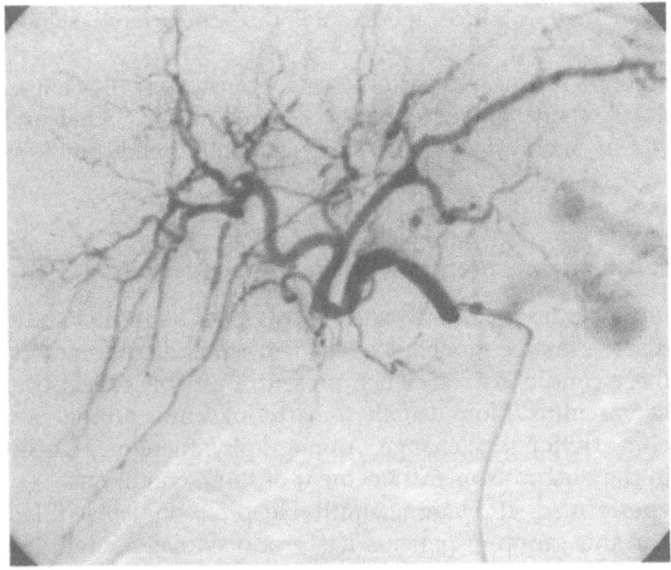

**Figure 8.2** Selective common hepatic arteriogram in a patient with PAN. Multiple small aneurysms are visible together with marked vascular irregularity and beading.

commonly abnormal with occlusive disease visible in 16–33% and aneurysms in 22–39% [3–6].

It is generally felt that the presence of aneurysms is associated with more severe disease and, subsequently, a worse prognosis. In a series of 17 patients reported by Travers *et al.* [3], 10 patients shown to have aneurysms were more likely to have features of loss of weight, severe hypertension, abdominal pain, cardiac involvement and central nervous system involvement. There were, however, no appreciable differences between those patients with and those without aneurysms in the incidence of nephritis, mononeuritis or polyarthritis.

In another series of 26 patients reported by Ewald *et al.* [5], the presence of aneurysms was associated with severe hypertension and clinically severe disease.

It is interesting to note that almost all patients with PAN associated with the presence of hepatitis B surface antigen will demonstrate visceral arterial aneurysms.

When typical arterial changes including aneurysms are present in the right clinical context, then classical PAN can be diagnosed with some confidence. Similar aneurysmal changes may, however, rarely occur in other vasculitides, such as Wegener's granulomatosis, systemic lupus erythematosus (SLE) and Churg–Strauss disease. Furthermore, other non-vasculitic conditions, such as atrial myxoma [7] and infective endocarditis may produce aneurysmal disease, arterial stenoses and occlusions.

In addition to its role in diagnosis, angiography may play a useful part in treatment when hemorrhage has occurred from a renal or hepatic aneurysm. In such circumstances per-catheter embolization may be life-saving.

### Systemic necrotizing arteritis with granulomatosis

This group includes conditions such as Wegener's arteritis and Churg–Strauss disease. As mentioned above, aneurysmal disease may rarely be seen in these conditions and, when present, probably reflects the intensity of the vasculitis. More usually the arteriogram is normal or there is non-specific arteriopathic change. Angiography, therefore, has little role to play in the diagnosis or management of these conditions.

Work performed at Hammersmith Hospital on labeled leukocyte scanning in this group of patients has produced some interesting findings [8]. Active disease is associated with markedly increased diffuse pulmonary uptake of granulocytes due to leukocyte margination. Increased activity is often also seen in organs involved by active disease, presumably as a result of white cell migration. Patients with systemic vasculitis frequently have increased nasal activity, and may have abnormal bowel and renal activity. This constellation of findings in a patient with unexplained fever should immediately suggest a diagnosis of systemic vasculitis.

The increased granulocyte margination in the lungs is not associated with overt evidence of lung damage in these patients; this probably also requires extravascular migration. It is not clear what triggers the subsequent extravascular migration of the marginating cells in the lung.

### SMALL-VESSEL VASCULITIS

This group includes conditions such as Henoch–Schönlein purpura, SLE and Sjögrens syndrome. Angiography has no role in this group.

### LARGE-VESSEL VASCULITIS

### Temporal arteritis

Clinical evidence of large artery involvement in this disease is present in 10–15% of patients, although there is autopsy evidence of large-vessel disease in as many as 60% of individuals [9, 10]. The supra-aortic vessels are those most commonly involved. Patients may complain of upper extremity claudication and clinical examination may reveal bruits over the carotid, subclavian, axillary and brachial arteries with diminished or absent pulses in the neck and arms.

The disease is generally patchy and angiographic findings are those of smooth-walled arterial stenoses or occlusions alternating with areas of normal or increased caliber, predominantly within the carotid, subclavian, axillary and brachial vessels. Irregular plaques and ulcerations are absent. Aortocoronary disease is reported and may be more common than is recognized [11]. Angiography plays little role in the diagnosis of this condition, but may be helpful in documenting the extent of disease in patients with ischemic symptoms in their upper limbs. Balloon angioplasty of arterial stenoses may be helpful in certain individuals who remain symptomatic after medical therapy.

Attempts have been made [12] to utilize angiography to localize a diseased segment of superficial temporal artery to guide biopsy. This has not proved useful because of the frequent presence of atheromatous disease in the branches of this vessel, which may mimic an arteritis, and because a normal arteriographic appearance may be seen in a histologically involved vessel. Doppler ultrasound has also been used to localize abnormal segments of vessel, but results are again not encouraging [13].

### Takayasu's arteritis

This disease is seen much more commonly in the Orient although it has a worldwide distribution. There is a marked female predominance (female to male 10:1) and the onset of disease generally occurs before the age of 40 years. The descending thoracic and abdominal aorta, the renal, superior mesenteric and supra-aortic vessels are commonly involved (Figures 8.3 and 8.4(c)). Most lesions are stenosing or occlusive, and take the form of proximally located stenoses or long segment occlusions; the latter may be the most striking feature of the disease (Figure 8.4(c)). Aneurysmal disease is seen, however, in approximately 10% of patients and this predominantly involves the descending thoracic aorta. Such aneurysms are much more commonly saccular than fusiform and are usually associated with long-standing severe hypertension.

There are interesting racial and geographical differences in the morphological pattern of involvement. In most countries, except Japan, the abdominal aorta is the most frequently involved vessel and renal and subclavian branches are the most commonly diseased aortic branches. In Japan, thoracic aortic and carotid arterial disease is more common. There is an unusually high incidence of aneurysmal disease in the Philippines, Thailand, Israel, Japan and western India [14]. Pulmonary arterial involvement is common, occurring in between 40 and 100% of patients,

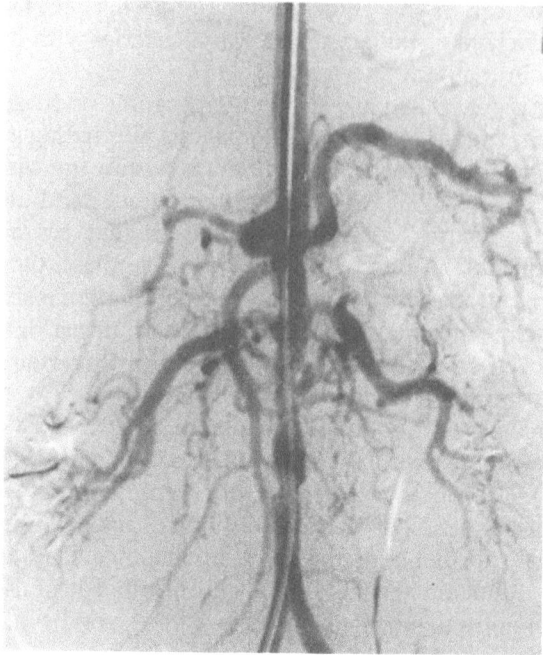

**Figure 8.3** Abdominal aortogram in Takayasu's arteritis. There is narrowing of the whole of the abdominal aorta with marked irregularity in its infrarenal portion. Bilateral proximal renal artery stenoses are also present.

and there the disease appears to show a predilection for the right upper lobe vessels [14,15]. Stenotic and occlusive disease is seen (Figure 8.4). Bronchial arterial hypertrophy and subsequent life-threatening hemoptysis may develop as a consequence of pulmonary infarction [16].

Angiography is useful for both the diagnosis and assessment of disease extent in this condition. In view of the often widespread vascular disease and frequent involvement of the aorta, intravenous digital subtraction angiography may be the best method of imaging. Magnetic resonance angiography may play more of a role in the future, both in diagnosis and follow-up [17, 18]. Stenotic disease involving the origins of visceral vessels may respond well to angioplasty. Treatment of life-threatening hemoptysis by embolization of hypertrophied bronchial arteries which have developed as a result of pulmonary infarction may be life-saving [16].

### Behçet's disease

This systemic vasculitis of unknown etiology was first described by Hulusi Behçet, Professor of Dermatovenereology in Istanbul, in 1937

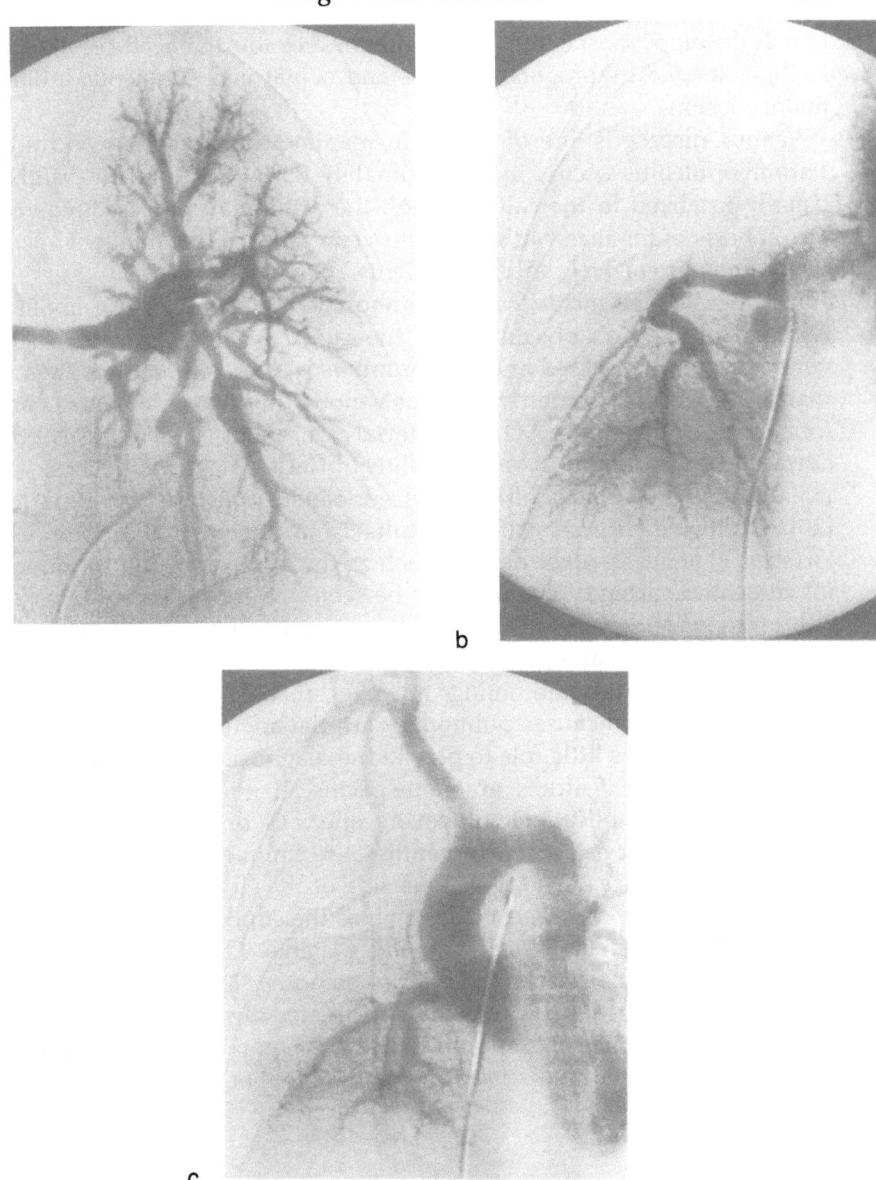

**Figure 8.4** Pulmonary artery involvement in Takayasu's arteritis. (a) Left pulmonary arteriogram demonstrates multiple tight stenoses within many of the pulmonary arterial branches. (b) Right pulmonary arteriogram in the same patient shows very severe disease with complete occlusion of all of the pulmonary arteries other than a single, tightly stenosed basal branch. (c) Later image from the right pulmonary arteriogram demonstrates the aorta and severe supraaortic vessel disease with occlusion of all branches except the right brachiocephalic artery.

[19]. It develops most commonly in the third or fourth decade of life and is characterized by orogenital ulcers and ocular and cutaneous inflammatory lesions.

Venous disease is one of the main manifestations of this condition. Thrombophlebitis occurs in approximately 25% of all patients, and is usually localized to the calf although the arms may also be involved. Thrombosis of the iliac veins and both vena cavae is less frequent and is seen almost exclusively in male patients.

Despite the high incidence of thrombophlebitis, pulmonary embolism is rare. This is probably related to the fact that long segments of the venous wall are diseased and the thrombus is, therefore, more adherent than in the more common variety of venous thrombosis which occurs, for example, after surgery. On the arterial side, aneurysmal disease of the aorta and carotid, femoral and popliteal arteries may occur. These are usually true aneurysms, although pseudoaneurysms are also reported [20]. Occlusive disease usually results from aneurysmal thrombosis. Arteritic changes in other organs, such as the liver, or within the superior mesenteric arterial territory may be seen but are extremely rare [21, 22]. Pulmonary disease, in the form of arterial aneurysms, arterial and venous thromboses and pulmonary infarcts are seen in fewer than 5% of patients. It is worthwhile noting, however, that Behçet's disease is the only arteritis which causes pulmonary arterial aneurysms.

Angiography has little role to play in the diagnosis of this condition as the arteriographic features are non-specific. It may be useful in the assessment of complications, however, especially on the venous side of the circulation. Per-catheter interventional techniques may be helpful in certain circumstances. On the venous side of the circulation, recanalization of occluded or stenosed veins, such as the superior vena cava, may be possible with metallic self-expanding stents. On the arterial side, angioplasty may be useful for arterial stenoses. Large aneurysms or pseudoaneurysms may occasionally be treated by embolization, perhaps as a temporizing measure in patients who are not surgical candidates because of active disease, and this treatment may be life-saving [23].

## VASCULITIS 'LOOK-ALIKES'

Arteries can respond in a limited number of ways to an insult by the formation of stenoses, occlusions or aneurysms, and it is therefore not surprising that there are a number of disorders that may mimic the arteritic changes seen in the foregoing diseases. Such conditions include chronic thromboembolism, atrial myxoma, septicemia, organ transplantation [24], infective endocarditis and ergotism. It is important that the radiologist is aware of the similarity of the arterial changes in some of these conditions to those in arteritic diseases so as to avoid misdiagnosis. In most cases, the correct diagnosis will be obvious from the patient's clinical presentation or will be suggested by the distribution of the arterial changes; for example, an arterial abnormality localized to one hand due to chronic thromboembolism from a subclavian artery stenosis. In some instances, however, the diagnosis will be less clear cut and will only become apparent after further investigation, which will often include tissue biopsy.

## CONCLUSION

Angiography plays a major role in the diagnosis and assessment of classical polyarteritis nodosa and Takayasu's arteritis, but it may also be useful in documenting the arterial and venous complications of Behçet's disease and in differentiating some of the 'look-alike' conditions from the true arteritides.

In addition to diagnosis, angiography may play a major part in treatment, either by embolization of a bleeding vessel (such as may occur in polyarteritis nodosa) or by recanalization of a stenosed or occluded vessel, as may be seen in Takayasu's arteritis or Behçet's disease. The role of radiolabeled leukocyte scanning in diagnosis and in the assessment of disease severity is, as yet, unknown.

## REFERENCES

1. Kussmaul, A. and Maier, R. (1866) Über eine bisher nicht beschreibene eigenthümliche Arterienerkrankung (Periarteritis nodosa), die mit Morbus Brightii und rapid fortschreitender allgemeiner Muskellähmung einhergeht. *Dtsch. Arch. Klin. Med.*, **1**, 484–518.
2. Fleming, R.J. and Stern, L.Z. (1965) Multiple intraparenchymal renal aneurysms in polyarteritis nodosa. *Radiology*, **84**, 100–3.
3. Travers, R.L., Allison, D.J., Brettle, R.P. and Hughes, G.R.V. (1979) Polyarteritis nodosa: A clinical and angiographic analysis of 17 cases. *Sem. Arth. Rheum.*, **8**, 184–9.
4. Hekali, P., Kajander, H., Pajari, R. *et al.* (1991) Diagnostic significance of angiographically observed visceral aneurysms with regard to polyarteritis nodosa. *Acta Radiologica*, **32**, 143–8.
5. Ewald, E.A., Griffin, D. and McCune, W.J. (1987) Correlation of angiographic abnormalities with disease manifestations and disease severity in polyarteritis nodosa. *J. Rheumatol.*, **14**, 952–6.
6. Verztman, L. (1980) Polyarteritis nodosa. *Clin. Rheum. Dis.*, **6**, 297–317.
7. Boussen, K., Moalla, M., Blondeau, P. *et al.* (1991) Embolization of cardiac myxomas masquerading as polyarteritis nodosa. *J. Rheumatol.*, **18**, 283–5.
8. Peters, A.M. (1994) The utility of [99m Tc] HMPAO-leukocytes for imaging infection. *Sem. Nuclear Med.*, **24**, 110–27.
9. Green, G.M., Lain, D., Sherwin, R.M. *et al.* (1986) Giant cell arteritis of the

legs. Clinical isolation of severe disease with gangrene and amputation. *Am. J. Med.*, **81**, 727–33.

10. Walz-Leblanc, B.A.E., Ameli, F.M. and Keystone, E.C. (1991) Giant cell arteritis presenting as limb claudication. Report and review of the literature. *J. Rheumatol.*, **18**, 470–2.

11. Mitnick, H.J., Tunick, P.A., Rotterdam, H. and Esposito, R. (1990) Antemortem diagnosis of giant aortitis. *J. Rheumatol.*, **17**, 708–11.

12. Sewell, J.R., Allison, D.J., Tarin, D. and Hughes, J.R.V. (1980) Combined temporal arteriography and selective biopsy in suspected giant cell arteritis. *Ann. Rheumat. Dis.*, **39**, 124–8.

13. Barrier, J., Potel, G., Renaut-Hovasse, H. *et al.* (1982) The use of Doppler flow studies in the diagnosis of giant cell arteritis. Selection of temporal biopsy site is facilitated. *J.A.M.A.* **248**, 2158–9.

14. Sharma, S., Rajani, M. and Talwar, K.K. (1992) Angiographic morphology in non-specific aortoarteritis (Takayasu's arteritis); a study of 126 patients from North India. *Cardiovasc. Intervent., Radiol.*, **15**, 160–5.

15. Yamada, I., Shibuya, H., Matsubara, O. *et al.* (1992) Pulmonary artery disease in Takayasu's arteritis: angiographic findings. *A.J.R.*, **159**, 263–9.

16. Lopez, A.J., Brady, A.J.B. and Jackson, J.E. (1992) Case report: Therapeutic bronchial arterial embolization in a case of Takayasu's arteritis. *Clin. Radiol.*, **45**, 415–7.

17. Yamada, I., Numano, F. and Suzuki, S. (1993) Takayasu's arteritis; evaluation with MR imaging. *Radiol.*, **188**, 89–94.

18. Oneson, S.R., Lewin, J.S. and Smith, A.S. (1992) MR angiography of Takayasu's arteritis. *J.C.A.T.*, **16**, 478–80.

19. Behçet, H. (1937) Über rezidivierende, aphthöse, durch ein Virus verursachte Geschwüren am Mund, am Auge und an den Genitalien. *Dermatol. Wochenschr.*, **105**, 1152–7.

20. Park, J.H., Han, M.C. and Bettman, M.A. (1984) Arterial manifestations of Behçet's disease. *A.J.R.*, **143**, 821–5.

21. Mathur, A.K., Maslow, J. and Urffer, P.A. (1989) Hepatic arteritis in Behçet's disease. *J. Rheumatol.*, **16**, 1516–17.

22. Gonzalez-Gay, M.A., Sanchez-Andrade, A., Pulpeiro, J.R. and Armesto, V. (1991) Hepatic arteritis in Behçet's disease. *J. Rheumatol.*, **18**, 152–3.

23. Hartnell, G.G. and Allison, D.J. (1987) Embolization as palliative treatment for aortic aneurysm complicating Behçet's disease. *J. Interven. Radiol.*, **2**, 109–12.

24. Hemingway, A.P. and Allison, D.J. (1980) Renal aneurysms in rejected renal transplants. *B.M.J.*, **281**, 1640–1.

# *Part Three*

## Primary and Systemic Vasculitides

# 9

# Polyarteritis nodosa: clinical aspects

*L. Guillevin, F. Lhote and P. Casassus*

## INTRODUCTION

Polyarteritis nodosa (PAN), first described by Kussmaul and Maier [1], is a well-known form of necrotizing angiitis whose main manifestations are: weight loss, fever, asthenia, peripheral neuropathy, renal involvement, musculoskeletal and cutaneous manifestations, hypertension, gastrointestinal (GI) tract involvement and cardiac failure. Churg–Strauss syndrome (CSS) [2] is a disorder characterized by hypereosinophilia and systemic necrotizing vasculitis similar to that of PAN and occurring in individuals with asthma and allergic rhinitis. The PAN group of vasculitides is heterogeneous and comprises classic PAN (c-PAN), microscopic polyangiitis (MPA), CSS and overlap syndrome. Considering the etiologies of PAN, primary and secondary vasculitides can also be distinguished, since PAN can be the consequence of hepatitis B virus (HBV) infection and sometimes of other etiological agents. Only c-PAN is addressed in this chapter; MPA and CSS are treated in detail elsewhere.

## CLASSIFICATION: THE POLYARTERITIS NODOSA GROUP

In Fauci's classification, this group comprised c-PAN, CSS and overlap angiitis. The detection of antineutrophil cytoplasm antibody (ANCA) detection and virological investigations, led to the further subdivision of this group.

*The Vasculitides.* Edited by B.M. Ansell, P.A. Bacon, J.T. Lie and H. Yazici.
Published in 1996 by Chapman & Hall, London. ISBN 978-0-412-64140-4.

## Classic polyarteritis nodosa

This vasculitis involves medium-sized arteries. Clinical manifestations are: fever, poor general condition, peripheral neuropathy, arthralgias, myalgias and visceral (GI tract, kidneys, heart, etc.) involvement. Kidney involvement is the consequence of renal vasculitis, and multiple renal infarcts can be responsible for renal insufficiency, sometimes leading to renal failure and malignant hypertension. Microaneurysms are often seen on abdominal and renal angiograms. ANCA are usually absent and we think that the coexistence of ANCA and medium-sized artery involvement is improbable. Indeed, a careful analysis of our personal data has shown that patients with c-PAN do not have ANCA. We previously reported [3] that ANCA were present in 10.7% of PAN with hepatitis B virus (HBV) markers and in 27.3% of patients with PAN of unknown etiology. Patients with PAN were selected according to pathology, clinical symptoms and the American College of Rheumatology classification criteria for PAN [4]. This classification is a useful tool for diagnosis but does not take into account MPA which seems to be an entity clearly distinguishable from c-PAN.

In France, HBV-related PAN formerly accounted for one-third (36%) of all cases of systemic PAN [5], but we have noted a progressive decrease in the number of cases since the development of vaccines against viral hepatitis. In 1992 and 1993, we observed that only 7% of the cases of c-PAN were attributable to HBV infection. The HBV-related vasculitis almost always takes the form of c-PAN. The clinical symptoms are the same as those observed in the other cases of PAN, except for orchitis which is more frequent in this group. In our male patients, 35% had orchitis and sometimes malignant hypertension. ANCA were rarely present (10%) [3]. Immune complexes seem to be the pathogenic mechanism involved in this disease [6].

## Microscopic polyangiitis

Formerly called microscopic polyarteritis nodosa, MPA is characterized histologically by the involvement of small-sized arteries and clinically by the presence of glomerulonephritis, in addition to the symptoms described above. Rapidly progressive glomerulonephritis (RPGN) is one of the major characteristics and leads to severe renal insufficiency and sometimes failure and dialysis. Other clinical symptoms, such as alveolar hemorrhage, when occurring in the context of PAN, can be attributed to MPA. The pulmonary–renal syndrome is a classic symptom of Goodpasture's syndrome but could be also due to MPA. However, we think that the symptoms of MPA are not exclusively pulmonary or renal and that some patients whose PAN appears to be 'classic' in fact have MPA. The latter diagnosis is highly probable when renal micro-aneurysms are absent and ANCA are present. ANCA are not found in every case of MPA [7, 8] but they have been reported in half of the patients. These antimyeloperoxidase antibodies give a perinuclear staining pattern in immunofluorescence labeling studies and are detectable by ELISA. In our experience, although ANCA are not present in every case of MPA, their association with the symptoms of PAN strongly suggests MPA. We have published that ANCA were found in some cases of c-PAN, but we now think that careful pathological examination will show a strong correlation between ANCA and small-sized vessel involvement. Nevertheless, this small-vessel vasculitis can be associated with that of medium-sized vessels, which would explain some conflicting results: in our opinion, in the case of concomitant medium- or large-sized vessel involvement or evidence of microaneurysms, arterial stenoses or renal infarcts, a diagnosis of classic PAN should be made, in contrast to the vasculitis nomenclature proposed by the Chapel Hill Group [9].

### Churg–Strauss syndrome

Initially, CSS was separated from PAN and characterized clinically by the presence of asthma and eosinophilia $> 1500/mm^3$, and histologically by the presence of medium-sized artery vasculitis, venous involvement and a granuloma. Small-sized arteries can also be affected. Recently, ANCA were detected in at least half the CSS patients tested [3, 7, 8]. This disease now assumes a new place in the PAN group and is certainly further from c-PAN than previously thought. Nevertheless, this syndrome shares with PAN a majority of symptoms, and the differences that are observed between c-PAN and CSS could reflect different etiologies and triggering factors and, consequently, other pathogenic mechanisms. We have shown [10] that CSS could occur after specific or non-specific immune stimulus, such as hyposensitization or vaccination. The syndrome is observed in patients who have pre-existing severe and corticodependent asthma, and these triggering factors are responsible for the exacerbation of the respiratory symptoms and the development of systemic vasculitis. When analyzing the differences between patients with c-PAN and CSS, we also noticed that none of them presented concomitant HBV infection. Based on epidemiological data and some anecdotal cases [10], we hypothesized that inhaled antigens were the cause of the disease and that triggering factors were necessary for its development. The pathogenic pathways have not been clearly elucidated and, if ANCA represent one of the most important mechanisms, the role of immune complexes [6] cannot be ignored.

## EPIDEMIOLOGY

PAN is a rare disease. In a study [11] that considered biopsy proven forms only, the annual incidence and prevalence of the disease were, respectively, 0.7/100 000 and 6.3/100 000 habitants. PAN affects men and women equally at every age, with a predominance between 40 and 60 years old.

## ETIOLOGY

In most cases, the etiology of PAN remains unknown. In some cases, PAN is the consequence of HBV infection. In 1984, this association was observed in France in 30–39% of the patients [12]. At present, we observe this association in fewer than 10% of them. The clinical characteristics of HBV-related PAN are similar to those of PAN without HBV infection. Nevertheless, we observed that orchitis, vascular nephropathy and GI involvement were more frequent in the case of HBV infection.

Other etiologies have been found in some PAN patients: human immunodeficiency virus (HIV) [13], cytomegalovirus (CMV), parvovirus B 19, human T lymphotropic virus type I, and hepatitis C virus (HCV) [14, 15].

## PATHOLOGY

PAN is an inflammatory vasculitis affecting segments of small- and medium-sized vessels. The vasculitic lesion often arises at arterial bifurcations [16, 17]. The acute phase of arterial wall inflammation is characterized by fibrinoid necrosis of the media associated with infiltration of pleiomorphic cells, mainly neutrophils; the arterial wall is destroyed; arterial aneurysms and thromboses can occur; perivascular inflammation is present. Arterial healing is characterized by fibrotic endarteritis; aneurysms usually regress [18]. Coexisting arterial lesions of different ages are present in the same organ [19]. In c-PAN only arteries are affected.

## CLINICAL ASPECTS

The clinical manifestations of PAN at the time of diagnosis described in several patient series reported in the literature [20–25] are summarized in Table 9.1. Initial symptoms vary from case to case [26]. Fever and weight loss are present in well over half the patients at the time of diagnosis, and myalgia and arthralgia in somewhat fewer. Fever has no specific characteristics; it can hover around 38°C or be higher than 40°C.

**Table 9.1** Main clinical and biological manifestations observed in polyarteritis nodosa (%)

| Manifestation | Mowrey and Lundberg 1954 [24] n = 607 | Nuzum and Nuzum 1954 [25] n = 175 | Frohnert and Sheps 1967 [20] n = 130 | Leib et al. 1979 [23] n = 64 | Guillevin et al. 1985 [21] n = 126 | Guillevin et al. 1992 [29] n = 182 |
|---|---|---|---|---|---|---|
| *Clinical* | | | | | | |
| Fever | 68 | 81 | 75 | 36 | 80 | 65 |
| Weight loss | 58 | 46 | 71 | – | – | – |
| Peripheral neuropathy | 66 | 54 | 72 | 36 | 70 | 70 |
| Myalgia | 50 | 35 | 30 | 73 | 55 | 54 |
| Skin involvement | 25 | 35 | 46 | 28 | 52 | 49 |
| Arthralgia | – | – | 58 | – | 63 | 46 |
| Kidney involvement | 83 | – | 67 | 63 | 45 | 36 |
| Hypertension | 58 | 54 | 56 | 55 | 38 | 33 |
| Abdominal pain | 48 | 62 | 14 | 16 | 24 | 26 |
| Lung involvement | 36 | 57 | 38 | 47 | 28 | 20 |
| Cardiac involvement | – | – | 8 | 17 | 20 | 9 |
| *Biological* | | | | | | |
| Erythrocyte sedimentation rate | – | 98 | 82 | 89 | 85 | – |
| Leukocytosis | 72 | 74 | 45 | 54 | 74 | – |
| HBV markers | – | – | – | 40 | 30 | – |

## Rheumatic symptoms

Myalgias are frequent (54.4%), can be intense and muscle tenderness is usual. Arthralgias are present in 46.2% of the cases and are associated with myalgia in most of them. Arthritis is asymmetric and predominantly present in the legs [27].

## Neurological symptoms

Mononeuritis multiplex is the most frequent symptom (70.3%) and it occurs early in the course of the disease [22]. Motor and sensory signs are asymmetric, affect predominantly the lower limbs, especially the sciatic nerve and its peroneal and tibial branches. Radial, cubital and median nerves are affected less frequently. Motor deficit occurs abruptly. Sensory signs are responsible for hypo- or hyperesthesia and pain in the area of neurological deficit, which is sometimes present before the sensory loss. In some PAN patients, it can be difficult to analyze the causes of pain: neurological, muscular and/or rheumatological involvement. The cerebrospinal fluid (CSF) is usually normal. Under treatment, mononeuritis multiplex regresses progressively and patients can recover without sequelae. However, when the latter are present, they are more sensory than motor. Palsies of cranial nerves are present in fewer than 2% of the cases, with the oculomotor (III, trochlear IV, abducent VI), facial (VII) or acoustic VIII nerves being affected. Sequelae of the VIIIth cranial nerve palsy are common.

Central nervous system (CNS) involvement is rare [28], but motor deficiencies, strokes and sometimes brain hemorrhages can be seen. Their etiology is not univocal: vasculitis of a cerebral artery or consequence of malignant hypertension, for example.

## Skin manifestations

Skin involvement is present in 25–60% of the patients with systemic PAN. In our previous study on 182 patients [29], we noted skin involvement in 48.9% of patients. Different types of skin lesions can be observed:

1. Vascular purpura is typically papulopetechial, sometimes bullous or vesiculous. Infiltration is not constant but, when present, represents the ideal site for biopsy.
2. Subcutaneous nodules are less frequent and consist of clusters of very small elements ranging from 0.5–2 cm. They are transient and should be biopsied as soon as possible.
3. Livedo reticularis is common.
4. Distal gangrene is the consequence of ischemia.

When a biopsy can be performed, it shows small-sized vessel and capillary vasculitis. When possible, the biopsy should include the derm in order to detect medium-sized vessels.

## Kidney involvement

In the literature, kidney involvement is present in 60–80% of the patients [26, 30], but was observed in only 35.7% of our patients [29] included in a prospective cohort. When autopsies were performed, kidney was the most frequently affected organ.

Kidney involvement is diverse, vascular or glomerular. According to the type of involvement, it is possible to diagnose two different diseases: MPA when glomerulonephritis is observed, and c-PAN when vascular nephropathy is responsible for renal failure.

Vascular nephropathy is the usual manifestation described in c-PAN. Rapid renal failure is the consequence of renal infarcts. When present, malignant hypertension is renin dependent. Oliguria or anuria can complicate the vascular nephropathy and can lead to irreversible dialysis during the acute phase of PAN or months or several years later. In such cases, renal angiography shows multiple microaneurysms and/or infarcts.

Glomerular nephropathy (RPGN) leads in a few days or weeks to renal failure, with proteinuria and microscopic hematuria. Renal angiograms are usually normal and the coexistence of vascular and glomerular nephropathy is rare. The presence of RPGN is characteristic of a disease distinct from c-PAN named MPA.

Ureteral involvement and renal or perirenal hematomas are observed [31, 34]. Ureteral stenosis is uni- or bilateral and can be responsible for anuria and renal insufficiency, or it can be asymptomatic. Ureteral stenosis is the consequence of periureteral vasculitis and secondary fibrosis. Renal or perirenal hematoma result from the rupture of microaneurysms [34, 35].

## Cardiovascular manifestations

Heart manifestations are limited to the myocardium. They are due to vasculitis of the coronary arteries or their branches, or to severe or malignant hypertension. Clinical, electrocardiographical or radiological abnormalities can be observed in 40% of PAN patients [29]. Cardiac failure involves the left heart more often than the right side. In our experience, angina is rare and coronary angiography is usually normal. Cardiomegaly can be seen in 25% of cases. The electrocardiogram can disclose signs that would be compatible with coronary ischemia. Atrioventricular block may be present. Pericarditis is rare.

Hypertension is present in 40% of the patients and it is usually mild; however, it should be kept in mind that it can be triggered or adversely affected by steroids. Malignant hypertension was detected in 7/165 of our cases [36]. Hypertension is more frequent in HBV-related PAN. Hypertension benefited from angiotensin converting enzyme inhibitors which improved its prognosis.

### GI involvement

GI tract involvement [37, 39] is one of the most severe manifestations of PAN. Abdominal pain has been noted in 34% of our patients. It is always difficult to establish the exact cause of the abdominal pain, but its presence can be the first symptom of GI vasculitis. In a majority of cases, ischemia is present in the small bowel, but rarely in the colon or stomach. Digestive bleeding and bowel perforation can occur [40]. Relapses after surgery or medical treatment are a sign of poor prognosis. Intractable abdominal pain associated with weight loss are the consequence of bowel ischemia. Digestive malabsorption and pancreatitis have been described [37]; their prognoses are extremely severe. Vasculitis of the appendix [41] or gallbladder is sometimes the first manifestation of PAN [39]; its significance is not unequivocal: it is sometimes a pathological curiosity without any other clinical, pathological and/or immunological involvement of vasculitis. In other cases, cholecystitis or appendicitis is another symptom of PAN. The prognoses for these manifestations seem different depending upon whether the cholecystitis or appendicitis is the first manifestation of PAN or a complication of a previously diagnosed and treated PAN: in the former situation, the prognosis remains good; however, when these symptoms occur during the course of PAN, they often precede other severe symptoms of GI involvement and the prognosis is poor.

Liver involvement, such as an infarction and hematomas, can exist even in the absence of HBV infection.

### Lung manifestations

A few cases of pneumopathy with infiltrates have been described. Pleural effusion was present in 5% of our patients.

### Miscellaneous

Orchitis is a classical symptom of PAN and is one of the ACR criteria for the diagnosis of PAN. This symptom is present in 36% of PAN related to HBV and is rarer in PAN without HBV [43]. An epididymal or testicular biopsy can provide the diagnosis of necrotizing vasculitis.

Breast and uterine PAN have also been described [44].

Ocular manifestations include retinal vasculitis, retinal detachment and cotton-wool spots. An ophthalmological examination should be performed systematically in PAN and, when necessary, should be complemented by retinal angiography.

## Clinical characteristics of PAN related to HBV infection

PAN related to HBV could be considered to be an immune complex (IC)-mediated disease [45], and vasculitis could be the consequence of excess antigen or an impairment of IC clearance by the reticuloendothelial system [46]. Nevertheless, viral antigen was rarely demonstrated in vessels or in IC. In France, we have observed that contamination due to an infected blood transfusion has now disappeared, but that intravenous drug use is becoming an important cause of HBV-related PAN. For the past several years, blood donors have been tested for the presence of anti-HBc antibodies (Ab) and transaminase levels and rejected when positive, even if they are not carriers of HBs or HBe antigen (Ag). The development of vaccines against HBV and their administration to people at risk also explain the dramatic decrease in the number of new cases observed since 1989. In the past few years, the frequency of PAN related to HBV has declined to 7.3%.

The immunological process responsible for PAN occurs early in the course of HBV infection. When it was possible to date the HBV contamination, we observed that, in most cases, PAN developed within less than six months after infection. Hepatitis is diagnosed sometimes, but is silent in most cases and PAN could be the first manifestation of HBV infection. In the group of PAN related to HBV, clinical data were roughly the same as those commonly observed in PAN, but we found some differences: malignant hypertension, renal infarction and orchiepididymitis were more often associated with HBV infection. Among another group of patients, severe malignant hypertension was observed in 26.8% of the cases, and orchiepididymitis was present in 25%. These manifestations occur in patients under 40 years of age. Their disease is acute and initially severe, but the outcome is excellent in most cases if adequate treatment is prescribed. Seroconversion usually leads to recovery. Sequelae are the consequence of vascular nephropathy but, even in patients with renal insufficiency, it is possible to obtain recovery with little residual impairment of renal function. Regarding other symptoms, we frequently observed abdominal manifestations (46.3%) and especially surgical emergencies (17%). In contrast to our previous results obtained in patients treated with steroids and cyclophosphamide, we did not observe in this study an increased mortality due to surgical emergencies [5].

In the study by Sergent *et al.* [47], two of the three deaths (among nine patients), were related to colon vasculitis. Among the Eskimo patients described by McMahon *et al.* [48], 31% of them died and one of the four early deaths was the consequence of bowel perforation. Digestive and renal manifestations resulted from ischemia, and angiography demonstrated the presence of microaneurysms and infarctions. It is interesting to note that control angiograms showed that the aneurysms had disappeared when PAN had been effectively treated. This disappearance could result from the thrombosis of aneurysms with evolution to fibrosis.

The prevalence of HCV in our patients was low, less than 10%, and confirms our previous results [14]. HCV does not seem to be an important etiological factor for PAN.

Hepatic manifestations of PAN during the course are clinically moderate. Hepatic cytolysis is moderate in most cases and cholestasis is minor or absent. When liver biopsies were performed, they frequently showed chronic hepatitis, even in PAN that occurred only a few months after HBV infection.

## BIOLOGICAL, IMMUNOLOGICAL AND ANGIOGRAPHIC INVESTIGATIONS

### Biological analyses

Inflammatory signs are present in the majority of cases: an erythrocyte sedimentation rate > 60 mm 1st hour, increased C reactive protein, high $\alpha$-2 globulin level, leukocytosis, sometimes hypereosinophilia > 1500/mm$^3$. IC are present in c-PAN [6]. ANCA are rare in c-PAN. By contrast ANCA have also been found in half of MPA. By immunofluorescence, ANCA give a perinuclear staining pattern; ELISA detects antimyeloperoxidase antibodies.

### Angiography

Microaneurysms are not pathognomonic but commonly present in c-PAN. Conversely, they are rarely seen in CSS. Aneurysms range in size from 1–5 mm and are predominantly seen in the kidneys, mesentery and liver; they regress under treatment [18]. Angiography is a useful diagnostic tool when other diagnostic examinations are negative. In our opinion, aneurysms are considered to classify PAN as c-PAN and, contrary to the conclusions of the Chapel Hill Nomenclature [9], could be considered exclusionary criteria for MPA.

**Biopsies**

The histological characteristics of PAN are detailed elsewhere.

## PROGNOSTIC FACTORS AND SEVERITY CRITERIA

Factors of poor prognosis have been identified in PAN and CSS and they are also relevant for other systemic necrotizing vasculitides. We have shown, based on a prospectively followed cohort of 334 patients with PAN or CSS, that mortality was increased when the following signs were present: renal symptoms, i.e. glomerulonephritis (when proteinuria is > 3 g/d; relative risk (RR) is × 3.6) and/or renal insufficiency (RR × 2); GI tract involvement characterized by bowel perforation and/or digestive hemorrhage and/or pancreatitis (RR × 2.7); cardiomyopathy (RR × 2); CNS involvement (RR × 1.75); and weight loss (RR × 1.75). In that prospective study, it was not possible to confirm that age was a poor prognostic factor as had previously been published [5].

However, we were able to construct a prognostic score for PAN based on five signs: cardiomyopathy, CNS involvement, severe GI tract symptoms, renal failure and high proteinuria (> 1 g/d). This score enabled us to determine the overall prognosis of the disease at the time of diagnosis, considering the presence of one, two or more factors or their total absence (Table 9.2), to choose the optimal treatment that seems to offer the best chance of success, and to avoid the occurrence of refractory disease.

## CONCLUSIONS

The clinical manifestations of PAN are well known, but different characteristics are observed according to the etiology of the disease and subclassification of PAN. When glomerulonephritis is present, MPA can be diagnosed and, conversely, renal vasculitis, which is responsible for renal failure and malignant hypertension, is characteristic of c-PAN.

**Table 9.2** Predictive mortality value of the five-factors score (FFS) in PAN

| FFS | Death (%) | Survival (%) | Relative risk (RR) | N |
|---|---|---|---|---|
| 0 | 12 | 88 | 0.63 | 217 |
| 1 | 26.25 | 73.75 | 1.38 | 80 |
| 2 | 45.95 | 54.05 | 2.4 | 37 |
| Total | 64 | 270 | 1 | 334 |

$p < 0.0001$ for scores 0 vs. 1 and 1 vs. 2.

These differences in the clinical expression of PAN show that the PAN group of vasculitides is not homogeneous.

## ACKNOWLEDGEMENTS

This study was based on prospective trials which were supported by grants from the Institut National pour la Santé et la Recherche Médicale (INSERM), the Caisse Nationale d'Assurance-Maladie des Travailleurs Salariés (CNAMTS) and the Association pour la Recherche sur les Angéites Nécrosantes (ARAN). Research was carried out with the help of the Cooperative Study Group for Polyarteritis Nodosa and the Société Nationale Française de Médecine Interne (SNFMI).

## REFERENCES

1. Kussmaul, A. and Maier, K. (1866) Über eine nicht bisher beschreibene eigenthümliche Arterienerkrankung (Periarteritis nodosa), die mit Morbus Brightii und rapid fortschreitender allgemeiner Muskelhämung einhergeht. *Dtsch. Arch. Klin. Med.*, **1**, 484–518.
2. Churg, A. (1983) Pulmonary angiitis and granulomatosis revisited. *Hum. Pathol.*, **14**, 868–83.
3. Guillevin, L., Visser, H., Noël, L. *et al.* (1993) Antineutrophil cytoplasm antibodies (ANCA) in systemic polyarteritis nodosa with and without hepatitis B infection and Churg–Strauss syndrome. 62 patients. *J. Rheum.*
4. Lightfoot, R.J., Michel, B.A., Bloch, D.A. *et al.* (1990) The American College of Rheumatology 1990 criteria for the classification of polyarteritis nodosa. *Arthritis Rheum.*, **33**, 1088–93.
5. Guillevin, L., Lê, T.H.D., Godeau, P. *et al.* (1988) Clinical findings and prognosis of polyarteritis nodosa and Churg–Strauss angiitis: a study in 165 patients. *Br. J. Rheumatol.*, **27**, 258–64.
6. Guillevin, L., Ronco, P. and Verroust, P. (1990) Circulating immune complexes in systemic necrotizing vasculitis of the polyarteritis nodosa group. Comparison of HBV-related polyarteritis nodosa and Churg–Strauss angiitis. *J. Autoimmun.*, **3**, 789–92.
7. Savage, C.O. and Lockwood, C.M. (1990) Antineutrophil antibodies in vasculitis. *Adv. Nephrol. Necker Hosp.*, **19**, 225–36.
8. Kallenberg, C.G., Mulder, A.H. and Tervaert, J.W. (1992) Antineutrophil cytoplasmic antibodies: a still-growing class of autoantibodies in inflammatory disorders. *Am. J. Med.*, **93**, 675–82.
9. Jennette, C.J., Falk, R., Andrassy, K. *et al.* (1994) Nomenclature of systemic vasculitides: Proposal for an international consensus conference. *Arthritis Rheum.*, **37**, 187–92.
10. Guillevin, L., Amouroux, J., Arbeille, B. and Boura, R. (1991) Churg–Strauss angiitis. Arguments favoring the responsibility of inhaled antigens *Chest*, **100**, 1472–3.
11. Scott, D., Bacon, P., Elliott, P. *et al.* (1982) Systemic vasculitis in a district general hospital 1972–1980: clinical and laboratory features, classification and prognosis in 80 cases. *Q. J. Med.*, **51**, 292–311.
12. Guillevin, L., Lê, T.H.D. and Gayraud, M. (1989) Systemic vasculitis of the

polyarteritis nodosa group and infections with hepatitis B virus: a study in 98 patients. *Eur. J. Intern. Med.*, **1**, 97–105.

13. Gherardi, R., Belec, L., Mhiri, C. *et al.* (1993) The spectrum of vasculitis in human immunodeficiency virus-infected patients. A clinicopathologic evaluation. *Arthritis Rheum.*, **36**, 1164–74.

14. Quint, L., Deny, P., Guillevin, L. *et al.* (1991) Hepatitis C virus in patients with polyarteritis nodosa. Prevalence in 38 patients. *Clin. Exp. Rheumatol.*, **9**, 253–7.

15. Carson, C.W., Conn, D.L., Czaja, A.J. *et al.* (1993) Frequency and significance of antibodies to hepatitis C virus in polyarteritis nodosa. *J. Rheumatol.*, **20**, 304–9.

16. Arkin, A. (1930) A clinical and pathological study of polyarteritis nodosa. *Am. J. Pathol.*, **6**, 401–2.

17. Lie, J.T. (1990) Illustrated histopathologic classification criteria for selected vasculitis syndromes. American College of Rheumatology Subcommittee on Classification of Vasculitis. *Arthritis Rheum.*, **33**, 1074–87.

18. Guillevin, L., Merrouche, Y., Ruel, M. *et al.* (1990) Regressing aneurysms in polyarteritis nodosa related to hepatitis B virus. *Eur. J. Intern. Med.*, **1**, 267–72.

19. Lie, J. (1990) Diagnostic histopathology of major systemic and pulmonary vasculitic syndromes. *Rheum. Dis. Clin. North Am.*, **16**, 269–92.

20. Frohnert, P. and Sheps, S. (1967) Long-term follow-up study of polyarteritis nodosa. *Am. J. Med.*, **48**, 8–14.

21. Guillevin, L., Fechner, J., Godeau, P. *et al.* (1985) Periartérite noueuse: étude clinique et thérapeutique de 126 patients suivis pendant 23 ans. *Ann. Med. Interne*, **136**, 6–12.

22. Guillevin, L., Lhote, F., Leon, A. *et al.* (1993) Treatment of polyarteritis nodosa related to hepatitis B virus with short-term steroid therapy associated with antiviral agents and plasma exchanges. A prospective trial in 33 patients. *J. Rheumatol.*, **20**, 289–98.

23. Leib, E., Restivo, C. and Paulus, H. (1979) Immunosuppressive and corticosteroid therapy of polyarteritis nodosa. *Am. J. Med..*, **67**, 941–7.

24. Mowrey, F. and Lundberg, E. (1954) The clinical manifestations of essential polyangiitis (periarteritis nodosa) with emphasis on the hepatic manifestations. *Ann. Intern. Med.*, **40**, 1145–64.

25. Nuzum, J. and Nuzum, J. (1954) Polyarteritis nodosa. Statistical review of 175 cases from literature and report of a typical case. *Arch. Intern. Med.*, **94**, 942–55.

26. Cupps, T. and Fauci, A. (1981) Systemic necrotizing vasculitis, in *The Vasculitides, volume XXI, Major Problems in Internal Medicine* (ed. L. H. Smith Jr), WB Saunders, Philadelphia.

27. Cohen, R., Con, D. and Ilstrup, D. (1980) Clinical features, prognosis and response to treatment in polyartentis. *Mayo Clin. Proc.*, **55**, 146–55.

28. Castaigne, P., Cambier, J., Escourolle, R. and Brunet, P. (1970) Les manifestations nerveuses centrales de la periartérite noueuse. A propos d'une observation anatomo-clinique. *Ann. Med. Interne*, **121**, 375–82.

29. Guillevin, L., Lhote, F., Jarrousse, B. and Fain, O. (1992) Treatment of polyarteritis nodosa and Churg–Strauss syndrome. A meta-analysis of 3 prospective controlled trials including 182 patients over 12 years. *Ann. Med. Interne*, **143**, 405–16.

30. Ronco, P., Mougenot, B., Kanfer, A. *et al.* (1989) Atteintes rénales des angéites nécrosantes. *Rev. Med. Interne*, **10**, 227–34.

31. Azar, N., Guillevin, L., Huong, D.L. *et al.* (1989) Symptomatic urogenital

manifestations of polyarteritis nodosa and Churg–Strauss angiitis: analysis of 8 of 165 patients. *J. Urol.*, **142**, 136–8.

32. Hachulla, E., Bourdon, F., Taieb, S. *et al.* (1993) Embolization of two bleeding aneurysms with platinum coils in a patient with polyarteritis nodosa. *J. Rheumatol.*, **20**, 158–61.

33. Lie, J.T. (1992) Retroperitoneal polyarteritis nodosa presenting as ureteral obstruction. *J. Rheumatol.*, **19**, 1628–31.

34. Smith, D.L. and Wernick, R. (1989) Spontaneous rupture of a renal artery aneurysm in polyarteritis nodosa: critical review of the literature and report of a case. *Am. J. Med.*, **87**, 464–7.

35. Ostrum, B and Soder, P. (1960) Periarteritis nodosa complicated by spontaneous perinephric hematoma roentgenographic findings in 3 cases and a review of the literature *Am. J. Roentgenol.*, **84**, 849–60.

36. Cohen, L., Guillevin, L., Meyrier, A. *et al.* (1986) Hypertension artérielle au cours de la périartérite noueuse. Incidence, paramètres clinicobiologiques et prognostiques dans une série de 165 cas. *Arch. Mal. Coeur Vaisseaux*, **79**, 773–8.

37. Desbazeille, F. and Soule, J. (1986) Manifestations digestives des vascularites. *Gastroentérol. Clin. Biol.*, **10**, 405–14.

38. Lê, T.H.D., Wechsler, B., Guillevin, L. *et al.* (1985) Les manifestations digestives de la périartérite noueuse dans une série de 120 cas. *Gastroentérol. Clin. Biol.*, **9**, 697–703.

39. Remigio, P. and Zaino, E. (1970) Polyarteritis nodosa of the gallbladder. *Surgery*, **67**, 427–31.

40. Williams, D.H., Kratka, C.D., Bonafede, J.P. and Katon, R.M. (1992) Polyarteritis nodosa of the gastrointestinal tract with endoscopically documented duodenal and jejunal ulceration. *Gastrointest. Endosc.*, **38**, 501–3.

41. Myoana, T. (1988) Necrotizing arteritis of the vermiform appendix. *Arch. Pathol. Lab. Med.*, **112**, 738–41.

42. Bookman, A., Goode, E., McLoughlin, M. and Cohen, Z. (1983) Polyarteritis nodosa complicated by a ruptured intrahepatic aneurysm. *Arthritis Rheum.*, **26**, 106–8.

43. Persellin, S.T. and Menke, D.M. (1992) Isolated polyarteritis nodosa of the male reproductive system. *J. Rheumatol.*, **19**, 985–8.

44. Piette, J.C., Bourgault, I., Legrain, S. *et al.* (1987) Systemic polyarteritis nodosa diagnosed at hysterectomy. *Am. J. Med.*, **82**, 836–8.

45. Trepo, C., Zuckerman, A., Bird, R. and Prince, A. (1974) The role of circulating hepatitis B antigen/antibody immune complexes in the pathogenesis of vascular and hepatic manifestations in polyarteritis nodosa. *J. Clin. Pathol.*, **27**, 863–8.

46. Lockwood, C., Worledge, S., Nicholas, A. *et al.* (1979) Reversal of impaired splenic function in patients with nephritis or vasculitis (or both) by plasma exchange. *N. Engl. J. Med.*, **300**, 524–30.

47. Sergent, J., Lockshin, M., Christian, C. and Gocke, D. (1976) Vasculitis with hepatitis B antigenemia: long-term observations in nine patients. *Medicine*, **55**, 1–18.

48. McMahon, B.J., Heyward, W.L., Templin, D.W. *et al.* (1989) Hepatitis B virus-associated polyarteritis nodosa in Alaskan Eskimos: clinical and epidemiologic features and long-term follow-up. *Hepatology*, **9**, 97–101.

# 10

# Microscopic polyarteritis (MPA)

*P.A. Bacon, C. Savage and D. Adu*

HISTORY

In the nineteenth century, following Kussmaul and Maier's [1] classic description, periarteritis nodosa was considered as a rare distinct disease entity. It was characterized by visible nodular lesions at autopsy involving the muscular arteries. Capillaries and venules were never involved. However, in the twentieth century, cases were described in which the diagnosis was made at microscopy – with consequent changes in the emphasis of disease descriptions. This led to an evolving concept of vasculitis and its classification. Initially, all the new descriptions were lumped together in an expanding unity. Subsequently, one entity after another was split off into distinct separate forms (Figure 10.1). Veszpremi and Jancso [2] were the first to describe a case with microscopic involvement of vessels in the absence of any gross evidence of periarteritis nodosa. This was felt at the time to be a most unusual case, and a decade later only five cases had been described with microscopic disease in the absence of gross lesions [3]. The chief impetus for the expanding interest in the microscopic forms of necrotizing vasculitis in the first half of the twentieth century (and for including them in the same category as classic PAN) was the recognition of the vascular manifestations of allergies. This could occur in relationship to classic allergens, such as foreign proteins, to bacteria, and to drugs. Rich [4] reported vascular lesions in both serum sickness and sulfonamide hypersensitivity. Shortly afterwards, he produced vascular lesions in rabbits, frequently in association with acute diffuse glomerulonephritis, by injection of horse serum [5]. The vascular lesions were found microscopically in arter-

*The Vasculitides*. Edited by B.M. Ansell, P.A. Bacon, J.T. Lie and H. Yazici.
Published in 1996 by Chapman & Hall, London. ISBN 978-0-412-64140-4.

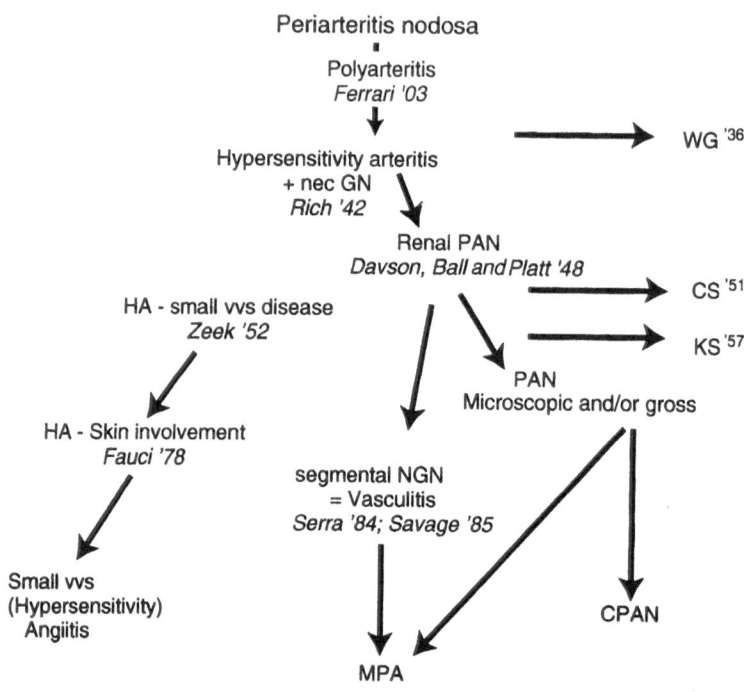

**Figure 10.1** Evolving definition of PAN.

ioles, capillaries and small muscular arteries in several organs, including the lungs.

In her classic review Zeek [6] divided the expanding cluster of necrotizing vasculitides into five main groups. The first of these was hypersensitivity vasculitis, which typically involved capillaries, arterioles and the small arteries as well as the venules. The organs involved were the kidney in particular, often with necrotizing glomerulonephritis, but the angiitis was usually widespread and pulmonary vessels were frequently involved.

## RENAL LIMITED VASCULITIS AND POLYARTERITIS

Renal lesions have been associated with polyarteritis since the comprehensive description by Kussmaul and Maier [1]. Most classic reviews have found the incidence of renal involvement to be around 80%. Gruber [7] had shown that multiple renal infarcts were commonly seen, but there were also cases in which diffuse changes in the renal parenchyma were present. These were usually described as a form of nephritis – 'Morbus Brightii' in Kussmaul and Maier's paper. Orphuls [8] believed it was an unrelated coincidental lesion, but others thought it was secondary to the arteritis [9] or due to the same causative (?toxic) factor. It is important to note that the lesions that were only detected microscopically were commoner than the aneurysmal type.

The morphological changes of polyarteritis nodosa – a term first used by Ferrari [10] – were described in detail by Arkin [11]. He pointed out that in cases without aneurysms the vasculitis may only be seen microscopically, and referred to such cases as microscopic PAN. This theme was taken up and expanded in a detailed clinicopathological study by Davson, Ball and Platt [12]. They linked the presence of the primary systemic vasculitis with the occurrence of segmental necrotizing glomerulonephritis in a subgroup of patients. They again suggested that such patients had a microscopic form of polyarteritis, which they thought was distinct from the periarteritis (or polyarteritis) nodosa described by Kussmaul and Maier.

They divided their patients into two groups, those with severe and widespread glomerular disease (9 out of 14) and those with minimal or no glomerular disease. Half of the first group had arterial (arcuate artery) involvement in addition to glomerular disease. They presented with a rather uniform clinical picture with pyrexia, leukocytosis and uremia in the presence of a normal blood pressure. The second group had variable clinical symptoms including elevated blood pressure, locomotor symptoms, abdominal pain, and heart and lung involvement. Pathology here showed lesions typical of the classic periarteritis and of malignant hypertension. They concluded that the microscopic form of polyarteritis appears different from the classic aneurysmal type. Importantly, they stated clearly that it was possible to separate microscopic polyarteritis from other forms of diffuse nephritis on both clinical and pathological grounds, establishing that the nephritis was not a coincidental or unrelated feature.

The idea of segmental necrotizing glomerulonephritis as a manifestation of vasculitis was pursued particularly in England. Serra *et al.* [13] described 53 patients with renal vasculitis by clinical and histological criteria. These were divided into two groups – those with vasculitis limited to the glomerular tuft, and those with extraglomerular vasculitic lesions in addition. Comparison of the clinical features of these two groups showed them to be indistinguishable. Although all had renal disease, and developed severe renal impairment, only a minority of them had presented initially with this. An extrarenal presentation was seen in two-thirds of their series. Systemic features, including fever, weakness, weight loss, episcleritis and musculoskeletal complaints, were common. The renal prognosis was poor, with only half surviving past the first year. Oliguria and rapid development of severe renal failure, in particular, carried a bad prognosis, emphasizing the urgent need for adequate nephrological assessment in microscopic polyarteritis. They concluded that such predominantly renal vasculitis did not easily fit the then current general classifications of the vasculitides.

This theme was pursued by Savage *et al.* [14] in a series of 34 patients with microscopic polyarteritis. They used the term to describe patients with clinical and histological evidence of small-vessel systemic vasculitis associated with focal segmental necrotizing glomerulonephritis. They excluded patients who had evidence of other distinct disease entities, such as Wegener's granulomatosis (WG) or a vasculitis associated with a connective tissue disease. All of them had clinical evidence of a systemic small-vessel vasculitis, which initially predominantly affected the skin and musculoskeletal systems and was accompanied by a focal segmental necrotizing glomerular nephritis. They emphasized that the distinction between microscopic polyarteritis and classic polyarteritis nodosa is justified by the different complications and prognosis of these two groups. All of their patients had impairment of renal function, which often developed rapidly and dominated the clinical course and both patient and kidney survival rates.

The importance of renal vasculitis in the overall prognosis of polyarteritis was emphasized again by a study from Birmingham [15] where patients aged over 50 were found to do much less well than younger patients. The higher mortality in older patients appeared to reflect the increasing severity of renal failure in these patients. Fifty-five per cent of patients with a serum creatinine in excess of 500 m$\mu$/l died as compared with only 6% of patients with a lower serum creatinine at presentation. Similarly, in the Serra *et al.* [13] study, 83% of oliguric patients died in contrast with 14% of non-oliguric patients. The Birmingham study showed that classic polyarteritis was often associated with an overt systemic presentation – with features including visceral infarction and peripheral neuritis – leading to early diagnosis and treatment. In contrast, the microscopic form had a more insidious clinical onset, so that severe renal impairment had often occurred before the true diagnosis was considered.

In the absence of aneurysms, there are no gross features in the kidney typical of polyarteritis. The renal lesion seen in these histological studies was a focal segmental necrotizing and thrombosing glomerulonephritis, which may be the only evidence of vasculitis within the kidney [12–14]. Extracapillary proliferation (crescents) was common in glomeruli affected by a segmental necrotizing and thrombosing glomerulonephritis. When extensive, it was usually associated with advanced renal failure, but overall there was often a poor correlation between the extent of glomerular involvement and the degree of renal impairment. The latter correlates closely with the degree of tubulointerstitial damage, but the mechanisms of this are unclear. It may be due to small-vessel vasculitis and ischemic damage. Glomerular immune deposits are usually sparse, often consisting of fibrin only, which is a major differentiating factor from conditions such as Henoch–Schönlein purpura and systemic lupus erythematosus (SLE) [16].

## DEFINITION OF MICROSCOPIC POLYANGIITIS

It is now widely accepted on both sides of the Atlantic that the microscopic form of necrotizing angiitis is a separate disease entity from the original descriptions of periarteritis or polyarteritis nodosa. The main question is that of nomenclature, and we favor that proposed by the Chapel Hill consensus conference [17]. This is important in view of the considerable preceding confusion over the names and definitions for necrotizing vasculitis affecting medium- and small-sized arteries. The preferred name is now microscopic polyangiitis (microscopic polyarteritis, MPA) which has to be distinguished from classic polyarteritis nodosa (c-PAN). The most important distinguishing feature is the presence of vasculitis in small vessels (arterioles, venules and capillaries) in MPA and its absence in c-PAN (Figure 11.2). It has to be emphasized that some overlap of the size of vessel involved is seen. By definition, MPA must have involvement of microscopic vessels, but it is recognized that it can also involve small- or medium-sized arteries (even with arterial aneurysms). In contrast, c-PAN must have no involvement of microscopic vessels and, therefore, no glomerulonephritis. By these definitions, MPA and c-PAN are differentiated by the presence or absence of small-vessel involvement, rather than by the involvement of medium-sized arteries which can occur in either. The definition of MPA also requires few or no immune deposits in the blood vessels, in order to allow differentiation from those variants of small-vessel vasculitis included in the hypersensitivity angiitis group that do have well-defined immune complexes. This is particularly important to allow distinction from Henoch–Schönlein purpura and cryoglobulonemic vasculitis, both with predominant cutaneous involvement. It also differentiates other forms of immune complex-mediated small-vessel vasculitis, such as that seen in SLE or serum sickness. It is recognized that identical glomerular lesions may be seen in WG, but there are other important distinctions from that condition examined below.

## COMPARISON WITH c-PAN AND WG

The clinical distinction between the microscopic and the classical aneurysmal form of polyarteritis has been building momentum since the first description of microscopic vessel involvement and the subsequent observation that, overall, renal vasculitis includes an arteritis and a necrotizing glomerulonephritis. However, the overlapping vessel size which occurs in MPA confuses interpretation of older series, and makes the interrelationship of the two syndromes difficult to interpret. For example, 25% of renal patients in our series [15], all of whom had focal segmental necrotizing glomerulonephritis, also had renal arteritis. There is a similarity in the clinical symptoms. Even in the renal series, the majority of patients first present with non-renal symptoms. However, previous experience in a district general hospital had emphasized the wide spectrum of severe organ involvement in c-PAN [18]. A comparison of several series of patients from renal units with those from general or overlapping departments confirmed that there are real differences [15]. The general series, which contained many cases of true PAN, showed a higher incidence of fever, heart, gastrointestinal and peripheral nerve involvement. The difference in vessel size involved appears to dictate not just the clinical pattern of organ damage, but also the prognosis. The visceral infarction or hemorrhage in c-PAN contributes prominently to the overall outcome [19]. The overt clinical symptoms often also contribute to an earlier diagnosis and thus institution of therapy in c-PAN.

In view of the preceding confusion, we have reanalyzed Birmingham patients seen over a decade to see how well the consensus definitions from Chapel Hill hold up in real practice (Table 10.1). The symptoms present when the patient is diagnosed emphasize the high frequency of systemic manifestations in both. However, in c-PAN there is an increased frequency of locomotor involvement, gut and neurological disease. In MPA, there is an increase in both renal and lung disease. Pulmonary involvement in MPA has been noted in previous series and contributes to a poor prognosis [14, 20]. Patients may present with overt or even massive pulmonary hemorrhage in association with renal failure – so-called pulmonary renal syndrome – and these patients tend to fair badly.

**Table 10.1** Features at presentation

|  | MPA (n = 95) | c-PAN (n = 14) | Renal WG (n = 31) |
|---|---|---|---|
| Systemic | 89 | 13 | 24 |
| LMS | 46 | 13 | 19 |
| Skin | 25 | 2 | 13 |
| Eye | 19 | 3 | 4 |
| ENT | 19 | 4 | 19 |
| Gut | 5 | 3 | 4 |
| Lung | 49 | 5 | 19 |
| Heart | 9 | 0 | 1 |
| CNS/PNS | 13 | 9 | 3 |
| Renal | 91 | 2 | 31 |

Less dramatic pulmonary involvement is even more frequent. The pathological lesion, a necrotizing vasculitis in small alveolar vessels, is essentially the same as in the kidney. The most striking difference between the two conditions in our series is the frequency of presentation. Eight MPA cases have been seen for every c-PAN case. This is not due to a biased collection from a renal center, since the Birmingham Vasculitis Group collects cases from rheumatological as well as renal clinics and from general physician referrals. It reflects the widespread experience in Europe now that MPA is common (and perhaps increasing in incidence) while PAN is rare (and perhaps decreasing). There is also a difference in outcome. Relapse rates are high in vasculitis and this was seen in two-fifths of our PAN cases, but in less than one-fifth of the MPA group. However, the overall prognosis is much worse in MPA, which in our series showed a 40% overall mortality over the decade, largely related to renal failure often associated with delayed referrals. In contrast, c-PAN responded well to therapy, with a 100% survival in our hands despite the relapses.

The advent of ANCA testing has provided another way to compare the two forms of angiitis. In MPA the majority of patients are ANCA positive. Antibody specificity is predominantly directed towards myeloperoxidase [21]. In contrast, in our patients with c-PAN, only one was ANCA positive. Others have also concluded that c-PAN, without evidence of small vessel involvement, is ANCA negative [22].

WG is the classic disease that is ANCA positive, although the antibody specificity is primarily to proteinase 3 rather than myeloperoxidase. An identical, pauci immune, necrotizing glomerulonephritis also occurs in this condition, but there is little to support the concept that MPA represents a renal limited form of WG. A comparison of the clinical features seen in our series is also shown in Table 10.1. and more detailed comparison of MPA with WG cases with renal involvement has also been made [23]. This shows the expected increased incidence of clinical evidence of upper plus lower airways involvement in WG. The real difference is at the pathological level, where granulomatous lesions are seen in WG but not in MPA. The prognosis appears better in WG also, even in the renal cases, with a lower frequency of progression to renal failure and death – despite a high relapse rate equivalent to that of c-PAN. This is not true in other series [24].

## THERAPY

Treatment of MPA is essentially the same as that of the other primary systemic vasculitides, such as WG and PAN. This is in part because treatment in these conditions is directed at the inflammatory consequences of the disease rather than at the primary mechanisms. The sole exception to this is antiviral treatment in hepatitis B induced PAN. Nevertheless, there are differences in emphasis, reflecting the increased severity and poorer prognosis in MPA. The mainstay of treatment initially is a combination of cyclophosphamide together with corticosteroid. There is strong circumstantial evidence to support the concept that the combination is more effective than either alone, although this has not been formally submitted to a trial in MPA. A difference between MPA and PAN is seen in the enhanced need for adjuvant therapy in the latter. Additional bolus doses of intravenous methylprednisolone, plasma exchange and, more recently, intravenous immunoglobulin, have all been used in this respect. In practice, they are added frequently, reflecting a high degree of perceived need for adjuvant therapy. Thus in our recent prospective study of two cyclophosphamide/ steroid regimes, escalation therapy was allowed as the clinician felt it was required. One of the above modalities was added in 60% of instances. The one-year survival in this prospective study was nearly 90% (reviewed in [25]). Thus the short-term prognosis in systemic vasculitis now appears very good, but the perceived need for escalation therapy suggests we still do not have the ideal treatment. The latter will only come from enhanced understanding of causes and mechanisms of vasculitis.

An important way to enhance the success of therapy in non-specialist centers is earlier recognition of disease leading to earlier institution of therapy. The renal lesion is the chief determinant of prognosis in the majority of series, particularly in the elderly. Thus early diagnosis, allowing institution of therapy before irreversible renal damage has occurred, would make a major difference. The chances of early diagnosis would be enhanced if a policy of urgent renal biopsy were pursued, both in cases where there are clinical signs of systemic vasculitis together with laboratory evidence of renal involvement, and in unexplained cases of renal failure with concomitant systemic symptoms, however mild. Hematuria is the commonest lesion in patients with MPA but, unfortunately, it is chiefly microscopic hematuria, not macroscopic [15]. Thus microscopic hematuria must be taken seriously and positive efforts taken to exclude the possibility of systemic vasculitis.

## CONCLUSION

The evolving concept of necrotizing vasculitis has now accepted the microscopic form of polyarteritis as a distinct entity from the classic aneurysmal form. The predominant renal involvement in the microscopic form is also clearly distinct from other forms of nephritis and relates to a primary vascular lesion. This led to the definition of microscopic polyangiitis as a separate entity, even though there is some overlap in both vessel size and clinical symptoms with c-PAN. Analysis of our data agrees with other series that MPA represents a clinically distinct entity that is both considerably more frequent and has a worse prognosis than c-PAN, related to the development of rapidly progressive renal failure. The relapse rate and the mortality confirmed the difference between the two conditions, as does the antibody status. Despite this, therapy is similar at present, but additional therapies, such as plasma exchange or pulse methoprednisolone, are more frequently required in MPA. There are thus good clinical, therapeutic and pathological reasons for accepting this as a separate entity which needs to be widely recognized, diagnosed early and treated aggressively before irreversible loss of renal function occurs.

## REFERENCES

1. Kussmaul, A. and Maier, R. (1866) Über eine nicht bisher beschreibene eigenthümliche Arterienerkrankung (Periarteritis nodosa), die mit Morbus Brightii und rapid fortschreitender allgemeiner Muskelhämung einhergeht. *Dtsch. Arch. Klin. Med.*, **1**, 484–518.
2. Veszpremi, D. and Jancso, M. (1903) Über einen Fall von Periarteritis nodosa. *Beitr. Path. Anat.*, **34**, 1–25.
3. Klotz, O. (1917) Periarteritis nodosa. *J. Med. Res.*, **XXXVII**, 1–49.
4. Rich, A.R. (1942) The role of hypersensitivity in periarteritis nodosa. *Bull. Johns Hopkins Hosp.*, **71**, 123–4. Additional evidence of the role of hypersensitivity in the etiology of periarteritis nodosa. *Ibid.*, 375–9.
5. Rich. A.R and Gregory, J.E. (1943) The experimental demonstration that periarteritis nodosa is a manifestation of hypersensitivity. *Bull. Johns Hopkins Hosp.*, **72**, 65–88.
6. Zeek, P.M. (1952) Periarteritis nodosa: a critical review. *Am. J. Clin. Path.*, **22**, 777–90.
7. Gruber, G.B. (1925) Zur Frage der Periarteritis nodosa, mit besonderer Berucksichtigung der gallenblassen und Nieren beteiligung. *Arch. path. Anat.*, **258**, 441–501.
8. Ophuls, W. (1923) Periarteritis acuta nodosa. *Arch. Int. Med.*, **32**, 870–98.
9. Gray, J. (1929) A case of periarteritis nodosa. *J. Path. Bact.*, **32**, 787–93.
10. Ferrari, E. (1903) Über Polyarteritis acuta nodosa (sogenannte Periarteriitis nodosa) und ihre Beiziehungen zur Polymyositis und Polyneuritis acuta. *Beitr. path. Anat.*, **34**, 1–25.
11. Arkin, A. (1930) A clinical and pathological study of periarteritis nodosa. *Am. J. Path.*, **6**, 401–26.
12. Davson, J., Ball, J. and Platt, R. (1948) The kidney in periarteritis nodosa. *Q. J. Med.*, **65**, 175–202.
13. Serra, A., Cameron, J.S., Turner, D.R. *et al.* (1984) Vasculitis affecting the kidney: presentation, histopathology and long term outcome. *Q. J. Med.*, **53**, 181–207.
14. Savage Caroline, O.S., Winearls, C.G., Evans, D.J. *et al.* (1985) Microscopic polyarteritis: presentation. *Q. J. Med.* **56**, 467–83.
15. Adu, D., Howie, A.J., Scott, D.G.I. *et al.* (1987) Polyarteritis and the kidney. *Q. J. Med.*, **62**, 221–37.

16. Rosen, S., Falk, R.J. and Jennette, J.C. Polyarteritis nodosa, including microscopic form and renal vasculitis, in *Systemic Vasculitides – Section Two: Idiopathic Vasculitides* (ed. A. Churg and J. Churg), Igaku-Shoin, New York, Tokyo.

17. Jennette, J.C., Falk, R.J., Andrassy, K. *et al.* (1994) Nomenclature of systemic vasculitides – Proposal of an International consensus conference. *Arthritis Rheum.*, **366**, 185–95.

18. Scott, D.G.I., Bacon, P.A., Elliott, P.J. *et al.* (1982) Systemic vasculitis in a district general hospital 1972–1980: clinical and laboratory features, classification and prognosis of 80 cases. *Q. J. Med.*, **51**, 292–311.

19. Sack, M., Cassidy, J.T. and Bole, G. (1975) Prognostic factors in polyarteritis. *J. Rheumatol.*, **2**, 411–20.

20. Haworth, S.J., Savage, C., Carr, D. *et al.* (1985) Pulmonary haemorrhaging complicating Wegener's granulomatosis and microscopic polyarteritis. *Br. Med. J.*, **290**, 1775–8.

21. Kallenburg, C.G.M. (1994) Anti-neutrophil cytoplasmic antibodies (ANCA): current perspectives. *Clin. Rheumatol.* In press.

22. Baranger, T.A.R., Audrain, M.A.P., Guillevan, L. and Esnault, V.L.M. (1993) ANCA are only found in polyarteritis nodosa when associated with microscopic polyangiitis features. *Clin. Exp. Immunol.*, **93**, 35.

23. Luqmani, R.A., Bacon, P.A., Beaman, M. *et al.* (1994) Classical versus non-renal Wegener's granulomatosis. *Q. J. Med.*, **87**, 161–7.

25. Luqmani, R.A., Pall, A., Adu, D. *et al.* (1996) (see Chapter 23 of this volume).

# 11

# New concepts in Wegener's granulomatosis

*W.L. Gross*

Wegener's granulomatosis (WG) is commonly thought to be part of the spectrum of disorders characterized by the presence of vasculitic lesions throughout the body. However, the pathological hallmark of WG is the coexistence of vasculitis and granuloma. Classic WG according to the original criteria of Godman and Churg [1] involves a triad of airway, lung and renal disease. Some forms of WG, however, may be confined to the upper respiratory tract or lung, and spare the kidneys and other organ systems. The initial descriptions of WG have been followed more recently by greater understanding of its nuances, pathogenesis and treatment.

## VARIOUS FORMS: LATENT, INCOMPLETE, CLASSIC AND FULL-BLOWN

Carrington and Liebow [2] first called attention to the limited form of WG: the term 'limited' was used to denote those cases with predominant involvement of the lungs in the absence of glomerulonephritis. The absence of nephritis appeared to augur a much more favorable prognosis than for patients with the classic 'WG triad'. DeRemee *et al.* [3] became aware of otherwise typical pathology and clinical signs and symptoms that were, however, limited in extent, in patients who nevertheless responded well to conventional regimens. Fienberg [4] demonstrated cases with indolent lesions, particularly in the upper respiratory tract, that in some instances evolved over a period of up to 18 years from

*The Vasculitides*. Edited by B.M. Ansell, P.A. Bacon, J.T. Lie and H. Yazici. Published in 1996 by Chapman & Hall, London. ISBN 978-0-412-64140-4.

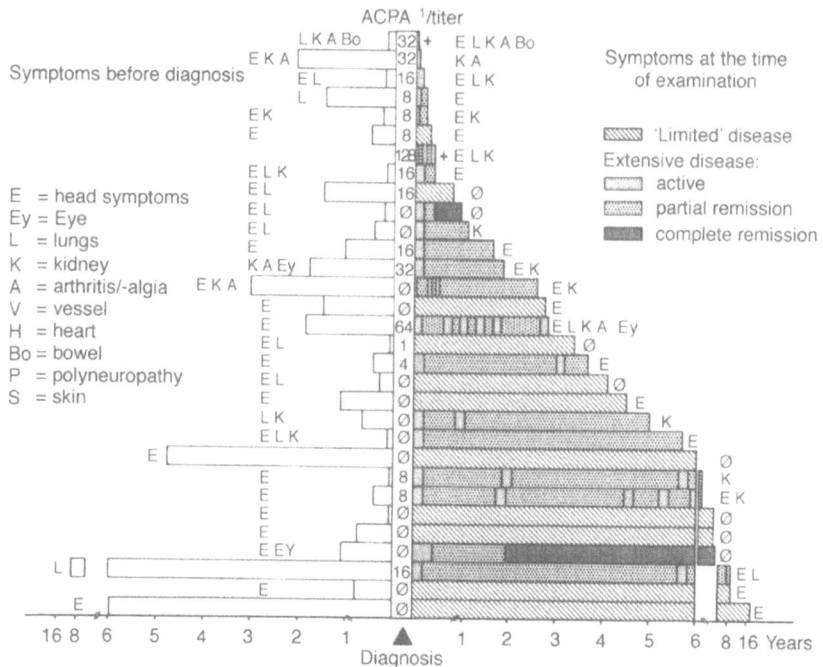

**Figure 11.1** Disease course prior to diagnosis (left side of the ordinate) and symptoms after diagnosis. Symptoms are given according to the modified ELK classification.

an incomplete form to a fully expressed 'WG triad'. Similar experiences with latent variants of WG having a more indolent course had been published earlier [5, 6]. The highly variable length of the latent course prior to the diagnosis of WG is demonstrated in Figure 11.1.

In their updated analysis of 158 patients with WG, Hoffman *et al.* [7] emphasized the variability of the disease prior to diagnosis: WG followed a confusing and indolent course in many cases (particularly in patients without renal manifestation) for up to 16 years before a definite diagnosis was established. The median and mean periods from the onset of symptoms to a diagnosis of WG were 4.7 and 15 months, respectively. Diagnosis of WG was made within three months after onset of symptoms in only 42% of patients; this is surprising because earlier studies had reported a median survival of only five months [8]. In the UK, 265 WG patients were observed between 1975 and 1985 [9]. The mean times from onset of symptoms to presentation and from presentation to diagnosis were both approximately seven months; correct diagnosis was often missed for many years (< 1–188 months). The mean survival of 4.2 years (shorter if renal disease was present) in patients receiving no drug treatment (10%) indicates that the variants must have been very mild. Furthermore, it is striking that the median survival of 72 patients treated with glucocorticoids alone exceeded 12 years. Unfortunately, the authors did not indicate the number of patients in the different treatment groups who had mild upper airway disease or disease that included renal involvement.

In the past, a major problem associated with the group of systemic vasculitides, including WG, was the lack of standardized diagnostic terms (names) and their definitions. The new classification criteria worked out by the American College of Rheumatology in 1990 [10], and the definitions given by the Chapel Hill Conference in 1992 [11] laid the foundation for the establishment of diagnostic criteria. In addition, the high specificity of c-ANCA (PR3-ANCA) for WG provides additional help in the diagnosis of even abortive forms of WG. Despite these advances, however, making a diagnosis of WG remains a problem in many patients with atypical variants not fulfilling the ACR criteria, or without typical histopathological and/or serological (ANCA) findings (Table 11.1).

These observations of latent, incomplete, or abortive variants of WG have led to the ELK concept developed by DeRemee [3]: E stands for symptoms in the upper respiratory tract, L the lung and K the kidneys. Implicit in this concept is a larger scope of clinical diversity and expression than was originally proposed. An extended ELK classification is now used as a 'disease extent index' in clinical studies [12–14] and as a supplement to activity indices: classic WG with the ELK triad represents the generalized (vasculitic = systemic) phase of WG that usually (~80%) begins with an initial phase characterized by symptoms mostly limited to E, L, EL as a locoregional restricted disease. Since more and more of these restricted forms of WG are being recognized early, it has been learned that the initial phase generally follows a rather subacute or chronic course of unpredictable duration before entering the generalized (systemic) phase. Whereas the latter is usually associated with high titers of c-ANCA, in the former c-ANCA can be lacking in about 50% of all cases. Classic WG is characterized, in addition to the granulomatous lesions, by systemic necrotizing vasculitis, usually with renal and pulmonary involvement. If untreated, generalized WG can transform into the fulminant form, which carries a devastating prognosis. Rheumatic complaints, eye or peripheral nerve involvement are frequently the ominous heralds of the onset of the generalized vasculitic phase, which is usually associated with constitutional (B) symptoms, such as malaise, weight loss, fever and night sweats, and highly elevated non-specific markers of inflammation, for example, ESR, CRP. c-ANCA (PR3-ANCA,

**Table 11.1** Names and definitions of ANCA-associated vasculitides (adopted by the Chapel Hill Consensus Conference on the Nomenclature of Systemic Vasculitis [12])*

*Small-vessel vasculitis*

| | |
|---|---|
| Wegener's granulomatosis | Granulomatous inflammation involving the respiratory tract, and necrotizing vasculitis affecting small to medium-sized vessels (e.g. capillaries, venules, arterioles, and arteries). *Necrotizing glomerulonephritis is common.* |
| Churg–Strauss syndrome | Eosinophil-rich and granulomatous inflammation involving the respiratory tract, and necrotizing vasculitis affecting small to medium-sized vessels, and associated with asthma and eosinophilia. |
| Microscopic polyangiitis | Necrotizing vasculitis, with few or no immune deposits, affecting small vessels (i.e. capillaries, venules, or arterioles). *Necrotizing arteritis involving small- and medium-sized arteries may be present. Necrotizing glomerulonephritis is very common. Pulmonary capillaritis often occurs.* |

* Large vessel refers to the aorta and the largest tranches directed toward major body regions (i.e. to the extremities and the hand and neck); medium-sized vessel refers to the main visceral arteries (e.g. renal, hepatic, coronary and mesenteric arteries); small vessel refers to venules, capillaries, arterioles and the intraparenchymal distal arterial radicles that connect with arterioles. Some small- and large-vessel vasculitides may involve medium-sized arteries, but large- and medium-sized vasculitides do not involve vessels smaller than arteries. Essential components are depicted in normal type; common, but not essential, components are shown in italics.
Note: WG may also occur as an inflammatory process without apparent vasculitis.

see below), which has a 90% specificity for WG, is now detectable in high titers in 95% of sera from generalized WG.

## DEFINITIONS AND CLASSIFICATION

In spite of substantial efforts by many investigators, the nomenclature for the various subsets of WG remains enigmatic. A major problem is the lack of standardized diagnostic terms and definitions. Thus, different names are used – for example, Wegener's granulomatosis, Wegener's vasculitis and microscopic polyarteritis or Wegener's type – for various forms of the same disease. In an attempt to address this problem, a proposal by an international consensus conference held in Chapel Hill

**Table 11.2** 1990 ACR-criteria for the classification of Wegener's granulomatosis (number of criteria present rule)*

| Criteria | Definition |
|---|---|
| 1. Nasal or oral inflammation | Development of painful or painless oral ulcers or purulent or bloody nasal discharge |
| 2. Abnormal chest X-ray | Roentgenogram of the chest showing the presence of nodules, fixed infiltrates, or cavities |
| 3. Urinary sediment | Microhematuria (over 5 red blood cells/hpf) or red cell casts in the urine sediment |
| 4. Granulomatous inflammation on biopsy | Histologic changes showing granulomatous inflammation within the wall of an artery or in the peri- or extravascular area (artery or arteriole) |

* For classification purposes, a patient shall be said to have WG if he/she has satisfied any two or more of these four criteria. This rule is associated with a sensitivity of 88.2% and a specifity of 92.0%.
Reproduced with permission from Leavitt *et al.* [10].

in 1992, gives both the names and definitions of vasculitides which resemble WG in many aspects [11] (Table 11.2).

WG was restricted by definition to patients with necrotizing granulomatous inflammation. As noted above, the inclusion of granulomatous inflammation in the definition of WG does not necessarily mean that histologic proof of granulomatous inflammation will ultimately be one of the classification or diagnostic criteria for the disease; non-invasive evaluations may some day be able to identify an abnormality that adequately predicts the presence of granulomatous inflammation without having to perform a histologic examination. The inclusion of granulomatous inflammation in the definition of WG, however, does exclude from this category patients who in fact have only non-granulomatous respiratory tract vasculitis, such as patients who have microscopic polyangiitis (synonym: microscopic polyarteritis, MPA) with pulmonary alveolar capillaritis.

Vessels of many types can be affected in WG, resulting in various clinical manifestations. Figure 11.2 illustrates the overlapping distribution of vessel involvement among different categories of vasculitis. It is important to realize that the overlapping features of WG and MPA, combined with the evolution of the inflammatory lesions over time (for example, glomerulonephritis followed by skin vasculitis followed

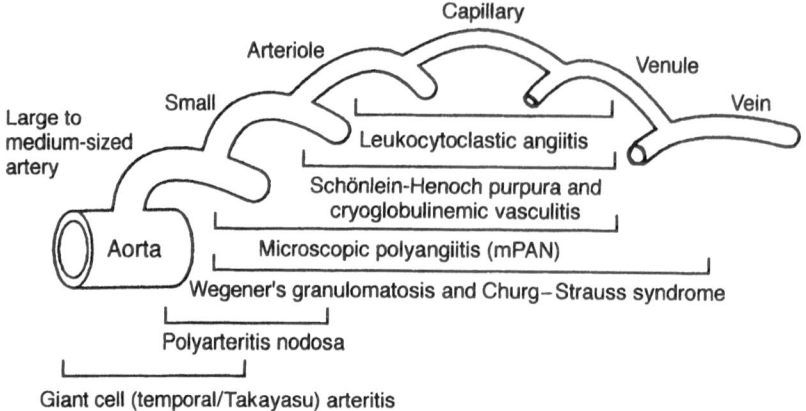

**Figure 11.2** Predominant range of vascular involvement by vasculitides as defined by the Chapel Hill Consensus Conference on the Nomenclature of Systemic Vasculitis. Note the substantial overlap among diseases. Large artery refers to the aorta and the largest branches directed toward major body regions (for example, to the extremities and the head and neck); medium-sized artery refers to the main visceral arteries (for example, renal, hepatic, coronary, and mesenteric arteries), and small artery refers to the distal intraparenchymal arterial radicles that connect with arterioles.

by pulmonary vasculitis), and the inability to immediately recognize clinically all aspects of systemic vasculitis in a given patient (especially the presence or absence of respiratory tract granulomatous inflammation), will inevitably mean that in some patients the name (diagnosis) ascribed to the vasculitis will change over time, for example a shift from an initial diagnosis of MPA to a diagnosis of WG.

Cytoplasmic autoantibodies (c-ANCA) with antigen specificity for proteinase 3 (PR3-ANCA) are a very sensitive and specific serologic marker for WG (for review, [15]). Patients with non-granulomatous pulmonary vasculitis (usually a hemorrhagic capillaritis) that is not caused by immune complexes or anti-basement membrane antibodies (that is, patients with MPA involving the lungs) may have perinuclear pattern ANCA (p-ANCA) with antigen specificity for myeloperoxidase (MPO-ANCA) or, less frequently, c-ANCA (PR3-ANCA). It should be noted that less common ANCA with other antigen specificities (for review, [15]) occur in patients with vasculitis, and that ANCA with other antigen specificities occur in non-vasculitic diseases, for example inflammatory bowel disease (for review, [16]).

There was much discussion at the 1992 Chapel Hill Conference about what diagnostic term to use for non-granulomatous small-vessel vasculitis affecting the upper and/or lower respiratory tracts, with or without necrotizing glomerulonephritis and without evidence of antibasement membrane antibody or immune complex mediation. Such patients usually also have a pauci-immune necrotizing and crescentic glomerulonephritis and thus typically manifest a pulmonary–renal vasculitic syndrome. The pulmonary vasculitis in such patients is predominantly a necrotizing alveolar capillaritis. Proposed names included Wegener's granulomatosis, Wegener's vasculitis, Wegener's syndrome, Wegener's disease, polyangiitis overlap syndrome, microscopic polyarteritis and microscopic polyangiitis. The final decision was to consider this as microscopic polyangiitis (microscopic polyarteritis) involving the respiratory tract.

Other causes of pulmonary–renal vasculitic syndrome must be excluded to make a diagnosis of microscopic polyangiitis with pulmonary involvement. This is accomplished in part by the requirement that there be few or no immune deposits, thus excluding immune complex-mediated and antibasement membrane antibody-mediated pulmonary and pulmonary–renal vasculitic syndromes. Most patients with such 'pauci-immune' pulmonary–renal vasculitic syndromes have either p-ANCA (MPO-ANCA) or c-ANCA (PR3-ANCA). The most difficult problem is differentiating between WG and microscopic polyangiitis with respiratory tract involvement.

The presence of ANCA in a patient with pulmonary–renal vasculitic syndrome speaks for the presence of WG or MPA. The ANCA in patients with WG as the cause of pulmonary–renal vasculitic syndrome are almost always of the cytoplasmic pattern (PR3-ANCA). The ANCA in patients with MPA may be p-ANCA (MPO-ANCA) or c-ANCA (PR3-ANCA). A patient with p-ANCA and pulmonary–renal vasculitic syndrome almost always has microscopic polyangiitis, that is, non-granulomatous pulmonary vasculitis, rather than WG, that is granulomatous pulmonary inflammation.

The conference participants realized that clinical differentiation between the ANCA associated pulmonary–renal vasculitic syndrome caused by WG and that caused by MPA may sometimes be difficult. In fact, some patients may initially have disease that is non-granulomatous, that is microscopic polyangiitis, and subsequently develop granulomatous inflammation, that is WG. This differentiation is less of a problem in patients with p-ANCA because they only very rarely develop granulomatous inflammation [17].

The Chapel Hill Conference did not address limited forms of WG, such as disease limited to the lungs. The definition proposed for WG in the context of providing a nomenclature for the vasculitides requires the presence of vasculitis; however, WG may also occur as an inflammatory process without apparent vasculitis.

In addition, to improve communication among physicians and to permit comparisons of research studies on systemic vasculitis performed at different centers, a subcommittee of the Diagnostic and Therapeutic Criteria Committee of the American Colle e of Rheumatology (ACR) recently developed classification rules for several forms of vasculitis, including WG [10], using an analytic approach based on prospectively gathered information from consecutive cases described elsewhere.

The classification criteria select the clinical findings and symptoms which both identify the disease (as one of a number of syndromes of vasculitis), and distinguish it from others (from all other vasculitides included in the data set). As a result, the classification criteria do not include the full spectrum of manifestations of the disease and are not meant to be used as diagnostic criteria. They provide a standard way to evaluate and describe groups of patients in therapeutic, epidemiological, or other investigations.

## CLINICAL MANIFESTATIONS AND THE TWO-PHASE COURSE

Clinical presentations vary widely, but the typical findings (Figure 11.3) in the upper respiratory tract, the lung and the kidney form a distinct clinicopathologic entity. Consequently, diagnosis is usually based on the combination of typical clinical features, histopathology, and PR3-ANCA findings. Recently, however, several new characteristics of WG have been discovered. From the clinical point of view, it is particularly important that the disease proceeds through two phases: in most cases WG begins with an initial stage that can last for many years without causing any life-threatening complications, but which can at any time turn into the classic systemic disease.

The clinical manifestations of WG exhibit a broad range with respect to clinical tempo and extent of expression. At one end of the spectrum is the indolent lesion, particularly in the upper respiratory tract (E according to the ELK format; 'initial phase'; 'purely granulomatous WG'), including nasal mucosal ulcerations ('rhinogenic granulomatosis' according to Friedrich Wegener), later on to sinusitis, mastoiditis. These lesions appear to move *per continuitatem* via the pharynx to the oral cavity leading to oral ulcerations, to the trachea and bronchi leading to (subglottic, bronchus) stenosis and to nodules in the lung, which tend to cavitate. Sometimes this granulomatous process moves up to the retro-orbital space leading to proptosis or, less frequently, to intracerebral granuloma. Because of the locoregional symptomatology confined mostly to the head, this initial phase of WG can be puzzling to the diagnostician; it often takes years until the true nature of the disease is suspected and histological diagnosis is finally made. Thus in its initial phase WG is usually not a life-threatening disease. Nevertheless, local complications, for example, subglottal larynx obstruction, can cause

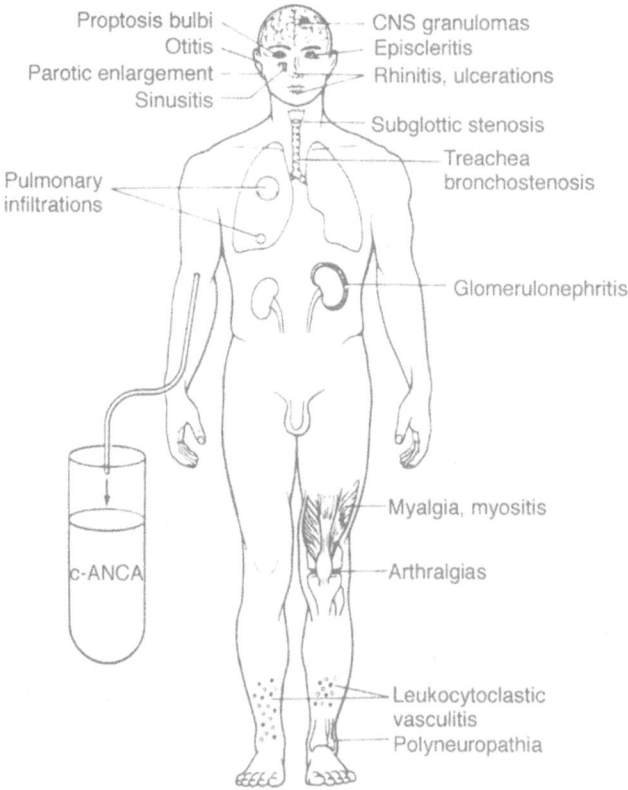

**Figure 11.3** Schematic demonstration of clinical symptoms in WG.

severe disease symptoms. On the other hand, at the time of first presentation of these locoregional restricted lesions it is absolutely unclear whether the disease will immediately transform into the generalized form of WG, progress very slowly, or – rarely – never progress but remain a *forme fruste* of WG frozen in the initial phase (stage). Clinically, these mono- or oligosymptomatic variants appear with a subglottic stenosis, cavitating nodule of the lung, proptosis, etc. The histology and the ANCA-findings can unmask these states as manifestations of WG (for review, [3, 5, 18]).

At the other extreme of the spectrum is fulminant classic WG. It appears as a true pulmonary–renal syndrome, and these two organ systems are then largely responsible for the fatal clinical course in untreated disease. Clinically, alveolar hemorrhage due to capillaritis of the lung and renal failure due to rapidly progressive necrotizing glomerulonephritis with crescent formation are the leading symptoms. The prognosis for patients with this disease expression is grave, and aggressive immunosuppressive therapy is urgently needed (see below). However, thorough anamnestic data obtained from patients with fulminant WG reveal that at least two out of three of them can remember prodromata in the ENT region, usually lasting for months and sometimes years! So, most cases of so-called 'fulminant WG' appear in reality to represent the culminating point of a formerly latent WG.

As always, however, the more typical course of WG lies between these two extremes. In fact, most cases of WG – even fulminant forms – begin in the ENT region with symptoms characteristic of but non-specific to the initial phase described above. This granulomatous process is usually not accompanied by clinically apparent vasculitis. The transition from the initial phase to the generalized (vasculitic) phase is, as mentioned already, always unpredictable, but it can be recognized clinically because it usually begins with constitutional symptoms common in systemic vasculitis (weight loss, fever, night sweats) and uncharacteristic rheumatic complaints (arthralgia, myalgia, frank arthritis, myositis, etc.). The true clinical face of systemic vasculitis then becomes more and more apparent: symptoms involving primarily the small vessels are 'red eye' (for example, episcleritis), skin rash (for example, purpura), peripheral neuropathy (for example, neuritis), and, most importantly, infiltration of the lung (capillaritis and renal lesion (glomerulonephritis)). If untreated, this stage can lead ultimately via rapidly progressive glomerulonephritis to oliguria and/or via increasing lung infiltration to cardiorespiratory insufficiency.

## HISTOLOGY AND HISTOGENESIS

It is widely believed that thoracotomy is necessary to obtain biopsy specimens adequate for histopathologic demonstration of WG. Since the majority of WG patients present first with symptoms involving the head, the diagnostic usefulness of nasal (and pronasal sinus) biopsies has been recognized, and criteria were recently proposed for diagnosis of WG based on biopsy specimens from the ENT region [19] (Plate 1).

The characteristic morphological picture in WG is a 'peculiar' granulomatous tissue with histiocytic epithelioid cell granulomas often containing giant cells of foreign body or Langhans' type. Additionally, there are many plasma cells, lymphocytes, leukocytes with polymorphs, sometimes eosinophils [20]. When the epithelioid cell granulomas disintegrate, the resulting picture is that of a necrotizing granulomatosis. Vessels can be involved in the form of granulomatous arteritis, phlebitis, perivasculitis or 'giant cell vasculitis' [21] as a secondary reaction to the primary necrotizing inflammation [8]. If vessels are not involved, Fienberg [22] speaks of an 'idiopneumonic pathergic granulomatosis' (for review, [5]).

In a study based on 35 open lung biopsies, Fienberg's group [23] sought to trace the histogenesis of the pulmonary lesions and were able to identify the early lesions. Micronecrosis, usually with neutrophils (microabscesses), constitutes the early phase in the development of pathognomonic organized palisading granuloma. There is then a progression of disease from micronecrosis to macronecrosis (= widespread necrosis), and then to fibrosis. Areas of macronecrosis surrounded by palisading histiocytes or diffuse granulomatous tissue indicate active disease. Arteritis and phlebitis, as described in classic WG, are present in most, but not all, cases. So, in this study palisading granuloma was virtually pathognomonic of WG and micronecrosis of collagen appeared to be the primary process in this disease.

Many years ago it was pointed out that the vascular lesions leading to vasculitis begin with damage to the endothelium [24, 25]. The vascular lesions may take many diverse forms [26] representing all of the pictures of necrotizing angiitis, not only those in classic panarteritis nodosa and its microscopically recognized form, but also vascular lesions with the appearance of allergic angiitis, angiitis granulomatosis or large-celled tuberculoid granulomatous arteritis. The arterial lesions are accompanied by lesions of veins, venules and capillaries, namely round-cell and histiocytic epithelioid cell infiltration of the intima and the wall, granulomatous, destructive foci of phlebitis, vascular miliary granulomas, endothelial lesions with aggregations of thrombocytes and local thromboses. In the course of the disease all organs can be involved via these vascular reactions.

The microscopic findings in the kidneys are correspondingly varied. Here, too, the above described vascular lesions and disseminated granulomas are occasionally found. Much more common, however, is proliferative focal glomerulitis [27] with necrosis and thrombosis of individual loops or larger segments of the glomerulus. Sometimes almost all glomeruli are destroyed, and this is accompanied by capsule adhesions and proliferations. The marked tendency toward the formation of granulomas is often stressed: periglomerulitis granulomatosa. Other WG cases show the picture of a diffuse, that is extracapillary, mesangioproliferative glomerulonephritis.

## LABORATORY INVESTIGATIONS

Normochromic, normocytic anemia, leukocytosis with mild (< 10%) eosinophilia, thrombocytosis and elevation of acute-phase proteins (CRP, etc.) are common laboratory findings in WG. However, it must be stressed that these laboratory abnormalities correlate with the phase and activity of the disease. Leukocyte counts, for example, can range from normal values in patients with initial phase disease to 'leukemoid' values in patients with fulminant WG. The same holds true for the whole range of acute reactants.

In contrast to the collagen vascular diseases, WG exhibits no (or only low titer) ANAs, no or only mild hypergammaglobulinemia, normal (or slightly elevated) levels of whole complement and no cryoglobulins. However, rheumatoid factor has been detected in more than 50% of WG patients and is noted most often in those with extensive disease.

PR3-ANCA have been shown to be a sensitive and specific marker for WG. PR3-ANCAs induce the classic (cytoplasmic) fluorescence staining pattern of c-ANCA and their presence can suggest a diagnosis of WG in patients with atypical or incomplete clinical manifestations. A clinico-pathological diagnosis, however, must be established, and therapy should not be initiated on the basis of positive ANCA-testing results alone. Since in many patients the ANCA titer correlates with clinical activity, ANCAs are helpful in monitoring patients during therapy, especially in distinguishing an infectious process from active disease.

## MONITORING OF DISEASE ACTIVITY AND DISEASE EXTENSION

Because generalized WG is a multisystem disease with a high frequency of relapse, it is especially important that the clinical aspect of disease activity and disease extension be monitored on a regular, objective and controlled basis by an interdisciplinary team of physicians, ENT surgeons, ophthal-mologists, neurologists (if appropriate) and radiologists. The value of erythrocyte sedimentation rate (ESR) and c-reactive protein (CRP) levels as – admittedly non-specific – indices of disease activity is generally accepted. In the initial phase of disease both ESR and CRP can be normal, while secondary and opportunisitic infections can lead to ele-vated levels of both. c-ANCA titers seem to correlate with disease activity in most patients [12]. Consequently, changes in c-ANCA titer remain one of the most important laboratory markers for alterations in disease activity, and the only one specific for WG. In contrast to ESR and CRP levels, the c-ANCA titer does not rise when there is a secondary infection. Highly elevated sIL-2R levels are found in patients with increased activity and may be an indicator of imminent relapse in patients in complete remission clinically [28] (Figure 11.4). The value of von Willebrand factor and other endothelial cell products (including thrombomodulin and ICAM-1) as indicators of disease activity has yet to be established.

Determination of disease extension by interdisciplinary clinical exam-inations is, in addition to disease activity, absolutely necessary for the planning of therapeutic strategies. Magnetic resonance imaging (MRI) scan of the head reveals disease activity at sites inaccessible by clinical investigation (sinuses: granulomatous lesions? Brain: white matter lesions?).

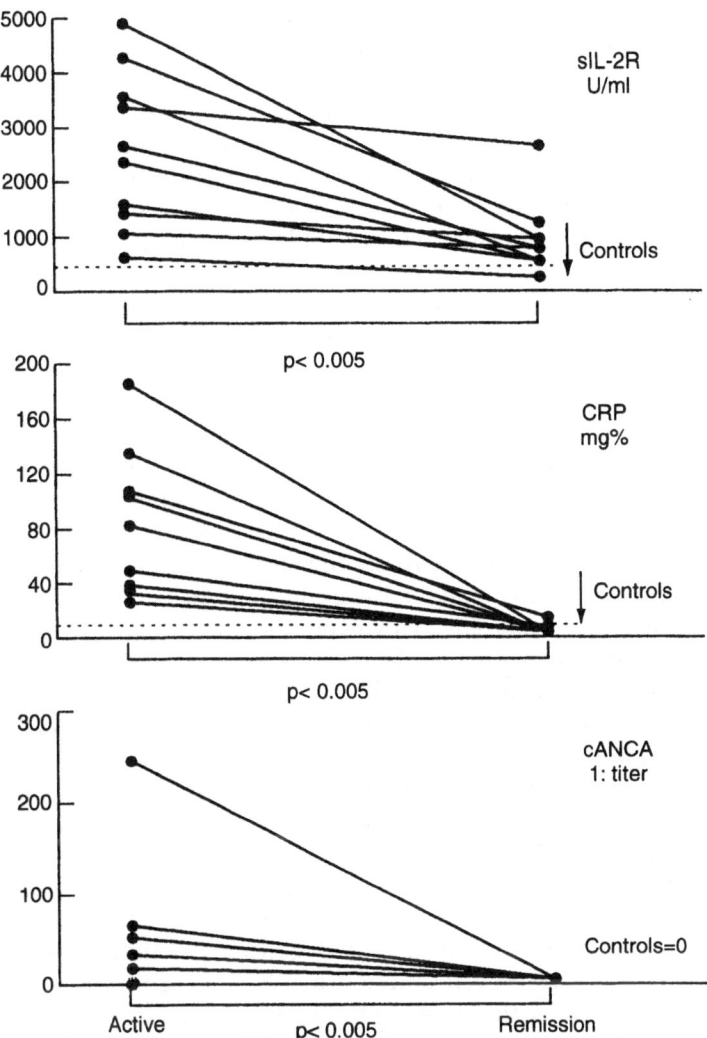

**Figure 11.4** Correlation of sIL-2R. C-reactive protein (CRP), and c-ANCA levels with clinical disease activity in 12 patients with generalized WG. In contrast to CRP and c-ANCA, sIL-2R levels remained elevated during remission.

## PATHOGENIC ASPECTS: GRANULOMA FORMATION AND VASCULITIS

As mentioned above, WG typically begins not as a vasculitic but as a granulomatous disease ('initial phase'). Granuloma was the name originally given to a chronic inflammatory mass resembling a tumor. The use of the term is now restricted to histological lesions largely composed of lymphocytes and cells of the myelomonocyte lineage. Granulomas are found in infectious diseases (for example, tuberculosis, lues), immuno-deficiency states (for example, hypogammaglobulinemia), in chronic granulomatous disease in children, and in diseases of unknown etiology (for example, Crohn's and Boeck's diseases). In chronic granulomatous disease, neutrophils do not develop a respiratory burst after phagocytosis, and the granuloma may represent an attempt by the cells to eliminate infectious agents. WG resembles chronic granulomatous disease in a few clinical aspects and in its response to treatment, for example with cotrimoxazole and corticosteroids [18]. It may or may not be induced by infectious agents. If PR3 is involved in defense mechanisms, for example against microbial infections, inhibition of PR3 by c-ANCA could result in an impairment of the defense system. Thus granuloma formation in WG, as in chronic granulomatous disease, could be a compensatory mechanism for eliminating infections, although this would not explain those cases of WG with granuloma formation which are ANCA negative in the initial phase (about 50%). The clinical response of WG to corticosteroids does not argue against this model, since in chronic granulomatous disease granulomas can be dramatically reversed by corticosteroids. However, the infectious agents are not eliminated by this treatment. Thus the granuloma can be reduced but generally will not be stopped [29].

Granuloma formation usually indicates a state of T cell hyperactivity and immunohistological studies in WG have revealed a predominance of $CD4^+$ cells in renal biopsies [30]. In cellular crescents of rapidly progressive glomerulonephritis and in the peripheral blood of WG patients, activated ($CD25^+$) T cells are also elevated. An increase in levels of serum cytokines, for example tumor necrosis factor alpha (TNF-$\alpha$) or IL-6, during the acute disease phase is an indirect indicator of T cell activation (for review, [31]). Elevated concentrations of soluble IL-2 receptor (sIL-2R) have been detected in WG sera, and the levels were shown to correlate with disease activity. It has been demonstrated that even in complete remission, sIL-2R levels tend to be elevated in WG [28]. However, T cell responses to neutrophil extract did not differ between vasculitis patients and controls. There were only low levels of antigen-specific proliferation, and these could not be amplified by *in vitro* selection [32]. In contrast to these findings, T cells from WG patients have been found to proliferate after exposure to highly purified PR3. The IgG subclass distribution of c-ANCA shows a high prevalence of IgG4-class antibody [30] together with increased total IgG4. This suggests repeated antigen stimulation in a T cell response and contrasts with the IgG subclass distribution of, for example, antinuclear antibodies, which are mainly restricted to IgG1 and IgG3 [33] (Table 11.3).

Recently, we presented a model in which ANCA and their target

**Table 11.3** Indications of the involvement of T cells in Wegener's granulomatosis

Granuloma formation in WG
Renal interstitial T cell infiltration
CD4+ T cells in the respiratory tract and kidneys
Elevated levels of CD25+ T cells
Elevated concentrations of sIL-2R
Prevalence of IgG4 subclass antibodies within c-ANCA
Proliferation of T cells after stimulation with PR3

antigens could be involved as a major pathogenic event in the induction of vasculitis in WG [34]. The most important phases in this model for ANCA-mediated vascular injury (ANCA-cytokine-sequence theory, Figure 11.5) are discussed below. The model is based on ANCA, cytokines and adhesion molecules, polymorphonuclear cells (PMN; granule proteins), and vascular endothelial cells. ANCA target antigens expressed on the cell surface in response to proinflammatory cytokines may interact with ANCA and lead to excessive activation of PMN (degranulation, generation of reactive oxygen species) and to cell lysis, and thus may induce necrotizing vasculitis.

In resting human PMN and monocytes, PR3 is mainly located intracellularly in MPO-positive granules [35]. Consequently, the target antigen of c-ANCA is not accessible to the circulating autoantibody *in vivo*. The following four stages may occur during the pathological process (for details and literature see [36, 37] as depicted in Figure 11.5).

First, infection (or an as yet unknown disease inherent factor) induces the production of proinflammatory cytokines (e.g. TNF-$\alpha$, IL-1, IL-8), which are elevated in systemic vasculitis [31]. *In vitro* exposure of PMN to TNF-$\alpha$/IL-8 leads to a time-dependent translocation of PR3 from their intragranular loci to the cell surface [38]. In this manner the autoantigen becomes accessible to the ANCA *in vivo*. In *ex vivo* studies, we have demonstrated that PR3 can be detected on the cell membrane of PMN from WG and sepsis patients [39]. Thus the presence of PR3 on PMN plasma membranes is not specific for ANCA-associated disorders, but it does enable circulating ANCA to bind to their target antigens.

Secondly, any pathogenic role of ANCA-activated PMN in systemic vasculitis demands a close connection of PMN to the endothelium, the target structure of vasculitis. Cytokine-induced expression of adhesion molecules (e.g. LFA-1 and ICAM-1) allows a close contact between PMN and the endothelial monolayer, with resultant shielding from $\alpha$-1AT. Immunohistological studies can demonstrate these molecules in tissue sections from these disease groups [40]. In addition, elevated levels of soluble ICAM-1 were demonstrated in active WG [41]. Recently, it was shown *in vitro* that Fab$_2$ fragments from ANCA increase neutrophil

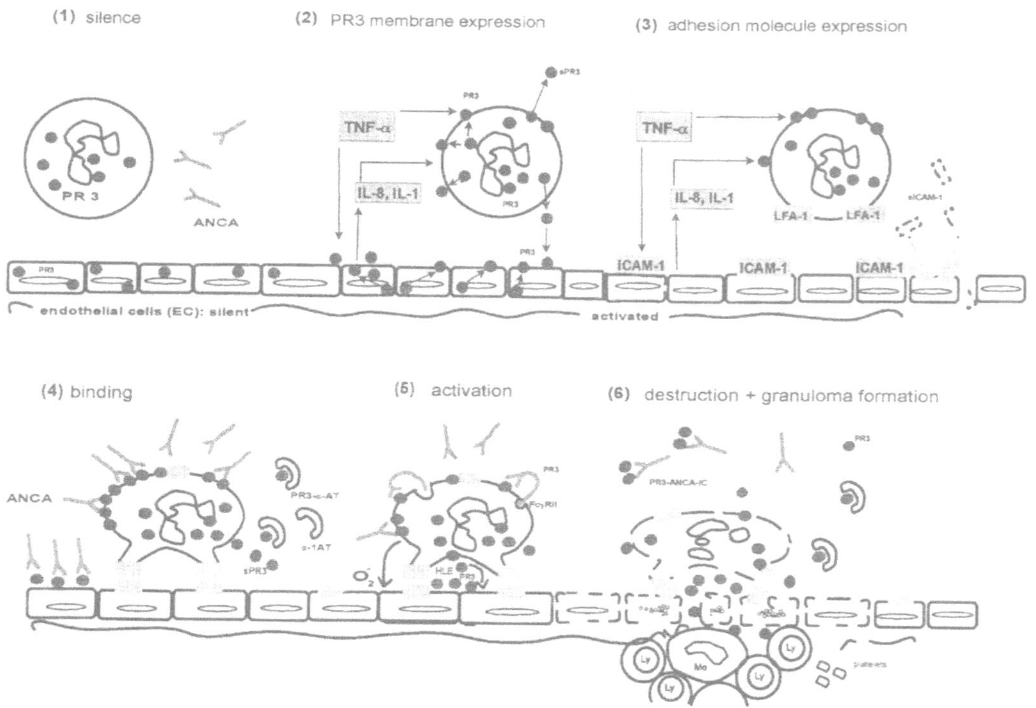

**Figure 11.5** Antineutrophil cytoplasmic antibody (ANCA)-cytokine sequence theory. (1) Resting neutrophil-proteinase 3 (PR3) mostly sequestered in azurophil granules. (2) Priming of polymorphonuclear leukocytes (PMN) by cytokines. Intracytoplasmic PR3 is translocated to the cell surface and becomes accessible to ANCA. Expression of adhesion molecules. (3) Adhesion of PMN to endothelial cells. (4–5) Interaction between ANCA and PR3 leads to activation of neutrophils with degranulation, generation of oxygen radicals and endothelial cell injury. (6) This finally results in intravascular lysis of PMN and necrotizing vasculitis.

adhesion to endothelial cells, augmented by prestimulation with TNF-α [42]. Interestingly, increased neutrophil adhesion to endothelium in patients with active vasculitis (WG and c-PAN) has been demonstrated [43].

Thirdly, binding of ANCA to granule proteins located on the cell surface of PMN enhances the PMN-endothelial cell interaction and further activates PMN to degranulate and to release toxic oxygen species into the immediate surroundings, for example into the isolated microenvironment between PMN and endothelial cells, which may contribute to enhanced tissue damage. Falk *et al.* [44] first demonstrated *in vitro* that ANCA activate neutrophils, as indicated by induction of a respiratory burst and degranulation. In addition, Lai and Lockwood [45] described how ANCA may selectively affect the signal transduction pathway (inositol phosphate generation and translocation of protein kinase C) in human PMN upon stimulation with a chemotactic peptide. Furthermore, it was found that in the presence of low levels of cytokines, ANCA stimulate PMN to damage endothelial cells *in vitro*, as measured by $^{51}$Cr and $^{111}$In release [46–48]. This might be caused by granule enzymes (for example, PR3, HLE, MPO), which may bind via charge interaction to the endothelial cell membrane. Here, they can lead to endothelial injury and subsequent necrotizing inflammatory lesions of the vessel wall. Deposition of neutrophil-derived cationic enzymes (e.g. MPO, PR3) on vascular endothelial cells was recently demonstrated *in vitro* [49, 50]. Furthermore, ANCA-antigen bound to endothelial cells can be recognized by ANCA and can enhance complement-dependent cytotoxicity [49]. This may result in further vascular damage. In a rat model, Kiser, Jenette and Falk [51] observed that intradermal injection of polyclonal rabbit anti-MPO antibodies resulted in increased vascular permeability. It was not determined whether this effect was direct or mediated by anti-MPO-induced leukocyte activation. The role of Fc receptors in this model has to be investigated.

Finally, cell lysis of ANCA-activated PMN occurs inside the vessel. This is in contrast to localized inflammation, which is characterized by transendothelial migration and chemotaxis of PMN to the inflammatory site. Donald, Edwards and McEvoy [52] found intravascular lysis of PMN in the vessels of WG lesions. In their report, organelles thought to originate from neutrophil cytoplasm were found to be coated by electron-dense material, indicating the presence of immunoglobulin on their surface. In this context, it is interesting that PMN from patients with active WG contain intracytoplasmic IgG as demonstrated by direct immunofluorescence [38].

In an alternative model proposed by Mayet *et al.* [53], vascular endothelial cells may be a direct target of ANCA. They demonstrated by laser scanning microscopy that PR3 is present in small amounts in the cytoplasm of cultured human umbilical vein endothelial cells. A time-dependent translocation of PR3 to the cell membrane was induced by stimulation with TNF-$\alpha$. This observation was recently confirmed using the polymerase chain reaction technique. In contrast, in immunohistological studies, PR3 does not appear to be expressed within endothelial cells in the vessels from WG and healthy tissue [54]. Furthermore, no immune deposits can be observed in ANCA-associated ('pauci-immune') vasculitis, which makes this highly interesting observation more doubtful with respect to its pathogenic role.

## PROGNOSIS AND TREATMENT

Because of the possibly life-threatening course of generalized WG, there has been general agreement that the 'standard' treatment consisting of daily 'low-dose' (2–4 mg/d) cyclophosphamide (CP) plus glucocorti-coids (GC) is of most benefit to the majority of patients. This view has been challenged by data indicating:

1. the variability of clinical course of the same disease,
2. continued morbidity despite CP plus GC therapy,
3. a high incidence of relapses,
4. the alarming range of toxic effects induced (for review, [55]).

Although the introduction of CP and GC therapy 30 years ago improved the prognosis in WG from a mean survival of five months [8] to a 93% chance of remission [29], an updated analysis [56] now presents less favorable data on both the efficacy of the protocol (remission rate 75%) and the attendant side-effects (serious infectious complications, 46%; CP-induced hemorrhagic cystitis, 43%; bladder cancer, 2.8%; mye-lodysplasia, 2%; etc.).

The knowledge that WG follows a two-phase course, together with the increasing identification of cases during the initial phase due to the diagnostic use of ANCA, provide a rationale for a stage adapted treatment of WG.

### Trimethoprim-sulfamethoxazole (Bactrim, T/S)

While T/S may have the capacity to induce remission in the 'initial phase' of WG [57, 58], the question of whether it may actually prolong the initial phase is currently being investigated in double-blind con-trolled studies. Recent experience does not indicate that T/S can pre-vent relapse of generalized WG in remission, and thus it does not appear suited for treatment of classic WG [14].

## Pulse CP

Inconsistent results were obtained in 12 patients with generalized WG treated with pulse CP [13]. Similarly, the NIH [59] reported only temporary benefit in 13 out of 14 patients treated with the same pulse CP regimen; lack of response and failure to sustain improvement or tolerate continued treatment was noted in 79% of patients. In contrast, two nephrologic groups [60, 61] reported the outcome after pulse CP therapy to be comparable with that of the 'standard protocol'. These differences may be ascribed to the fact that all of the nephrologic patients had renal involvement, whereas only 50% and 32% of the patients investigated in the former studies showed signs of renal disease. In addition, the group studied in Chapel Hill [61] was heterogeneous: only 19 out of 37 WG patients had c-ANCA and a majority of the other patients had p-ANCA, suggesting that cases of microscopic polyangiitis were included. Therefore, in order to find clinical and/or immunological markers affecting the response to pulse CP, we studied its efficacy in 43 c-ANCA-positive and biopsy proven WG patients (Table 11.4). Fifty-eight percent of the patients did not respond to pulse CP treatment. Collectively, a poor response was seen in patients with generalized disease involving more than four organ systems (mainly the heart, nervous system, eyes, skin) with constitutional symptoms and with a high c-ANCA titer [62].

## Methotrexate (MTX)

Hoffman *et al.* [56] recently used 'low-dose' (20–30 mg/weekly) MTX plus GC in 29 WG patients without 'immediately life-threatening disease', although 12 patients (41%) did have active glomerulonephritis. This produced a marked improvement in 76% and remission in 69% of the patients. However, 3 out of 29 (10%) suffered from *pneumocystis carinii* pneumonia during the first month of treatment (including prednisone 60 mg/day).

## Cyclosporine A (CYA)

Until now CYA has been used only as a back-up treatment in WG. In four patients with poor response to (oral) CP or with intolerable side-effects, CYA was able to control disease activity at dosages between 5 and 10 mg/kg/day [63]. Recently, Allen *et al.* [64] presented data on patients with active WG undergoing combined CYA and low-dose GC therapy. CYA was effective at initial doses of up to 5 m/kg/day, but mild flare-ups occurred when this was lowered to 2 mg/kg/day.

Schmitt *et al.* [65] reported on 20 patients who received renal trans-

**Table 11.4** Synopsis of clinical, laboratory and immunological findings in responders and nonresponders to pulse cyclophosphamide treatment (pulse-Cy)

| | Responder group (n = 18) | Non-responder group (n = 25) | |
|---|---|---|---|
| Male/female | 5/13 | 15/10 | |
| Median age | 47 (17–73) | 47 (23–68) | |
| Sympt. before diagn. was established (months) | 3 (0–168) | 6 (0–168) | |
| Daily Cy/Prd therapy prior to study entry | 13 (72%) | 18 (72%) | |
| Duration of therapy (months) | 10 (4–45) | 9 (1–34) | |
| Number of pulse-Cy given during the study | 6 (6–15) | 4 (1–15) | |
|   dosage of Cy (mg/m$^2$) | 637 (412–833) | 689 (350–909) | |
| Oral Prd dosage during pulse-Cy therapy | 13 (72%) | 14 (56%) | |
|   daily dosage (mg) | 4.5 (0–30) | 5.0 (0–20) | |
| Disease extension score | | | |
|   at diagnosis | 9 (2–11) | 11 (3–15) | |
|   start of pulse-Cy | 4 (0–11) | 4 (0–11) | |
|   end of pulse-Cy* | 0 (0–5) | 6 (2–11) | $p < 0.05$ |
| With only E and/or L at the start of pulse-Cy | 7 (39%) | 2 (8%) | $p < 0.05$ |
| Median c-ANCA titers | | | |
|   at diagnosis** | 1:16 (0–256) | 1:128 (0–512) | |
|   start of pulse-Cy** | 1:8 (0–128) | 1:16 (0–512) | |
|   end of pulse-Cy | 0 (0–128) | 1:64 (0–2048) | $p < 0.05$ |
| c-ANCA titer of 1:64 or higher | | | |
|   at diagnosis** | 5 (36%) | 13 (68%) | |
|   start of pulse-Cy** | 1 (6%) | 10 (42%) | $p < 0.05$ |
|   end of pulse-Cy | 1 (5%) | 14 (56%) | $p < 0.05$ |
| Median sIL-2R levels (U/ml)[+] | | | |
|   start of pulse-Cy*** | 1126 (415–2428) | 812 (241–2699) | |
|   end of pulse-Cy*** | 707 (321–3025) | 891 (355–2804) | |
| Median c-reactive protein (mg/dl)[##] | | | |
|   start of pulse-Cy[#] | 1.1 (0–7.7) | 0.0 (0–11.3) | |
|   end of pulse-Cy[#] | 0.0 (0–1.4) | 2.1 (0–26) | $p < 0.05$ |

* Data on organ involvement at the end of pulse-Cy in one non-responder missing.
** c-ANCA titers were determined at diagnosis in 33/43, at the start of pulse-Cy in 41/43 patients.
*** Serum levels of sIL-2R were measured prior to pulse-Cy in 41/43 patients.
[#] Serum levels of c-reactive protein were measured at the start and end of pulse-Cy in 41/43 patients.
[+] median value of healthy blood donors: 556 U/ml.
[##] normal: < 0.8 mg/dl.
Reproduced with permission from Reinhold-Keller *et al.* (1993).

plants (Tx) between 1982 and 1993. Treatment before Tx of oral CP and GC in 18 patients, and of pulse CP and GC or azathioprine plus GC in one patient each. At the time of Tx, six patients showed symptoms of active disease. Nevertheless, 18 patients are still alive with functioning grafts (mean creatinine, 1.7 mg/%) and 16 are in complete remission. These data demonstrate that the rate of survival and graft function in patients with WG is comparable to that of other transplant recipients, so patients with active WG need not be excluded from transplantation. Frequencies of relapse after transplant are given in Figure 11.2. These data indicate that immunosuppressive therapy after transplantation must not necessarily include CP and that CYA is effective under such circumstances.

## High dose intravenous immunoglobulin (IVIG)

The ability of IVIG to diminish vasculitic features was described recently [66, 67]. IVIG was given to 26 patients (WG, 14; m-PAN, 11; rheumatoid vasculitis, one) requiring an escalation of therapy, but mostly refractory to or intolerant of GC and cytotoxic drugs. They received Sandoglobulin 2 g/kg over five days; 26 out of 26 appeared to improve after IVIG, with 13 out of 26 achieving full remission. The benefit was sustained in 18 out of 26 at 12 months. Allowing for changes in other medications, 19 out of 26 were in full remission after one year, six in partial remission, while one had died of septic complications. In another study, IVIG was given to eight patients with WG and one patient with systemic p-ANCA-associated vasculitis, none of whom had reached complete remission under 'standard' therapy [68]. The response was measured by blind interdisciplinary clinical assessment (ENT, etc.), cranial MRI, and immunodiagnostic analyses. All nine patients were treated with Venimmune (0.4 g/kg/day for five days). Sixty per cent of the patients responded to therapy by showing improvement of single disease manifestations, but complete remission was not experienced by any of the patients. IVIG may thus play a role as an adjuvant to conventional therapy in patients with progressive disease.

## Monoclonal antibody therapy

This was introduced for treatment of vasculitis in 1990 [69]. A case of severe vasculitis treated several times with an anti-CDw52 antibody (Campath 1H) responded with significant, but only transient, improvement. Remission lasting for more than two years, however, was achieved by following anti-CD52 with injection of an anti-CD4 antibody. The same group has since reported on four patients (microscopic polyangiitis, Sjögren's syndrome, Behçet's disease) who were unresponsive to immunosuppressive drugs but who received 'substantial and sustained benefit' from Campath 1H and hIgG1CD4 [70]. Similar results have been obtained in a few patients with systemic WG [Lockwood, personal communication]. Two other groups have reported on the successful treatment of two severe and intractable cases of relapsing polychondritis with vasculitis using a mouse and chimeric anti-CD4 monoclonal antibody [71, 72].

## REFERENCES

1. Godman, G.C. and Churg, J. (1954) Wegener's granulomatosis. Pathology and review of the literature. *Arch. Pathol. Lab. Med.*, **58**, 533–53.
2. Carrington, C.B. and Liebow, A.A. (1966) Limited forms of angiitis and granulomatosis of Wegener's type. *Am. J. Med.*, **41**, 497.
3. DeRemee, R.A., McDonald, T.J., Harrison, E.G. *et al.* (1976) Wegener's granulomatosis. Anatomic correlates, a proposed classification. *Mayo Clin. Proc.*, **51**, 777–81.
4. Fienberg, R. (1981) The protracted superficial phenomenon in pathergic (Wegener's) granulomatosis. *Hum. Pathol.*, **12**, 458–67.
5. Gross, W.L. (1989) Wegener's granulomatosis: New aspects of the disease course, immunodiagnostic procedures, and stage-adapted treatment. *Sarcoidosis*, **6**, 15–29.
6. Cordier, J.F., Valeyre, D., Guillevin, L. *et al.* (1990) Pulmonary Wegener's granulomatosis. A clinical imaging study of 77 cases. *Chest*, **97**, 906–12.
7. Hoffman, G.S., Kerr, G.S., Leavitt, R.Y. *et al.* (1992) Wegener's granulomatosis: An analysis of 158 patients. *Ann. Intern. Med.*, **116**, 488–98.
8. Walton, E.W. (1958) Giant cell granuloma of the respiratory tract (Wegener's granulomatosis). *B.M.J.*, 265–270.
9. Anderson, G., Coles. E., Crabe, M. *et al.* (1992) Wegener's granuloma: A series of 265 British cases seen between 1975 and 1985. A report by a sub-committee of the British Thoracic Society Research Committee. *Q. J. Med.*, **83**, 427–38.
10. Leavitt, R.Y., Fauci, A.S., Bloch, D.A. *et al.* (1990) The American College of Rheumatology 1990 Criteria for the Classification of Wegener's Granulomatosis. *Arthritis Rheum.*, **33**, 1101–6.
11. Jenette, J.C., Falk, R.J., Andrassy, K. *et al.* (1994) Nomenclature of systemic vasculitides. Proposal of an international conference. *Arthritis Rheum.*, **37**, 187–92.
12. Nölle, B., Specks, U., Lüdemann, J. *et al.* (1989) Anticytoplasmic autoantibodies. Their immunodiagnostic value in Wegener's granulomatosis. *Ann. Int. Med.*, **111**, 28–40.
13. Deguchi, Y., Shibata, N. and Kishimoto, S. (1990) Enhanced expression of the tumur necrosis factor/cachectin gene in peripheral blood mononuclear cells from patients with systemic vasculitis. *Clin. Exp. Immunol.*, **81**, 311–14.
14. Reinhold-Keller, E., Kekow, J., Schnabel, A. *et al.* (1994) Influence of disease manifestation and cANCA titer on the response to pulse cyclophosphamide therapy in Wegener's granulomatosis. *Arthritis Rheum.*, **37**, 919–24.
15. Gross, W.L., Hauschild, S. and Mistry, N. (1993) The classical relevance of ANCA in vasculitis. *Clin. Exp. Immunol.*, **93** (Suppl. 1), 7–11.

16. Peter, H.H., Metzger, D., Rump, A. and Rother, E. (1993) ANCA in diseases other than systemic vasculitis. *Clin. Exp. Immunol.*, **93**(Suppl. 1), 12–14.

17. Ulmer, M., Rautmann, A. and Gross, W.L. (1992) Immunodiagnostic aspects of autoantibodies against myeloperoxidase. *Clin. Nephrol.*, **37**, 161–8.

18. Boudes, P.J. (1989) Acquired chronic granulomatous disease and Wegener's granulomatosis (letter). *Br. J. Rheumatol.*, **28**, 361–2.

19. Travis, W.D., Hoffman, G.S., Leavitt, R.Y. *et al.* (1991) Surgical pathology of the lung in Wegener's granulomatosis. *Am. J. Surg. Pathol.*, **15**, 315–33.

20. Wegener, F. (1967) Die pneumogene allgemeine Granulomatose – sog. Wegener'sche Granulomatose, *Lehrbuch der Speziellen Pathologischen Anatomie* (eds M. Staemmler and E. Kaufmann), De Gruyter, Berlin, pp 225.

21. Klinger, H. (1931) Grenzformen der Panarteriitis nodosa. *Z. Path.*, **42**, 455.

22. Fienberg, R. (1955) Pathergic granulomatosis. *Am. J. Med.*, **19**, 829.

23. Mark, E.J., Matsubara, O., Tau-Lin, N.S. and Fienberg, R. (1988) The pulmonary biopsy in the early diagnosis of Wegener's (pathergic) granulomatosis. *Hum. Pathol.*, **19**, 1065–71.

24. Altman, H.W. and Schiche, H. (1959) Ein Beitrag zur Histologie und zur Einordnung der Wegener'schen Granulomatose. *Beitr. Pathol. Anat.*, **121**, 211.

25. Letterer, E. (1958) Die Morphologie der allergischen Gefäbreaktionen. Ber Zusammenkunft Dtsch *Ophthalmol. Ges.*, **61**, 35.

26. Wegener, F. (1968) About the so-called Wegener's granulomatosis with special reference to the generalized vascular lesions, Morgagni, p. 5.

27. Zollinger, H.U. (1966) Die Herdglomerulitis beim Wegener-Syndrom in Niere und ableitenden Harnwegen, *Spezielle Pathologische Anatomie* (eds W. Doerr and E. Uhlinger), Springer, Berlin, p. 386.

28. Schmitt W.H., Heesen, C. and Csernok, E. (1992) Elevated serum levels of soluble interleukin-2-receptor (sIL-2R) in Wegener's granulomatosis (WG): asssociation with disease activity. *Arthritis Rheum.*, **35**, 108–10.

29. Fauci, A.S., Barton, H., Katz, P. *et al.* (1983) Wegener's granulomatosis: prospective clinical and therapeutic experience with 85 patients for 21 years. *Ann. Intern. Med.*, **98**, 76–85.

30. Brouwer, E., Cohen Tervaert, J.W., Weening, J.J. *et al.* (1991) Immunohistology of renal biopsies in Wegener's granulomatosis (WG): clues to its pathogenesis. *Kidney Int.*, **39**, 1055.

31. Kekow, J., Szymkowiak, Ch. and Gross, W.L. (1994) Involvement of cytokines in granuloma formation within primary systemic vasculitis, in *Cytokines: Basic Principles and Clinical Applications* (ed. S. Romagnani), Raven Press, New York.

32. Mathieson, P.W., Lockwood, C.M. and Oliveira, D.B.G. (1992) T and B cell responses to neutrophil cytoplasmic antigens in systemic vasculitis. *Clin. Immunol. Immunopathol.*, **63**, 135–41.

33. Kallenberg, C.G.M., Cohen Tervaert, J.W., van der Woude, F.J. *et al.* (1991) Autoimmunity to lysosomal enzymes: new clues to vasculitis and glomerulonephritis. *Immunol. Today*, **12**, 61–4.

34. Gross, W.L., Csernok, E. and Schmitt, W.H. (1991) Antineutrophil cytoplasmic autoantibodies: immunobiological aspects. *Klin. Wochenschr.*, **69–73**, 558–66.

35. Csernok, E., Lüdemann, J. and Gross, W.L. (1990) Ultrastructural localization of proteinase 3, the target antigen of anti-cytoplasmic antibodies circulating in Wegener's granulomatosis. *Am. J. Pathol.*, **137**, 1113–20.

36. Gross, W.L. Schmitt, W.H. and Csernok, E. (1993) ANCA and associated

diseases: immunodiagnostic and pathogenic aspects. *Clin. Exp. Immunol.*, **91**, 1–12.

37. Gross, W.L. Csernok, E. and Flesch, B.K. (1993) 'Classic' anti-neutrophil cytoplasmic antibodies 'Wegener's autoantigen' and their immunopathogenic role in Wegener's granulomatosis. *J. Autoimmun.*, **6**, 171–84.

38. Csernok, E., Schmitt, W.H., Martin, E. *et al.* (1993) Membrane surface proteinase 3 expression and intracytoplasmatic immunoglobulin on neutrophils from patients with ANCA-associated vasculitides, in *ANCA-Associated Vasculitides* (ed. W.L. Gross), Plenum Press, New York and London, pp. 45–50.

39. Csernok, E., Ernst, M. Schmitt, W.H. *et al.* (1991) Translocation of PR-3 on the cell surface of neutrophils: association of disease activity in Wegener's granulomatosis. *Arthritis Rheum.*, **34**, 79.

40. Müller, G.A., Marrovic-Lipkowski, J. and Müller, C.A. (1991) Intercellular adhesion molecule-1 expression in human kidneys with glomerulonephritis. *Clin. Nephrol.*, **36**, 203–8.

41. Hauschild, S., Schmitt, W.H. and Kekow, J. (1992) Hohe Serumspiegel von ICAM-1 bei der aktiven generalisierten Wegener'schen Granulomatose. *Immun. Infekt.*, **20**, 84–5.

42. Keogan, M.T., Rifkin, I., Ronda, N. *et al.* (1993) Antineutrophil cytoplasm antibodies (ANCA) increase neutrophil adhesion to cultured human endothelium, in *ANCA Associated Vasculitides: Immunological and Clinical Aspects* (ed. W.L. Gross), Plenum Press, New York and London, pp. 115–19.

43. Parida, S., Adu, D., Taylor, C.M. *et al* (1991) Increased neutrophil adhesion to resting endothelium to vasculitis. *Nephrol. Dial. Transplant*, **6**, 899.

44. Falk, R.J., Hogan, S., Carey, T.S. *et al.* (1990) Clinical course of anti-neutrophil cytoplasmic autoantibody-associated glomerulonephritis and systemic vasculitis. *Ann. Intern. Med.*, **113**, 656–63.

45. Lai, K.N. and Lockwood, C.M. (1991) The effect of anti-neutrophil cytoplasm autoantibodies on the signal transduction in human neutrophils. *Clin. Exp. Immunol.*, **85**, 396–401.

46. Ewert, B.H., Jenette, J.C. and Falk, R.J. (1991) The pathogenetic role of antineutrophil cytoplasmic autoantibodies. *Am. J. Kidney Dis.*, **18**, 188–95.

47. Savage, C., Gaskin, G., Pusey, C.D. and Pearson, J.D. (1993) Myeloperoxidase binds to vascular endothelial cells, is recognized by ANCA and can enhance complement dependent cytotoxicity, in *ANCA-Associated Vasculitides* (ed. W.L. Gross), Plenum Press, New York, pp. 121–3.

48. Ewert, B.H., Becker, M., Jenette, C. *et al.* (1994) Antimyeloperoxidase antibodies stimulate neutrophils to adhere to human umbilical vein endothelial cells, in *Abstractbook for 3rd International Workshop on ANCA*, held in Washington, USA, 1990.

49. Savage, C., Gaskin, G., Pusey, C.D. and Pearson, J.D. (1993) Myeloperoxidase binds to vascular endothelial cells, is recognized by ANCA and can enhance complement dependent cytotoxicity, in *ANCA-Associated Vasculitides: Immunolgical and Clincal Aspects* (ed. W.L. Gross), Plenum Press, New York and London, pp. 121–3.

50. Varagunam, M., Adu, D., Taylor, C.M. *et al.* (1993) Endothelium, myeloperoxidase, interaction in vasculitides, in *ANCA-Associated Vasculitides: Immunological and Clinical Aspects* (ed. W.L. Gross), Plenum Press, New York and London, pp. 419–22.

51. Kiser, M.A., Jenette, J.C. and Falk, R.J. (1993) Vascular permeability changes induced by antibodies to myeloperoxidase, in *ANCA-Associated Vasculitides:*

*Immunolgical and Clinical Aspects* (ed. W.L. Gross), Plenum Press, New York and London, pp. 125–7.

52. Donald, K.J., Edwards, R.L. and McEvoy, J.D.S. (1976) An ultrastructural study of the pathogenesis of tissue injury in limited Wegener's granulomatosis. *Pathol.*, **8**, 161–9.

53. Mayet, W.J., Hermann, E., Csernok, E. *et al.* (1993) In vitro interactions of cANCA (antibodies to proteinase 3) with human endothelial cells, in *ANCA Associated Vasculitides: Immunological and Clinical Aspects* (ed. W.L. Gross), Plenum Press, New York and London, pp. 109–13.

54. Braun, M.G., Csernok, E., Gross, W.L. *et al.* (1991) Proteinase 3, the target antigen of anticytoplasmic antibodies circulating in Wegener's granulomatosis. *Am. J. Pathol.*, **139**, 831–8.

55. Gross, W.L. (1994) New developments in the treatment of systemic vasculitis. *Curr. Opin. Rheumatol.*, **6**, 11–19.

56. Hoffman, G.S., Leavitt, R.Y., Kerr, G.S. *et al.* (1992) The treatment of Wegener's granulomatosis with glucocorticoids and methotrexate. *Arthritis Rheum.*, **35**, 1322–9.

57. Reinhold-Keller, E., Beigel, A., Duncker, G. *et al.* (1993) Trimethoprim-sulfamethaxozole (T/S) in the long-term treatment of Wegener's granulomatosis (WG). *Clin. Exp. Immunol.*, **93**(Suppl. 1), **38** (abstract).

58. DeRemee, R.A. (1993) Wegener's granulomatosis, in *Sarcoidosis and other granulomatous disorders* (ed. D.J. James), Marcel Dekker Inc., New York, pp. 657–80.

59. Hoffman, G.S., Leavitt, R.Y., Fleisher, T.A. *et al* (1990) Treatment of Wegener's granulomatosis with intermittent high-dose cyclophosphamide. *Am. J. Med.*, **89**, 403–10.

60. Haubitz, M., Frei, U., Rother, U. *et al.* (1991) Cyclophosphamide pulse therapy in Wegener's granulomatosis. *Nephrol. Dial. Transplant*, **6**, 531–5.

61. Falk, R.J., Terrel, R.S., Charles, L.A. *et al.* (1990) Anti-neutrophil cytoplasmic autoantibodies induce neutrophils to degranulate and produce oxygen radicals in vitro. *Proc. Nat. Acad. Sci. USA*, **87**, 4115–19.

62. Reinhold-Keller, E., Handrock, K., Mertens, J. *et al.* (1994) Rezidive bei Patienten mit einer ehemals generalisierten Wegener'schen Granulomatose unter Cotrim vs. ohne Therapie. *Med. Klin.*, **89**(Suppl. 1), 487–9.

63. Schollmeyer, P. and Grotz, W. (1990) Cyclosporin in the treatment of Wegener's granulomatosis (WG) and related diseases. APMIS (*Copenhagen*), **19**(Suppl.), 54–5.

64. Allen, N.B. Caldwell, D.S., Rice, J.R. and McCullum, R.M. (1993) Cyclosporin A therapy for Wegener's granulomatosis, in *ANCA-Associated Vasculitides* (ed. W.L. Gross), Plenum Press, London, 473–6.

65. Schmitt, W.H., Haubitz, M., Mistry, N. *et al.* (1993) Renal transplantation in Wegener's granulomatosis (Letter). *Lancet*, **342**, 860.

66. Jayne, D.R.W. and Lockwood, C.M. (1993) Pooled immunoglobulin in the management of systemic vasculitis, in *ANCA-Associated Vasculitides* (ed. W.L. Gross), Plenum Press, London, pp. 469–72.

67. Tuso, P., Moudgil, A., Hay, J. *et al.* (1992) Treatment of antineutrophil cytoplasmic autoantibody-positive systemic vasculitis and glomerulonephritis with pooled intravenous gammaglobulin. *Am. J. Kidney Dis.*, **20**, 504–8.

68. Richter, C., Schnabel, A., Csernok, E. *et al.* (1993) Treatment of Wegener's granulomatosis with intravenous immunoglobulin, in *ANCA-Associated Vasculitides* (ed. W.L. Gross), Plenum Press, London, pp. 487–9.

69. Mathieson, P.W., Cobbold, S.P., Hale, G. *et al.* (1990) Monoclonal antibody therapy in systemic vasculitis. *N. Engl. J. Med.*, **323**, 250–4.
70. Lockwood, C.M., Thiru, S., Isaacs, J.D. *et al.* (1993) Long-term remission of intractable systemic vasculitis with monoclonal antibody therapy. *Lancet*, **341**, 1620–2.
71. Van der Lubbe, P.A., Miltenburg, A.M. and Breedveld, F.C. (1991) Anti-CD4 monoclonal antibody for relapsing polychondritis. *Lancet*, **337**, 1349.
72. Choy, E.H.S., Chikaza, J.C., Kingsley, G.H. *et al.* (1991) Chimeric anti-CD4 monoclonal antibody for relapsing polychondritis. *Lancet*, **338**, 450.

# 12

# Giant cell arteritis

*B.-Å. Bengtsson*

## HISTORICAL BACKGROUND

In the Tadikivat of Ali Ibn Isa [1], from the tenth century, removal of the temporal artery was recommended in order to treat 'not only migraine and headache in those patients that are subject to chronic eye disease, but also acute, sharp, catarrhal affections including those showing heat and inflammation of the temporal muscles. These diseased conditions may terminate in loss of sight'.

The modern history of giant cell arteritis started over 100 years ago, when Hutchinson, a British surgeon, observed an 80-year-old patient with 'red streaks on his head' which were painful and prevented him from wearing his hat [2]. He named the disease 'arteritis of the aged'. Biopsy findings showing granulomatous inflammation were first described by Horton, Magath and Brown [3]. The name giant cell arteritis (GCA) was first used by Gilmour [4], a pathologist, who discovered that temporal arteritis (TA) was a manifestation of a generalized vascular disease.

Whereas GCA was considered a rare disease 50 years ago [3], it is now known to be an important and significant cause of morbidity in elderly people. As the clinical manifestations of the disease have been recognized, GCA has come to be one of the first rather than one of the last diagnoses considered in elderly people who display symptoms such as proximal muscle pain and stiffness, headache, deteriorating general condition, fever of unknown etiology and an unexplained increase in the erythrocyte sedimentation rate (ESR). The etiology of GCA is still unknown, although an autoimmune pathogenesis seems probable. The

*The Vasculitides.* Edited by B.M. Ansell, P.A. Bacon, J.T. Lie and H. Yazici.
Published in 1996 by Chapman & Hall, London. ISBN 978-0-412-64140-4.

predominance in the white population, familial aggregation and association with the HLA-DR4 antigen also favour a genetic predisposition.

## EPIDEMIOLOGY

GCA affects the white population almost exclusively. Patients with GCA from other ethnic backgrounds have rarely been reported. Most reports originate from northern Europe [5–7] and parts of the northern USA [8–10] with populations of the same ethnic background. GCA affects elderly people and is seldom diagnosed below the age of 50.

A population-based study from Olmsted County, Minnesota, USA, between 1950 and 1985, focused on patients showing clinical signs of TA [10]. Temporal artery biopsy revealed arteritis in 88 of 94 diagnosed patients. The age and sex adjusted incidence rate per 100 000 inhabitants aged 50 years or more was 17.0. The age adjusted rates were approximately three times higher in women than in men. Furthermore, a significant increase in total incidence rate was observed over the period of the study.

An epidemiological study of GCA was performed in Gothenburg, Sweden, covering the years 1977 to 1986. [7]. Only biopsy-proven cases were included in the study. Potential cases of GCA were identified using the register of the Department of Pathology which contains all temporal artery biopsies performed in the Gothenburg area. A total of 2307 temporal artery biopsies were performed during the 10-year period. The diagnosis of GCA was made in 284 of them (12.5%); 220 were women (77%) and 64 were men (23%), and their mean age at diagnosis was 72 years, with a range of 39–90 years. In the group of patients aged 50 years and older, the average incidence was 18.3, values for women and men were 25 and 9.4 per 100 000 inhabitants, respectively. The incidence figures for biopsy-proven GCA are similar to those reported both from Ribe in Denmark [6] and Olmsted County [10].

The incidence rate for women in the Gothenburg study increased significantly between 1977 and 1986. In males there was a tendency towards a declining incidence rate, although this was not significant. The incidence rate observed during the 1980s for women was almost twice that which we observed in Gothenburg 10 years earlier [5]. The cause of this increase in the incidence rate and the sex difference is obscure. Increased awareness and improved diagnostic skill cannot be the only explanation because comparable changes in both sexes would then be expected. Furthermore, the proportion of biopsies revealing GCA did not change significantly, neither in the study in Olmsted County nor in the study in Gothenburg. The biopsy rate increased in both studies, probably reflecting an increasing number of patients presenting with symptoms compatible with GCA. Thus, taking our data as well as the results obtained from Olmsted County into consideration, it appears that there is an increase in the incidence of GCA among women.

That the abnormal events leading to arteritic lesions are much more common than hitherto believed, and are not always manifested in life as a disease recognized by us as GCA, cannot be excluded. The findings of Östberg [11] are provocative in this respect. In an investigation comprising 889 autopsies performed in Malmö, Sweden, during a period of six months, she found a 1.7% incidence of arteritic lesions in the temporal artery or aorta. In a retrospective analysis of more than 20 000 autopsies, 79 (0.4%) were found to have GCA involving the aorta. The conclusion was that GCA is far more common than is indicated by the clinical diagnosis.

ETIOLOGY

There has been much conjecture about the cause of GCA. An infectious process is suggested by the fever, leukocytosis, elevated ESR and generalized illness of the patients, but no organisms have ever been isolated and no inclusions or particles have been found in ultrastructural studies of biopsy specimens from patients with GCA. Hepatitis B was suggested as a trigger for GCA by Bacon, Doherty and Zuckerman [12], but other studies have not confirmed this observation. Even if the etiology is still obscure, there is an increasing number of data suggesting both genetic predisposition and autoimmune pathogenesis.

The predominance in the white population, familial aggregation [13] and the association with HLA-DR4 antigen [14] favor genetic predisposition of the disease. Cid *et al.* [15] observed that the DR4 antigen was observed significantly more frequent in biopsy-proven patients who displayed symptoms of polymyalgia rheumatica (PMR) than in those without PMR. Immunogenetic analysis has demonstrated that GCA and rheumatoid arthritis (RA) are associated with two different domains of the HLA-DR4 molecule. Recently, Weyand *et al.* [16] showed that the distribution of HLA-DRB1 alleles in PMR resembles that found in GCA. Their data support that similar pathogenic mechanisms involving the HLA-DRB1 gene are functional in PMR and GCA and that they are distinct from the contribution of the HLA-DRB1 polymorphism in the pathogenesis of RA.

The vascular damage in GCA could be a result of either deposition of circulating immunocomplexes, with subsequent activation of the complement system, or an immune response directed against the vascular wall itself or its constituents. Deposition in the arterial wall of immunoglobulin, as well as increased levels of circulating immune complexes, have been reported in some studies [17], but not in others [18]. The striking localization of the inflammation around the internal elastic membrane in the vessel wall observed by Kimmerstiel, Gilmour and Hodges in 1952 [19] has given rise to a hypothesis of an autoimmune attack directed toward the elastic membrane. With the development of monoclonal antibodies, the functional subpopulations of the lymphoid cells can be studied *in situ*. Andersson *et al.* [20] found that infiltration of T lymphocytes and macrophages dominated in the arteritic lesions in GCA. The T helper/inducer (CD4) subset dominated over the T suppressor/cytotoxic (CD8) subset, and only a low number of B lymphocytes and natural killer cells were detected. Similar results have been observed by Banks *et al.* [21] and Cid *et al.* [22]. Further studies by Andersson *et al.* [23] showed that 16–42% of all T lymphocytes in the arteritic lesions expressed HLA-DR antigen, suggesting that the T lymphocytes were immunologically activated. In contrast, only 6% of T lymphocytes in peripheral blood expressed HLA-DR antigen, which is the same as that reported for healthy persons. This observation suggests that T lymphocytes are activated locally in the arterial wall. Andersson also demonstrated that smooth muscle cells did not express HLA-DR antigen. Further support for local activation of the T cells was the finding of interleukin-2 receptors on the lymphocytes. *In vitro* experiments have indicated that corticosteroids suppress interleukin-2 expression. Cid *et al.* [22] also demonstrated the existence of interdigitating reticulum cells in the arteritic lesions. These cells have previously been found in lesions in which local stimulus for immune response is suspected, but not in degenerative conditions or processes mediated by immunocomplexes.

Present data support the notion that T cell activation occurs in the arteritic lesion rather than systemically. This could represent an activation and expansion of one or a few antigen-specific clones. Alternatively, polyclonal activation may be induced by multiple antigenic epitopes or perhaps by an immunologically non-specific polyclonal mitogen. Recently, it was found that lymphocytes in the GCA lesions were of polyclonal origin and expressed intergrin-type adhesion receptors supporting the hypothesis that GCA involves an inflammatory response, during which polyclonal T cells adhere to arterial tissue components and accumulate in the developing lesions [24].

## MORPHOLOGY

Recently, Nordborg, Bengtsson and Nordborg [25] reported in great detail the histopathological findings in temporal artery biopsy specimens from patients with a clinical diagnosis of PMR and TA. In patients with TA, two different patterns with inflammatory reactions were encountered. A focal foreign body giant cell reaction affecting only part of the circumference was seen in some specimens, which were otherwise markedly atrophic, mainly affecting the medial smooth muscle with large calcifications. The giant cells apparently attacked and incorporated calcified portions of the internal elastic membrane. In other specimens quite a different picture was seen. Macrophages with some admixture of lymphocytes and plasma cells invaded the vessel wall, generally in the entire circumference of the vessel. Langhans' giant cells were often found in great abundance, but they did not attack or include calcified elastic fragments. They were generally located at the outer border of the internal elastic membrane. The focality of the foreign body giant cell reaction and its striking spatial relationship with lymphocytes indicate an early, more specific phase of inflammation which is highly suggestive of an interaction between those two types of cells. The diffuse macrophage attack, on the other hand, is compatible with a later, non-specific inflammatory stage.

Analysis of the non-inflamed vascular segments revealed a significantly greater atrophy of the medial smooth muscle in the PMR group than in corresponding control subjects. In addition, calcifications of the internal elastic membrane, when encountered, were significantly larger than in the controls. These observations with morphologic similarities in terms of arterial wall atrophy and calcification, as well as inflammatory reaction, indicate that PMR and TA represent different degrees or stages of the same disease. The arterial disease in PMR and TA may initially be due to a metabolic disturbance in the arterial wall, which leads to medial smooth muscle atrophy as well as to the degeneration and dystrophic calcification of the internal elastic membrane. PMR and TA patients are prone to foreign body giant cell reactions to the degenerated, calcified internal elastic membrane. Foreign body giant cells, possibly of smooth muscle cell origin, present antigens to lymphocytes, which produce cytokines. A massive and diffuse macrophage invasion ensues and causes major damage to the arterial wall. Secondary giant cells (Langhans' type), which do not attack calcified internal elastic membrane, are formed by the fusions of macrophages.

## CLINICAL PRESENTATION

The modern history of GCA runs along two paths: that of TA and that of PMR. During the 1950s, the connection between TA and PMR had been established [26–28]. The terms PMR and TA are used in this review to define a typical clinical syndrome, regardless of the temporal artery biopsy findings. The term 'giant cell arteritis' is used by us to define the group as a whole. The typical clinical symptoms in GCA include headache, scalp tenderness, jaw claudication, PMR and systemic symptoms. GCA can cause ischemic lesion in different organs, including the heart and the brain, giving rise to heart infarction and to a great variety of neurological defects. Brain infarctions caused by arteritic involvement of the internal carotid or the vertebrobasilar arterial system are well known.

PMR is a syndrome characterized by ache and morning stiffness in the neck, shoulder, upper arms, hipgirdle and thighs, Usually, the ache and stiffness are symmetric. In view of the clinical similarities between biopsy-proven and biopsy-negative patients with PMR, many authors favor the concept that PMR is an expression of an underlying GCA. Morphological similarities further support this hypothesis [29–32]. Mild and transient synovitis is seen in both biopsy-proven and biopsy-negative patients with PMR, but joint destruction does not develop. It is questionable, however, if this synovitis has anything to do with the clinical symptoms of the patient. Because of the synovitis, the relationship between PMR and RA has been discussed, but the percentage of patients with PMR who eventually develop seropositive RA is small, not higher than one could expect in a normal population.

## OPHTHALMIC COMPLICATIONS

By compromising the blood supply to the prelaminar and lamina cribosa regions, arteritic lesions in the short posterior ciliary arteries or in the ophthalmic artery result in an infarction of the optic nerve head (anterior ischemic optic neuropathy [33]). This is the most common cause of blindness in GCA, and accounts for 80–90% of all cases of permanent blindness. If the arteritic process strikes the central retinal artery, the retina is damaged. Besides these complications, there is a diverse spectrum of seldom-appreciated neuro-ophthalmic complications in GCA, such as transient ischemic attacks, opthalmopareses and cortical blindness.

The overall incidence of visual loss observed in older series of GCA was as high as 30–60% [34, 35]. The remarkable efficacy of steroid treatment in preventing such catastrophies appeared after the advent of cortisone therapy, and in modern series, much lower figures have been reported. Today, around 10% of patients with biopsy-proven GCA suffer permanent loss of vision [5]. Rarely, visual acuity has been shown to improve after institution of steroid treatment.

## DIAGNOSIS

Biopsy of the temporal artery is still the only way to make a firm diagnosis. The histological picture is characteristic and rarely offers any problems to an experienced pathologist. Opinions differ on the usefulness of performing a temporal artery biopsy, because a negative biopsy cannot exclude arteritis [36]. In our opinion, patients who display symptoms of PMR and have a negative temporal artery biopsy require further investigations in order to exclude diseases that could mimic PMR. The need for such investigations will be obviated if the biopsy is positive, and long-term corticosteroid treatment can be started if histological proof is obtained [13]. Furthermore, a higher yield of positive biopsies can be obtained by increasing the length of the biopsies and the number of sections that are being examined [Nordborg, personal communication].

## TREATMENT

The aim of treatment of GCA is to relieve the patient of troublesome symptoms and prevent ischemic catastrophies. The use of corticosteroids in the treatment of biopsy-verified GCA is not controversial. When a negative biopsy is obtained, some authors are reluctant to prescribe corticosteroids to patients with a pure 'PMR' syndrome. [37]. The main controversy today, however, concerns the initial dose of corticosteroids to be used and the length of treatment required. Some researchers advocate one month of high initial doses of steroids for biopsy-verified patients and low initial doses for biopsy-negative patients. Others, including the author, initiate corticosteroid treatment regardless of the biopsy findings, with a single morning dose of an average 30–40 mg of prednisolone followed by weekly decrements. A single morning dose reduces such adverse effects as corticosteroid-induced suppression of the adrenal glands.

The mean duration of treatment varies in different series. By using a life-table method, Andersson [38] estimated the mean duration of treatment to be five years. Similarly to Chuang *et al.* [9], Andersson observed a shorter duration of treatment for men than for women. Corticosteroids relieved the symptoms of GCA, but there was no evidence to suggest that they shortened the duration of the disease. No laboratory test can predict when therapy should be discontinued. The author gradually decreases the doses of prednisolone to about 2.5 mg/day, and then, if no clinical symptoms have recurred after about one month, discontinues therapy. Using this method, 50% of the patients will not need corticosteroids again, while the rest will usually relapse during the following month and need prolonged treatment [39]. If the relapse is detected early, it is often enough to start treatment with prednisolone 10–15 mg/day and reduce gradually to 2.5–5 mg/day within one to two months.

## ADVERSE EFFECTS

The most commonly reported adverse effects of corticosteroids are vertebral crush fractures, hip fractures, diabetes mellitus and cataracts. The prevalence of these complications vary widely in different studies. In addition, these conditions are common in elderly people. Therefore, it is difficult to attribute such events to the corticosteroid treatment. In

Gothenburg, the incidence of diabetes mellitus and hip fractures among patients with GCA was about the same as in the general population at corresponding ages [39]. In a retrospective study, we evaluated the bone mineral content of the calcaneus and the signs of osteoporosis on spinal X-ray in 26 patients with GCA who were receiving prednisolone for an average period of five years [40]. We did not detect any additional osteoporosis that could be attributed to the cortisone treatment compared with the general population of the same age. In another prospective study of the bone mineral content of the third lumbar vertebra using dual photon absorptiometry, no significant reduction of bone mineral content was found during the first two years of treatment [41].

## PROGNOSIS

Although GCA affects the arteries, the statistical survival rate of patients with GCA has been shown to be the same as that of the general population [8, 40]. These series have been comparatively small, however, and possibly not large enough to detect minor differences in survival rates. In the study from Gothenburg of 284 biopsy-proven patients with GCA, Nordborg and Bengtsson [42] observed an increased number of deaths from vascular causes (cerebrovascular, myocardial infarction, cardiac failure and rupture of dissecting aneurysm of the aorta) during the first four months after diagnosis. Thereafter, the observed number of deaths equals that of the general population. Furthermore, in the majority of patients who died within the first four months, treatment had probably been insufficient to suppress the disease activity either because of a too low dose of corticosteroids or premature discontinuation of treatment, or simply because there was too little time to suppress disease activity before death occurred. Autopsy was performed in some of these patients, and the microscopic findings showed arteritic changes affecting various parts of the arterial tree. The fatal outcome in patients with GCA has been poorly recognized, and previous reports have generally been case reports or autopsy studies. The autopsy rate is low in many countries, and microscopic examinations of arteries in patients dying from vascular disorders are not routine, especially not in elderly people. Hence, GCA may be concealed among the cases currently diagnosed as ischemic catastrophies due to arteriosclerosis.

## REFERENCES

1. Ali Ibn Isa (1904) *Erinnerungsbuch für Augenärzte aus arabischen Handschriften, übersetzt und erläutert von J Hirschberg und J Lippert*, Verlag von Veit, Leipzig.
2. Hutchinson, J. (1890) Diseases of the arteries. *Arch. Surg. (London)*, **1**, 323.

3. Horton, B.T., Magath, T.B. and Brown, G.E. (1937) An undescribed form of arteritis of the temporal vessels. *Proc. Staff Meet. Mayo Clin.*, **7**, 700.

4. Gilmour, J.R. (1941) Giant-cell chronic arteritis. *J. Path.*, **53**, 263.

5. Bengtsson, B.-Å. and Malmvall, B.E. (1981) The epidemiology of giant cell arteritis and polymyalgia rheumatica – incidences of different clinical presentations and eye complications. *Arthritis Rheum.*, **24**, 899–904.

6. Boesen, P. and Sörensen, S.F. (1987) Giant cell arteritis, temporal arteritis and polymyalgia rheumatica in a Danish County. A prospective investigation 1982–1985. *Arthritis Rheum.*, **30**, 294–9.

7. Nordborg, E. and Bengtsson, B.-Å. (1990) Epidemiology of biopsy-proven giant cell arteritis (GCA). *J. Int. Med.*, **227**, 233–6.

8. Huston, K.A., Hunder, G.G., Lie, J.T. *et al.* (1978) Temporal arteritis. A 25-year epidemiologic, clinical and pathologic study. *Ann. Int. Med.*, **88**, 162–7.

9. Chuang, T.Y., Hunder, G.G., Ilstrup, D.M. and Kurland, L.T. (1982) Polymyalgia rheumatica. A 10-year epidemiologic and clinical study. *Ann. Int. Med.*, **97**, 672–80.

10. Machado, E.B.V., Michet, C.J., Ballard, D.J. *et al.* (1988) Trends in incidence and clinical presentation of temporal arteritis in Olmsted County, Minnesota 1950–1985. *Arthritis Rheum.*, **31**, 745–9.

11. Östberg, G. (1973) On arteritis with special reference to polymyalgia arterica. *Acta Path. Microbiol. Scand.*, **237** (Sect A, Suppl.).

12. Bacon, P.A., Doherty, S.M. and Zuckerman, A.J. (1975) Hepatitis B antibody in polymyalgia rheumatica. *Lancet*, **2**, 476–8.

13. Bengtsson, B.-Å. and Malmvall, B.E. (1982) Giant cell arteritis. Incidence of giant cell arteritis. *Acta Med. Scand.*, **658** (Suppl.), 15–17.

14. Barrier, J., Bignon, J.-D., Souliou, J.-P. and Grolleau, J. (1981) Increased prevalence of HLA-DR4 in giant-cell arteritis. *New Engl. J. Med.*, **305**, 104–5.

15. Cid, M.C., Ercilla, G., Vilaseca, J. *et al.* (1988) Polymyalgia rheumatica: a syndrome associated with HLA-DR4 antigen. *Arthritis Rheum.*, **31**, 678–82.

16. Weyand, C.M., Hunder, N.N.H., Hicock, K.C. *et al.* (1994) HLA-DRB1 alleles in polymyalgia rheumatica, giant cell arteritis, and rheumatoid arthritis. *Arthritis Rheum.*, **37**, 514–20.

17. Papaioannou, C.C., Gupta, R.C., Hunder, G.G. and McDuffie, F.C. (1980) Circulating immune complexes in giant cell arteritis and polymyalgia rheumatica. *Arthritis Rheum.*, **23**, 1021–5.

18. Malmvall, B.E., Bengtsson, B.-Å., Nilsson, L.-Å. and Bjursten, L.-M. (1981) Immune complexes, rheumatoid factors, and cellular immunological parameters in patients with giant cell arteritis. *Ann. Rheum. Dis.*, **40**, 276–80.

19. Kimmerstiel, P., Gilmour, M.T. and Hodges, H.H. (1952) Degeneration of elastic fibers in granulomatous giant cell arteritis (temporal arteritis). *Arch. Pathol.*, **54**, 157–68.

20. Andersson, R., Jonsson, R., Tarkowski, A. *et al.* (1987) T-cell subsets and expression of immunological activation markers in the arterial walls of patients with giant cell arteritis. *Ann. Rheum. Dis.*, **46**, 915–23.

21. Banks, P.M., Cohen, M.D., Ginsburg, W.W. and Hunder, G.G. (1983) Immunohistologic and cytochemical studies of temporal arteritis. *Arthritis Rheum.*, **26**, 1201–7.

22. Cid, M.C., Campo, E., Ercilla, G. *et al.* (1989) Immunohistochemical analysis of lymphoid and macrophage cell subsets and their immunologic activation markers in temporal arteritis. Influence of corticosteroid treatment. *Arthritis Rheum.*, **32**, 884–93.

23. Andersson, R., Hansson, G.K., Söderström, T. *et al.* (1988) HLA-DR expres-

sion in the vascular lesion and circulating T lymphocytes of patients with giant cell arteritis. *Clin. Exp. Immunol.*, **73**, 82–7.

24. Schaufelberger, C., Stemme, S., Andersson, R. and Hansson, G.K. (1993) T lymphocytes in giant cell arteritic lesions are polyclonal cells expressing αβ type antigen receptors and VLA-1 integrin receptors. *Clin. Exp. Immunol.*, **91**, 421–8.

25. Nordborg, E., Bengtsson, B.-Å. and Nordborg, C. (1991) Temporal artery morphology and morphometry in giant cell arteritis. *Acta Pathol. Microbiol. Immunol. Scand.*, **99**, 1013–23.

26. Porsman, V.A. (1951) *Proc. II Congr. Europ. Rheum.*, (editorial scientia, Barcelona) p. 479.

27. Paulley, J.W. (1956) Anarthritic rheumatoid disease. *Lancet*, **2**, 946.

28. Paulley, J.W. and Hughes, J.P. (1960) Giant-cell arteritis, or arteritis of the aged. *Br. Med. J.*, **2**, 1562.

29. Bengtsson, B-Å. and Malmvall, B.E. (1981) The epidemiology of giant cell arteritis including temporal arteritis and polymyalgia rheumatica. *Arthritis Rheum.*, **24**, 899–904.

30. Fauchald, P., Rygvold, O. and Oystese, B. (1972) Temporal arteritis and polymyalgia rheumatica. Clinical and biopsy findings. *Ann. Int. Med.*, **77**, 845–52.

31. Hamrin, B. (1992) Polymyalgia arterica. *Acta Med. Scand.*, **192** (Suppl.), 86–99.

32. Soelberg-Sörensen, P. and Lorentzen, I. (1977) Giant cell arteritis, temporal arteritis and polymyalgia rheumatica. *Acta Med. Scand.*, **201**, 207–13.

33. Hayreh, S.S. (1991) Ophthalmic features of giant cell arteritis. *Baillière's Clin. Rheumat.*, **5**(3), 431–69.

34. Birkhead, N.C., Wagener, H.P. and Shick, R.M. (1957) Treatment of temporal arteritis with adrenal corticosteroids. Results in fifty-five cases in which lesion was proved at biopsy. *J. Am. Med. Ass.*, **163**, 821.

35. Hollenhorst, R.W., Brown, J.R.,Wagener, H.P. and Schick, R.M. (1960) Neurologic aspects of temporal arteritis. *Neurology*, **10**, 490.

36. Hall, S., Lie, J.T. and Kierland, L.T. (1983) The therapeutic impact of temporal artery biopsy. *Lancet*, **ii**, 12217–20.

37. Hunder, G.G. and Michet, C.J. (1985) Giant cell arteritis and polymyalgia rheumatica. *Clin. Rheum. Dis.*, **11**, 471–83.

38. Andersson, R. (1988) Giant cell arteritis – immunology, long-term corticosteroid treatment and mortality, PhD Thesis, University of Gothenburg, Gothenburg.

39. Andersson, R., Malmvall, B.E. and Bengtsson, B.-Å. (1986) Long-term corticosteroid treatment in giant cell arteritis. *Acta Med. Scand.*, **220**, 465–9.

40. Andersson, R., Rundgren, A., Rosengren, K. *et al.* (1990) Osteoporosis after long-term corticosteroid treatment of giant cell arteritis. *J. Int. Med.*, **227**, 391–5.

41. Nordborg, E., Hansson, T., Jonsson, R. *et al.* (1993) Bone mineral content of the third vertebra during 18 months of prednisolone treatment for giant cell arteritis. *Clin. Rheumatol.*, **12**, 455–60.

42. Nordborg, E. and Bengtsson, B.-Å. (1990) Epidemiology of biopsy-proven giant cell arteritis (GCA). *J. Int. Med.*, **227**, 233–6.

# 13

# Takayasu arteritis: a current update

*J.T. Lie*

Takayasu arteritis has many synonyms (Table 13.1), among them are aortic arch syndrome, pulseless disease, primary arteritis, non-specific aortoarteritis, idiopathic aortitis and occlusive thromboarteriopathy [1–25]. Of interest, it was probably an English surgeon, Savory, in 1856 [26], who first described the entity now known as Takayasu arteritis. Takayasu [27], a Japanese ophthalmologist, had merely reported 'a case with peculiar changes of the central retinal vessels' in a young woman who apparently had pulseless disease of which Takayasu was totally unaware, and he certainly did not describe an arteritis of the aorta or the aortic arch vessels then in 1908, or later.

Takayasu arteritis is a chronic, non-arteriosclerotic, inflammatory disease of the aorta and its brachiocephalic branches, the pulmonary arteries and, less commonly, the coronary, renal, visceral, upper and lower limb arteries. This arterial disease of unknown cause has a world-wide distribution but with greater prevalence in the Orient, Africa, Latin America and eastern European countries, and it typically affects young women between 15 and 45 years of age six to nine times more commonly than men in any age group. Four types of Takayasu arteritis is recognized according to its anatomic distribution:

- type I involves the ascending aorta, aortic arch and arch vessels with or without aneurysm formation;
- type II involves the thoracoabdominal aorta;
- type III involves the aortic arch and thoracoabdominal aorta; and
- type IV involves the pulmonary arteries (Figure 13.1).

*The Vasculitides.* Edited by B.M. Ansell, P.A. Bacon, J.T. Lie and H. Yazici.
Published in 1996 by Chapman & Hall, London. ISBN 978-0-412-64140-4.

**Table 13.1** Selected synonyms of Takayasu arteritis

| Reference | Authors (year) | Synonym |
|---|---|---|
| 1 | Beneke (1925) | Sclerosing aortitis |
| 2 | Griffin (1939) | Reverse coarctation |
| 3 | Frövig (1946) | Aortic arch syndrome |
| 4 | Shimizu and Sano (1951) | Pulseless disease |
| 5 | Caccamise and Whitman (1952) | Takayasu's disease |
| 6 | Learmonth (1952) | Obliterative arteritis |
| 7 | Ross and McKusick (1953) | Young female arteritis |
| 8 | Martorell and Fabre (1954) | Martorell syndrome |
| 9 | DeBes, Lucas and Barcons (1955) | Takayasu syndrome |
| 10 | Barker and Edwards (1955) | Primary arteritis |
| 11 | Kalmansohn and Kalmansohn (1957) | Non-specific arteritis |
| 12 | Gibbons and King (1957) | Obliterative brachycephalic arteritis |
| 13 | Isaacson (1964) | Idiopathic aortitis |
| 14 | Sen *et al.* (1962) | Stenosing aortitis |
| 15 | Inada, Shimizu and Yokoyama (1962) | Atypical coarctation |
| 16 | Judge *et al.* (1962) | Takayasu arteritis |
| 17 | Sen *et al.* (1963) | Middle aortic syndrome |
| 18 | Strachan (1964) | Takayasu's arteriopathy |
| 19 | Nasu and Mamiya (1966) | Truncoarteritis productiva obliterans |
| 20 | Maekawa and Ishikawa (1966) | Occlusive thromboarteriopathy |
| 21 | Domingo *et al.* (1967) | Acquired aortoarteritis |
| 22 | Committee Report (1968) | Aortitis syndrome |
| 23 | Marquis *et al.* (1968) | Idiopathic medial aortoarteriopathy |
| 24 | Sen *et al.* (1973) | Non-specific aortoarteritis |

Reproduced with permission from [25].

Involvement of the pulmonary arteries and aneurysmal disease of the ascending aorta resulting in aortic insufficiency occur in 10–30% of cases [28, 29]. Abdominal Takayasu aortitis is an important cause of renovascular hypertension in children, adolescents and young adults [30, 31]. Reversal of hypertension after corticosteroid therapy occurs only rarely [32].

## CLINICAL FEATURES AND DIAGNOSTIC CRITERIA

The stereotype of the patient with Takayasu arteritis is a young woman of Oriental descent. Among the 52 patients reported from the USA [33, 34], 36 were North American Caucasian, five were Oriental, four were

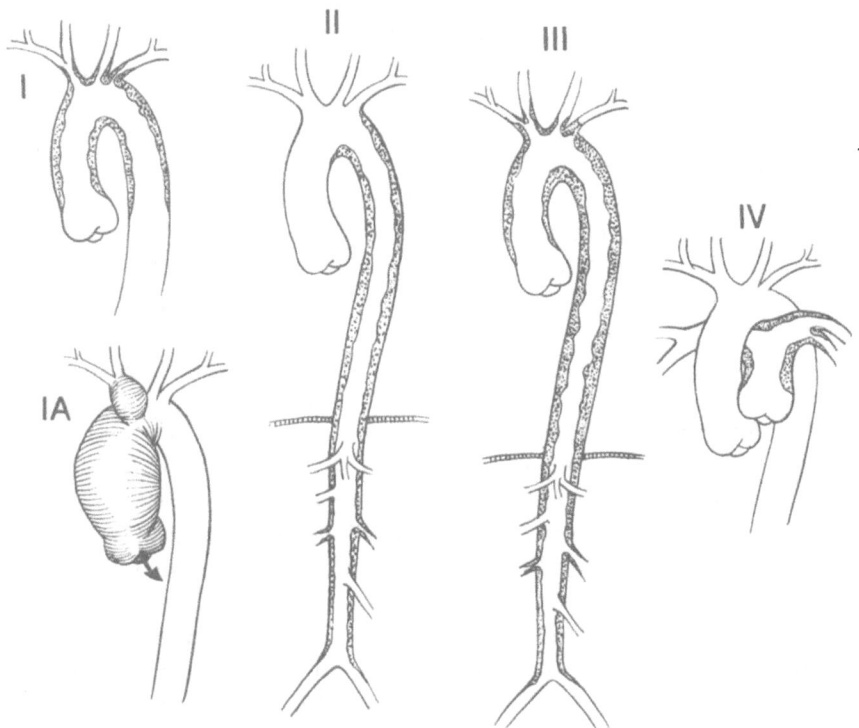

**Figure 13.1** Topographic classification of Takayasu arteritis. In type I (Shimizu–Sano) the disease is limited to the aortic arch and its branches; type IA is the aneurysmal disease subgroup, often with annuloaortic ectasia and aortic regurgitation. In type II (Kimoto) the disease is confined to the descending thoracic and abdominal aorta. Type III (Inada) has combined features of types I and II. Type IV (Lupi–Herrera) donotes pulmonary artery involvement. (Reproduced with permission from [25].)

African American, four were Mexican, and one each Native American, Filipino and Middle Eastern descent. In general, the disease is not immediately recognized in these patients from North America. Only two out of 32 patients reported from the Mayo Clinic series in Minnesota, USA, were suspected of having Takayasu arteritis [33]. A population study of the Olmsted County, Minnesota residents over a 13-year period yielded an incidence rate of 2.6 cases per million per year [33], which is less than one-tenth of that reported from Japan [35].

The clinical features of Takayasu arteritis at presentation vary according to the duration of the disease before the patient seeks medical attention [35–38]. An earlier study from Japan found a striking reciprocal relationship between elevation of the erythrocyte sedimentation rate

(ESR) and duration of the disease [36]. According to more recent studies from Japan [37, 38], the criteria proposed for the diagnosis of Takayasu arteritis were based on clinical and angiographic data from 108 patients: 96 with Takayasu arteritis and 12 with another disease of the aorta. The criteria consist of one obligatory criterion of age under 40 years; two major criteria (left and right midsubclavian artery lesions); and nine minor criteria (elevated ESR, hypertension, common carotid artery tenderness, aortic insufficiency or annuloectasia, and lesions of the pulmonary artery, left common carotid artery, distal brachiocephalic trunk, thoracic aorta and abdominal aorta). In addition to the obligatory criterion, the presence of two major, or one major plus two or more minor, or four or more minor criteria would indicate a high probability of Takayasu arteritis. The criteria had an 84% sensitivity in patients with the disease [38].

The American College of Rheumatology classification criteria for Takayasu arteritis are: age of onset before 40 years of age; upper and/ or lower limb claudication; brachial artery pressure deficit of greater than 10 mmHg; subclavian artery or aortic bruit; and an abnormal or diagnostic arteriography. The presence of any three or more of these five criteria has a diagnostic sensitivity of 90.5% and specificity of 97.8% [39].

Aside from clinical evidence of vascular insufficiency, symptoms of myalgia, arthralgia and synovitis are common, particularly early in the course of the disease. Takayasu arteritis also has been reported to occur

**Table 13.2** Takayasu arteritis

| Distribution of vascular lesion | 1958–73 (n=76) | | 1975–84 (n=115) | |
|---|---|---|---|---|
| Ascending aorta | 78 | | 73 | |
|   coronary artery | | 11 | | 45* |
| Aortic arch | 86 | | 74 | |
|   innominate artery | | 82 | | 57 |
|   right subclavian artery | | 75 | | 45 |
|   right common carotid artery | | 82 | | 50 |
|   left common carotid artery | | 87 | | 62 |
|   left subclavian artery | | 88 | | 57 |
| Thoracic aorta | 71 | | 81 | |
| Abdominal aorta | 59 | | 67 | |
|   celiac artery | | 13 | | 9 |
|   superior mesenteric artery | | 15 | | 5 |
|   renal artery | | 24 | | 40* |
|   inferior mesenteric artery | | 8 | | 2 |
| Common iliac artery | 3 | | 27* | |
| Pulmonary artery trunk | 45 | | 35 | |
|   intrapulmonary branches | | 28 | | 17 |

Adapted from Hotchi [35].
*Indicates significant change in frequency of involvement between the two periods.

in patients with adult or juvenile rheumatoid arthritis and spondyloar-thropathy [40]. Non-specific constitutional symptoms, such as anemia, weight loss, malaise and fever of unknown origin, are not uncommon, especially in the 'prepulseless phase' of the disease. Infrequent, but clinically important, manifestations of Takayasu arteritis include cuta-neous disease, coronary arterial lesions, interstitial lung disease, pul-monary hypertension and glomerulonephritis [25, 40–48].

In Japan, where Takayasu arteritis has been studied extensively, comparison of patient survival and autopsy data in two different time periods, 1958–73 and 1975–84, totaling 191 patients [35], shows that the age distribution at onset of disease now extends into the seventh decade with the age distribution at death improved by three decades, from 50–59 to 80–89 years. This trend probably indicates improved and earlier diagnosis and more effective treatment of the disease. With a few exceptions, the anatomic distribution of arterial lesions has changed very little over the years (Table 13.2). Among the patients with Takayasu arteritis reported from North America [33, 34], about 20% were 40 years of age or older at onset of disease, but survival data comparable to those of the Japanese series [35] are unavailable.

## PATHOLOGICAL SPECTRUM

Because Takayasu arteritis is a chronic, recurrent, inflammatory vascular disease that runs a clinical course of latency, exacerbation, quiescence and recrudescence over a period of months or years, the histopathology of affected arteries accordingly exhibits a spectrum of changes in concert with the inflammatory activity of the arteritis [25, 49–52]. For a correct interpretation of the histopathology of biopsy or autopsy material, it is imperative to have at hand all pertinent clinical and angiographic data of the patient, including any prior drug treatment. Biopsy is more likely to reveal the diagnostic active-phase histopathology of Takayasu arteritis, and a negative biopsy does not always rule out the disease because of the inherent sampling problem of the site and size selection in biopsies.

The characteristic gross changes of segmental tapering and often irregular narrowing of the affected artery are produced by the undu-lated intimal proliferation and fibrotic contraction of the media and adventitia. These arteritic changes with or without any coexisting throm-bosis account for the typical stenosis, coarctation and post-stenotic dilations visualized on arteriography (Figure 13.2) [53–55]. Aneurysm formation and arterial dissection are less well known complications of Takayasu arteritis [23, 49–52]. Annuloaortic ectasia and aortic aneurysm, with or without clinically significant aortic insufficiency, occur in 10–30% of patients with Takayasu's arteritis [25, 29, 56]. Saccular and/or fusiform aneurysm can also occur in the abdominal aorta (Figure 13.3).

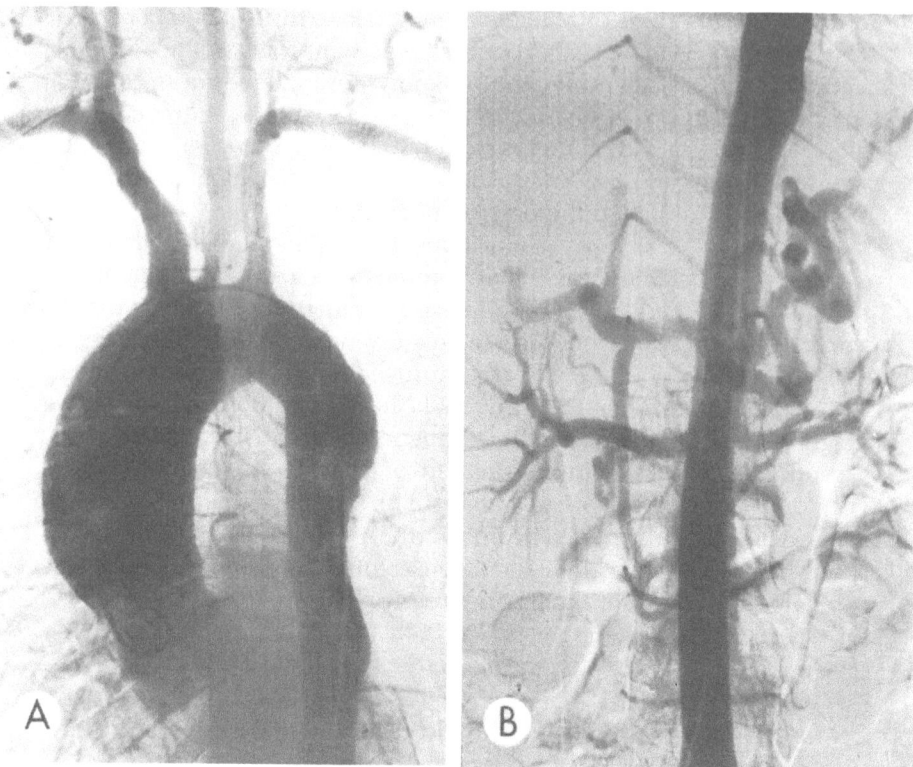

**Figure 13.2** The aortic arch (A) and abdominal aorta (B) angiography in Takayasu arteritis showing irregular aortic arch vessel stenoses, aneurysm of the ascending aorta, focal obstruction of the descending thoracic aorta and long pipe-like narrowing of the abdominal aorta. (Reproduced with permission from [25].)

Rarely, ruptured aneurysm or dissection, or perforation of a localized arteritis may occur (Figure 13.4) [57].

In the active or prepulseless phase of the disease, as seen in surgical biopsies (Figure 13.5), the vascular lesion is characterized by a granulomatous giant cell arteritis. The inflammatory infiltrate is predominantly or exclusively lymphoplasmacytic with variable numbers of eosinophils and histiocytes (Figure 13.6). Both the Langhans' and foreign body type giant cells may be present, and myogenic giant cells derived from the medial smooth muscle have also been described [49]. Contrary to the often repeated statement from an untraceable source, the arterial inflammation begins and is most pronounced in the media, rather than in the adventitia. The intensity of the cell infiltrate and the number of giant cells, however, are highly variable. The arterial inflammation is invari-

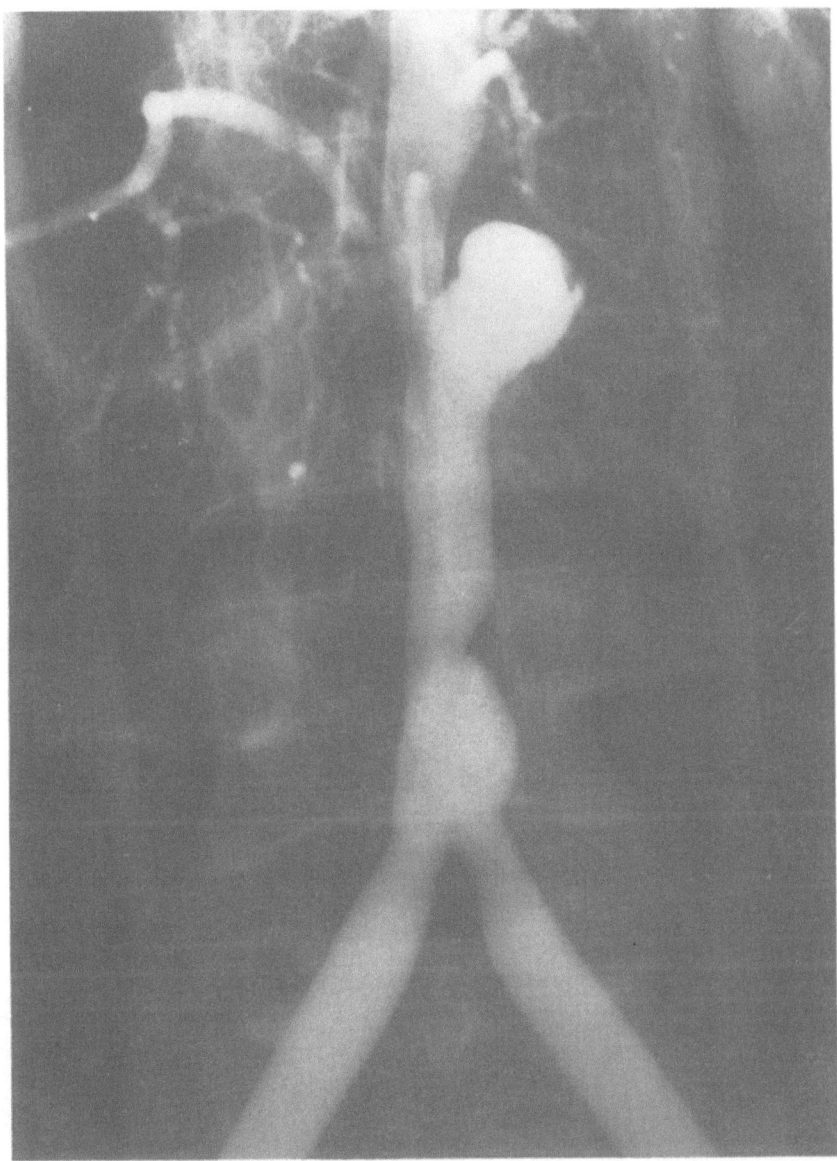

**Figure 13.3** Abnormal aortogram in Takayasu arteritis shows a high-grade narrowing of the abdominal aorta beginning at the level of the origin of the right renal artery, involving the origin of the left renal artery, extending down to the aortic bifurcation and ending with a small fusiform aneurysm. There is also a saccular aneurysm arising from the left side of the aorta at the narrowest level. (Reproduced with permission from [25].)

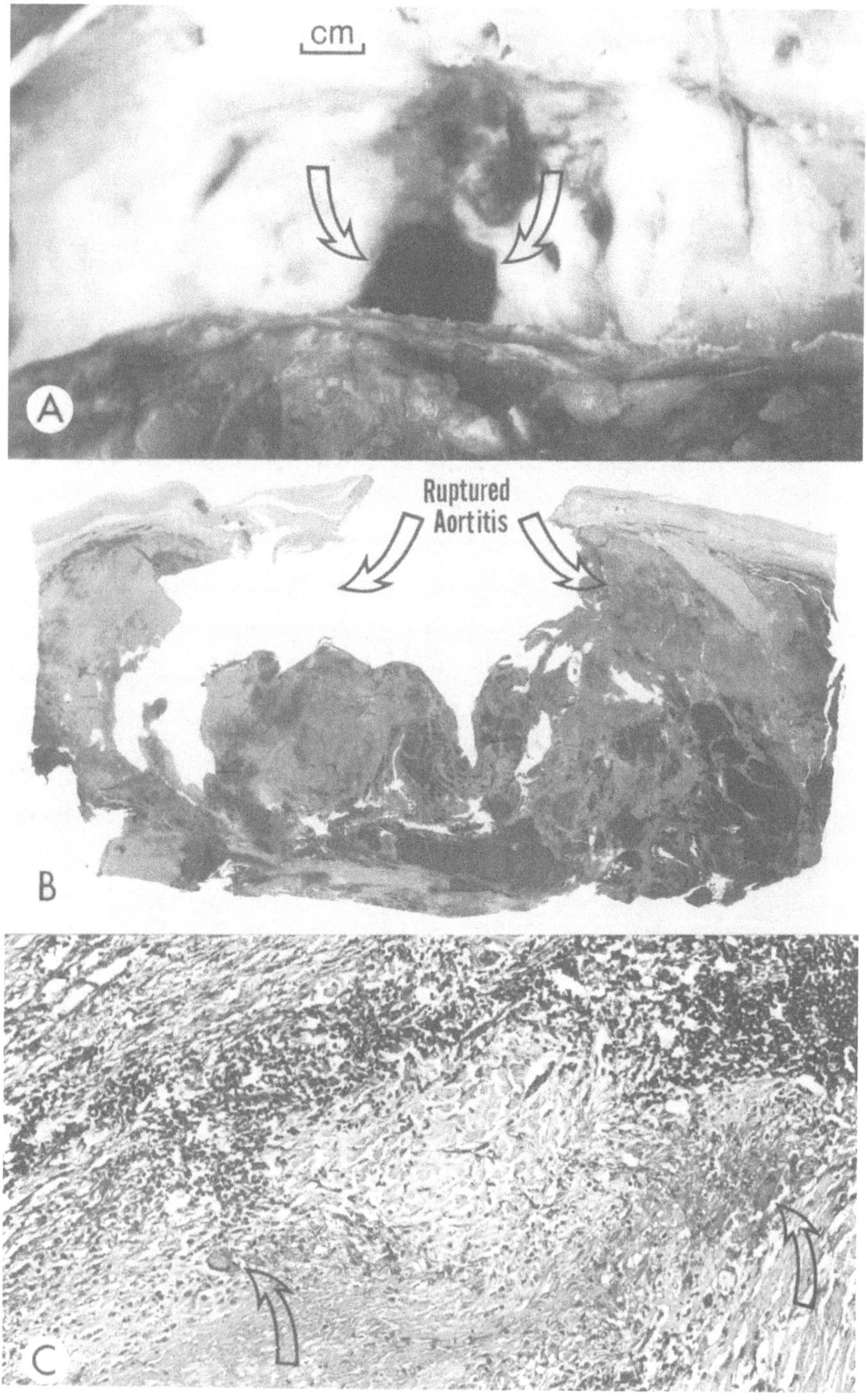

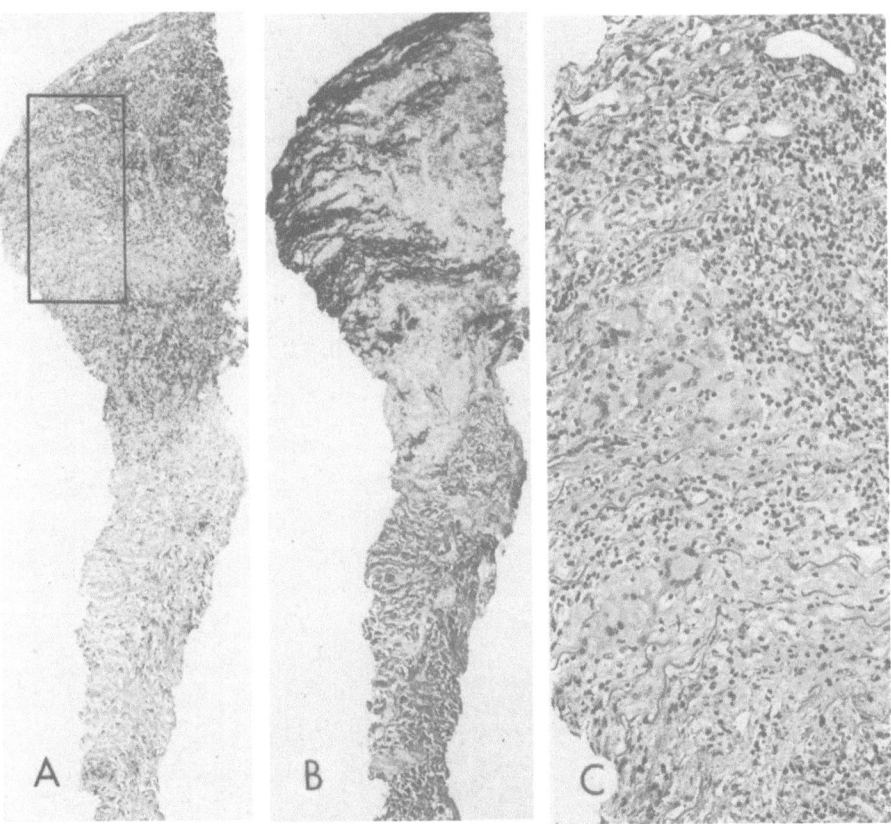

**Figure 13.5** Photomicrograph of active phase Takayasu arteritis as seen in a surgical biopsy of aorta. Matching H&E (A) and elastic stain (B) sections show a granulomatous aortitis with adventitial fibrosis (lower-half) and patchy destruction of the medial musculoelastic lamellae (upper-half). (C) Close-up view of granulomatous inflammation and giant cells in the boxed area of (A). (Hematoxylin-eosin: A and B × 40; C × 160.) (Reproduced with permission from [25].)

**Figure 13.4** Gross photograph (A) of the infrarenal abdominal aorta in Takayasu arteritis with a sharply punched defect (arrows) indicating the site of localized aortitis with rupture. Whole-mount histologic section (B) of the ruptured aortitis with thrombus. Representative photomicrograph of aortitis (C) taken from an area adjacent to the ruptured site, with lymphoplasmacytic infiltrate and multi-nucleated giant cells (arrows). (Hematoxylin-eosin: B × 4; C × 60.) (Reproduced with permission from [25].)

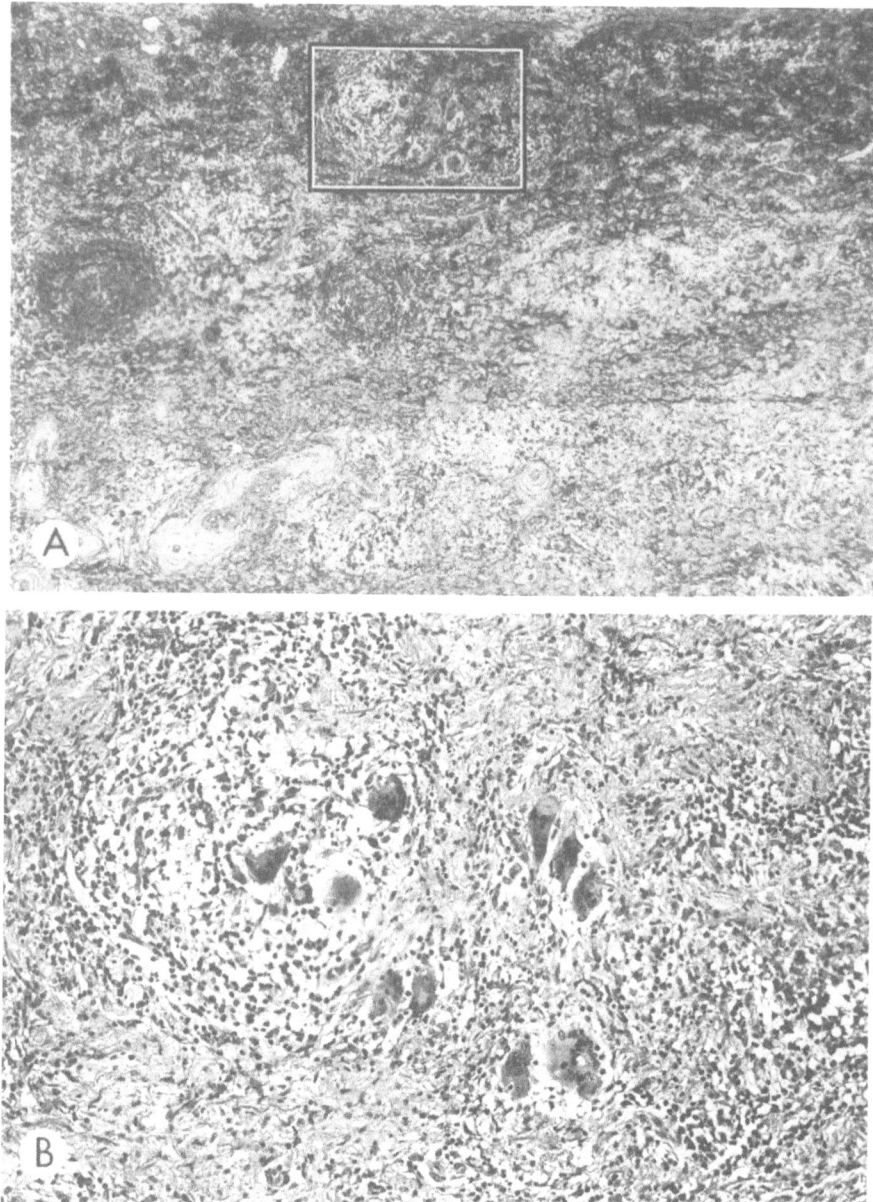

**Figure 13.6** Active phase Takayasu aortitis histopathology (A) in a patient who had the disease diagnosed five years earlier and received corticosteroid therapy. Close-up view (B) of granulomatous inflammation with giant cells. Note adventitial fibrosis and the persistent intense inflammation in the media with patchy destruction of medial musculoelastic lamellae. (Elastic stain: A × 40. Hematoxylin-eosin: B × 160.) (Reproduced with permission from [25].)

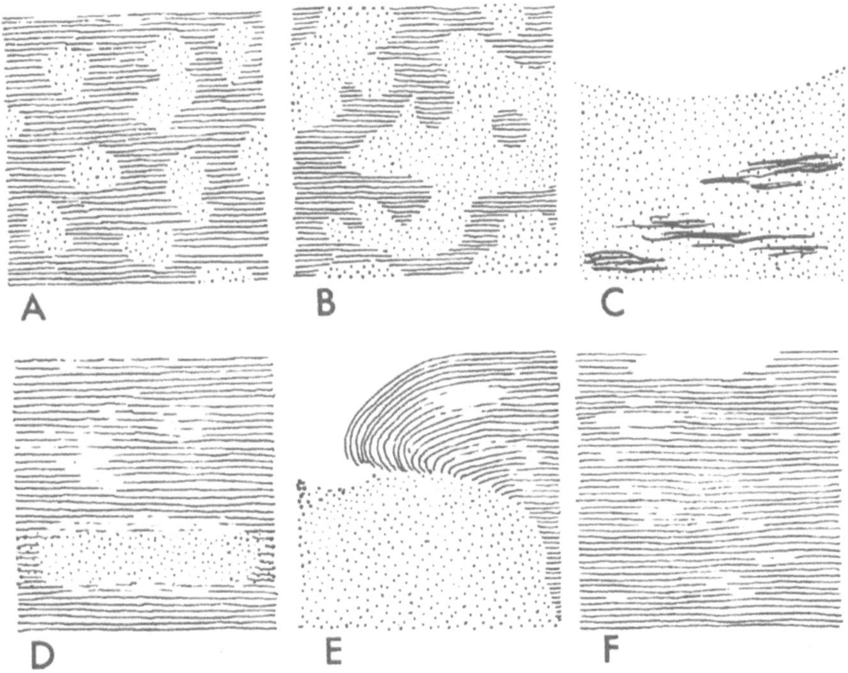

A        B        C

D        E        F

**Figure 13.7** Schematic representations of cross-sections of the media illustrating the various types of lesions encountered in Takayasu arteritis. The normal musculoelastic tissue is illustrated by horizontal wavy lines. The dotted areas represent the zones of medial destruction. (A) The typical well-defined more or less cuboidal foci of destruction. (B) Confluent foci of destruction. (C) Complete destruction with thinning of the media. (D) A band of destruction in the adventitial part of the media. (E) Total dissolution (tear) of the intima and media with an overhanging cliff of normal media. (F) Hyaline necrosis (fine dots) at the junction of intima with media. (Reproduced with permission from [25].)

ably accompanied by focal destruction of the medial musculoelastic lamellae and, with time, progressive replacement fibrosis (Figure 13.7). A band of infarct-like focal aseptic necrosis of the media (Figure 13.8), as described by Heggtveit, Hanningar and Morrione [58] in panaortitis, may also be observed occasionally.

Interestingly enough, the active phase histopathology of arteritis can persist for an extended period, even when the patient has been treated or is still on immunosuppressive therapy. With recurrence or exacerbation of the disease, there is also a return of the active-phase histopathology of granulomatous arteritis (Figure 13.6) even after a clinically successful suppression of inflammatory activity by long-term corticosteroid therapy. The same granulomatous arteritis is observed in visceral, coronary

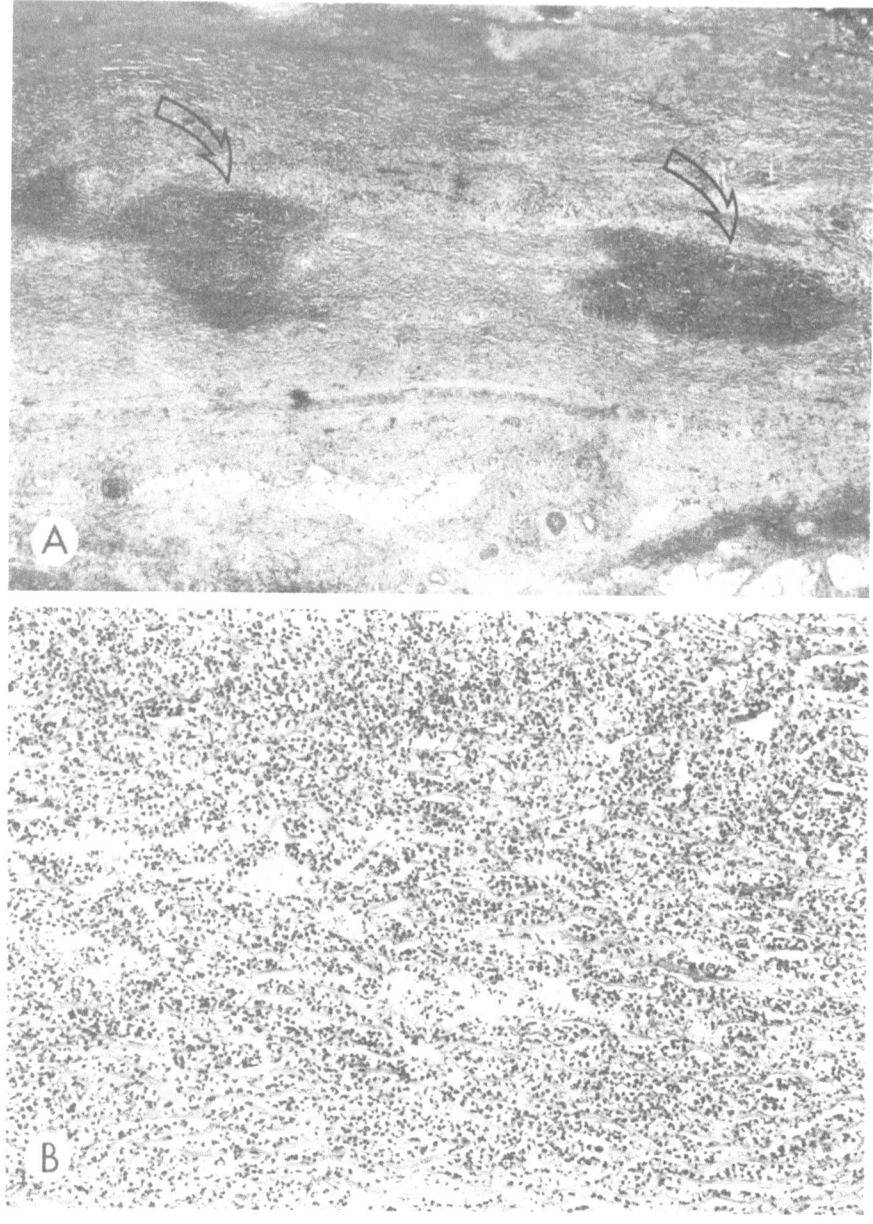

**Figure 13.8** Photomicrograph (A) of two bands of medial infarction (arrows) in Takayasu aortitis. Close-up view (B) of infarction with intense polymorpho-nuclear infiltration in the necrotic media. (Hematoxylin-eosin: A × 40; B × 160.) (Reproduced with permission from [25].)

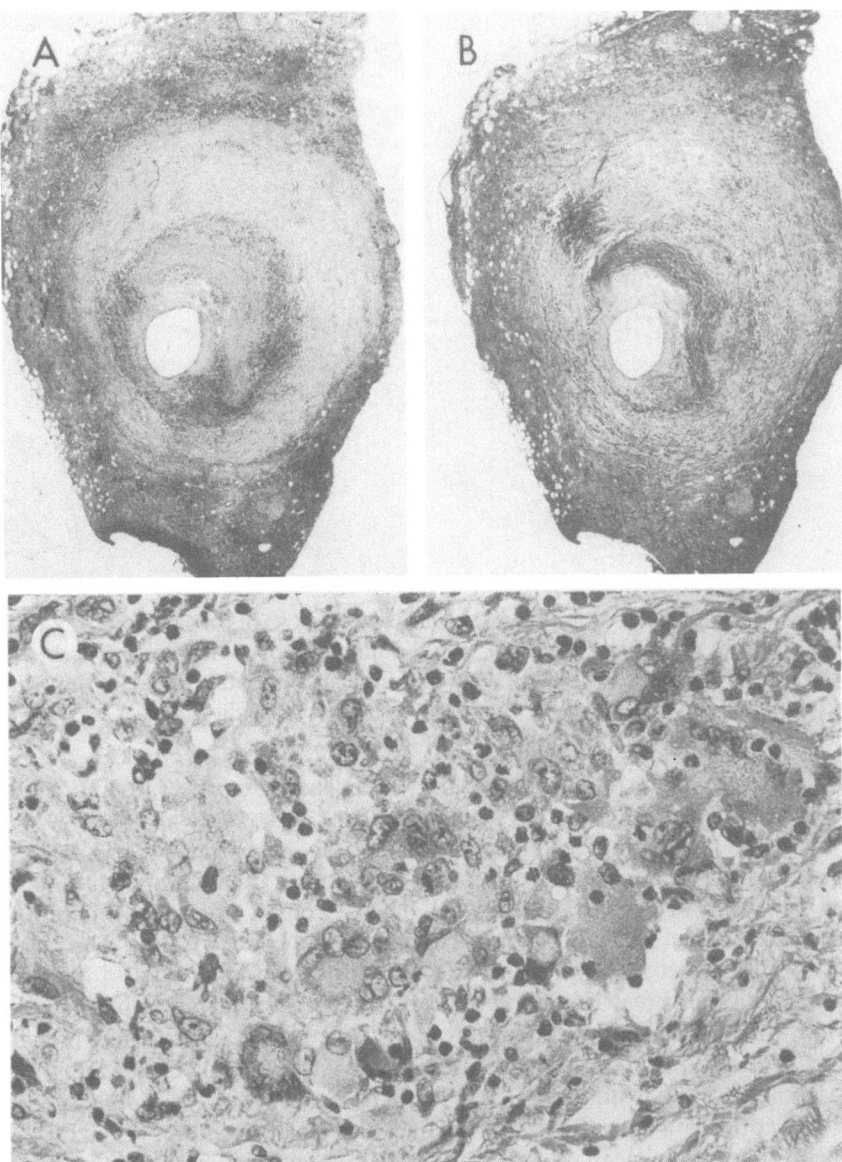

**Figure 13.9** Matching H&E (A) and elastic stain (B) sections of the coronary artery involved with Takayasu arteritis. Note transmural fibrosis and inflammatory infiltrate in the media. Close-up view (C) of granulomatous inflammation with giant cells in the media of the coronary artery. (Hematoxylin-eosin: A × 16; C × 400. Elastic stain: B × 16.) (Reproduced with permission from [25].)

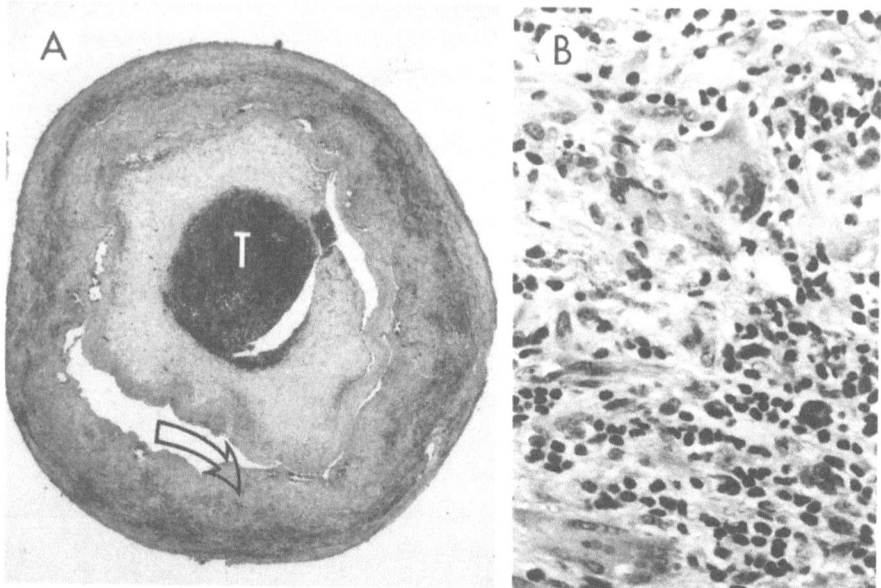

**Figure 13.10** Endarterectomized Takayasu arteritis (A) of the carotid artery with transmural inflammation (arrow), marked luminal narrowing and an occlusive thrombus (T). Close-up view (B) of an area indicated by arrow in (A), to show the granulomatous inflammation and giant cells. (Hematoxylin-eosin: A × 16; B × 400.) (Reproduced with permission from [25].)

and pulmonary arteries affected by the disease (Figure 13.9), and in an endarterectomized occluded artery (Figure 13.10).

In the late or pulseless phase of the disease, transmural sclerosis with scanty or no inflammatory cell infiltrate becomes the histologic hallmark of Takayasu arteritis (Figure 13.11). The degree of intimal proliferation and adventitial fibrosis is proportional to the duration and severity of the arteritis. Secondary thrombosis occurs not infrequently and it poses an additional threat to complete occlusion of the affected artery. It is in the subacute phase and burnt-out stage that a Takayasu's arteritis may escape recognition and be mistaken for arteriosclerosis obliterans by a pathologist who is unfamiliar with the disease or who relies solely on the gross appearance for the diagnosis of a fibrotic or occluded artery when examining a surgical biopsy, endarterectomy, amputation or autopsy specimen.

## CONCLUSION

In diseases of large and medium-sized arteries, if one excludes arteriosclerosis or atherosclerosis, syphilitic arteritis and large-vessel arteritis of

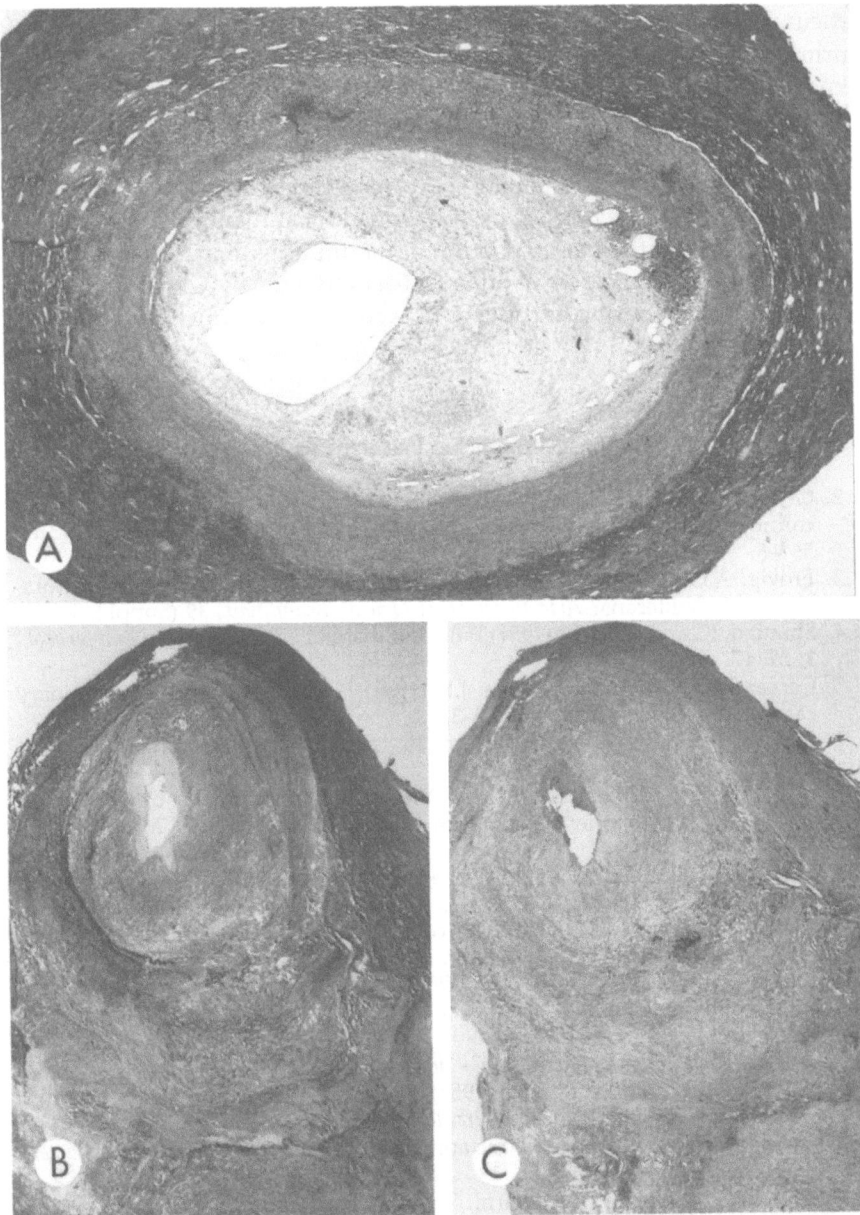

**Figure 13.11** Histopathology of subacute phase (A) and end-stage (B, C) Takayasu arteritis. (A) shows occlusive intimal proliferation and adventitial fibrosis with residual inflammatory infiltrate in the media. Matching elastic stain (B) and H&E (C) sections of burnt-out arteritis with the artery reduced to a fibrous cord. (Hematoxylin-eosin: A and C × 16. Elastic stain: B × 16.) (Reproduced with permission from [25].)

rheumatic–rheumatoid disorders, one is left with a peculiar type of primary granulomatous aortitis and arteritis of obscure origin that is known as Takayasu arteritis. The disease has a worldwide distribution, but it occurs most commonly in the Far East, India, Africa, South-East Asia, eastern Europe and South America. Women of childbearing age are afflicted with the disease about six to nine times more frequently than men of any age. The disease has protean clinical manifestations and no diagnostic laboratory tests. Angiographic findings may be highly suggestive, but confirmation requires a tissue diagnosis which should be made with deliberation and the appropriate clinical correlation.

## REFERENCES

1. Beneke, R. (1925) Ein eigentümlicher Fall schwieliger Aortitis. *Virchows Arch. Pathol. Anat.*, **254**, 722–33.
2. Griffin, H.M. (1939) Reverse coarctation with vasomotor gradient: report of cardiovascular anomaly with symptoms of brain tumor. *Proc. Mayo Clin.*, **14**, 561–5.
3. Frövig, A.G. (1946) Bilateral obliteration of common carotid artery. Thromboangiitis obliterans? *Acta Psychiat. et Neurol. Scandinav.*, **39** (Suppl.), 4–79.
4. Shimizu, K. and Sano, K. (1951) Pulseless disease. *J. Neuropathol. Clin. Neurol.*, **1**, 37–47.
5. Caccamise, W.C. and Whitman, J.F. (1951) Pulseless disease: a preliminary case report. *Am. Heart J.*, **44**, 629–33.
6. Learmonth, B.C. (1952) Report of 16 cases of sudden aortic occlusion. *Edinburgh Med. J.*, **59**, 65–93.
7. Ross, R.S. and McKusick, V.A. (1953) Aortic arch syndromes: diminished or absent pulses in arteries arising from arch of aorta. *Arch. Inter. Med.*, **92**, 701–40.
8. Martorell, F. and Fabre, J. (1954) The syndrome of obliteration of the supra-aortic branches. *Angiology*, **5**, 39–42.
9. DeBes, L.T., Lucas, J.G.S. and Barcons, F.B. (1955) The case of Takayasu syndrome: the pulseless disease. *Br. Heart J.*, **17**, 484–8.
10. Barker, N.W. and Edwards, J.E. (1955) Primary arteritis of the aortic arch. *Circulation*, **11**, 486–92.
11. Kalmansohn, R.B. and Kalmansohn, R.W. (1957) Thrombotic obliteration of the branches of the aortic arch. *Circulation*, **15**, 237–44.
12. Gibbons, T.B. and King, R.L. (1957) Obliterative brachiocephalic arteritis: pulseless disease of Takayasu. *Circulation*, **15**, 845–9.
13. Isaacson, C. (1961) An idiopathic aortitis in young Africans. *J. Pathol. Bacteriol.*, **81**, 69–79.
14. Sen, K.P., Kinare, S.G., Kulkarni, T.P. and Parulkar, G.B. (1962) Stenosing aortitis of unknown etiology. *Surgery*, **51**, 317–25.
15. Inada, K., Shimizu, K. and Yokoyama, T. (1962) Pulseless disease and atypical coarctation of the aorta with special reference to their genesis. *Surgery*, **52**, 433–43.
16. Judge, R.C., Currier, R.D., Graci, W.A. and Figley, M.M. (1962) Takayasu's arteritis and the aortic arch syndrome. *Am. J. Med.*, **32**, 379–92.
17. Sen, P.K., Kinare, S.G., Engineer, S.D. and Parulkar, G.B. (1963) The middle aortic syndrome. *Br. Heart J.*, **25**, 610–18.

18. Strachan, R.W. (1964) The natural history of Takayasu's arteriopathy. *Q. J. Med.*, **33**, 57–69.
19. Nasu, T. and Mamiya, N. (1966) Pathogenesis of truncarteritis productive obliterans; so-called pulseless disease or aortic arch syndrome. *Jap. Circ. J.*, **30**, 68–71.
20. Maekawa, M. and Ishikawa, K. (1966) Occlusive thromboarteriopathy. *Jap. Circ.*, **30**, 79–85.
21. Domingo, R.T., Maramba, T.P., Torres, L.F. and Wesolowski, S.A. (1967) Acquired aortoarteritis. A worldwide vascular entity. *Arch. Surg.*, **95**, 780–90.
22. Committee Report (1968) Clinical and pathological studies of aortitis syndrome. *Jap. Heart J.*, **9**, 76–87.
23. Marquis, Y., Richardson, J.B., Ritchie, A.C. and Wigle, E.D. (1968) Idiopathic medial aortopathy and arteriopathy. *Am. J. Med.*, **44**, 939–54.
24. Sen, P.K., Kinare, S.G., Kelkar, M.D. and Paulkar, G.B. (1973) *Nonspecific Aorto-arteritis. A Monograph Based on the Study of 101 cases*, Tata-McGraw Hill, Bombay.
25. Lie, J.T. (1990) Takayasu arteritis, in *Systemic Vasculitides* (eds A. Churg, A. Churg and J. Churg), Igaku Shoin, New York, pp. 159–79.
26. Savory, W.S. (1856) Case of a young woman in whom the main arteries of both upper extremities and of the left side of the neck were throughout completely obliterated. *Med. Chir. Trans. London*, **39**, 205–11.
27. Takayasu, M. (1908) A case with unusual changes of the central vessels in the retina. *Acta Soc. Ophthalmol. Jap.*, **35**, 750–4.
28. Sharma, S., Kamalakar, T., Rajani, M. *et al.* (1990) The incidence and patterns of pulmonary artery involvement in Takayasu's arteritis. *Clin. Radiol.*, **42**, 177–81.
29. Sharma, S., Rajani, M., Kamalakar, T. *et al.* (1990) The association between aneurysm formation and systemic hypertension in Takayasu's arteritis. *Clin. Radiol.*, **42**, 182–7.
30. Kanaraj, T.J. and Wong, H.O. (1959) Primary arteritis of abdominal aorta in children causing bilateral stenosis of renal arteries and hypertension. *Circulation*, **20**, 856–63.
31. Wiggelinkhuizen, J. and Cremin, B.J. (1978) Takayasu arteritis and renovascular hypertension in childhood. *Pediatrics*, **62**, 209–17.
32. Kulkarni, T.P., D'Cruz, I.A., Gandhi, M.J. and Dadhich, D.S. (1974) Reversal of renovascular hypertension caused by nonspecific aortitis after corticosteroid therapy. *Br. Heart J.*, **36**, 114–16.
33. Hall, S., Barr, W., Lie, J.T. *et al.* (1985) Takayasu arteritis: A study of 32 North American patients. *Medicine*, **64**, 89–99.
34. Shelhamer, J.H., Volkman, D.J., Parrillo, J.E. *et al.* (1985) Takayasu arteritis and its therapy. *Ann. Intern. Med.*, **103**, 121–6.
35. Hotchi, M. (1993) Pathology of Takayasu arteritis, in *Intractable Vasculitis Syndromes* (ed. T. Tanabe), Hokaido University Press, Sapporo, pp. 85–92.
36. Nakao, K., Ikeda, M. and Kimatra, S. (1967) Takayasu arteritis: clinical report of eighty-four cases and immunological study of seven cases. *Circulation*, **35**, 1141–55.
37. Ishikawa, K. (1986) Patterns of symptoms and prognosis in occlusive thromboartopathy (Takayasu disease). *J. Am. Coll. Cardiol.*, **8**, 1041–6.
38. Ishikawa, K. (1988) Diagnostic approach and proposed criteria for the clinical diagnosis of Takayasu's arteriopathy. *J. Am. Coll. Cardiol.*, **12**, 964–72.
39. Arend, W.P., Michel, B.A., Bloch, D.A. *et al.* (1990) The American College of

Rheumatology 1990 criteria for the classification of Takayasu arteritis. *Arthritis Rheum.*, **33**, 1129–34.

40. Hall, S. and Buchbinder, R. (1990) Takayasu's arteritis. *Rheum. Dis. Clin. North Am.*, **16**, 411–22.

41. Perniciaro, C.V., Winkelmann, R.K. and Hunder, G.G. (1987) Cutaneous manifestations of Takayasu's arteritis: a clinicopathologic correlation. *J. Am. Acad. Dermatol.*, **17**, 998–1012.

42. Hass, A. and Steihm, E.R. (1986) Takayasu's arteritis presenting as pulmonary hypertension. *Am. J. Dis. Child.*, **140**, 372–4.

43. Koumi, S., Endo, T., Okumura, H. *et al.* (1990) A case of Takayasu's arteritis associated with membranoproliferative glomerulonephritis and nephrotic syndrome. *Nephron.* **54**, 344–6.

44. Koyabu, S., Isaka, N., Yada, T. *et al.* (1993) Severe respiratory failure caused by recurrent pulmonary hemorrhage in Takayasu's Arteritis. *Chest*, **104**, 1905–6.

45. Cipriano, P., Silverman, J., Perbroth, M. *et al.* (1977) Coronary ostial narrowing in Takayasu's arteritis. *Am. J. Cardiol.*, **39**, 744–50.

46. Tanaka, A., Fukayama, M., Funata, N. *et al.* (1988) Coronary arteritis and aortoarteritis in the elderly male. *Virchows Arch. Pathol. Anat.*, **414**, 9–14.

47. Lande, A. and Bard, R. (1976) Takayasu arteritis: An unrecognized cause of pulmonary hypertension. *Angiology*, **27**, 114–21.

48. Chauvaud, S., Mace, L., Brunewald, P. *et al.* (1987) Takayasu arteritis with bilateral pulmonary artery stenosis: Successful surgical correction. *J. Thorac. Cardiovasc. Surg.*, **94**, 246–50.

49. Nasu, T. (1963) Pathology of pulseless disease: A systematic study and critical review of twenty-one autopsy cases reported in Japan. *Angiology*, **14**, 225–42.

50. Vinijchaikul, K. (1967) Primary arteritis of the aorta and its main branches (Takayasu arteriopathy): A clinicopathologic autopsy study of cases. *Am. J. Med.*, **43**, 15–27.

51. Restrepo, C., Tejeda, C. and Correa, P. (1969) Nonsyphilitic aortitis. *Arch. Pathol.*, **87**, 1–12.

52. Rose, A.G. and Sinclair-Smith, C.C. (1980) Takayasu arteritis: A study of 16 autopsy cases. *Arch. Pathol. Lab. Med.*, **104**, 231–7.

53. Lande, A. and Gross, A. (1972) Total aortography in the diagnosis of Takayasu arteritis. *Am. J. Radiol.*, **116**, 165–78.

54. Deutch, V., Wexler, L. and Deutch, H. (1974) Takayasu arteritis: An angiographic study with remarks on ethnic distribution in Israel. *Am. J. Radiol.*, **122**, 13–28.

55. Lie, Y.Q. and Du, J.H. (1982) Aortoarteritis: A collective angiographic experience in 244 cases. *Internat. Angiol.*, **3**, 487–97.

56. Ueda, H., Sugiura, M., Ito, I. *et al.* (1967) Aortic insufficiency associated with aortitis syndrome. *Jap. Heart J.*, **8**, 107–20.

57. Lie, J.T. (1987) Segmental Takayasu (giant cell) aortitis with rupture and limited dissection. *Hum. Pathol.*, **18**, 1183–5.

58. Heggtveit, H., Hanningar, G.R. and Morrione, T.G. (1963) Panaortitis. *Am. J. Pathol.*, **42**, 151–72.

# 14

# Behçet's syndrome with emphasis on clinical genetics and mortality

*H. Yazıcı, B. Çiçek, G. Başaran, V. Hamuryudan,*
*N. Çakir, I. Dimitriatris, S. Yurdakul, F. Moral,*
*Y. Özyazgan, M.C. Mat and Y. Tüzün*

Behçet's syndrome (BS) is a systemic vasculitis of unknown causes(s), which characteristically involves the skin, the oral and the genital regions, the eye and the central nervous sytem. Involvement of the kidneys and the peripheral nervous system are distinctly rare. Among the vasculitides, BS is notable for its predilection for the venous side of the vascular tree [1].

On the arterial side, aneurysm formation in both the systemic and pulmonary trees (Figure 14.1) is not uncommon and is an important cause of morbidity and mortality. Table 14.1 gives a list of the more important manifestations of BS. The pathergy reaction (Plate 2) is the non-specific hyperreactivity of the tissues to a trauma and is quite unique to BS. When produced in the skin, it can be used as an aid to diagnosis. Curiously it is quite rare among patients from northern Europe [2]. We have recently shown that it can be suppressed by surgical cleaning of the skin, which in turn implies that more than disruption of the integrity of the epidermis and dermis is involved in its formation [3]. The augmented skin response to intradermally injected monosodium urate crystals is similar, however, between Turkish and British patients [4].

The main vascular pathology in BS is an intense inflammation of the whole vessel wall (Figure 14.2). Thus when thrombi are formed they are quite adherent to the endothelium.

*The Vasculitides.* Edited by B.M. Ansell, P.A. Bacon, J.T. Lie and H. Yazici.
Published in 1996 by Chapman & Hall, London. ISBN 978-0-412-64140-4.

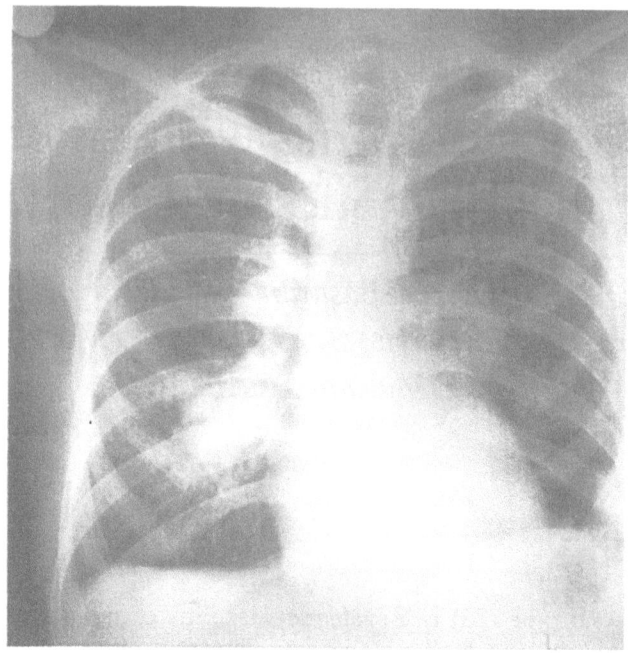

**Figure 14.1** Pulmonary arterial aneurysms.

**Table 14.1** Clinical findings in Behçet's syndrome

| Lesion | Frequency |
|---|---|
| Oral aphthous ulceration | 97–99% |
| Genital ulcerations | 80% |
| Skin lesions | 80% |
| Erythema nodosum | 50% |
| Folliculitis | 80% |
| Pathergy | 60% (Mediterranean countries and Japan) |
| Eye lesions | 59% |
| Arthritis | 40–50% |
| Arterial occlusion/aneurysm | 4% |
| Thrombophlebitis | 25% |
| Neurological involvement | 5% |
| Gastrointestinal involvement | 0–25% (Japan) |
| Epididymitis | 5% |

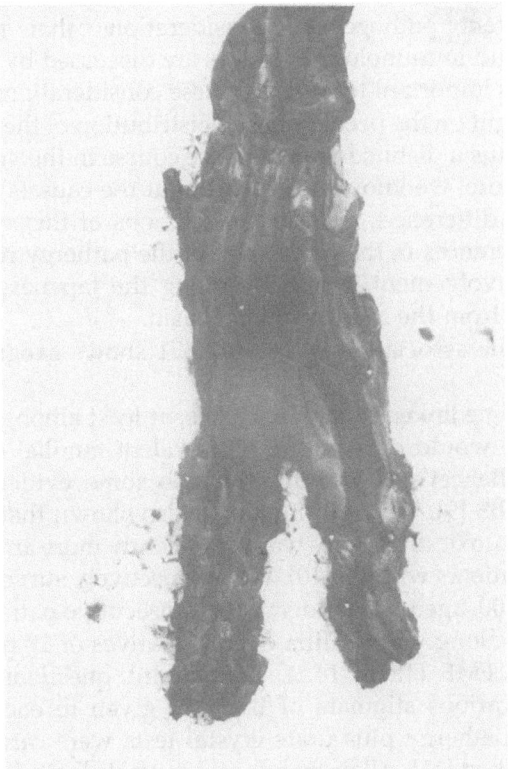

**Figure 14.2** Thrombosis of inferior vena cava.

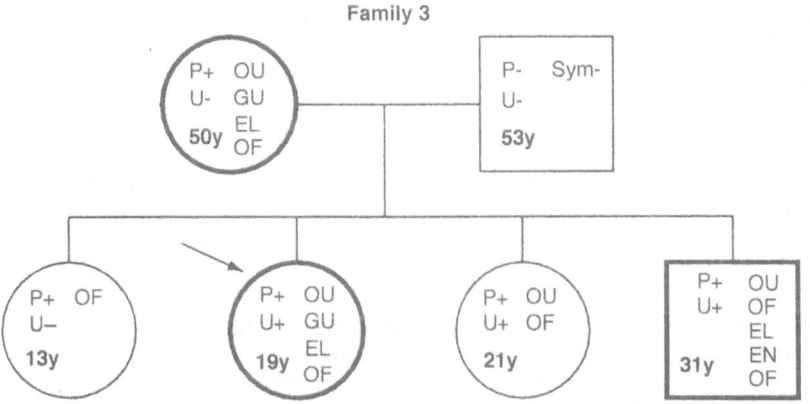

**Figure 14.3** The pathergy and urate crystal tests in family members (Sym: symptoms of BS; P: pathergy reaction; U: urate reaction; OU: oral ulceration; GU: genital ulceration; EL: eye lesions; EN: erythema nodosum; OF: ostiofolliculitis; arrow: index case with BS; dark lines: patients with BS).

Some current pathogenetic considerations that mainly concern endothelial and immunological factors are discussed by Kansu in Chapter 15 [5]. It is important to note that these considerations thus far do not shed much light on the peculair tissue distribution of the pathology in BS and why it runs a distinctly more severe course in the young adult male [6]. Furthermore, we know very little about the cause(s) for the marked geographical differences in the manifestations of this syndrome. Apart from the differences in the prevalence of the pathergy reaction as noted above, gut involvement, common among the Japanese, is rarely seen among those from the Mediterranean basin.

Similarly, the association with HLA B51 shows geographic variation [7].

In a syndrome linked to an HLA allele, at least among its more severe cases [8], one would expect a more prevalent familial occurrence than one sees in Behçet's syndrome. There is some evidence for familial clustering in BS [9]. We had previously also shown that the prevalence of lymphocytotoxic antibodies was significantly more among the healthy children of patients with BS [10]. We prospectively surveyed first degree relatives (n=200, aged 10 or older) of 62 consecutive patients with BS in a blind fashion along with 67 first degree relatives of 17 patients with AS and 17 with FMF (Table 14.2). A standard questionnaire about the presence of various stigmata of BS were given to each proband and control, and pathergy plus urate crystal tests were carried out as previously described [11]. All probands and controls lived in Istanbul. There were five families (5/62 = 8.0%) in whom there was at least another patient with BS in the family. In these families, positivity in pathergy and/or the urate crystal test did not always go in parallel with the presence of the full syndrome. Such a family is shown in Figure 14.3.

Table 14.3 shows the cardinal findings of Behçet's syndrome among the relatives of probands and the controls. It is to be noted there were no significant differences in the prevalences apart from in arthritis, uveitis and thrombophlebitis which, in fact, were significantly more common among the relatives of patients in AS and FMF. However, when we analyzed our data for the possible association of the pathergy and/or urate crystal test positivity and the prevalence of the various clinical

**Table 14.2** Age and sex of patients with Behçet's syndrome (BS), ankylosing spondylitis (AS) and familial mediterranean fever

|     | n  | M/F   | Age ($\times \pm$ SD) |
| --- | --- | ----- | --------------------- |
| BS  | 62 | 36/26 | $30 \pm 9$            |
| AS  | 17 | 13/4  | $41 \pm 4$            |
| FMF | 17 | 10/7  | $21 \pm 9$            |

**Table 14.3** Prevalence of pathergy and MSU positivity and the clinical findings of Behçet's syndrome among the first-degree relatives of probands and controls

|  | Behçet's relatives | | Control relatives | |
|---|---|---|---|---|
| *n* | 200 | | 67 | |
| Pathergy (+) | 13[1] | (7) | 1 | (1) |
| MSU(+) | 30 | (15) | 7 | (10) |
| Oral ulceration (OU) | 67 | (34) | 23 | (34) |
| OU within the past 6 months | 62 | (26) | 10 | (15) |
| Genital ulceration | 19 | (5) | 3 | (4) |
| Inflammation of the eye | 20 | (10) | 13 | (19)[2] |
| Ostiofolliculitis | 43 | (22) | 19 | (28) |
| Erythema nodosum | 7 | (4) | 2 | (3) |
| Arthritis | 7 | (4) | 9 | (13)[3] |
| Thrombophlebitis | 2 | (1) | 6 | (9)[4] |

Parentheses indicate percentages. [1] nine were mothers, two were sisters, one was a daughter and one was a father. [2] $p < 0.05$. [3] $p < 0.01$. [4] $p < 0.001$.

**Table 14.4** The prevalence of aphthae among the pathergy and/or MSU test positive relatives of patients with BS*

|  | Pathergy (+) $n=13$ | MSU (+) $n=30$ | Pathergy and/or MSU (+) $n=37$ |
|---|---|---|---|
| Aphthae | 8 (61.5%) | 20 (66.6%) | 23 (62.2%) |
| Odds ratio | 1.95 | 2.43 | 2.30 |
| Significance | $0.10 > p > 0.05$ | $p < 0.001$ | $p < 0.01$ |

*$n=200$; overall prevalence of aphthae: $67/200 = 33.5\%$.

manifestations of BS, we found that the prevalence of aphthae was significantly increased among the pathergy and/or the urate crystal test positive individuals (Table 14.4). It was also noteworthy (Table 14.3) that among the 10 parents who were pathergy positive, nine were mothers. This indicates maternal preference in genetic transmission. This issue has been debated in the past and our Japanese colleagues had indicated a maternal preference for the inheritance of HLA B5 [12]. However, in a subsequent study from our group, this could not be substantiated [13].

Thus the three main conclusions from our family survey were:

1. There is indeed familial clustering of BS.
2. The presence of aphthae and pathergy positivity and/or urate crystal test positivity are, to some degree, linked.

3. The 'maternal preference in genetic transmission' concept should be reassessed.

The impact of BS on survival had never been adequately assessed. In a previous 52 ± seven month follow-up of 51 patients, we had no reported deaths [6]. We have lately looked at the 10-year mortality in BS. This was mainly prompted by our recent awareness of the rather high ($\approx$ 50%) mortality of our patients with pulmonary arterial aneurysms, despite treatment [14]. There were 143 new patients (92 males and 51 females) registered in our BS outpatient clinic in 1982. They were called back 10 years later for reassessment. Thirty-five were still attending our clinic (seen within one year). Another 79 were contacted by letter or telephone. Of these, 29 could be brought to the hospital for reassessment 38 (25%) were lost to follow-up.

There were six deaths among the whole group. All were male and the probable causes of death are given in Table 14.5. In one instance (ZS) we do not know the immediate cause of death. He had a relentless course of a very severe organic brain syndrome with multiple infarcts that had led to dementia. Using the Turkish mortality tables the maximum number of deaths were calculated that could be expected in the general population for the specified age brackets (Table 14.6). For those patients that were in

**Table 14.5** Causes of death (all male)

| Name | Age first seen | Age of death | Probable cause of death |
|------|------|------|------|
| MS | 43 | 52 | PAA |
| YÖ | 43 | 48 | GI hemorrhage (VCIS) |
| MC | 24 | 29 | Hemoptysis (VCSS) |
| FS | 49 | 52 | CVA |
| ZS | 20 | 28 | CNS involvement? |
| AM | 23 | 28 | PAA |

**Table 14.6** Minimum 10-year mortality among 92 male patients with Behçet's syndrome*

| Age bracket | Number of patients | Number of deaths | Maximum expected number of deaths ($\alpha < 0.01$) |
|------|------|------|------|
| 15–24 | 16 | 2 | 1.49 |
| 25–34 | 34 | 1 | 3.20 |
| 35–44 | 19 | 2 | 3.60 |
| 45–54 | 6 | 1 | 2.84 |

*Data retrieval: 75/92 (81%).
For CSO 1953–58 ($\alpha < 0.05$) = 1.87.
For ADST 1949–51 ($\alpha < 0.05$) = 1.56.

the 15–24 age bracket when they first saw us, the observed mortality of two was indeed well above the expected (p < 0.01). No such differences were observed for the other age brackets. This observed minimum (albeit high) mortality, especially among the male and young patients with BS, is quite in line with our previous observations [6].

These observations bring to discussion early aggressive treatment in this age group. Hamuryudan *et al.* [16] recently studied the long-term (53 ± 28 SD months) outcome of our patients who had initially been the subjects of our azathioprine study [15]. The data are still being analyzed. However, there seems to be a definite trend for a better long-term prognosis for those patients who had received azathioprine early in the course. We also have some promising data on the systemic use of α2b-interferon. In patients without eye disease in a controlled setting, systemic α2b-interferon had beneficial effects on several manifestations of the syndrome. In the cases of arthritis and the erythrocyte sedimentation rate, these assumed statistical significance [16].

There remain two very important sites of organ involvement in BS for which nobody has adequate data on treatment. These are central nervous system involvement and the major vessel (both arterial and venous) involvement. For better control of these and the other manifestations we definitely need more understanding of the pathogenetic mechanisms.

## REFERENCES

1. Lie, J.T. (1992) Vascular involvement in Behçet's disease. Arterial and venous and vessels of all sizes. *J. Rheumatol.*, **19**, 341–3.
2. Yazici, H., Chamberlain, M.A., Tüzün, Y. *et al.* (1984) A comparative study of the pathergy reaction among Turkish and British patients with Behçet's disease. *Ann. Rheum. Dis.*, **43**, 74–5.
3. Fresko, İ., Yazici, H., Bayramiçli, M. *et al.* (1993) Effect of surgical cleaning of the skin on the pathergy phenomenon in Behçet's syndrome. *Ann. Rheum. Dis.*, **52**, 619–20.
4. Çakir, N., Yazici, H., Chamberlain, M.A. *et al.* (1991) Response to intradermal injection of monosodium urate crystals in Behçet's syndrome. *Ann. Rheum. Dis.*, **50**, 634–6.
5. Kansu, E. (1995) Endothelial cell function in Behçet's syndrome (see Chapter 15 of this book).
6. Yazici, H., Tüzün, Y., Pazarli, H. *et al.* (1984) Influence of age of onset and sex on the prevalence and severity of manifestations of Behçet's syndrome. *Ann. Rheum. Dis.*, **43**, 783–9.
7. Yazici, H., Chamberlain, M.A., Schreuder, I. *et al.* (1980) HLA Antigens in Behçet's disease: a reappraisal by a comparative study of Turkish and British patients. *Ann. Rheum. Dis.*, **39**, 344–8.
8. Yurdakul, S., Günaydin, I., Tüzün, Y. *et al.* (1988) The prevalence of Behçet's syndrome in a rural area of Northern Turkey. *J. Rheumatol.*, **15**, 820–2.

9. Dündar, S.V., Gencalp, U. and Simsek, H. (1985) Familial cases of Behçet's disease. *Br. J. Dermatol.*, **113**, 319–21.

10. Günaydin, I., Yazici, H., Özbakir, F. *et al.* (1992) Lymphocytotoxic antibodies in the healthy children of patients with Behçet's syndrome. *J. Rheumatol.*, **19**, 1966–98.

11. Tüzün, Y., Yazici, H., Pazarli, H. *et al.* (1979) The usefulness of the nonspecific skin hyperreactivity (the pathergy test) in Behçet's disease in Turkey. *Acta Derm. (Stockholm)*, **59**, 77–9.

12. Ohno, S., Sugiura, S., Ohguchi, M. and Aoki, K. (1979) Close association of HLA-B5 with Behçet's disease, in *Immunology and Immunopathology of the Eye* (eds A.M. Silverstein and G.R. O'Connor), Mason, New York, Chapter 4.

13. Yazici, H., Chamberlain, M.A., Schreuder, I. *et al.* (1983) HLA-B5 and Behçet's disease. *Ann. Rheum. Dis.*, **42**, 602–3.

14. Hamuryudan, V., Yurdakul, S., Moral, F. *et al.* (1994) Pulmonary arterial aneurysms in Behçet's syndrome: A report of 24 cases. *Br. J. Rheum.*, **33**, 48–51.

15. Yazici, H., Pazarli, H., Barnes, C.G. *et al.* (1990) A controlled trial of azathioprine in Behçet's syndrome. *N. Eng. J. Med.*, **322**, 281–5.

16. Hamuryudan, V., Moral, F., Yurdakul, S. *et al.* (1994) Systemic interferon α2b treatment in Behçet's syndrome. *J. Rheumatol.*, **21**, 1098–100.

# 15

# Endothelial cell dysfunction in Behçet's disease

*E. Kansu*

## INTRODUCTION

Behçet's disease (BD) was first described by Hulusi Behçet in 1937 with a classic triad of uveitis, and recurrent oral and genital ulcerations [1]. Since then, the entity has been defined as a multisystem disorder with other systemic manifestations including skin lesions, vascular involvement of both arteries and veins, arthritis, pulmonary, gastrointestinal and central nervous system findings. The diagnosis can be made using the International Study Group Criteria for Behçet's Disease [2].

Vascular involvement can be seen at the onset of the disease. There have been several studies reporting the frequency and types of the vascular involvement in BD [3–5]. The reports on vascular complications in BD have indicated prevalences ranging from 4% to 30%, some of which include venous system involvement only. Vascular involvement includes superficial or deep vein thrombosis, arterial aneurysms, Budd–Chiari syndrome and arterial occlusions [5]. In recent analysis, we have reported on 137 Turkish patients with BD, of whom 38 had vascular involvement (a prevalence of 27.7%) with a clear preponderance of venous lesions (88%) compared with arterial involvement (12%) (Table 15.1) [5].

The etiology of BD is not known. Viruses, streptococcal infection, autoimmune mechanisms and endothelial cell dysfunction have been postulated and to date no definite cause has yet been identified [6, 7]. An increased prevalence of HLA-B5 antigen reported in BD from Turkey,

*The Vasculitides.* Edited by B.M. Ansell, P.A. Bacon, J.T. Lie and H. Yazici. Published in 1996 by Chapman & Hall, London. ISBN 978-0-412-64140-4.

**Table 15.1** Behçet's disease – vascular complications (4–30%)

Thrombophlebitis
Budd–Chiari syndrome
Vena cava obstruction
Arterial occlusions
Arterial aneurysms

Israel and Japan was considered to represent a possible genetic suscept-ibility in this entity [8, 9].

Although the exact cause is not known, vasculitis is observed as the common pathology in all the systems involved. In our previous studies, we demonstrated dense perivascular infiltration of lymphocytes around the small venules. Endothelial cells appeared to be normal under light microscope. Using appropriate monoclonal antibodies for T lymphocyte subpopulations, we demonstrated that the majority of mononuclear cells located at the base of the oral ulcers and within the perivascular infil-trates were CD3+-CD4+ T lymphocytes (Figure 15.1) [10]. Our studies also showed perivascular HLA-DR+ cells which indicated the presence of activated lymphocytes.

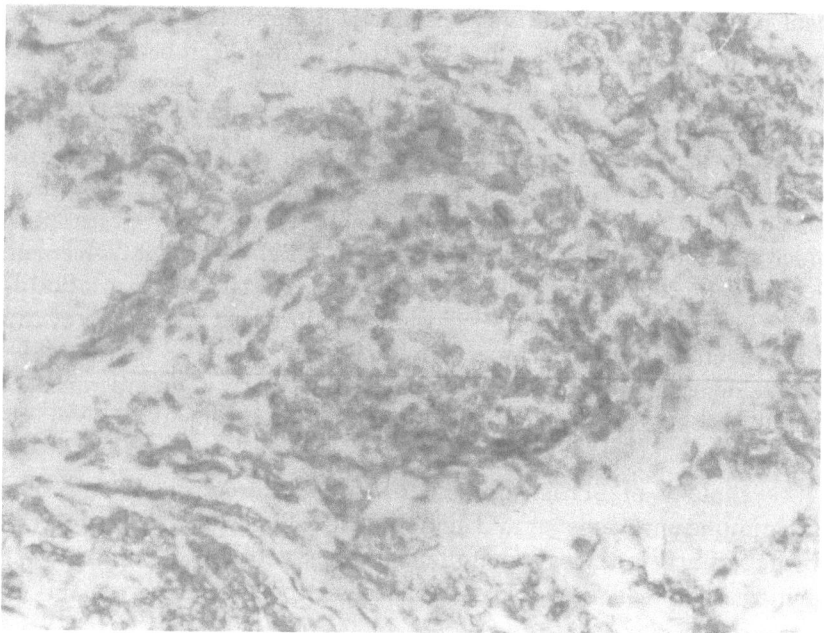

**Figure 15.1** Immunoperoxidase staining of CD3+-CD4+ helper T cells in the perivascular infiltrate (counterstained with hematoxylin, × 460) [10].

**Table 15.2** Functions of endothelium

---

- Hemostatic
    procoagulant
    antithrombotic
    fibrinolytic
- Vascular tone
- Leukocyte interaction
- Permeability/leakage
- Angiogenesis

---

In view of our findings in patients with BD, it appears very likely that the inflammatory infiltrate present at the perivascular space could represent an 'immune vasculitis' due to T cell mediated cellular immune response without significant activation and alteration of peripheral blood lymphocyte subpopulations, except natural killer (NK) cells [10].

In order to elucidate the pathogenesis of the vascular involvement and thrombophlebitis commonly seen in BD, we designed several studies to investigate the endothelial cell functions which could enhance and alter the thrombus formation within the vascular structures. In this chapter we will try to summarize the functional data on the endothelium obtained from several laboratories, including our own.

## FUNCTIONS OF VASCULAR ENDOTHELIUM

Endothelial cells carry out various physiological functions, which are briefly summarized in Table 15.2. In this section, functional features of the normal vascular endothelium and corresponding findings reported to date in patients with Behçet's disease will be summarized under appropriate headings and subtitles.

## HEMOSTATIC FUNCTIONS

Endothelial cells synthesize various active substances to maintain the fluidity of blood and the patency of the blood vessels. Circulating unactivated blood does not normally interact with intact vascular endothelium. Hemostatic functions of vascular endothelium include procoagulant, antithrombotic and fibrinolytic activities.

### Procoagulant activities

Molecules synthesized by vascular endothelial cells that exhibit known procoagulant effects include: von Willebrand factor (vWF), tissue factor

(TF), plasminogen activator-inhibitor-1 and 2 (PAI-1, PAI-2), surface binding of clotting factors, factor-V and platelet-activating factor (PAF).

vWF is synthesized by endothelium, stored in multimeric form in Weibel–Palade endothelial cell secretory granules and released as multimeric molecules of 1–20 million daltons. vWF is an important adhesive molecule for both platelet adhesion and aggregation. Enhanced release from vascular endothelial cells is induced by thrombin and increased plasma levels are found in several processes characterized by inflammation.

Increased plasma levels of vWF were detected in patients with BD and vascular involvement in 1987 by Yazici and co-workers [12]. Studies done in the UK by Chamberlain *et al.* showed higher levels of vWF:antigen in 15 patients with BD compared with patients with seronegative arthritis [11]. Pivetti-Pezzi and co-workers found raised levels of vWF:antigen in three out of 35 Italian BD patients having retinal vasculitis with systemic involvement [13]. Elevated levels of vWF were found in seven out of 84 BD patients (8.3%) in the studies reported by Aydintuǧ *et al.* [14], and Direskeneli found significantly high levels of Factor VIII antigen in BD patients and in patients with systemic vasculitis compared with control subjects [15]. All these results showing elevated plasma levels of vWF and vWF:antigen (vWF:Ag) in patients with BD and in patients with different types of vasculitis may reflect abnormalities or damage to the vascular endothelium due to inflammatory processes.

Vascular homeostasis results from the regulated interaction of the coagulation and fibrinolytic systems. Any imbalance in the system may lead to an increased risk for thrombosis or tendency to bleeding diathesis. The balance between plasminogen activators (PAs) and plasminogen activator-inhibitor (PAI-1) appears to have a very critical importance in the fibrinolytic activity. PAI-1 is a 52 kD protein generated and released by endothelial cells to maintain a significant circulating concentration. PAI-1 is a rapid and specific inhibitor of both tissue-type and urokinase-type plasminogen activators. Recent studies using *in situ* hybridization techniques indicate that the primary site of synthesis of PAI-1 in the mouse is the vascular endothelial cells. Elevated PAI-1 during septicemia and atherosclerosis may inhibit normal fibrin clearance mechanisms and may contribute to thrombotic events. Studies performed by Chamberlain and co-workers showed that subjects with Behçet's disease had depressed plasma fibrinolytic activity with reduced plasminogen activator activity due to increased tissue plasminogen activator-inhibitor (tPA-I) and PAI:antigen [11]. PAI: antigen levels were increased in the acute phase response and high levels were detected in BD patients which was considered to be a reflection of an inflammatory vasculitis. Similar results were also obtained in patients with malignancy, postmyocardial infarction and postoperative situations [11]. Özcebe and co-workers were not able to demonstrate any major abnormalities in prothrombin time, aPTT, tPA-Ag and PAI-1 levels pre- and post-1-deamino-8-D-arginine vasopressin (DDAVP) infusion in 14 patients with active BD compared with 10 healthy volunteers [16].

## Antithrombotic and antiaggregant activities

Several molecules are synthesized by vascular endothelial cells that exhibit antithrombotic activities. Molecules with known antithrombotic effects include: prostacyclin ($PGI_2$), nitric oxide (NO), thrombomodulin (TBM), heparin-like glycosamino-glycans and tissue plasminogen activator (tPA).

$PGI_2$ is a metabolite of arachidonic acid with very potent antiaggregatory and vasodilatory properties. $PGI_2$ is synthesized by the endothelial cells, secreted extracellularly and acts on platelets as well as vascular smooth muscle cells. $PGI_2$ is unstable in aqueous solutions of neutral pH with half-lives of 2–3 minutes. In contrast, for the balance of hemostasis, $PGI_2$ is converted to thromboxane $A_2(TxA_2)$ in platelets. $TxA_2$ induces platelet activation, constricts smooth muscle cells and promotes platelet aggregation [17].

In our previous studies, we investigated the biosynthesis of $PGI_2$ by vascular endothelium and plasma $TxB_2$ levels in patients with active BD [18]. The plasma 6-keto $PGF_{1\alpha}$, which is a stable metabolite of $PGI_2$, was measured by a radioimmunoassay and the levels were significantly reduced ($58.6 \pm 10.8$ pg/ml) in 37 patients with BD compared with normal subjects ($110.3 \pm 22.4$ pg/ml). $PGI_2$ levels in patients with thrombophlebitis were not significantly different from those without vascular involvement [18].

Plasma $TxB_2$ levels in BD patients averaged $80.6 \pm 10.8$ pg/ml and in the group without vascular thrombosis the average $TxB_2$ value was $65.4 \pm 5.7$ pg/ml. These values were significantly increased compared with normal control subjects with an average value of $49.7 \pm 10.6$ pg/ml (range of 20.4–148.3 pg/ml). These data suggested that the cyclooxygenase pathway in the platelets was intact and the impairment of prostanoid synthesis in patients with BD could be confined to the vessel walls or endothelium.

In order to evaluate the low plasma $PGI_2$ levels and increased $TxB_2$ values further, we obtained forearm vein segments from the patients and markedly impaired $PGI_2$ biosynthesis was demonstrated *in vitro* [19]. In three patients, no biosynthesis of $PGI_2$ by the fresh vein segments was observed during 30 minutes of incubation, while six normal subjects synthesized increasing amounts of $PGI_2$ during the same period (Figure 15.2). In patients with BD, these data suggested a possible interaction between inflammatory perivascular cellular infiltrate secreted cytokines

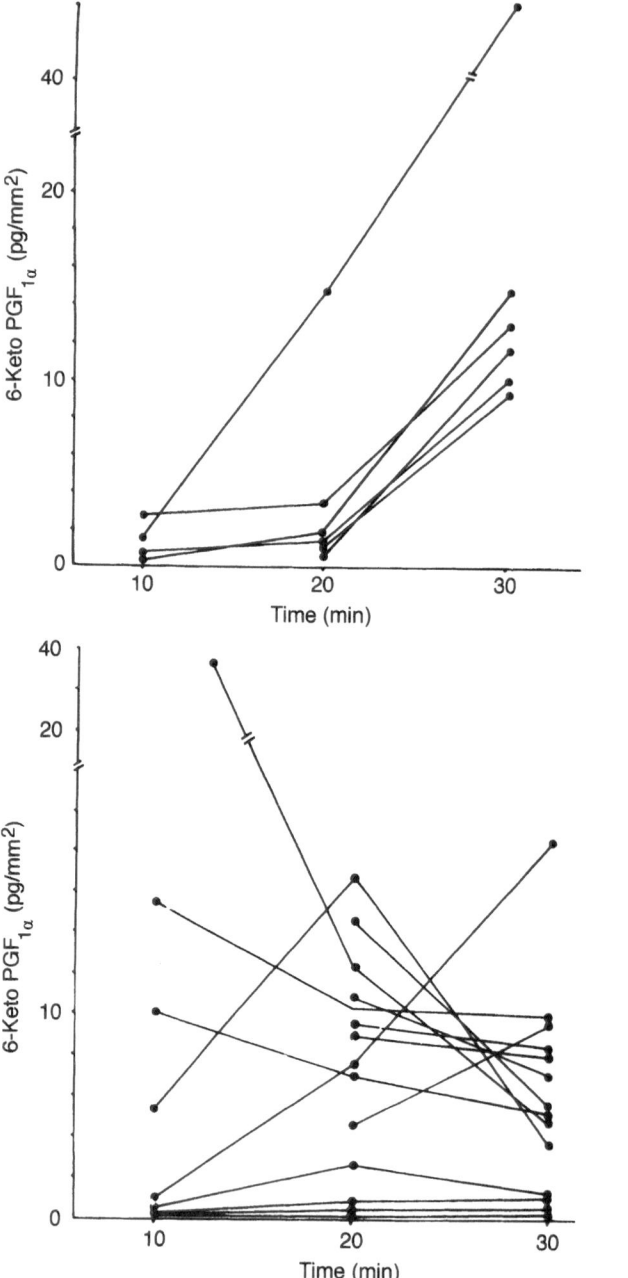

**Figure 15.2** Production of prostacyclin during a 30 minute incubation period by venous tissue from normal control subjects (upper) and from patients with Behçet's disease (lower). 6-keto-prostaglandin F1-α levels are expressed as pg/mm$^2$ of the endothelial surface of the vein segment [18].

and the cells of the vessel wall resulting in significant functional impairment or dysfunction of the endothelial cells. In addition, reduced biosynthesis of $PGI_2$ by the vascular endothelial cells could have a contributory role in the pathogenesis of thrombotic complications seen in BD patients.

Thrombomodulin (TBM) is a cell-surface 35 kD integral glycoprotein mainly present on the luminal surface of endothelial cells. It comprises six epidermal growth factor-like domains. TBM acts as a 'thrombin receptor' and neutralizes its procoagulant activities. When thrombin is produced, it rapidly binds epidermal-growth factor domains in thrombomodulin and forms a termolecular complex with protein-C facilitated by repeats of thrombomodulin. TBM also acts as a co-factor for thrombin catalyzed activation of protein-C [20]. Activated protein-C catalytically degrades coagulation co-factors Va and VIIIa bound to membrane surfaces, interrupting subsequent thrombin production. TBM has been accepted as a major anticoagulant proteoglycan present on the luminal surfaces of the vascular endothelial cells.

Increased TBM levels had been reported in disseminated intravascular coagulation (DIC), adult respiratory distress syndrome (ARDS) and pulmonary thromboembolism, thrombotic thrombocytopenic purpura and diabetic microangiopathy [21]. In the recent International Conference on BD, Tribout and co-workers, using a two-site ELISA, reported significantly low levels of TBM in patients with BD compared with controls, but thrombotic events were not statistically associated with lower TBM levels [22]. Ohdama *et al.* published the plasma TBM concentrations in patients with systemic lupus erythematosus (SLE), rheumatoid arthritis, juvenile rheumatoid arthritis, Sjögren's syndrome, systemic sclerosis, polymyositis, Wegener's granulomatosis and Behçet's disease [23]. Plasma TBM levels in BD, systemic sclerosis, polymyositis and Wegener's granulomatosis were significantly higher than those in the control group. The mean value of TBM for six active BD patients was significantly higher than that of the controls but the difference was not remarkable [23]. Two patients with vascular-type BD showed the highest plasma TBM among six patients with active BD. All the elevated values were decreased along with the amelioration of the findings following treatment. The authors suggested that TBM could be helpful for evaluating vascular injury in collagen diseases.

In view of conflicting plasma thrombomodulin results in patients with BD, further studies in a large number of patients are needed to clarify this issue.

## Fibrinolytic activities

Hypofibrinolysis leading to decreased removal of fibrin deposits would be a prime candidate for a role in the development of atherothrombosis.

Recent experimental data have shown that increased plasma PAI-1 levels do indeed have a prothrombotic effect. Increased PAI-1 decreases endogenous and exogenous fibrinolysis and increases the extension of thrombosis.

Normal or elevated levels of t-PAI, tPA:antigen and PAI-1 as well as decreased levels of $\alpha$-2-plasmin inhibitor had been reported in patients with BD. No consistent data is available to date and we need more well-standardized studies to have better information on the fibrinolytic system in BD.

## VASCULAR TONE

Vascular endothelial cells have an important metabolic function with respect to vasoactive substances. The outer surface of the endothelial cell contains angiotensin-converting enzyme which catalyzes the formation of the vasoconstrictor angiotensin II from its inactive precursor angiotensin I. The same enzyme inactivates bradykinin which is well-known as a potent vasodilator. Endothelial cells can also generate $PGI_2$, nitric oxide (NO or EDRF) and endothelin-1 for the physiologic regulation of the vascular tone and maintaining the liquidity of blood. The physiologic role of $PGI_2$ and its importance in BD was reviewed above.

Nitric oxide (NO), formerly known as endothelial derived relaxing factor (EDRF), is liberated by the action of nitric oxide synthase on L-arginine in endothelial cells. NO can be generated by various cells and by diverse stimuli to cause vasorelaxation [24]. NO is an essential modulator of vascular smooth muscle tone, systemic vascular resistance and regional blood flow. It has been shown that nitric oxide and prostacyclin synergize with each other as inhibitors of platelet aggregation and inducers of disaggregation [25]. Recent studies suggest that impaired generation of NO contributes to the vascular and thrombotic complications in atherosclerosis [25]. Research on the biology of NO and its role in several vascular disorders including hypertension, thrombosis, septic shock and atherosclerosis is in progress. To date, no scientific data is available in BD, but it holds promise as a very fascinating research problem.

Endothelin (ET) is a novel vasoconstrictor peptide which is synthesized by the vascular endothelial cells [26]. It is the most potent vasoconstrictor agent known to date and human ET has the same structure as porcine ET. ET is a family consisting of three isopeptides named ET-1, ET-2 and ET-3 [26]. We studied plasma levels of ET-1,2 in patients with active and inactive BD with and without vascular complications. Healthy individuals and patients with systemic lupus erythematosus (SLE) and vasculitis in the active and inactive periods of their disease were included as the control groups. Plasma ET-1,2 levels were analyzed in 49 patients with BD, 16 patients with vasculitis and in 23 healthy subjects. The highest mean concentration of ET-1,2 was found in patients with active vascular BD (72.5 ± 17.4 fmol/ml) compared with patients with inactive BD without vascular involvement (32.3 ± 2.6 fmol/ml) and healthy subjects (29.8 ± 2.5 fmol/ml). The mean concentration of ET-1,2 was also elevated in patients with active vasculitis including SLE [27].

Recent studies suggest that ET may serve as a plasma indicator of endothelial cell damage due to vascular injury. Increased plasma ET levels were reported in patients with pulmonary hypertension, primary Raynaud's phenomenon and in those undergoing surgical procedures [28].

Elevated plasma levels of ET could be a secondary event either due to an increased biosynthesis or release of the preformed peptide from the damaged endothelium due to vasculitis. Our data may suggest that a rise in ET-1,2 levels could reflect the endothelial cell injury induced by 'immune-cell mediated' vasculitis. In addition, further studies are warranted to identify whether ET could serve as a novel parameter in the entities characterized with vasculitis.

## LEUKOCYTE INTERACTION

Vascular endothelium is now recognized as an active participant in a variety of pathophysiological processes, including inflammation and tumor metastasis. In recent years, intense research projects focused on endothelial cell surface molecules that support the adhesion of leukocytes. Leukocytes also express specific counter-receptors for one or more of the endothelial adhesive glycoproteins [29]. Four families of endothelial-leukocyte adhesion molecules have been identified and characterized. These include selectins, integrins, immunoglobulin-gene superfamily and cadherins.

Selectins are involved in the early, low-affinity binding interactions and their lectin-binding sites interact with oligosaccharides presented on the surfaces of reactive cells. There are three well-described members of the selectins: platelet-selectin (P-selectin), endothelial-selectin (E-selectin) and leukocyte-selectin (L-selectin) [30]. Integrins are a family of membrane glycoproteins that mediate cell adhesion. Integrins appear to be the primary mediators of cell–extracellular matrix adhesion and also serve as one of the many families of molecules active in cell–cell adhesion. Members of the integrin receptor family include LFA-1 (CD11a/CD18), Mac-1 (CD11b/CD18), p150,95 and VLA-1 through to VLA-6 [30]. Three members of the immunoglobulin gene superfamily are well identified: ICAM-1 (CD54), ICAM-2, VCAM-1 and NCAM. Endothelial cells also express CD31, an immunoglobulin-like molecule

**Table 15.3** Families of adhesion molecules and their counter-receptors

| Adhesion molecule | Counter-receptor |
| --- | --- |
| Ig-superfamily | |
| CD2 | LFA-3 |
| LFA-3 (CD58) | CD2 |
| ICAM-1 (CD54) | LFA-1 |
| ICAM-2 | LFA-1 |
| VCAM-1 | VLA-4 |
| Integrin family | |
| LFA-1 (CD11a/CD18) | ICAM-1 |
| Mac-1 (CD11b/CD18) | ICAM-1 |
| p150,95 (CD11c/CD18) | Unknown |
| VLA-1 (CD48a/CD29) | Laminin, collagen |
| VLA-5 (CD49c/CD29) | Fibronectin |
| Selectin family | |
| E-selectin (ELAM-1) | Sialyl–Lewis antigen |
| L-selectin (LECAM-1) | Vascular addressins |
| P-selectin | Unknown |

that is concentrated at opposing endothelial cell borders. Endothelial cells can also express LFA-3 (CD58), and class-1 and class-2 MHC antigens which are immunoglobulin-related molecules. ICAM-1 is widely distributed on endothelial cells and its surface expression can be increased by exposure of endothelial cells to various cytokines or to endotoxin. Families of adhesion molecules and their counter-receptors involved in leukocyte–endothelial cell interactions are briefly summarized in Table 15.3 [31].

Özgün and co-workers studied the effect of sera obtained from BD patients on the adherence of normal peripheral blood neutrophils to human vascular endothelial cells *in vitro*. Sera from the patients induced an increase in surface expression of CD11a molecule on normal neutrophils; however, CD11b molecule did not show any significant increments. ICAM-1 expression was also found to be increased on the human vascular endothelial cells following incubation with the sera of patients with BD. The CD11a-positive neutrophils were higher (95%) in patients with BD compared with healthy control subjects (54%) [32].

Inaba and co-workers studied L-selectin-Mac-1 (CD11b/CD18) and CD44 expression on the peripheral blood leukocytes in 25 Japanese patients with BD who had active ocular lesions. L-selection (LECAM-1) and CD44 expression was dramatically reduced from polymorphonuclear leukocytes upon ocular attacks and CD11b/CD18 expression was not significantly altered [33]. There were no considerable changes on the surfaces of T lymphocytes concerning the expression of these adhesion molecules during active ocular inflammation.

Studies are in progress in several laboratories researching the role of the adhesion system to elucidate the pathogenesis of pathergy test, chemotaxis of neutrophils to the lesion sites and vasculitis in BD.

## Antibodies to endothelial cells

Surface molecules on endothelial cells carry epitopes which could serve as targets for autoantibodies or alloantibodies. Antiendothelial cell antibodies (AECA) have been detected by their ability to bind to cultured human vascular endothelial cells *in vitro*. AECA may damage endothelial cells via complement-mediated or antibody-dependent cellular cytotoxic mechanisms. AECA have been described in different entities characterized by vasculitis and/or thrombosis. Aydintuğ and co-workers showed that circulating antibodies directed against surface antigens on human endothelial cells were present in 18.1% of patients with BD but not in healthy controls [34]. There was no evidence for cytotoxicity induced by AECA in BD patients. The prevalance of acute thrombotic events and retinal lesions was found to be significantly higher in AECA-positive BD patients than in the AECA-negative group. Anticardiolipin and antineutrophil cytoplasmic antibodies were negative in all. Direskeneli *et al.* detected AECA in 29% of 70 BD patients; arthritis had a positive relationship with the presence of AECA [15]. There was no relationship with other organ manifestations, pathergy test, HLA-B5 and AECA levels.

The role of AECA in the pathogenesis of vascular endothelial damage is not well described and the antigenic specificity of AECA is not yet clear. These antibodies not being cytotoxic for the endothelial cells may not easily explain the changes observed in endothelial cell functions. Although the presence of AECA in patients with BD and other rheumatic diseases has been demonstrated, the findings to date only suggest a possible role of AECA rather than a cause–effect relationship. More studies are needed to clarify the mechanisms and relationship of endothelial cell damage in vasculitis and appearance of AECA.

Histopathologic and immunohistochemical studies of the biopsy lesions obtained from patients with Behçet's disease reveal typical features of 'vasculitis' with very dense perivascular infiltrations of T lymphocytes, B lymphocytes and neutrophils. Secretion of inflammatory cytokines from these cells including tumor necrosis factor-alpha (TNF-α), interleukin-1 (IL-1) and interleukin-6 (IL-6) may have injury, derangement and/or modulating effect on the vascular endothelial cells. IL-1 and TNF modulate both qualitative and quantitative changes in the secretion and surface expression of many vasoactive and hemostatic mediator molecules. The mechanism of endothelial cell damage in BD where vascular thrombosis frequently appears, could be related to direct endothelial damage by the cellular infiltrate or indirect cytopathic action of the cytokines (notably IL-1 and TNF).

Increased levels of plasma vWF:antigen [11, 12], impaired biosynthesis of $PGI_2$ in the vessel walls [19], increased levels of endothelin-1,2 [27] and thrombomodulin [21] and the presence of antiendothelial cell antibodies [34] are highly suggestive indicators for endothelial cell dysfunction in patients with BD. These findings may also have a contributory role in the pathogenesis of thrombotic vascular complications seen in BD patients. The use of some endothelial cell markers in BD or in patients

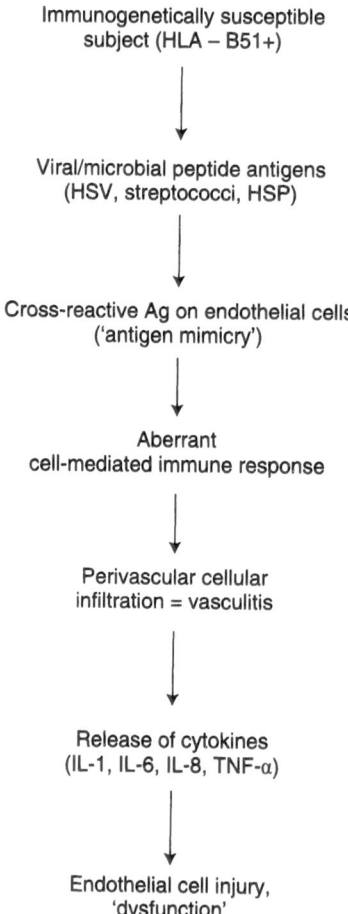

**Figure 15.3** Possible mechanisms for endothelial cell dysfunction in Behçet's disease.

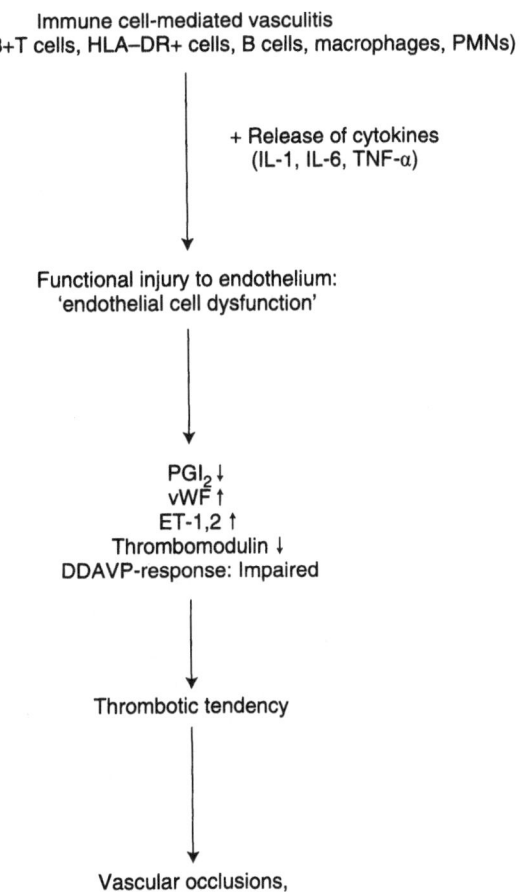

Immune cell-mediated vasculitis
(CD4+ and CD8+T cells, HLA–DR+ cells, B cells, macrophages, PMNs)

+ Release of cytokines
(IL-1, IL-6, TNF-α)

Functional injury to endothelium:
'endothelial cell dysfunction'

PGI$_2$ ↓
vWF ↑
ET-1,2 ↑
Thrombomodulin ↓
DDAVP-response: Impaired

Thrombotic tendency

Vascular occlusions,
Local lesions

**Figure 15.4** Possible mechanisms for vascular occlusion in Behçet's disease.

with other forms of vasculitis may also be beneficial in identifying and monitoring effective therapy.

In view of the available information to date, the possible mechanisms for endothelial cell dysfunction and vascular occlusions are briefly proposed in Figures 15.3 and 15.4.

## REFERENCES

1. Behçet, H. (1937) Über rezidivierende aphtose, durch ein Virus verursachte Geschwüren am Mund, am Auge und an den Genitalien. *Dermatol. Wochenschr.*, **105**, 1152–7.

2. International Study Group for Behçet's Disease (1990) Criteria for diagnosis of Behçet's Disease. *Lancet,* **335**, 1078–80.

3. Kansu, E., Özer, F.L., Akalin, E. *et al.* (1972) Behçet's syndrome with obstruction of the venae cavae: A report of seven cases. *Q.J. Med.,* **41**, 151–68.

4. Müftüoğlu, A.Ü., Yurdakul, S., Yazici, H. *et al.* (1986) Vascular involvement in Behçet's disease-a review of 129 cases, in *Recent Advances in Behçet's disease* (eds T. Lehner and C.G. Barnes), Royal Society of Medicine Services International Congress and Symposium Series No. 103, London, pp. 255–60.

5. Koç, Y., Güllü, I., Akpek, G. *et al.* (1992) Vascular involvement in Behçet's disease. *J. Rheumatol.,* **19**, 402–10.

6. Lehner, T. (1986) The role of disorder in immunoregulation associated with herpes simplex virus type 1 in Behçet's disease, in *Recent Advances in Behçet's disease* (eds T. Lehner and C.G. Barnes), Royal Society of Medicine Services, No. 103, London, pp. 31–6.

7. Ohno, S., Yoshikawa, K., Sasamoto, Y. *et al.* (1991) Immune responses against strepococcus sanguis in Behçet's disease, in *Behçet's disease – Basic and Clinical aspects* (eds J.D. O'Duffy and E. Kokmen), Marcel Dekker Inc., New York, pp. 395–404.

8. Ohno, S., Ohguchi, M., Hirose, S. *et al.* (1982) Close association of HLA-BW51 with Behçet's disease. *Arch. Opthalmol.,* **100**, 1455–8.

9. Yazici, H., Akokan, G., Yalçin, B. and Müftüoğlu, A. (1979) The high prevalence of HLA-B5 in Behçet's disease. *Clin. Exp. Immunol.,* **30**, 259–61.

10. Celenligil, H., Kansu, E., Ruacan, Ş. and Eratalay, K. (1991) Characterization of peripheral blood lymphocytes and immunohistological analysis of oral ulcers in Behçet's disease, in *Behçet's Disease – Basic and Clinical Aspects* (eds J.D. O'Duffy and E. Kökmen), Marcel Dekker Inc., New York, pp. 487–96.

11. Chamberlain, M.A., Hampton, K.K., Menon, D. and Davies, J.A. (1991) Coagulation and fibrinolytic activity in Behçet's disease, in *Behçet's Disease – Basic and Clinical Aspects* (eds J.D. O'Duffy and E. Kökmen), Marcel Dekker Inc., New York, pp. 541–4.

12. Yazici, H., Hekim, N., Ozbakir, F. *et al.* (1987) Von Willebrand factor in Behçet's syndrome, *J. Rheumatol.,* **14**, 305–6.

13. Pivetti-Pezzi, P., Priori, R., Catarinelli, G. *et al.* (1992) Markers of vascular injury in Behçet's disease associated with retinal vasculitis. *Ann. Ophthalmol.,* **24**, 411–4.

14. Aydintuğ, A.O., Tokgöz, G., Jedryka-Goral, A. *et al.* (1993) Von Willebrand factor antigen in patients with Behçet's disease. *Rev. Méd. Int.,* **14** (Suppl. 1), 132s, Abstract 157.

15. Direskeneli, H., Keser, G., D'Cruz, D. *et al.* (1993) Anti-endothelial cell antibodies and endothelial cell dysfunction in Behçet's disease. *Rev. Méd. Int.,* **14** (suppl 1), 134s, Abstract 162.

16. Özcebe, O., Özdemir, O., Dündar, S. and Kirazli, S. (1993) Impaired F. VIII: C release to intravenous 1-Deamino-8-D-arginine vasopressin in Behçet's disease: a clue for endothelial dysfunction? *Rev. Méd. Intern.,* **14** (Suppl. 1) 27s, Abstract CO8.

17. Fitzgerald, G.A., Oates, J.A., Hawiger, J. *et al.* (1983) Endogenous biosynthesis of prostacyclin and thromboxane and platelet function during chronic administration of aspirin in man. *J. Clin. Invest.,* **71**, 676–88.

18. Kansu, E., Şahin, G., Şahin, F. *et al.* (1986) Impaired prostacyclin synthesis by vessel walls in Behçet's disease. *Lancet,* **2**, 1154.

19. Kansu, E., Sivri, B., Şahin, G. *et al.* (1991) Endothelial cell dysfunction in

Behçet's disease, in *Behçet's disease – Basic and Clinical Aspects* (eds J.D. O'Duffy and E. Kokmen), Marcel Dekker Inc., New York. pp. 523–30.

20. Esmon, C.T. (1989) The roles of protein C and thrombomodulin in the regulation of blood coagulation. *J. Biol. Chem.*, **264**, 4743–6.
21. Takano, S., Kimura, S., Ohdama, S. and Aoki, N. (1990) Plasma thrombomodulin in health and diseases. *Blood*, **76**, 2024–9.
22. Tribout, B., Huong-Du, L.T., Gozin, D. *et al.* (1993) Low levels of circulating thrombomodulin in patients with Behçet's disease. *Rev. Méd. Int.*, **14** (Suppl. 1), 27, Abstract CO9.
23. Ohdama, S., Takano, S., Miyake, S. *et al.* (1994) Plasma thrombomodulin as a marker of vascular injuries in collagen vascular diseases. *Am. J. Pathol.*, **101**, 109–13.
24. Palmer, R.M., Ashton, D.S. and Moncada, S. (1988) Vascular endothelial cells synthesize nitric oxide from L-arginine. *Nature*, **333**, 664–6.
25. Radomski, M.W. and Moncada, S. (1993) Regulation of vascular homeostasis by nitric oxide. *Thromb. Haemost.*, **70**, 36–41.
26. Inoue, A., Yanigasawa, M., Kimura, S. *et al.* (1989) The human endothelin family: three structurally and pharmacologically distinct isopeptides predicted by three separate genes. *Proc. Nat. Acad. Sci. USA*, **86**, 2863–7.
27. Koç, Y., Kansu, E., Koray, Z. *et al.* (1993) Endothelin-1,2 levels in Behçet's disease, in *Behçet's Disease* (eds P. Godeau and B. Wechsler), Elsevier Science Publishers, pp. 97–101.
28. Vane, J.R., Anggard, E.E. and Botting, R.M. (1990) Regulatory functions of the vascular endothelium. *N. Engl. J. Med.*, **323**, 27–36.
29. Granger, D.N., and Kubes, P. (1994) The microcirculation and inflammation: Modulation of leukocyte-endothelial cell adhesion. *J. Leukocyte Biol.*, **55**, 662–75.
30. Ruoslahti, E. (1991) Integrins. *J. Clin. Invest.*, **87**, 1–5.
31. Pardi, R., Inverardi, L. and Bender, J.R. (1992) Regulatory mechanisms in leukocyte adhesion: flexible receptors for sophisticated travelers. *Immunol. Today*, **13**, 224–30.
32. Özgün, S., Akoğlu, T., Direskeneli, H. and Yurdakul, S. (1993) Neutrophil adhesion to endothelial cells in patients with Behçet's disease. *Rev. Méd. Int.*, **14** (Suppl. 1), 133s, Abstract 159.
33. Inaba, G., Kaku, H., Ujihara, H. *et al.* (1993) Leukocyte adhesion molecules in active ocular lesions of Behçet's disease. *Rev. Méd. Int.*, **14** (Suppl. 1), 24s, Abstract CO4.
34. Aydintuğ, A.O., Tokgöz, G., D'Cruz, D.P. *et al.* (1993) Antibodies to endothelial cells in patients with Behçet's disease. *Clin. Immunol. Immunopathol.*, **67**, 157–62.

# 16

# Pulmonary vasculitides

*L. Mouthon, F. Lhote and L. Guillevin*

## INTRODUCTION

Pulmonary vasculitides include various conditions whose sole or predominant histological feature is inflammation of pulmonary vessels, and refer only to vascular damage related to a primary pathogenetic mechanism and of proven or probable immunological origin. Vasculitis classifications available in the literature are based on the size of the vessels involved, presumed pathophysiological mechanisms of the diseases [1], or histopathological characteristics of the vasculitis. Histological findings usually reflect clinical symptoms of suspected diseases, including organs preferentially involved, radiographic and computed tomography (CT) scan findings, immunological profile, and evolution of the disease process under appropriate therapy. Considering all these elements, most lung vasculitides can be identified as well-defined clinicopathological entities (Table 16.1): Wegener's granulomatosis (WG), Churg–Strauss syndrome (CSS), Behçet's disease, Takayasu's arteritis, temporal arteritis, etc. Alveolar hemorrhage, when present, can be isolated and considered to be idiopathic, related to Goodpasture's syndrome or to other systemic diseases involving both lungs and kidneys.

Exclusion of secondary vascular involvement is sometimes difficult and inflammatory conditions, such as infectious diseases, can lead to vasculitis simply by extension of the parenchymal disease process to the vascular wall. Therefore, it is essential to perform appropriate cultures of biopsy specimens to eliminate an infectious etiology. Moreover, confident diagnosis of primary vasculitis requires evidence of vascular damage, such as necrosis or an inflammatory infiltrate in the vessel

*The Vasculitides.* Edited by B.M. Ansell, P.A. Bacon, J.T. Lie and H. Yazici.
Published in 1996 by Chapman & Hall, London. ISBN 978-0-412-64140-4.

**Table 16.1** Clinicopathological entities of lung vasculitides

| Disease | Size and type of vessels involved | | |
| --- | --- | --- | --- |
| | Large pulmonary arteries | Small- and medium-sized arteries | Capillaries, veins, venules |
| Wegener's syndrome | | Parenchymal necrosis | Condensation and intra-alveolar hemorrhage |
| Churg–Strauss syndrome | | | Lung condensation |
| Behçet's syndrome | Aneurysms and stenosis | | Rare lung condensation and intra-alveolar hemorrhage |
| Takayasu's arteritis | Stenosis | | |
| Temporal arteritis | | Parenchymal necrosis | Rare lung condensation |
| Microscopic polyangiitis | | | Intra-alveolar hemorrhage |
| Miscellaneous | | | |

wall, thrombosis, aneurysm or disruption of the elastic lamina of larger vessels.

Although deposition of immune complexes within the vessel wall has been extensively investigated as the major immunopathological process leading to lung vasculitis, no experimental data have been obtained that could support such a hypothesis. On the other hand, autoantibodies directed against neutrophil cytoplasmic antigens (ANCA) have recently been detected in the sera of patients with systemic vasculitides [2]. Although ANCA frequency varies depending upon the disease, the presence of c-ANCA diffuse cytoplasmic staining pattern favors a diagnosis of WG, while that of p-ANCA perinuclear staining pattern indicates CSS and antibodies to myeloperoxidase (MPO). ANCA are of pathophysiological interest because they support the hypothesis of an autoantibody mediated vascular injury. Herein clinical, pathological, radiological and immunological aspects and outcome of the major lung vasculitides, especially WG, CSS and microscopic polyangiitis (MPA), are reviewed.

DIAGNOSIS OF LUNG VASCULITIDES

Typical clinical symptoms that enable a correct diagnosis of lung vasculitis can occur late in the course of the disease. Although difficult, to obtain early diagnosis would result in early treatment and improve the prognosis of the disease. Since lung vasculitis often develops in conjunction with systemic vascular disorders, attention and complementary examinations must be directed at extrapulmonary signs and symptoms, such as cutaneous vasculitis, upper respiratory tract involvement, glomerulonephritis and peripheral nervous system involvement. Poor general condition is observed frequently and is associated with biological findings that are consistent with an inflammatory syndrome. When evidence for extrapulmonary vascular involvement is not found, or a particular condition exists, for example hypereosinophilia and sometimes p-ANCA in CSS, or anti-MPO ANCA by ELISA and immunofluorescence (c-ANCA), and anti-PR3 ANCA in WG syndrome, a surgical pulmonary biopsy should be considered.

### Alveolar hemorrhage and Goodpasture's syndrome

Alveolar hemorrhage can be observed in a large number of diseases, and can be idiopathic or related to Goodpasture's syndrome or other systemic diseases involving both lungs and kidneys (pulmonary–renal syndrome). Independently of the underlying disease process, typical clinical findings of pulmonary hemorrhage include: hemoptysis, alveolar opacities on chest X-ray, anemia, dyspnea and hypoxemia [3]. Therefore, the diagnosis is highly dependent upon clinical, biological and pathological extrapulmonary findings. According to Thomas and Irwin, diffuse intra-alveolar hemorrhage can be classified into three major groups on the basis of immunological studies [4]: in the first group, antiglomerular basement membrane (GBM) antibodies are present and their ability to recognize and alter the basement membrane of alveoli causes diffuse pulmonary hemorrhage [5]; in the second group, deposition of immune complexes leads to pulmonary and renal damage; in the third group, diffuse pulmonary hemorrhage without apparent extrapulmonary disease and without evidence of an immunological mechanism may represent idiopathic pulmonary hemosiderosis [3, 6]. This classification must be reviewed taking into consideration ANCA detection, which has greatly improved our understanding of the pathogenetic processes of systemic vasculitides.

#### Goodpasture's syndrome

Goodpasture's syndrome is an autoimmune disease caused by anti-GBM antibodies, which react specifically with the basement membranes of the glomeruli, alveoli and renal tubules, leading to alveolar hemorrhage and glomerular damage [5]. In most cases, alveolar hemorrhage occurs in patients presenting with renal involvement, although it can be the initial event. Anti-GBM antibodies can be detected in serum by radioimmunoassay (RIA) or ELISA. Neither episodes of alveolar hemorrhage nor their severity correlate with the level of anti-GBM antibodies. In most cases, renal biopsy reveals either focal or diffuse glomerulonephritis, with crescents and necrosis. Immunofluorescence reveals linear deposits of IgG and often C3 along the capillary walls [5]. Treatment combines plasma exchanges and steroids. In rare cases the association of ANCA and anti-GBM antibodies has been found.

### Pulmonary hemorrhage

Pulmonary hemorrhage may also occur without anti-GBM antibodies. Many diseases other than Goodpasture's syndrome can lead to intraalveolar hemorrhage, such as systemic lupus erythematosus (SLE), WG, other systemic necrotizing vasculitides, cryoglobulinemia, immune complex-mediated glomerulonephritis and idiopathic crescentic glomerulonephritis with negative immunofluorescence. Most of these diseases are addressed in other chapters and herein we focus on systemic necrotizing vasculitides.

## LUNG VASCULITIDES

### Wegener's syndrome

WG was first described in 1936 [7] and is characterized by involvement of lung (nodules and infiltrates), ear, nose and throat (ENT) and kidneys (RPGN); the vasculitis involves medium- and small-sized arteries, but also veins and capillaries; extravascular granulomas are present. c-ANCA are the immunological marker of the disease and ELISA detects antibodies that are directed against proteinase 3 (PR3). ANCA are present in 80% of patients with systemic manifestations involving ENT, lung and kidneys, and in 40% of those with mono- or paucisymptomatic disease.

The lung is the most frequently involved organ and it may even be the only one affected. Usually, the initial symptoms are rhinitis, sinusitis or other ENT signs. Pulmonary signs usually occur later [8], but can be the initial symptoms. Glomerulonephritis can be observed at any time during the course of the disease.

*Pathology*

Pathological lesions are well known. Macroscopically pulmonary nodules are white necrotic masses that are sometimes excavated. The three pathological criteria of WG are necrosis, vasculitis and granuloma [9, 10].

At the early stage of the disease, microabscesses with neutrophils [10] are found in intra-alveolar septae, vessel walls, pleura and bronchioles. Necrosis can spread and progressively destroy larger areas of lung. Necrotic zones are surrounded by histiocytes and a palisadic granuloma. Fibrosis is the ultimate stage of necrosis.

Vasculitis can be present in arteries, veins and capillaries; necrosis of the vessel wall, infiltrates with lymphocytes, neutrophils and histiocytes are observed. Healing is characterized by stenosis of the vessel lumen and fibrosis of vessel wall.

The granuloma of WG is characterized by histiocytes which surround the center of the granuloma. Giant cells can also be found. Leukocytoclastic capillaritis is frequent, mostly in patients who simultaneously present alveolar hemorrhage. During the acute phase of the disease, exudative alveolitis, alveolar hemorrhage and capillaritis are seen. A bronchocentric granuloma can be seen [11]. Bronchial involvement is characterized by necrosis, micro-abscesses of the bronchial mucosa and palisadic granuloma. Bronchiolitis has also been observed: follicular, granulomatous and sometimes with organizing pneumonia. Fibrosis is common in WG and present most frequently under cyclophosphamide treatment. Pleuritis also occurs often, secondary to lung nodules or as the consequence of pleural vasculitis.

*Clinical aspects*

WG symptoms are not specific: cough, fever, pleural effusion, pneumothorax, etc. [12]. The main respiratory and radiological symptoms are described in Tables 16.2 and 16.3. Tracheobronchial lesions are frequent in WG [12]: bronchial inflammation, mucosal ulcers, stenosis and pseudotumor can be seen [13]. Tracheal stenosis has been observed alone or associated with other respiratory signs of WG [12, 14–16]. Pulmonary radiographs and CT scans show nodules ranging from 5–100 mm in diameter, regularly round [8, 12, 17], but fewer than 10 nodules are found in both lungs. Cavities with or without liquid are seen.

Lung infiltrates can be the consequence of alveolar hemorrhage [12], or can also be due to pneumonia. Less frequently, interstitial lung disease has been described.

Atelectasia is rare and due to endobronchial stenosis which is often

**Table 16.2** Clinical manifestations in 77 patients with pulmonary Wegener's granulomatosis [12]

| Manifestation* | No. (%) |
|---|---|
| Lung + kidney + URT + other | 42 (55) |
| Lung + kidney + other | 8 (10) |
| Lung + URT + other | 7 (9) |
| Lung only | 7 (9) |
| Lung + URT | 5 (7) |
| Lung + kidney | 4 (5) |
| Lung + URT + kidney | 3 (4) |
| Lung + other | 1 (1) |

*URT, upper respiratory tract; other, any clinical manifestation in any organ other than the lung, kidney or URT.

**Table 16.3** Features of pleuropulmonary imaging in 77 patients with pulmonary Wegener's granulomatosis [12]

| Type of opacity | No. of cases (%) | Cavitary opacities |
|---|---|---|
| Nodules | 53 (69) | 26/53 (49%) |
|    unilateral | 14 (26) | |
|    bilateral | 39 (74) | |
| Infiltrates | 41 (53) | 7/41 (17%) |
|    unilateral | 15 (37) | |
|    bilateral | 26 (63) | |
| Pleural opacity | 19 (12) | |
| Atelectasis | 3 (4) | |

associated with underlying pneumonia. Pleural effusion is rarely present alone [12, 18].

CT scans are more sensitive than X-rays. They can show mediastinal nodes, tracheobronchial stenosis and calcifications [17, 19], although X-rays can also show diaphragmatic palsy (Figures 16.1 and 16.2).

X-rays and CT scans can also reveal symptoms of infectious pneumonia complicating steroid and immunosuppressive treatments. Interstitial lesions can be the consequence of *Pneumocystis carinii* pneumonia. Bacterial pneumonia and tuberculosis are frequent complications of WG therapy [20].

Alveolar hemorrhage was described above and it can complicate or be the first symptom of WG. Its frequency is difficult to evaluate and could range from 5–13% of cases [21]. Its symptoms are often severe: dyspnea, cough, fever, thoracic pain and hemoptysis but, in most cases, they are minor.

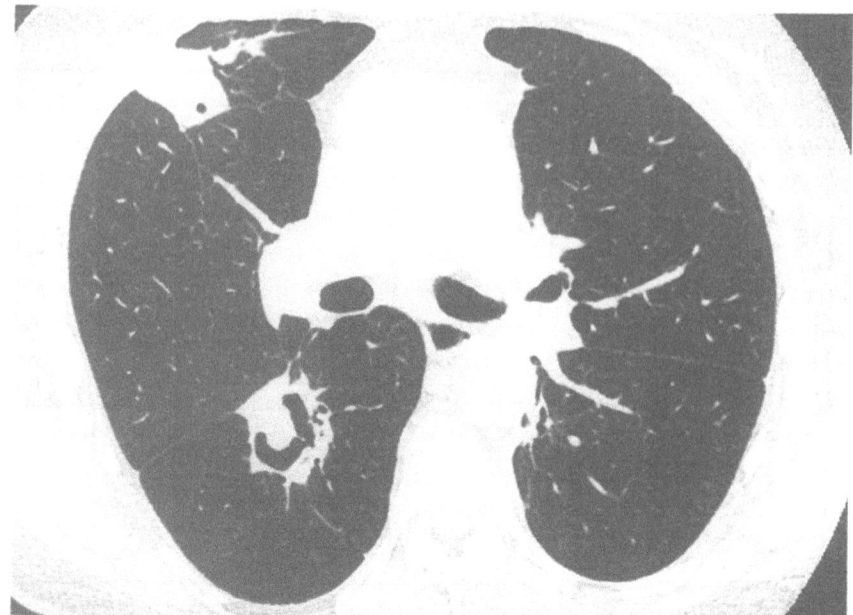

a

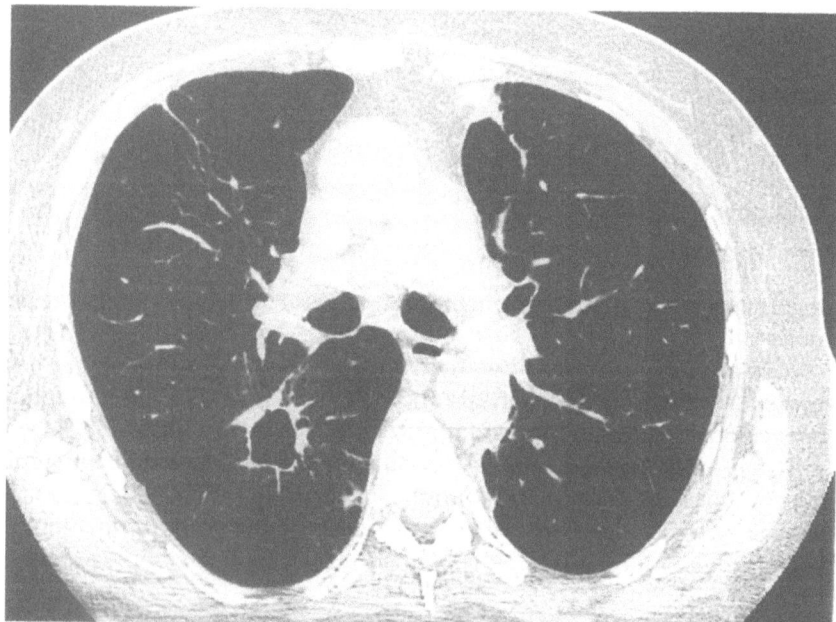

b

**Figure 16.1 (a)** Lung computerized tomography. Nodule of the upper lobe of the right lung and cavitary opacity of the right inferior lobe. **(b)** After 30 months' treatment: disappearance of the upper lobe nodule and cystic modification of the inferior lobe nodule.

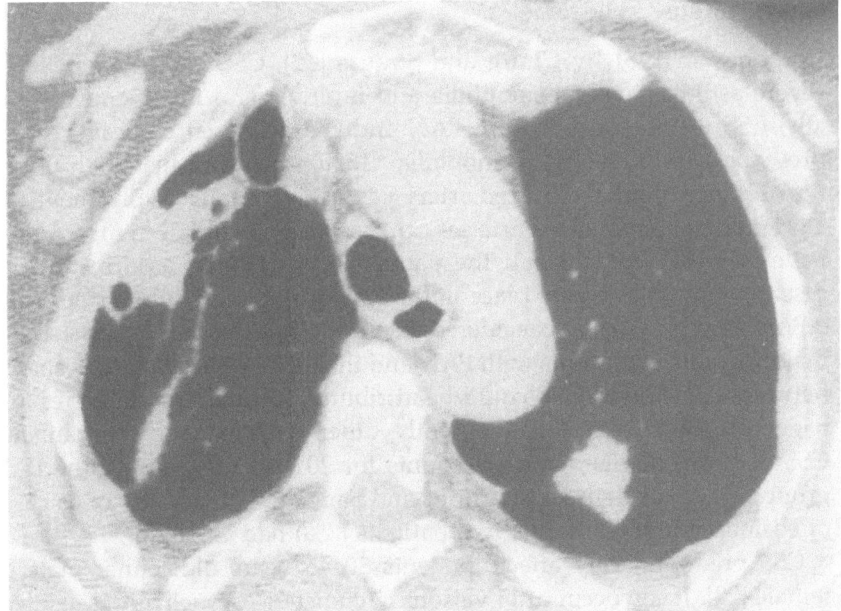

a

b

**Figure 16.2 (a)** Lung computerized tomography. Parenchymal condensation and cavitation of a nodule of the upper right lobe and upper left lobe nodule. **(b)** Disappearance of the right condensation with fibrous opacity.

**Churg–Strauss syndrome**

Described in 1951 by Churg and Strauss [22], CSS is characterized by severe asthma, hypereosinophilia and multivisceral involvement. Initially, CSS was separated from PAN and characterized clinically by the presence of asthma and eosinophilia >1500/mm$^3$, and histologically by the presence of medium-sized artery vasculitis, venous involvement and a granuloma. Small-sized arteries can also be involved. Recently, ANCA have been detected in half the patients [23, 24] and, as a result, this disease now takes a new place in the PAN group, certainly farther from c-PAN than previously considered. Nevertheless, this syndrome shares a majority of its symptoms with PAN and the differences that are observed between c-PAN and CSS could be attributed to different etiologies and triggering factors and consequently other pathogenetic mechanisms. CSS is a rare disease which accounts for 20% of the cases of the PAN group [25, 26] as defined by Cupps and Fauci [27]. In a study of 65 cases of eosinophilic pneumonia, two patients (3%) had CSS [28].

CSS predominantly affects patients 30–45 years old, either male or female [29]. It can occur after various circumstances, such as desensitization, vaccination, inhaled antigens and abrupt stoppage of steroid treatment [30, 31]. We have shown with others [30, 31] that CSS could occur after specific or non-specific immune stimulation, such as hyposensitization or vaccination. The syndrome is observed in patients with long-standing severe and corticodependent asthma, and the triggering factors are responsible for the exacerbation of the respiratory symptoms and the development of systemic vasculitis.

The diagnosis is based on clinical symptoms that can be associated with vasculitis of medium- and small-sized vessels, with or without concomitant extra- or perivascular granuloma. Clinical manifestations of CSS are detailed in Table 16.4.

*Asthma*

Asthma precedes the systemic manifestations in 95% of cases. It is corticodependent with continuous dyspnea, and worsens when the CSS develops. Other manifestations of allergy are sometimes found in these patients and in their families. Radiological signs include: uni- or bilateral lung infiltrates, which regress spontaneously or under steroids; pulmonary nodules or interstitial signs are more rare; a pleural effusion is present in 25–65% of cases [32–34].

The other clinical manifestations are the same as those observed in PAN.

**Table 16.4** Clinical manifestations of CSS (237 cases)

| Authors [Reference] | Review of the literature [29] (1984) | Lanham et al. [29] (1984) | Guillevin et al. [30] (1987) | Gaskin, Clutterbuck and Pusey [36]* (1991) | Haas et al. [32] (1991) |
|---|---|---|---|---|---|
| n | 138 | 16 | 43 | 21 | 16 |
| Sex (M/F) | 72/66 | 12/4 | 24/19 | 14/7 | 12/4 |
| Age   Mean | 38 | 38 | 43.2 | 46.5 | 42.5 |
| Range | | | (7–66) | (23–69) | (17–74) |
| Asthma | 100% | 100% | 100% | 100% | 100% |
| Poor condition | | | 72% | | 100% |
| Lung infiltrates | 74% | 72% | 77% | 43% | 62% |
| Rhinitis | 69% | 70% | 21% | | 10% |
| Mononeuritis multiplex | 64% | 66% | 67% | 70% | 75% |
| GI tract involvement | 62% | 59% | 37% | 58% | 56% |
| Cardiomyopathy | 52% | 47% | 49% | 15% | 56% |
| Arthralgia, polyarthritis | 46% | 51% | 28% | 43% | 31% |
| Skin manifestations | | | | 50% | 68% |
| purpura | 46% | 48% | 28% | | 25% |
| nodules | 33% | 30% | 21% | | 25% |
| Renal impairment | 42% | 49% | 16% | 80% | 31% |
| Pleurisy | 29% | 29% | | | 25% |

* Nephrologic study.

### Rheumatic symptoms

Myalgia and arthralgia are present in half the cases. Arthritis is asymmetrical and predominantly present in the legs.

### Neurological symptoms

Mononeuritis multiplex is the most frequent symptom (more than 60%); it occurs early in the course of the disease. Motor and sensory signs are asymmetrical, affecting predominantly the lower limbs, especially the sciatic nerve and its tibial and peroneal branches. Radial, cubital and median nerves are less frequently involved. The motor deficit develops abruptly. Sensory signs are responsible for hypo- or hyperesthesia, and pain in the region of neurological defect. Pain is sometimes present before the deficiency. In some PAN patients, it can be difficult to analyze the cause of their pain, whether neurological, muscular or rheumatological in origin. The cerebrospinal fluid (CSF) is usually normal. Under treatment, mononeuritis multiplex patients improve progressively and can recover without sequelae. When the latter are present, they are more sensory than motor. Palsies of cranial nerves are present in fewer than 2% of cases and affect the oculomotor (III), trochlear (IV), abducens (VI), facial (VII) or acoustic (VIII) nerves. Sequelae of the VIIIth cranial nerve palsy is common within this 2%. Although deficits, strokes and sometimes brain hemorrhages can be seen, central nervous system (CNS) involvement is rare [35]; its etiology is not univocal: vasculitis of a brain artery or consequence of malignant hypertension, for example.

*Skin manifestations*

Skin involvement, present in 70% of CSS patients, occurs more frequently than in PAN and includes vascular purpura, which is typically papulo-petechial, and sometimes bullous or vesiculous. Infiltration is not a constant feature but, when present, should be the ideal site for a biopsy; subcutaneous nodules are rarer and consist of small elements ranging in size from 0.5 cm to 2 cm. They are transient and should be biopsied as soon as possible. A granuloma can be found; livedo racemosa or reticularis is common; distal gangrene is the consequence of ischemia.

Biopsies show small-sized vessel and capillary vasculitis. When possible, the tissue sample should include the derm in order to assess medium-sized vessel involvement accurately.

*Kidney involvement*

According to the literature, kidney involvement is present in 42% of patients [29].

Kidney manifestations are diverse, glomerular and can be severe, leading to hemodialysis. Unlike c-PAN, vascular nephropathy is rare. In patients with glomerulonephritis, p-ANCA are often present [36].

*Cardiovascular manifestations*

Heart manifestations are limited to the myocardium and are due to coronary vasculitis or hypertension. Clinical, electrocardiographical or radiological anomalies can be observed in half the patients. Cardiac failure involves the left heart more than the right side. Angina is rare and coronarography is usually normal. Cardiomyopathy can be seen in 25% of cases [29].

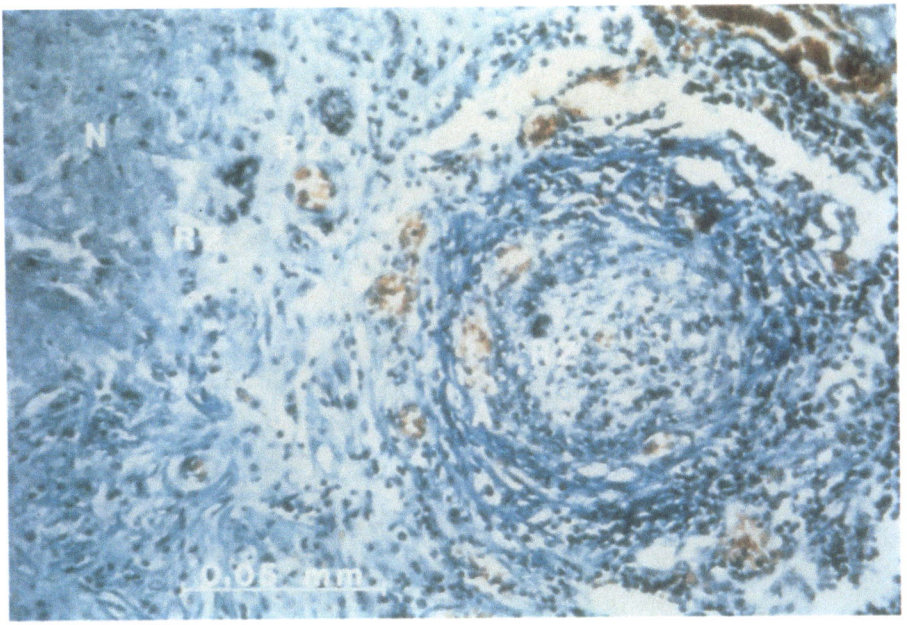

Plate 1 Histopathology of Wegener's granulomatosis. Nasal biopsy. RZ: giant cell, N:necrosis. (Prof. Müller-Hermelink, Würzburg).

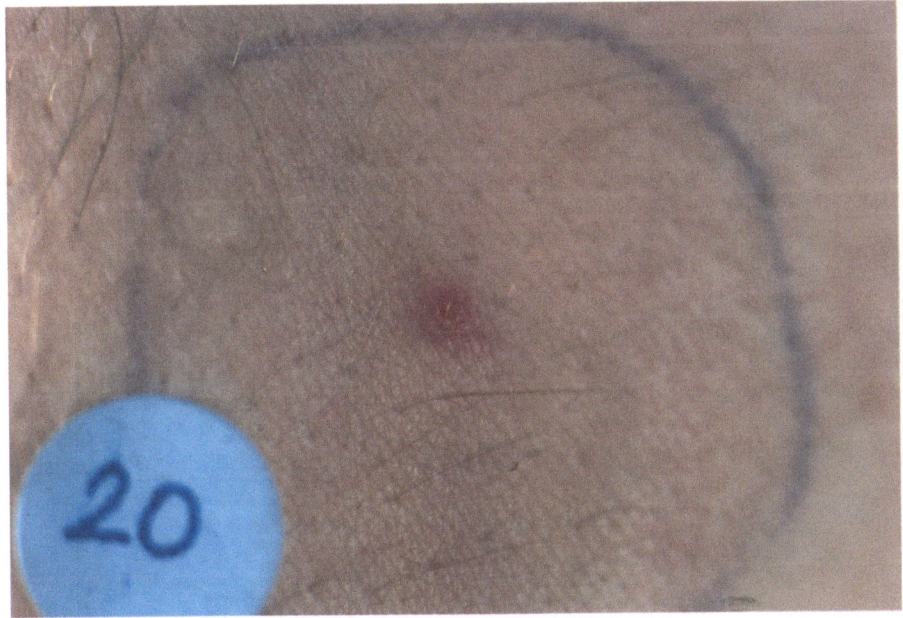

Plate 2 The pathergy reaction.

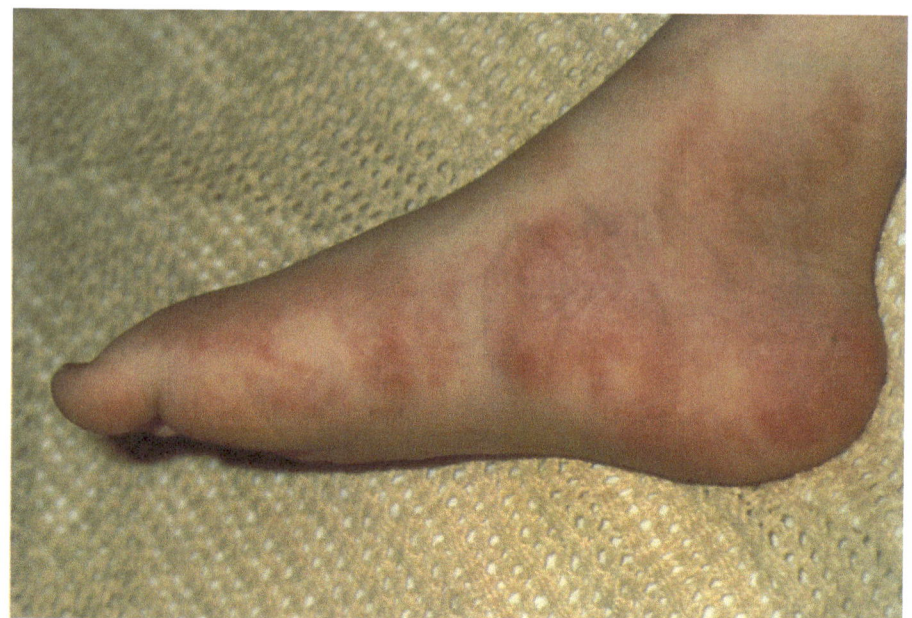

Plate 3 Cutaneous vasculitis.

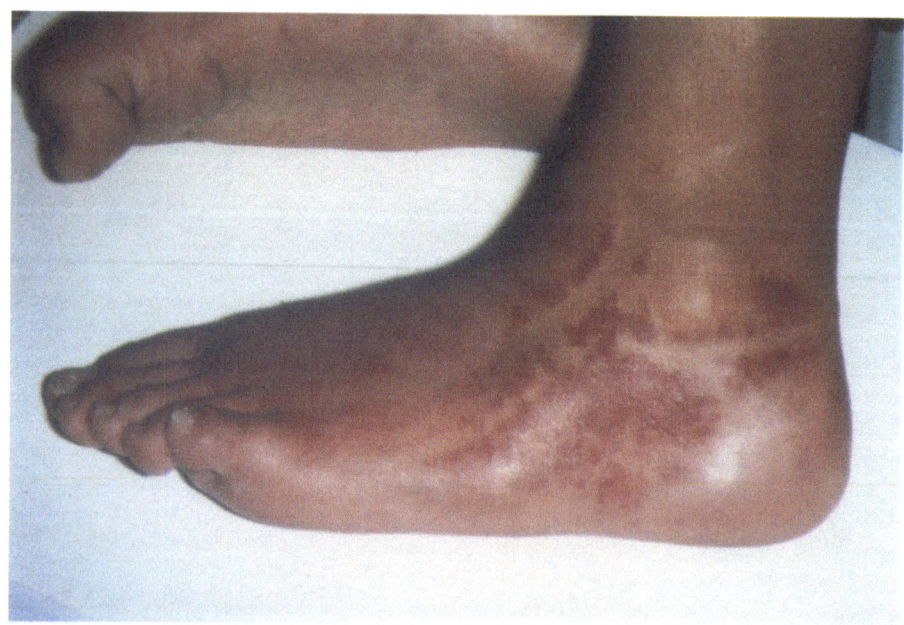

Plate 4 Henoch–Schönlein purpura in an adolescent child with FMF.

*Gastrointestinal involvement*

GI tract involvement is one of the most severe manifestations of CSS, and abdominal pain was noted in 40–60% of our patients [29]. It is always difficult to determine the exact origin of the abdominal pain, but its presence can be the first symptom of GI vasculitis. In a majority of cases, ischemia is present in the small bowel but rarely in the colon or stomach. Digestive bleeding and bowel perforation can be seen [29]. Relapses after surgery or medical treatment are a sign of poor prognosis. Intractable abdominal pain associated with weight loss are the consequence of bowel ischemia. Digestive malabsorption has also been described [37]. The prognosis for patients with these symptoms is extremely poor. Vasculitis of the appendix [38] or gallbladder is sometimes the first manifestation [39].

*Biological and angiographic investigations*

Hypereosinophilia is always greater than $1500/mm^3$, and sometimes reaches $50\ 000/mm^3$. Eosinophilia can also be detected in bronchoalveolar lavage fluids. Eosinophil numbers return to normal under steroids. The IgE level is increased in 75% of cases. ANCA were present in 66.7% of our patients and were anti-MPO in most cases [40]; in rare cases, anti-PR3 were found. Persistence of ANCA is sometimes associated with relapses, but the level of ANCA cannot predict the occurrence of relapes [41]. The pathogenetic mechanism is not clearly known and, if ANCA are one of the most important mechanisms, the responsibility of immune complexes [24] cannot be forgotten.

Abdominal and renal angiographies are usually normal.

## Giant cell arteritis

Giant cell (temporal) arteritis and Takayasu's arteritis share similar histological findings, but are usually quite different in their presentations and courses. For example, pulmonary involvement is relatively frequent in Takayasu's arteritis, whereas it is rare in the course of temporal arteritis.

*Takayasu's arteritis (pulseless disease)*

Takayasu's arteritis is a giant cell segmental arteritis that involves principally the aorta, its main branches and pulmonary arteries. This disease is rare, develops predominantly in women before 40 years of age and occurs worldwide, although most affected individuals are Japanese. The etiology and pathogenesis remain unknown. A mycobacterial origin has been suspected since a majority of patients had evidence of active or remote tuberculosis, and granulomas have been found in arterial walls. A hypersensitivity reaction to mycobacteria has also been suggested, but Takayasu's arteritis appears to be more probably an immune-complex mediated disease.

Upon histological examination, most changes are limited to the larger elastic vessels and indicate a patchy panarteritis. The inflammatory process in Takayasu's arteritis shows a predilection for arteries with abundant elastic tissue in their walls, such as pulmonary arteries, which are frequently affected by this disease process [42]. Granulomatous arteritis probably represents the active phase of the disease. The inflammatory process is confined to the media and adventitia, and the cell infiltrate consists predominantly of lymphocytes, plasma cells and histiocytes, with varying numbers of giant cells. Chronicity of the disease leads to sclerosing arteritis. Disruption of the elastic lamellae and thrombosis and aneurysm formation can be observed in both the active and chronic forms of the disease [42].

The condition is classified into four types: disease confined to the aortic arch and its branches (type I), disease limited to the descending thoracic and abdominal aorta (type II), a mixed form that combines types I and II (type III), and involvement of the pulmonary arterial tree (type IV) [43]. The course of Takayasu's arteritis can be divided into two stages. The first, or 'prepulseless' stage, is characterized by nonspecific inflammatory features, such as fever, night sweats, malaise, arthralgia, myalgia and skin rash, which usually precede signs and symptoms of vascular insufficiency by a prolonged period of several years [44]. Then, clinical manifestations arise secondary to narrowing and occlusion or dilatation of the aorta and its major branches. However, a lot of patients have inflammatory features at the time of diagnosis.

The frequency of pulmonary artery involvement in Takayasu's vasculitis varies from one study to another [43]. Absent in two studies, pulmonary arteritis was reported in 40–100% of cases based on autopsy findings [42]. A few studies of pulmonary angiography findings reported an intermediate frequency of pulmonary artery involvement [45–47]. Pulmonary artery involvement revealing pulseless disease has rarely been observed [48–50].

The intimal, medial and adventitial fibroses are responsible for extensive vascular stenosis, resulting in the pulmonary artery hypertension that is the principal manifestation of Takayasu's arteritis affecting the lungs. However, small branches can also be involved, and four major types of pulmonary vascular injury have been distinguished [48].

Proximal pulmonary artery involvement, which can lead to pulmonary hypertension and hemoptysis, has been reported on multiple occasions [43, 48]. Chest X-ray can show enlargement of the pulmonary artery trunk or decreased vasculature of one lung or lobe [51]. The diagnosis can be derived from angiographic findings, which can evidence pulmonary artery stenosis or occlusion of one of its branches [46]. Distal pulmonary artery stenosis is a common feature of Takayasu's arteritis. Usually associated with proximal pulmonary–artery involvement, when restricted to the distal pulmonary artery, pulseless disease is difficult to distinguish from congenital peripheral pulmonary stenosis [48]. Disease progression is usually marked by pulmonary hypertension. Rare cases involving microscopic arteries were diagnosed based upon histological examination. Two of the three cases reported by Lupi-Herrera, Sanchez-Torres and Murcushamer [43] were not associated with proximal pulmonary artery involvement. A few cases of coronary-pulmonary artery, or coronary-bronchial–pulmonary artery communication have been reported [48].

Pulmonary artery involvement is often diagnosed late at the sclerosis stage that no longer responds to steroid therapy. However, dramatic improvement has been obtained surgically for very proximal stenoses of the pulmonary artery [52]. Steroid treatment would probably be beneficial if pulmonary involvement in pulseless disease could be diagnosed early.

## Temporal arteritis

Temporal arteritis is a relatively common systemic vascular disease that predominantly affects individuals between 70 and 80 years old. Vasculitis usually develops in the larger vessels of the head and neck, involving mainly branches of the external carotid artery, but evidence of involvement of the aorta and its major branches has also been noted in a small portion of cases [53].

Histological findings in temporal artery biopsies usually consist of giant cell segmental panarteritis. Classical symptoms include headache, jaw claudication and blindness; occasionally, there may be high fever, anemia and polymyalgia rheumatica.

Pulmonary involvement is rare in temporal arteritis. However, an isolated chronic cough of unobvious cause that resolved quickly under steroid therapy has been reported [54, 55]. In addition, the same authors reported respiratory tract symptoms and signs including sore throat, hoarseness, choking sensation and chest pain, sometimes accompanied by nodular or interstitial infiltrates on X-rays, as initial findings of temporal arteritis [54]. A few cases of pulmonary nodules and interstitial infiltrates have also been published [56]. In these reports, histological examination of lung biopsies revealed interstitial, non-caseating granuloma [57] and areas of infarction, fibrosis and features of giant cell arteritis of the pulmonary artery [58]. Aneurysms and narrowing of pulmonary arteries were reported separately in two cases [59, 60]. Pleuritis may also occur in the course of temporal arteritis. In most of the cases reported so far, pulmonary involvement usually regressed with corticosteroid therapy [57].

Several cases of a visceral disseminated form of giant cell arteritis involving the lungs have been reported [61]. However, none of those cases had temporal artery involvement nor did they present the clinical picture of the polymyalgia–temporal arteritis syndrome [61]. Taken together, these features stress the fact that temporal arteritis represents a spectrum of diseases whose manifestations vary from local symptomatology of the temporal arteries to multisystemic disease involving many organs including the lungs.

## Miscellaneous pulmonary vasculitis

### Cryoglobulinemia

Essential mixed cryoglobulinemia is characterized by the presence in the serum of globulins that precipitate upon exposure to cold temperatures (cryoglobulins). These proteins are composed of monoclonal or polyclonal IgG and/or IgM, and can be associated with various conditions. The symptomatology seems to derive from the deposition of circulating immune complexes within the vascular wall leading to a secondary inflammatory response.

Although circulating immune-complex vasculitis can occur during the course of cryoglobulinemia, it usually involves skin and kidneys, and pulmonary involvement is infrequent. One study of 23 patients showed evidence of small airway obstruction and increased alveolar arterial gradient in nine cases and interstitial fibrosis on chest radiographs in 18 patients with mixed essential cryoglobulinemia [62]. One case of lymphocytic alveolitis [63] and one case of respiratory distress syndrome have also been described [64]. However, pathological features have rarely been documented and pulmonary involvement in cryoglobulinemias needs to be specified.

### Henoch–Schönlein (or Schönlein–Henoch) purpura

Henoch–Schönlein purpura is a clinical entity characterized by diffuse, necrotizing vasculitis involving the joints, skin, GI tract and kidneys. Pulmonary involvement is extremely rare in this condition and can lead to alveolar hemorrhage. In a review of 77 adults with Henoch–Schönlein purpura, respiratory tract involvement was found in only four of them, and none of them underwent lung biopsy [65]. When autopsy was performed in fatal cases, pathological examination revealed leukocyto-clastic vasculitis and extensive alveolar hemorrhage [66]. In one case, immunohistochemical staining showed IgA deposition along the alveolar septae [66]. Although the prognosis of Henoch–Schönlein purpura is generally good under steroid therapy, severe renal involvement and unusual alveolar hemorrhage can occasionally result in death.

*Behçet's syndrome*

Behçet's syndrome is a systemic vasculitis affecting predominantly the eyes, mouth and deep veins. In some cases, the lung can also be involved. Diagnostic criteria consist of the association of relapsing oral aphthosis with two of the following symptoms: genital ulcers, uveitis, pathergic skin test and other skin manifestations [67].

Involvement of large pulmonary vessels is responsible for aneurysm development. Histological examination of pulmonary vessels can detect vasculitis of various-sized vessels, arteries and veins and the vasovasorum of larger vessels. Vessel inflammation or fibrinoid necrosis can be seen. Thrombosis is a major element of vasculitis with both entities being responsible for pulmonary infarcts, aneurysms and hemoptysis. The latter can be the consequence of capillaritis, aneurysm rupture or pulmonary infarcts.

*Pulmonary artery aneurysms*

These are rarely seen but can be responsible for a patient's death. Their frequency is low, 1/1731 cases in a Japanese study [68], 1/316 for Benamour *et al.* [69], 3/196 for Wechsler *et al.* [70], 2/450 for Hamza, Ayed and Zribi [71].

They develop on large pulmonary arteries and less frequently on segmental arteries. They are bilateral and usually $</=$ 3. They can be asymptomatic, but are often diagnosed in conjunction with hemoptysis, which can be severe. Aneurysm rupture can be preceded by thoracic pain. X-rays, CT scans, magnetic resonance imaging and angiography can all contribute positively to making the diagnosis, and investigations should be performed before treatment. In association with the medical treatment of Behçet's disease, surgery or embolization of aneurysms is necessary to avoid rupture (Figure 16.3). The association of pulmonary artery aneurysms and venous thromboses has been described as the Hughes–Stovin syndrome [72]. This syndrome was initially described without Behçet's syndrome, but is now considered to be strongly associated with it [73].

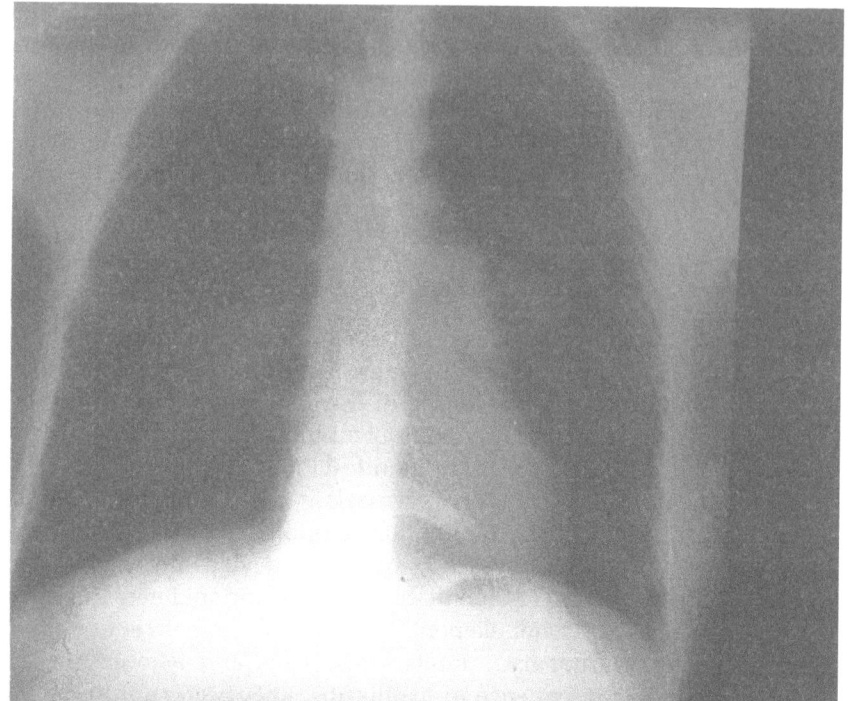

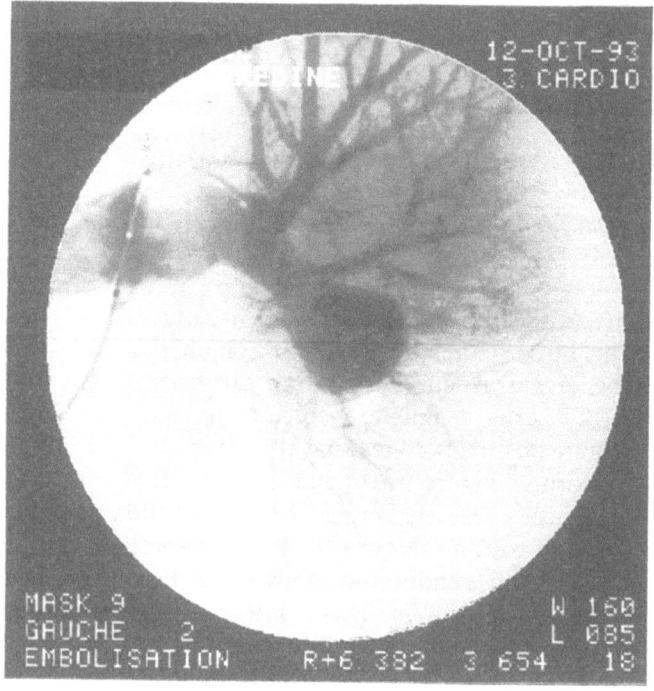

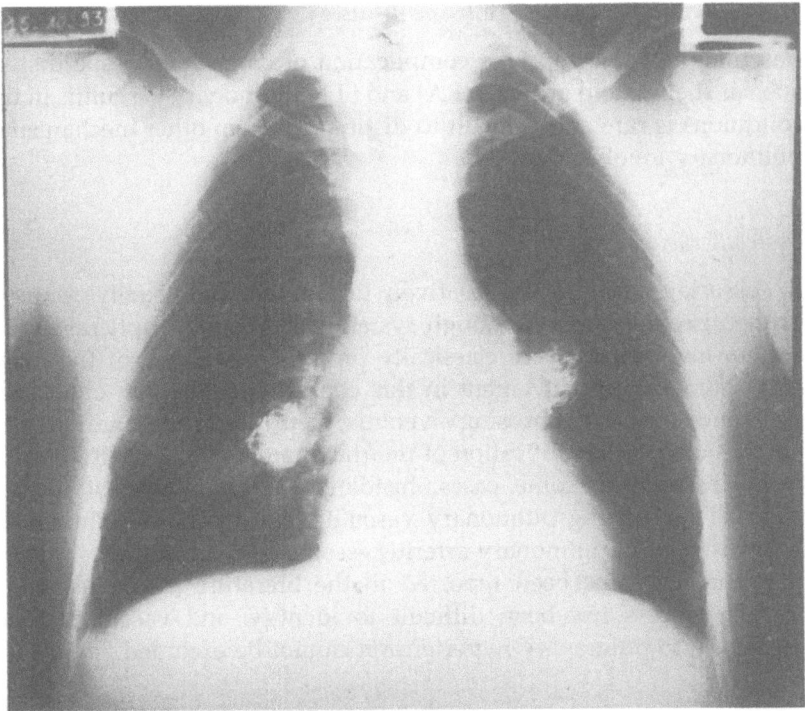

c

**Figure 16.3 (a)** Bilateral pulmonary aneurysms. **(b)** Angiography showing right pulmonary artery aneurysm. **(c)** Coil embolization of the pulmonary aneurysms.

### Thrombosis

Vessel thromboses with or without pulmonary infarcts are frequent. They represent a local phenomenon which is the consequence of stenosis, local inflammation and aneurysm thrombosis. In most cases, they are observed downstream from aneurysms. Pulmonary hypertension can occur [74, 75].

Clinical symptoms are the same as those observed in the case of pulmonary embolism.

Thromboses of other vessels are frequent, especially the vena cava and other mediastinal veins.

### Other pulmonary manifestations

Lung infiltrates can be the consequence of incomplete pulmonary infarcts or vasculitis. An obstructive syndrome has sometimes been described. Lung infections can also occur and be attributable to the treatments given for vasculitis (steroids, cytotoxic agents, etc.).

*Lung vasculitis in connective tissue diseases*

Vasculitis can develop as a complication of connective tissue diseases, such as rheumatoid arthritis (RA) and SLE. Pulmonary vasculitis in these conditions is rare and difficult to distinguish from other mechanisms of pulmonary involvement.

*Systemic rheumatoid vasculitis*

RA-associated vasculitis is relatively uncommon and usually occurs late in the disease process, although systemic, skin and peripheral nervous system involvements are classically prominent features of the disease [76]. Pulmonary involvement in this context is present in one-third of cases and consists of fibrosing alveolitis, pleuritis or pulmonary nodules, but pathological identification of the rheumatoid origin is often missing [76]. However, in some cases, histological examination of the lung revealed necrotizing pulmonary vasculitis and interstitial fibrosis [77]. A few reports of pulmonary arteritis associated with primary hypertension and RA have been reported in the literature [78]. The primary disease process has been difficult to identify, and vasculitic lesions secondary to pulmonary hypertension cannot be excluded.

*Systemic lupus erythematosus*

SLE vasculitis usually consists of microvascular lesions that are prominent in the skin; large-vessel involvement is rare [79]. As for rheumatoid vasculitis, pulmonary arteritis has been observed in association with primary hypertension in SLE patients, although vasculitis can involve the lungs even without evidence of pulmonary hypertension [80]. Alveolar hemorrhage resulting from microvasculitis has been reported to occur in patients suffering from SLE [81]. It was extensive in all the cases reported and most of the patients died within a few days [3].

*Other connective tissue diseases*

Very few cases of pulmonary vasculitis have been reported during the course of inflammatory muscular diseases.

*Inflammatory bowel disease*

Although patients with inflammatory bowel disease can develop inflammatory complications in the lung, interstitial pulmonary involvement is rare in this condition. Very few cases of cavitating lung nodules have been reported, apparently unrelated to WG since none of these patients had c-ANCA [82]. Anecdotal reports have described vasculitic lesions of lung biopsy specimens [83]. More patients having concomitant lung vasculitis will have to be identified before these lesions can be considered part of the spectrum of inflammatory bowel disease.

*Hypocomplementemic urticarial vasculitis*

This vasculitis associates urticarial skin lesions, low serum complement levels, and two of the following features: dermal venulitis, arthritis, glomerulonephritis, episcleritis or uveitis, recurrent abdominal pain and C1q precipitin in plasma [84]. Chronic obstructive pulmonary disease has been reported in hypocomplementemic urticarial vasculitis, even in patients with a low pack–year cigarette smoking history [85]. Histological findings associate lung venulitis and immunoglobulin deposits, suggesting the possible interaction of smoking and immunologically mediated processes in the pathogenesis of the disease.

## REFERENCES

1. Lie, J. (1977) Nosology of pulmonary vasculitides. *Mayo Clin. Proc.*, **52**, 520–2.
2. van der Woude, F., Rasmussen, N., Lobatto, S. *et al.* (1985) Autoantibodies against neutrophils and monocytes: tool for diagnosis and marker of disease activity in Wegener's granulomatosis. *Lancet*, **1**, 425–9.
3. Leatherman, J., Davies, S. and Hoidal, J. (1984) Alveolar hemorrhage syndromes: diffuse microvascular lung hemorrage in immune and idiopathic disorders. *Medicine*, **63**, 343–61.
4. Thomas, H. and Irwin, R. (1975) Classification of diffuse intrapulmonary hemorrage. *Chest*, **68**, 483.
5. Hudson, B., Wieslander, J., Wisdom, B. and Noelken, M. (1989) Biology of disease. Goodpasture's syndrome: molecular architecture and function of basement membrane antigen. *Lab. Invest.*, **61**, 256–69.
6. Soergel, K. and Sommers, S. (1962) Idiopathic pulmonary hemosiderosis and related syndromes. *Am. J. Med.*, **32**, 499.
7. Wegener, F. (1936) Über generalisierte septische Gefäßerkrankungen. *Verh. Dtsch. Pathol. Ges.*, **29**, 202.
8. Walton, E. (1958) Giant-cell granuloma of respiratory tract (Wegener's granulomatosis). *Br. Med. J.*, **2**, 265.
9. Travis, W., Hoffman, G., Leavitt, R. *et al.* (1991) Surgical pathology of the lung in Wegener's granulomatosis. Review of 87 open lung biopsies from 67 patients. *Am. J. Surg. Pathol.*, **15**, 315–33.
10. Mark, E.J., Matsubara, O., Tan, L.N. and Fienberg, R. (1988) The pulmonary biopsy in the early diagnosis of Wegener's (pathergic) granulomatosis: a study based on 35 open lung biopsies. *Hum. Pathol.*, **19**, 1065–71.
11. Yousem, S.A. (1991) Bronchocentric injury in Wegener's granulomatosis: a report of five cases. *Hum. Pathol.*, **22**, 535–40.
12. Cordier, J.F., Valeyre, D., Guillevin, L. *et al.* (1990) Pulmonary Wegener's granulomatosis. A clinical and imaging study of 77 cases. *Chest*, **97**, 906–12.
13. Hellmann, D., Laing, T., Petri, M. *et al.* (1987) Wegener's granulomatosis: isolated involvement of the trachea and larynx. *Ann. Rheum. Dis.*, **46**, 628–31.

14. Lampman, J., Querubin, R. and Kondapalli, P. (1981) Subglottic stenosis in Wegener's granulomatosis. *Chest*, **79**, 230–2.

15. McDonald, T., Welland, L. and DeRemee, R. (1989) Head and neck involvement in Wegener's granulomatosis. *Semin. Respir. Med.*, **10**, 133–5.

16. Strange, C., Halstead, L., Baumann, M. and Sahn, S.A. (1990) Subglottic stenosis in Wegener's granulomatosis: development during cyclophosphamide treatment with response to carbon dioxide laser therapy. *Thorax*, **45**, 300–1.

17. Aberle, D.R., Gamsu, G. and Lynch, D. (1990) Thoracic manifestations of Wegener's granulomatosis: diagnosis and course. *Radiology*.

18. England, D.M. and Unger, J.M. (1987) Pleural-based mass in an elderly man with arthralgias. *Chest*, **91**, 603–4.

19. Maguire, R., Fauci, A., Doppman, J. and Wolf, S. (1978) Unusual radiographic features of Wegener's granulomatosis. *A.J.R.*, **130**, 233–8.

20. Jarrousse, B., Guillevin, L., Bindi, E. *et al.* (1993) Increased risk of *Pneumocystis carinii* pneumonia in patients with Wegener's granulomatosis. *Clin. Exp. Rheumatol.*, **11**, 615–21.

21. Colby, T. (1989) Diffuse pulmonary hemorrhage in Wegener's granulomatosis. *Semin. Respir. Med.*, **10**, 136–40.

22. Churg, J. and Strauss, L. (1951) Allergic granulomatosis, allergic angiitis and periarteritis nodosa. *Am. J. Path.*, **27**, 277–94.

23. Savage, C.O. and Lockwood, C.M. (1990) Antineutrophil antibodies in vasculitis. *Adv. Nephrol. Necker Hosp.*, **19**, 225–36.

24. Guillevin, L., Ronco, P. and Verroust, P. (1990) Circulating immune complexes in systemic necrotizing vasculitis of the polyarteritis nodosa group. Comparison of HBV-related polyarteritis nodosa and Churg–Strauss angiitis. *J. Autoimmun.*, **3**, 789–92.

25. Guillevin, L., Lhote, F., Jarrousse, B. and Fain, O. (1992) Treatment of polyarteritis nodosa and Churg–Strauss syndrome. A meta-analysis of 3 prospective controlled trials including 182 patients over 12 years. *Ann. Méd. Interne*, **143**, 405–16.

26. Jarrousse, B. and Guillevin, L. (1990) L'angéite de Churg et Strauss. *S.T.V.*, **2**, 481–6.

27. Cupps, T. and Fauci, A. (1981) Systemic necrotizing vasculitis, in *The Vasculitides, Vol. XXI, Major Problems in Internal Medicine* (ed. L.H. Smith Jr.), W.B. Saunders, Philadelphia.

28. Capewell, S., Chapman, B.J., Alexander, F. *et al.* (1992) Pulmonary eosinophilia with systemic features: therapy and prognosis. *Respir. Med.*, **86**, 485–90.

29. Lanham, J., Elkon, K., Pusey, C. and Hughes, G. (1984) Systemic vasculitis with asthma and eosinophilia: a clinical approach to the Churg–Strauss syndrome. *Medicine*, **63**, 65–81.

30. Guillevin, L., Guittard, T., Blétry, O. *et al.* (1987) Systemic necrotizing angiitis with asthma: causes and precipitating factors in 43 cases. *Lung*, **165**, 165–72.

31. Guillevin, L., Amouroux, J., Arbeille, B. and Boura, R. (1991) Churg–Strauss angiitis. Arguments favoring the responsibility of inhaled antigens. *Chest*, **100**, 1472–3.

32. Haas, C., Geneau, C., Odinot, J.M. *et al.* (1991) Angéite allergique et granulomateuse: le syndrome de Churg–Strauss. Une étude rétrospective de 16 cas. *Ann. Méd. Interne*, **142**, 335–42.

33. Harkavy, J. (1943) Vascular allergy. *J. Allergy*, **14**, 507–37.

34. Lhote, F., Cohen, P., Chemlal, K. *et al.* (1992) Les manifestations pleurales au

cours de la périartérite noueuse, de la maladie de Wegener et du lupus érythémateux systémique. *Ann. Méd. Interne*, **143**, 228–32.

35. Castaigne, P., Cambier, J., Escourolle, R. and Brunet, P. (1970) Les manifestations nerveuses centrales de la periartérite noueuse. A propos d'une observation anatomo-clinique. *Ann. Méd. Interne*, **121**, 375–82.

36. Gaskin, G., Clutterbuck, E.J. and Pusey, C.D. (1991) Renal disease in the Churg–Strauss syndrome. Diagnosis, management and outcome. *Contrib. Nephrol.*, **94**, 58–65.

37. Desbazeille, F. and Soule, J. (1986) Manifestations digestives des vascularites. *Gastroentérol. Clin. Biol.*, 405–14.

38. Myoana, T. (1988) Necrotizing arteritis of the vermiform appendix. *Arch. Pathol. Lab. Med.*, **112**, 738–41.

39. Remigio, P. and Zaino, E. (1970) Polyarteritis nodosa of the gallbladder. *Surgery*, **67**, 427–31.

40. Guillevin, L., Visser, H., Noël, L. *et al.* (1993) Antineutrophil cytoplasm antibodies (ANCA) in systemic polyarteritis nodosa with and without hepatitis B infection and Churg-Strauss syndrome. 62 patients. *J. Rheumatol.*, **20**, 1345–9.

41. Baril, L., Guillevin, L., Lhote, F. *et al.* (1992) Persistance des anticorps anticytoplasme des polynucléaires neutrophiles (ANCA) chez des patients asymptomatiques atteints de périartérite noueuse et de syndrome de Churg et Strauss. Suivi de 54 patients. *Rev. Méd. Interne*, (Suppl.), 402 (Abstract).

42. Saito, Y., Hirota, K., Ito, I. *et al.* (1972) Clinical and pathological studies of five autopsied cases of aortitis syndrome. *Jpn. Heart J.*, **13**, 20–3.

43. Lupi-Herrera, E., Sanchez-Torres, G. and Murcushamer, Jea. (1997) Takayasu's arteritis. Clinical study of 107 cases. *Am. Heart J.*, **93**, 94.

44. Hall, S., Barr, W., Lie, J. *et al.* (1985) Takayasu's arteritis. A study of 32 North American patients. *Medicine*, **64**, 89–99.

45. Sharma, S., Kamalakar, T., Rajani, M. *et al.* (1990) The incidence and patterns of pulmonary artery involvement in Takayasu's arteritis. *Clin. Radiol.*, **42**, 177–81.

46. Yamada, I., Shibuya, H., Matsubara, O. *et al.* (1992) Pulmonary artery disease in Takayasu's arteritis: angiographic findings. *A.J.R.*, **159**, 263–9.

47. Yamato, M., Lecky, J., Hiramatsu, K. and Kohda, E. (1986) Takayasu arteritis: radiographic and angiographic findings in 59 patients. *Radiology*, **161**, 329–34.

48. Bletry, O., Kieffer, E., Herson, S. *et al.* (1991) Formes artérielles pulmonaires graves de l'artérite de Takayasu: trois observations et revue de la littérature. *Arch. Mal. Coeur*, **84**, 817–22.

49. Haas, A. and Stiehm, E. (1986) Takayasu's arteritis presenting as pulmonary hypertension. *Am. J. Child. Dis.*, **140**, 372–4.

50. Hayashi, K., Nagasaki, M., Matsunaga, N. *et al.* (1986) Initial pulmonary artery involvement in Takayasu arteritis. *Radiology*, **159**, 401–3.

51. Berkmen, Y. and Lande, A. (1975) Chest roentgenography as a window to the diagnosis of Takayasu's arteritis. *Am. J. Roentgenol.*, **125**, 842–6.

52. Jakob, H., Volb, R., Stangl, G. *et al.* (1990) Surgical correction of a severely obstructed pulmonary artery bifurcation in Takayasu's arteritis. *Eur, J. Cardiothorac. Surg.*, **4**, 456–8.

53. Vayssairat, M. and Housset, E. (1991) Maladie de Horton, in *Maladies Systémiques* (eds M.F. Kahn and A.P. Peltier), Flammarion, Paris.

54. Larson, T., Hall, S., Hepper, N. and Hunder, G. (1984) Respiratory tract as a clue to giant cell arteritis. *Ann. Intern. Méd.*, **101**, 594–7.

55. Desmet, G., Knockaert, D. and Bobbaers, J. (1990) Temporal arteritis: the silent presentation and delay in diagnosis. *J. Intern. Med.*, **227**, 237–40.
56. Kramer, M., Melzer, E., Nesher, G. and Sonnenblick, M. (1987) Pulmonary manifestations of temporal arteritis. *Eur. J. Respir. Dis.*, **71**, 430–3.
57. Karam, G. and Fulmer, J. (1982) Giant cell arteritis presenting as interstitial lung disease. *Chest*, **82**, 781–4.
58. Doyle, L., McWilliam, L. and Hasleton, P. (1988) Giant cell arteritis with pulmonary involvement. *Br. J. Dis. Chest*, **82**, 88–92.
59. Dennison, A., Watkins, R. and Gunning, A. (1985) Simultaneous aortic and pulmonary artery aneurysms due to giant cell arteritis. *Thorax*, **40**, 156–7.
60. Wagenaar, S., Westerman, C. and Corrin, B. (1981) Giant cell arteritis limited to large elastic pulmonary arteries. *Thorax*, **36**, 876–7.
61. Lie, J. (1978) Disseminated giant cell arteritis. Histologic description and differentiation from other granulomatous vasculitides. *Am. J. Clin. Pathol.*, **69**, 299–305.
62. Bombardieri, S., Paoletti, P., Ferri, C. *et al.* (1979) Lung involvement in essential mixed cryoglobulinemia. *Am. J. Med.*, **66**, 748–56.
63. Bertorelli, G., Pesci, A., Manganelli, P. *et al.* (1991) Subclinical pulmonary involvement in essential mixed cryoglobulinemia assessed by bronchoalveolar lavage [letter]. *Chest*, **100**, 1478–9.
64. Stagg, M., Lauber, J. and Michalski, J. (1989) Mixed essential cryoglobulinemia and adult respiratory distress syndrome: a case report. *Am. J. Med.*, **87**, 445–8.
65. Cream, J., Gumpel, J. and Peachey, R. (1970) Schönlein–Henoch purpura in the adult. A study of 77 adults with anaphylactoïd or Schönlein–Henoch purpura. *Q. J. Med. New Series*, **39**, 461.
66. Kathuria, S. and Cheifec, G. (1982) Fatal pulmonary Henoch-Schönlein syndrome. *Chest*, **182**, 654.
67. International Study group for Behçet's disease (1990) Criteria for diagnosis of Behçet's disease. *Lancet*, **335**, 1078–80.
68. Shimizu, T., Ehrlich, G., Inaba, G. and Hayashi, K. (1979) Behçet's disease. *Semin. Arthritis Rheum.*, **8**, 223–60.
69. Benamour, S., Zeroual, B., Bennis, R. *et al.* (1990) Maladie de Behçet. 316 cas. *Presse Med.*, **19**, 1485–9.
70. Wechsler, B., Le, T.H.D., De Gennes, C. *et al.* (1989) Manifestations artérielles de la maladie de Behçet. *Rev. Med. Interne Paris*, **10**, 303–11.
71. Hamza, M., Ayed, K. and Zribi, A. (1991) Maladie de Behçet, in *Les Maladies systémiques* (eds M.F. Kahn, A.P. Peltier, O. Meyer, and J.C. Piette), Editions Flammarion, Paris, pp. 917–47.
72. Hughes, J. and Stovin, P. (1959) Segmental pulmonary artery aneurysms with peripheral venous thrombosis. *Br. J. Dis. Chest*, **53**, 19–27.
73. Durieux, P., Bletry, O., Huchon, G. *et al.* (1981) Multiple pulmonary arterial aneurysms in Behçet's disease and Hughes–Stovin syndrome. *Am. J. Med.*, **71**, 736–41.
74. Efthimiou, J., Johnston, C., Spiro, S. and Turner-Warwick, M. (1986) Pulmonary disease in Behçet's syndrome. *Q. J. Med.*, **58**, 259–80.
75. Raz, I., Okon, E. and Chajek-Shaul, T. (1989) Pulmonary manifestations in Behçet's syndrome. *Chest*, **95**, 585–9.
76. Scott, D., Bacon, P. and Tribe, C. (1981) Systemic rheumatoid vasculitis: a clinical and laboratory study of 50 cases. *Medicine*, **60**, 288–97.
77. Lakhampal, S., Conn, D. and Lie, J. (1994) Clinical and prognostic signifi-

cance of vasculitis as an early manifestation of connective tissue disease syndrome. *Ann. Intern. Méd.*, **101**, 743–8.

78. Morikawa, J., Kitamura, K., Habuchi, Y. *et al.* (1988) Pulmonary hypertension in a patient with rheumatoid arthritis. *Chest*, **93**, 876–8.

79. Piette, J., Chapelon, C., Boussen, K. *et al.* (1987) Vascularities lupiques. *Ann. Intern. Med. Paris*, **138**, 425–36.

80. Fayemi, A. (1976) Pulmonary vascular disease in systemic lupus erythematosus. *Am. J. Clin. Pathol.*, **65**, 284–90.

81. Myers, J. and Katzenstein, A. (1986) Microangiitis in lupus-induced pulmonary hemorrage. *Am. J. Clin. Pathol.*, **85**, 552–6.

82. Camus, P., Piard, F., Ascroft, T. *et al.* (1993) The lung in inflammatory bowel disease. *Medicine*, **72**, 151–83.

83. Collins, J., Bendig, D. and Taylor, W. (1979) Pulmonary vasculitis complicating childhood ulcerative colitis. *Gastroenterol. Clin. Biol.*, **77**, 1091–3.

84. McDuffie, F., Sams, W.J., Maldonaldo, J. *et al.* (1973) Hypocomplementemia with cutaneous vasculitis and arthritis: possible immune complex syndrome. *Mayo Clin. Proc.*, **48**, 340–8.

85. Schwartz, H., McDuffie, F., Black, L. *et al.* (1982) Hypocomplementemic urticarial vasculitis. Association with chronic obstructive pulmonary disease. *Mayo Clin. Proc.*, **57**, 231–8.

# 17

# Angiitis of the central nervous system

*J.T. Lie*

Angiitis, or vasculitis, is defined as inflammation of the blood vessels that may result in damage to the vessel walls or vascular occlusion, and causes ischemic injury to the organ tissues served by the affected blood vessels. Angiitis may occur without a known cause (**primary angiitis**) or it may occur as a manifestation of a diverse group of underlying disease (**secondary angiitis**).

The central nervous system (CNS) may be the seat of a wide variety of vasculitides, all but one of which are associated with a known underlying disease (Table 17.1). Because in most patients cerebral vasculitis is often diagnosed clinically without a biopsy confirmation, one should be aware of a wide variety of non-vasculitic conditions that may simulate vasculitis clinically, angiographically, or both.

## PRIMARY GRANULOMATOUS ANGIITIS OF THE CENTRAL NERVOUS SYSTEM

Primary granulomatous angiitis of the central nervous system (GANS) was probably first described in 1922 by Harbitz, among the 'unknown forms of arteritis' [1]. This condition later became known as 'non-infectious granulomatous angiitis involving the central nervous system' [2], 'granulomatous angiitis of the nervous system' [3], or 'isolated angiitis of the central nervous system' [4]. Identical lesions have also been reported in the literature as 'giant-cell arteritis involving

*The Vasculitides.* Edited by B.M. Ansell, P.A. Bacon, J.T. Lie and H. Yazici. Published in 1996 by Chapman & Hall, London. ISBN 978-0-412-64140-4.

**Table 17.1** Angiitis of the central nervous system

*Primary*
Granulomatous angiitis of the nervous system

*Secondary*
Infective angiitis (bacterial, fungal, mycoplasmal, protozoal, rickettsial, viral)
Systemic diseases:
- giant cell arteritis (temporal arteritis, Takayasu arteritis)
- polyarteritis nodosa
- Wegener's granulomatosis
- Churg–Strauss syndrome
- hypersensitivity angiitis
- systemic lupus erythematosus and other rheumatologic disorders
- Behçet's syndrome
- Cogan syndrome
- necrotizing sarcoid granulomatosis and sarcoid angiitis
- inflammatory bowel disease

Drug-induced or drug-related:
- allopurinol
- amphetamines
- cocaine
- ephedrine
- heroin

Malignancy related:
- Hodgkin's lymphoma
- non-Hodgkin's lymphoma
- lymphomatoid granulomatosis
- leukemia
- metastatic small cell lung cancer

*Vasculitis simulators*
Fibromuscular dysplasia
Moyamoya disease
Antiphospholipid syndrome
Vasospasm or acute arterial hypertension
Thrombotic thrombocytopenic purpura
Cardiac myxoma embolism
Sickle cell anemia
Radiation vasculopathy
Acute meningoencephalitis
Acute angioendotheliomatosis

small meningeal and intracerebral vessels' [5]. Although GANS is usually restricted to the CNS, autopsies in some of the previously reported cases have shown a variable degree of involvement of extracranial arteries, including pulmonary and abdominal visceral angiitis

[6, 7]. Thus, the designation of 'primary angiitis' of the central nervous system is more appropriate than 'isolated angiitis'.

Primary GANS is uncommon and, to date, fewer than 120 cases have been reported in the English language literature [8, 9]. The disease occurs in patients of all ages (mean age, 44; range, 3–75 years) with a 5:3 or 4:3 male preponderance [10]. Although the definitive diagnosis of GANS requires biopsy confirmation, most of the cases reported were diagnosed clinically and angiographically, or at autopsy [7, 10]. Biopsy, on the other hand, may not always be conclusive because of the focal nature of the distribution of GANS and the inherent sampling problem for this type of lesion [11]. Such is the dilemma that also applies to the biopsy diagnosis of virtually all other systemic and isolated vasculitides [12].

## Diagnosis

Primary GANS is a distinct nosologic entity with well-characterized clinical, angiographic, and histopathologic diagnostic features (Table 17.2). In about two-thirds of the reported cases, the presenting clinical features include headache, confusion, altered mental status and progressive intellectual deterioration. Diplopia, blurred vision, nystagmus, pupillary abnormalities and dysarthria are indications of possible cranial nerve involvement. Hemiparesis and focal cerebral manifestations are not uncommon, whereas extrapyramidal signs and brainstem strokes occur infrequently.

**Table 17.2** Diagnostic features of primary granulomatous angiitis of the nervous system

*Clinical*
Unexplained, acute, progressive encephalopathy without evidence of extracranial or systemic vasculitis; sometimes associated with history of herpes zoster viral infection, Hodgkin's lymphoma, or illicit drug use; mean age 44 (range 3–75) years with a 4:3 male preponderance.

*Angiographic*
Multifocal, segmental irregularity, stenosis, and dilatation of small- and medium-sized intracranial arteries; alterations in blood flow or associated mass effects on imaging; rarely normal cerebral arteriography.

*Histopathologic*
Focal segmental, lymphocytic, necrotizing or granulomatous angiitis of leptomeningeal and cortical parenchymal small blood vessels, commonly with skipped lesions and coexistence of morphologically acute, healing and healed lesions.

Angiographic findings in cerebral vasculitis are not entirely specific but they are highly characteristic [13, 14]. Stenosis of the affected arteries may be focal or diffuse, partial or complete, circumferential or eccentric, and smooth or irregular. Dilatation of the blood vessels tends to be discrete and aneurysmal. Changes in blood flow may be accelerated or diminished. Distribution of arterial alterations may be localized or widespread, usually affecting the smaller leptomeningeal and intraparenchymal blood vessels.

Although angiography may correctly portray the presence of a cerebral vasculitis, it is incapable of revealing the histologic type of vasculitis, which only a positive biopsy can provide. Acute arterial hypertension can mimic the angiographic appearance of cerebral vasculitis [15]. Unusual distribution involving the larger anterior and posterior cerebral arteries, as well as both carotid siphons, has also been reported [16]. In addition to its diagnostic usefulness, angiography is the only means currently available to monitor disease activity and the outcome of drug treatment in cerebral vasculitis [17]. Conversely, GANS may mimic a cerebral mass lesion in magnetic resonance imaging [18].

Pathologically, GANS affects the small- and medium-sized arteries and only rarely the veins of the brain and meninges. It is characterized by a segmental necrotizing or granulomatous vasculitis, with or without identifiable giant cells, often accompanied by thrombosis. Skipped lesions are common. The inflammatory infiltrate is predominantly lymphocytic with a variable number of neutrophils, eosinophils, histiocytes and plasma cells (Figure 17.1). Both Langhans' and foreign body-type giant cells may be present and in variable numbers (Figure 17.2). Intimal fibrosis signifies the results of healing. Morphologically acute, healing and healed lesions frequently coexist in different segments of the same artery or adjacent arteries.

## Etiology

The etiology of primary GANS is unknown. The association of GANS with lymphoma or leukemia in about 25% of patients, many of whom also have a prior or concurrent herpes zoster infection, may be more than coincidental [10]. Among patients with GANS and herpes zoster infection, herpes virus-like particles have been identified in the glial cell [19] or in the blood vessels affected by GANS [20]. There have also been reports of GANS associated with isolation of human T lymphotrophic virus type III from the CNS [21], and in a patient with acquired immune deficiency syndrome-related complex [22].

Mycoplasmal infection has also been linked to GANS. A necrotizing cerebral vasculitis, histologically similar to GANS, was produced in young turkeys by intravenous injection of *Mycoplasma gallisepticum*

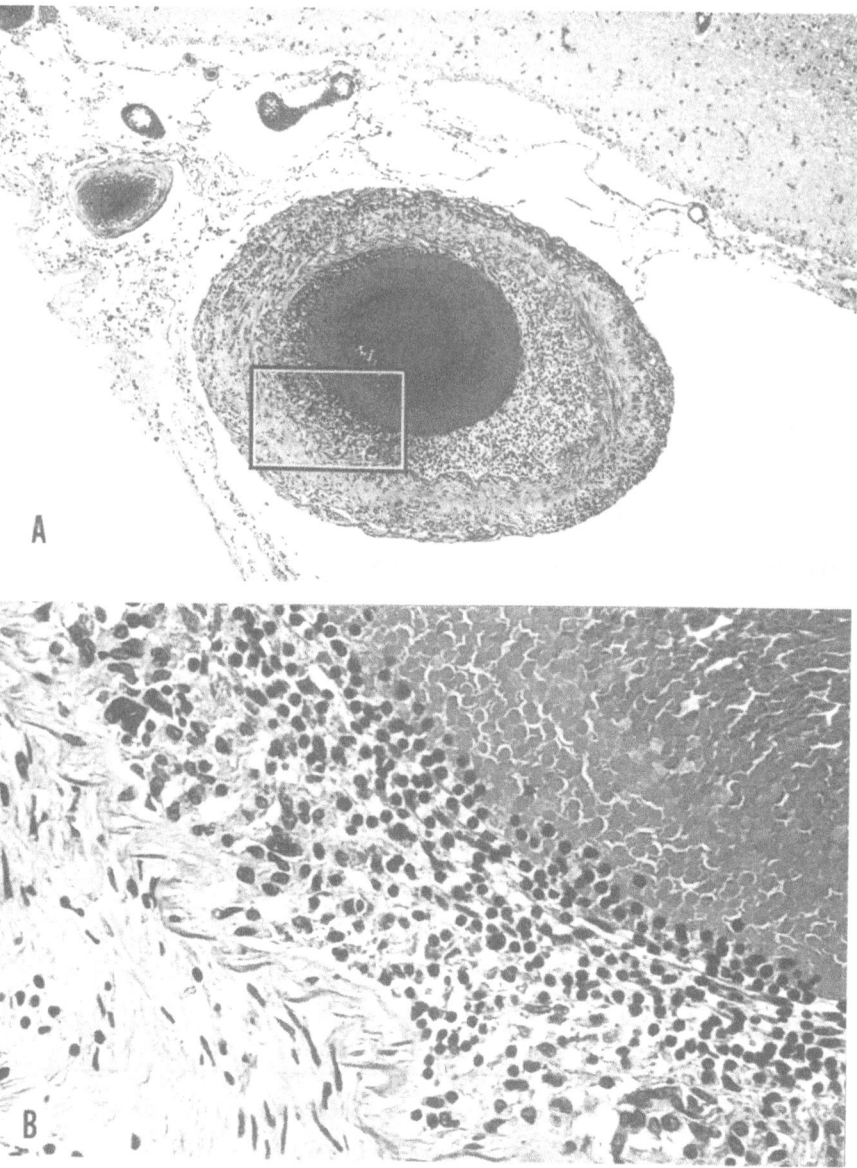

**Figure 17.1** Non-granulomatous primary angiitis (A) of a leptomeningeal artery. Close-up view (B) of boxed area in (A) to show lymphoplasmacytic infiltrate without giant cells. (Hematoxylin-eosin: A × 64, B × 400.)

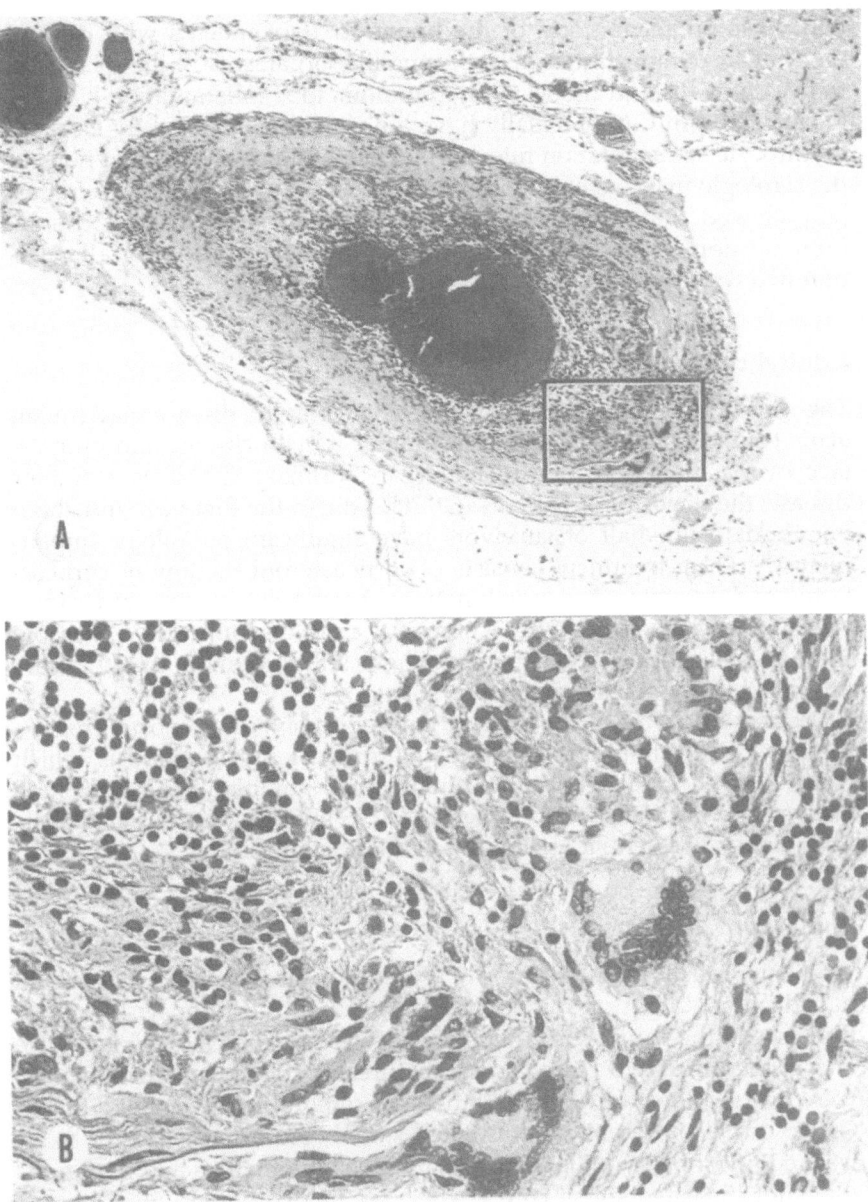

**Figure 17.2** Granulomatous primary angiitis (A) of a leptomeningeal artery of the same patient shown in Figure 17.1. Close-up view (B) of boxed area in (A) to show multinucleated giant cells. (Hematoxylin-eosin: A × 64, B × 400.)

[23]. Electron microscopy of the brains in two patients who died of GANS had demonstrated mycoplasma-like organisms in the glial cells, the vessel walls and tissues with granulomatous inflammation [24].

Patients with GANS usually have a normal or only slightly elevated erythrocyte sedimentation rate, normal hematologic studies and none of the serologic markers of autoimmune disease. An intriguing case report [25] showed that primary GANS may develop and progress despite immunosuppressive therapy in doses sufficient to prevent graft rejection in a renal transplant.

### Clinical course

The clinical outcome of GANS is unpredictable; the disease may run an acute course of three days to six weeks or remain chronic and progressive over a period of one to five years. Primary GANS is a serious disease; the majority of patients (60–70%) die in the first year and about one-third to one-half of survivors have significant neurologic impairment [7]. Drug treatment consists of corticosteroid therapy or corticosteroid and cytotoxic drug combinations. Long-term survival of patients with GANS is possible with aggressive immunosuppressive therapy; there is a case report of one patient who is still alive more than 12 years after the biopsy-proven diagnosis [26].

Because of the need for prompt and aggressive immunosuppressive therapy, GANS should be differentiated from cerebral sarcoidosis, which could also be interpreted as granulomatous angiitis of the CNS [27]. Conversely, cases suspected of being GANS might in fact be sarcoidosis [28]. Biopsy is essential for the differential diagnosis.

### Spinal cord GANS

Three previously reported cases of Hodgkin's disease-associated GANS had documentation of spinal cord involvement at autopsy. In one of these cases, the spinal cord involvement was not clinically evident [29], and in the other two cases clinical features of spinal cord involvement were overshadowed by manifestations of cerebral involvement [30, 31]. GANS can also present as an isolated spinal cord arteritis [32] or as an isolated myelopathy [33]. Clinically, both granulomatous and non-granulomatous isolated spinal cord GANS may be diagnosed with a biopsy to exclude cord involvement by Hodgkin's disease or non-Hodgkin's lymphoma that mimics GANS (Figures 17.3 and 17.4) [34].

## SECONDARY ANGIITIS OF THE CNS

Secondary angiitis of the CNS may be associated with infections, systemic vasculitides, rheumatologic disorders, drug hypersensitivity, illicit

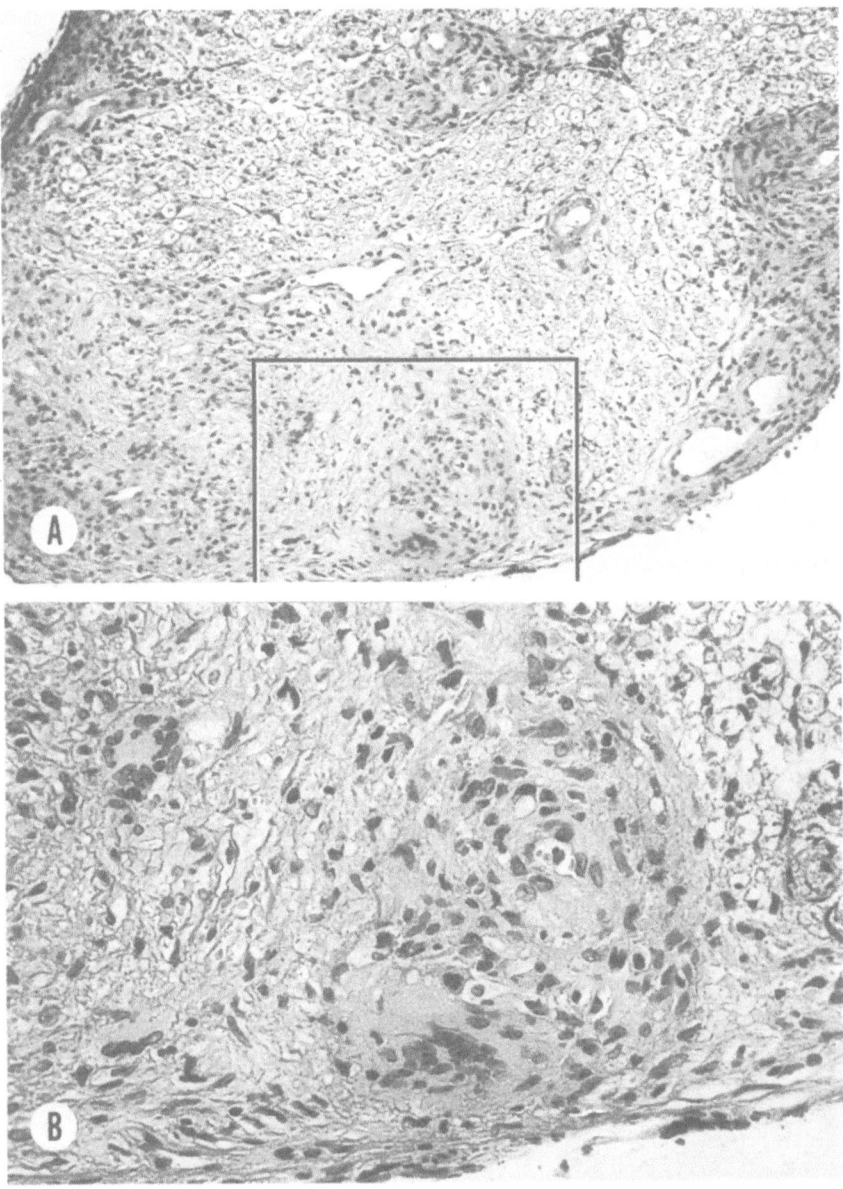

**Figure 17.3** Granulomatous angiitis (A) of the spinal cord. Close-up view (B) of the boxed area in (A). (Hematoxylin-eosin: A × 160, B × 400.)

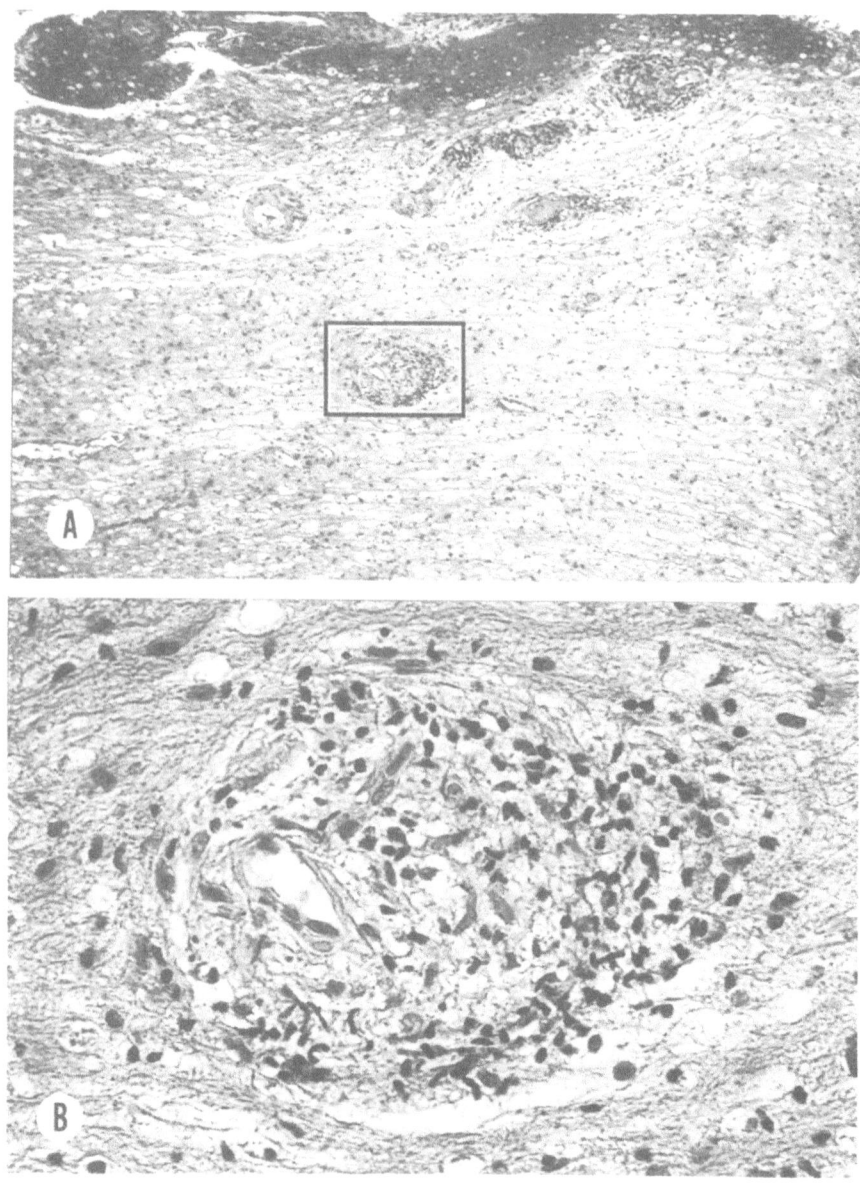

**Figure 17.4** Non-granulomatous angiitis (A) of the spinal cord. Close-up view (B) of boxed area in (A) without giant cells in the infiltrate. Biopsy of the same patient shown in Figure 17.3. (Hematoxylin-eosin: A × 160, B × 400.)

drug abuse and malignancy (Table 17.1). Only selected examples of secondary angiitis of the central nervous system are briefly reviewed here.

## Giant cell arteritis

The giant cell arteritis group comprises temporal arteritis and Takayasu arteritis, two distinct syndromes that have in common histologically a granulomatous inflammation of small-, medium- and large-sized arteries [35]. Although neurologic dysfunction and peripheral neuropathy are not uncommon in both temporal arteritis [36] and Takayasu arteritis [37], angiitis of the intracerebral and peripheral nerve blood vessels are distinctly rare.

## Polyarteritis nodosa and Churg–Strauss syndrome

Neurologic abnormalities occur in 25–50% of patients with polyarteritis nodosa. CNS disease is second only to renal disease as a cause of death in polyarteritis nodosa [38]. The onset of CNS disease tends to be later in the course of polyarteritis nodosa than that of peripheral neuropathies [39]. Peripheral neuropathy is more common in systemic necrotizing vasculitis than physical evaluation alone suggests. Using abnormal electrophysiologic tests as a guide, the diagnostic yield of sural nerve biopsy will be greatly enhanced [40]. Neurological abnormalities of Churg–Strauss syndrome, including cerebral vasculitis, are similar in their frequency and pattern to those in polyarteritis nodosa.

## Wegener's granulomatosis

Based on clinical and pathological criteria, the frequency of CNS involvement in Wegener's granulomatosis (WG) varies from 22% [41] to 54% [42]. Necrotizing granulomatosis and focal vasculitis occur independently more often than in combination. Cyclophosphamide therapy has reduced the incidence of neurologic complications of WG, but treatment failure still occurs [43].

## Hypersensitivity angiitis

The incidence of neurologic complication in hypersensitivity angiitis is unknown. As is exemplified by the classic serum sickness and Schönlein–Henoch purpura [44], the CNS abnormalities include encephalopathy, seizures, coma and peripheral neuropathies. Cerebral vasculitis is uncommon, but peripheral nerve necrotizing vasculitis frequently accompanies cutaneous vasculitis or hypersensitivity angiitis.

## Systemic lupus erythematosus and other rheumatologic disorders

CNS disease may be part of the clinical presentation of systemic lupus erythematosus (SLE) in 25–75% of cases, of which about 30% are non-inflammatory vasculopathy of SLE and only 12.5% true vasculitis [45].

CNS disease is uncommon in progressive systemic sclerosis. Cranial neuropathy was found in 4% of 442 consecutive cases of progressive systemic sclerosis in one series [46], none of which could be attributed to vasculitis. Among the three cases of progressive systemic sclerosis presumed to have angiitis of the CNS, two had angiographic findings compatible with cerebral angiitis [47, 48], and the third was found to have only a right thalamic hemorrhage at post-mortem examination [49].

While vasculitis-related peripheral neuropathy is common in rheumatoid arthritis, angiitis of the CNS is exceedingly rare. A survey of the literature up to 1987 shows that there are only 12 reported cases of rheumatoid cerebral vasculitis [10]. These 12 cases had documented rheumatoid arthritis 1–30 years prior to the clinical onset of neurologic complications.

Equally rare is systemic necrotizing vasculitis involving cerebral blood vessels in rheumatic heart disease, dermatopolymyositis, primary Sjögren's syndrome [50] and relapsing polychondritis [51].

## Behçet's syndrome

Behçet's syndrome is a systemic disease with vasculitis as the common basis of its protean clinical manifestation. Behçet's syndrome is generally a benign disease, but the presence of neurologic involvement worsens the prognosis. About 30% of patients with Behçet's syndrome have neurologic abnormalities that include meningoencephalitis, cortico-spinal tract involvement, cerebral ataxia, pseudotumor, pseudobulbar palsy, paraplegia, sensory disturbances and seizures [52, 53].

The CNS pathologic findings in Behçet's syndrome include venous thrombosis, meningeal inflammation, perivascular cell loss and demyelination, and non-specific small-vessel lymphocytic vasculitis. Angiographically, resolution of cerebral vasculitis can be demonstrated after the successful treatment of Behçet's syndrome with combined cytotoxic agents and corticosteroids [54].

## Cogan's syndrome

Cogan's syndrome is an uncommon disease characterized by non-syphilitic interstitial keratitis associated with vestibuloauditory dysfunction, usually in young adults and with an equal sex distribution. At least 130 cases have been reported in the literature [55]. The prevalence of systemic vasculitis has been estimated to range from 15–72% [56]. Small-vessel vasculitides in Cogan's syndrome are histopathologically indistinguishable from the polyarteritis nodosa-type necrotizing vasculitis, and aortitis is the most common form large-vessel vasculitis, often associated with aortic insufficiency. Apart from retinal vasculitis, involvement of intracerebral blood vessels is rare.

### Sarcoidosis

Sarcoidosis affects the CNS in about 5% of patients. Because the majority of these patients have clear evidence of systemic sarcoidosis, this condition is seldom confused with primary GANS. The most common form of CNS sarcoidosis is a basal, aseptic, granulomatous meningitis. A systemic or pulmonary granulomatous necrotizing angiitis has been reported in many patients with sarcoidosis [57], but angiographic changes suggestive of cerebral vasculitis have been reported in CNS sarcoidosis only very occasionally [58].

### Inflammatory bowel disease

The association of vasculitis and inflammatory bowel disease is infrequent and intriguing. To date, there has been only a single case report of cerebral vasculitis in ulcerative colitis [59] and one of cerebral vasculitis in celiac disease [60].

### Cerebral vasculitis associated with drug use and drug abuse

Illicit drug-induced systemic necrotizing angiitis was first reported in 1970 [61, 62]. This was soon followed by a description of 19 drug-abuse patients, most of whom used intravenous methamphetamines, who had tissue-proven cerebral vasculitis or small-vessel occlusive disease on cerebral angiograms [63]. Since then, cerebral vasculitis has been reported in patients who are polydrug abusers and those who use heroin, ephedrine or cocaine exclusively [64].

### Miscellaneous conditions

The CNS may be involved in a variety of systemic diseases, and some of these cases have clinical or angiographic features that mimic cerebral vasculitis. Lymphomatoid granulomatosis (LGM) is not a vasculitis, but a peripheral T cell lymphoma with angioinvasive and angiodestructive features that mimic vasculitis. CNS dysfunction occurs in 20% of patients and peripheral neuropathy in 15% of patients. There has been a report of one patient in whom clinical evidence of LGM was confined

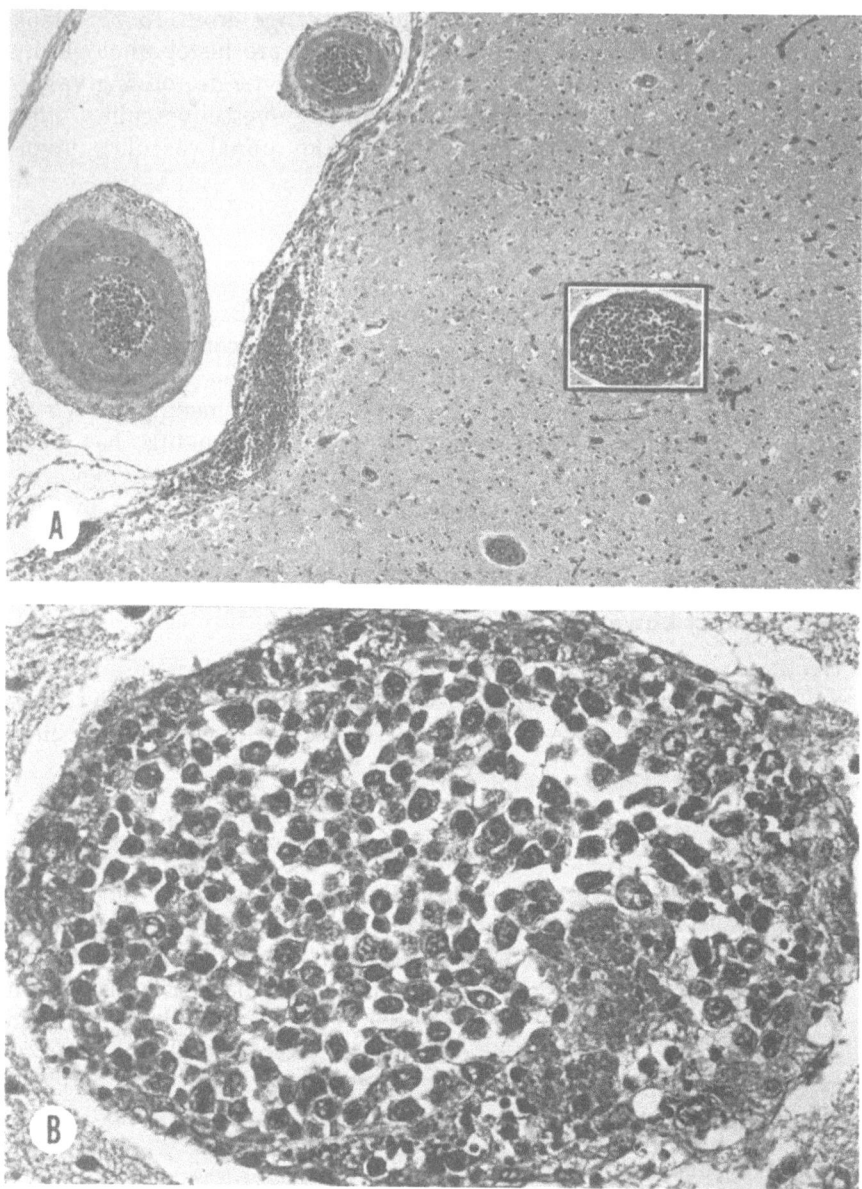

**Figure 17.5** Malignant angioendotheliomatosis (A) involving leptomeningeal and cortical small blood vessels. Close-up view (B) of boxed area in (A) to show intravascular lymphoma cells occluding the vessel lumen. (Hematoxylin-eosin: A × 64, B × 400.)

to the CNS [65]. Malignant angioendotheliomatosis, likewise, may simulate primary GANS (Figure 17.5) [66].

Fibromuscular dysplasia may mimic vasculitis angiographically but it rarely affects the intracerebral arteries. Moyamoya disease, most prevalent in Japan where it was first described, is a peculiar clinical entity that is associated with transient ischemic attacks, stroke or intracerebral hemorrhage in young adults. The name is coined (after a Japanese expression for something hazy, just like 'a puff of cigarette smoke drifting in the air') for the wispy appearance of collaterals demonstrated by cerebral angiography [67]. Histopathologic changes in the brain in autopsy cases do not suggest vasculitis; these changes have been described as thickening of the affected vessels with hyperplasia of the intima, disorganization of the internal elastic lamina and medial fibrosis [68].

Antiphospholipid syndromes are diseases associated with the presence of autoantibodies in the blood against the negatively charged phospholipids, principally lupus anticoagulant and anticardiolipin antibodies. Antiphospholipid syndromes may be primary (not associated with a known underlying disease) or secondary (associated with SLE, lupus-like disease or another connective tissue disease). The syndromes have a wide spectrum of clinical manifestations, including recurrent arterial and venous thrombosis, recurrent fetal loss, thrombocytopenia, hemolytic anemia, livedo reticularis, transient ischemic attacks and strokes. However, coagulopathy rather than vasculitis appears to be the pathogenesis of occlusive vascular disease in antiphospholipid syndromes [69, 70].

Cardiac myxoma embolism may occur in systemic or cerebral circulation, and neurologic abnormalities are the presenting manifestation in about one-third of patients with cardiac myxomas [71, 72]. Cerebral angiography often reveals microaneurysms that mimic vasculitis, which may progress after cardiac surgery for the removal of myxomas, as may the neurologic deficits.

CONCLUSION

There are many causes of cerebral angiitis, and many non-vasculitic conditions may simulate cerebral angiitis. Primary GANS, though relatively uncommon, is clinically the most important type of cerebral angiitis because of the overall unfavorable prognosis. Cerebral angiitis is usually suspected clinically and recognized by angiography, but a definitive diagnosis still requires tissue documentation of the presence of a true vasculitis. A positive biopsy is diagnostic for the disease demonstrated, but a single negative biopsy does not necessarily exclude primary or secondary cerebral vasculitis.

REFERENCES

1. Harbitz, F. (1922) Unknown forms of arteritis with special reference to their relation to syphilitic arteritis and periarteritis nodosa. *Am. J. Med.*, **163**, 250–72.
2. Newman, W. and Wolf, A. (1952) Non-infectious granulomatous angiitis involving the central nervous system. *Trans Am. Neurol. Assoc.*, **77**, 114–7.
3. Budzilovich, G.N., Feigin, I. and Siegel, H. (1963) Granulomatous angiitis of the nervous system. *Arch. Pathol. Lab. Med.*, **76**, 250–6.
4. Cupps, T.R., Morre, P.M. and Fauci, A.S. (1983) Isolated angiitis of the central nervous system. *Am. J. Med.*, **74**, 97–106.
5. McCormick, H.M. and Neuberger, K.T. (1958) Giant-cell arteritis involving small meningeal and intracerebral vessels. *J. Neuropathol. Exp. Neurol.*, **17**, 471–8.
6. Burger, P.C., Burch, J.G. and Vogel, F.S. (1977) Granulomatous angiitis. *Stroke*, **8**, 29–35.
7. Calabrese, L.H. and Mallek, J.A. (1988) Primary angiitis of the central nervous system: report of 8 new cases, review of the literature, and proposal for diagnostic criteria. *Medicine*, **67**, 20–39.
8. Moore, P.M. (1989) Diagnosis and management of isolated angiitis of the central nervous system. *Neurology*, **39**, 167–73.
9. Calabrese, L.H., Furlan, A.J., Gragg, L.A. and Ropos, T.J. (1992) Primary angiitis of the central nervous system: diagnostic criteria and clinical approach. *Cleve. Clin. J. Med.*, **59**, 293–306.
10. Sigal, L.H. (1987) The neurologic presentation of vasculitis and rheumatologic syndromes: a review. *Medicine*, **66**, 157–80.
11. Sabharwal, U.K., Keogh, L.H., Weisman, M.H. and Zvaifler, N.J. (1982) Granulomatous angiitis of the nervous system: case report and review of the literature. *Arthritis Rheum.*, **25**, 342–5.
12. Lie, J.T. (1989) Systemic and isolated vasculitis: a rational approach to classification and pathologic diagnosis. *Pathol. Annu.*, **24**(1), 25–114.
13. Ferris, E.J. and Levine, J. (1973) Cerebral arteritis: classification. *Radiology*, **109**, 327–41.
14. Sole-Llenas, J., Mercader, J.M. and Mirosa, F. (1978) Cerebral arteritis. *Angiology*, **29**, 713–8.
15. Garner, B.F., Burns, P., Bunning, R.D. and Laureno, R. (1990) Acute blood pressure elevation can mimic arteriographic appearance of cerebral vasculitis. *J. Rheumatol.*, **17**, 93–7.
16. Husein, A.M.A. and Haq, N. (1990) Cerebral arteritis with unusual distribution. *Clin. Radiol.*, **41**, 353–4.
17. Stein, R.L., Martino, C.R., Weinert, D.M. *et al.* (1987) Cerebral angiography as a guide for therapy in isolated central nervous system vasculitis. *J.A.M.A.*, **257**, 2193–5.
18. Johnson, M., Maciunas, R., Dutt, P. *et al.* (1989) Granulomatous angiitis masquerading as a mass lesion. *Surg. Neurol.*, **31**, 49–53.
19. Reyes, M.G., Fresco, R., Chokroverty, S. and Salud, R.Q. (1976) Virus-like particles in granulomatous angiitis of the central nervous system. *Neurology*, **26**, 292–9.
20. Linnermann, C.C. Jr. and Alvira, M.M. (1980) Pathogenesis of varicella-zoster angiitis in the CNS. *Arch. Neurol.*, **37**, 239–40.
21. Yanker, B.A., Skolink, P.R., Shoukimas, G.M. *et al.* (1986) Cerebral granulo-

matous angiitis associated with isolation of human T-lymphotropic virus type II from the central nervous system. *Ann. Neurol.*, **20**, 362–4.

22. Vinters, H.V., Guerra, W.F., Eppolito, L. and Keith III, P.E. (1988) Necrotizing vasculitis of the nervous system in a patient with AIDS-related complex. *Neuropathol. Appl. Neurobiol.*, **14**, 417–24.

23. Thomas, L., Davidson, M. and McCluskey, T.R. (1966) Studies of PPLO Infection. I. The production of cerebral polyarteritis by *Mycoplasma gallisepticum* in turkeys; the neurotoxic property of the mycoplasma. *J. Exp. Med.*, **123**, 897–912.

24. Arthur, G. and Margolis, G. (1977) Mycoplasma-like structures in granulomatous angiitis of the central nervous system. Case reports with light and electron microscopies studies. *Arch. Pathol. Lab. Med.*, **101**, 382–7.

25. Rosenberg, R.J. (1985) Isolated angiitis of the brain. *Am. J. Med.*, **79**, 629–32.

26. Zimmerman, R.S., Young, H.F. and Hadfield, M.G. (1990) Granulomatous angiitis of the nervous system: a case report of long-term survival. *Surg. Neurol.*, **33**, 206–12.

27. Bottcher, E. (1959) Disseminated sarcoidosis with a marked granulomatous arteritis. *Arch. Pathol.*, **68**, 419–23.

28. Peison, B. and Padleckas, R. (1964) Granulomatous angiitis of the central nervous system. *Ill. Med. J.*, **126**, 330–4.

29. Rottino, A. and Hoffman, G. (1950) A sarcoid form of encephalitis in a patient with Hodgkin's disease: case report with review of the literature. *J. Neuropathol. Exp. Neurol.*, **9**, 103–8.

30. Rewcastle, N.G. and Tom, M.I. (1962) Non-infectious granulomatous angiitis of the nervous system associated with Hodgkin's disease. *J. Neurol. Neurosurg. Psychiatry*, **25**, 57–8.

31. Rosenblum, W.I. and Hadfield, M.G. (1972) Granulomatous angiitis of the nervous system in cases of herpes zoster and lymphosarcoma. *Neurology*, **22**, 348–54.

32. Feasby, T.E., Ferguson, G.G. and Kaufman, J.C.E. (1975) Isolated spinal cord arteritis. *Can. J. Neurol. Sci.*, **2**, 143–6.

33. Rawlinson, D.G. and Braun, C.W. (1981) Granulomatous angiitis of the nervous system first seen as relapsing myelopathy. *Arch. Neurol.*, **38**, 129–31.

34. Inwards, D.J., Piepgras, D.G., Lie, J.T. *et al.* (1991) Granulomatous angiitis of the spinal cord associated with Hodgkin's disease. *Cancer*, **68** 1318–22.

35. Lie, J.T. (1987) The classification and diagnosis of vasculitis in large- and medium-sized blood vessels. *Pathol. Annu.*, **22(Part 1)**, 125–62.

36. Caselli, R.J., Hunder, G.G. and Whisnant, J.P. (1988) Neurologic disease in biopsy-proven giant cell (temporal) arteritis. *Neurology*, **38**, 352–9.

37. Currier, R.D., DeJong, R.N. and Bole, G.G. (1954) Pulseless disease: central nervous system manifestations. *Neurology*, **4**, 818–30.

38. Travers, R.L., Allison, D.J., Brettle, R.P. and Hughes, G.R.V. (1979) Polyarteritis nodosa: a clinical and angiographic analysis of 17 cases. *Semin. Arthritis Rheum.*, **8**, 184–99.

39. Moore, P.M. and Fauci, A.S. (1981) Neurologic manifestations of systemic vasculitis: a retrospective and prospective study of the clinicopathologic features and responses to therapy in 25 patients. *Am. J. Med.*, **71**, 517–24.

40. Wees, S.J., Sunwoo, I.N. and Oh, S.J. (1981) Sural nerve biopsy in systemic necrotizing vasculitis. *Am. J. Med.*, **71**, 525–32.

41. Fauci, A.S., Haynes, B.F., Katz, P. and Wolff, S.M. (1983) Wegener's granulomatosis: prospective clinical and therapeutic experience with 85 patients for 21 years. *Ann. Intern. Med.*, **98**, 76–85.

42. Drachmann, D.A. (1963) Neurological complications of Wegener's granulomatosis. *Arch. Neurol.*, **8**, 145–55.
43. Kroneman III, O.C. and Pevzner, M. (1986) Failure of cyclophosphamide to prevent cerebritis in Wegener's granulomatosis. *Am. J. Med.*, **80**, 526–7.
44. Lewis, I.C. and Philpott, M.G. (1956) Neurological complications in the Schönlein–Henoch syndrome. *Arch. Dis. Child.*, **31**, 369–71.
45. Johnson, R.T. and Richardson, E.P. (1986) The neurological manifestations of systemic lupus erythematosus: a clinical pathological study of 24 cases and review of the literature. *Medicine*, **47**, 337–69.
46. Farrell, D.A. and Medsger Jr., T.A. (1982) Trigeminal neuropathy in progressive systemic sclerosis. *Am. J. Med.*, **73**, 57–62.
47. Lee, J.L. and Haynes, J.M. (1967) Carotid arteritis and cerebral infarction due to scleroderma. *Neurology*, **17**, 18–22.
48. Estey, E., Lieberman, A., Pinto, R. *et al.* (1979) Cerebral arteritis in scleroderma. *Stroke*, **5**, 595–7.
49. Wise, T.N. and Ginzler, E.M. (1975) Scleroderma cerebritis, an unusual manifestation of progressive systemic sclerosis. *Dis. Nerv. System*, **36**, 60–2.
50. Andonopoulos, A.P., Lagos, G., Drosos, A.A. and Moutsopoulos, H.M. (1990) The spectrum of neurological involvement in Sjögren's syndrome: *Br. J. Rheumatol.*, **29**, 21–3.
51. Stewart, S.S., Ashizawa, T., Dudley Jr., A.W. *et al.* (1988) Cerebral vasculitis in relapsing polychondritis. *Neurology*, **38**, 150–2.
52. Chajek, T. and Fainaru, M. (1975) Behçet's disease: report of 41 cases and a review of the literature. *Medicine*, **54**, 179–96.
53. Shimizu, T., Ehrlich, G.E., Inaba, G. and Hayashi, K. (1979) Behçet's disease (Behçet's syndrome). *Semin. Arthritis Rheum.*, **8**, 223–60.
54. Zelenski, J.D., Caparo, J.A., Holden, D. and Calabrese, L.H. (1989) Central nervous system vasculitis in Behçet's syndrome: angiographic improvement after therapy with cytoxic agents. *Arthritis Rheum.*, **32**, 217–20.
55. Bielory, L., Conti, J. and Frohman, L. (1990) Cogan's syndrome. *J. Allergy Clin. Immunol.*, **85**, 808–15.
56. Vollertsen, R.S., McDonald, T.J., Younge, B.R. *et al.* (1986) Cogan's syndrome: 18 cases and a review of the literature. *Mayo Clin. Proc.*, **61**, 344–61.
57. Lie, J.T. (1989) Classification of pulmonary angiitis and granulomatosis: histopathologic perspectives. *Semin. Respir. Med.*, **10**, 111–21.
58. Lawrence, W.P., El Gammal, T., Pool Jr., W.H. and Apter, L. (1974) Radiological manifestations of neurosarcoidosis: report of three cases and review of the literature. *Clin. Radiol.*, **25**, 343–8.
59. Nelson, J., Barron, M.M., Riggs, J.E. *et al.* (1986) Cerebral vasculitis and ulcerative colitis. *Neurology*, **36**, 719–21.
60. Rush, P.J., Inman, R., Bernstein, M. *et al.* (1986) Isolated vasculitis of the central nervous system in a patient with celiac disease. *Am. J. Med.*, **81**, 1092–4.
61. Gocke, D.J., Hsu, K., Morgan, C. *et al.* (1970) Association between polyarteritis and Australia antigen. *Lancet*, **ii**, 1149–53.
62. Citron, B.P., Halperin, M. and McCarron, M. (1970) Necrotizing angiitis associated with drug abuse. *N. Engl. J. Med.*, **283**, 1003–11.
63. Rumbauch, C.L., Bergeron, R.T., Fang, H.C.H. and McCormack, R. (1971) Cerebral angiographic changes in the drug abuse patient. *Radiology*, **101**, 335–44.
64. Kaye, B.R. and Fainstat, M. (1987) Cerebral vasculitis associated with cocaine abuse. *J.A.M.A.*, **258**, 2104–6.

65. Kokmen, E., Billman Jr., K. and Abell, M.R. (1977) Lymphomatoid granulomatosis clinically confined to the CNS. *Arch. Neurol.*, **34**, 782–4.
66. Lie, J.T. (1992) Malignant angioendotheliomatosis (intravascular lymphomatosis) simulating primary angiitis of the central nervous system. *Arthritis Rheum.*, **35**, 831–4.
67. Suzuki, J. and Takaku, A. (1969) Cerebrovascular 'Moyamoya' disease. *Arch. Neurol.*, **20**, 288–99.
68. Yamashita, M., Oka, K. and Tanaka, K. (1983) Histopathology of the brain vascular network in Moyamoya disease. *Stroke*, **14**, 50–8.
69. Lie, J.T. (1989) Vasculopathy in the antiphospholipid syndromes *J. Rheumatol.*, **16**, 713–15.
70. Lie, J.T. (1994) Vasculitis in antiphospholipid syndrome: culprit or consort? *J. Rheumatol.*, **21**, 397–9.
71. Wold, L.E. and Lie, J.T. (1980) Cardiac myxomas: a clinicopathologic profile. *Am. J. Pathol.*, **101**, 219–39.
72. Roeltgen, D.P., Weimer, G.R. and Patterson, L.F. (1981) Delayed neurologic complications of left atrial myxoma. *Neurology*, **31**, 8–13.

# Part Four

## Connective Tissue and Dermal Vasculitides

# 18

# Rheumatoid vasculitis

*P.A. Bacon, R.A. Luqmani and D.G.I. Scott*

HISTORY

Vascular inflammation in rheumatoid arthritis (RA) was described well before vasculitis was recognized as an integral part of the rheumatoid disease process. The first description was probably that of Bannatyne in 1898, a physician in Bath, England, who also had an interest in pathology. In the post-mortem section of a peripheral nerve he described 'infiltration of small round cells in the nerve sheath, around the blood vessels' [1]. This was in a rheumatoid patient who had had evidence of motor neuropathy during life. In 1947, Bywaters noted a rheumatoid patient with nail edge and nail fold lesions who later went on to develop gangrene of all four limbs and the gut. This started him on his extensive investigations into vascular involvement in RA [2]. Ellman and Ball [3] described inflammatory changes in the heart, lungs and kidney of one patient and were the first to coin the term 'rheumatoid disease'. However, they went on to suggest that such vasculitis represented the concurrence of polyarteritis and RA in a single patient [4]. It was Sokoloff, Wilens and Bunim in 1951 who were the first to recognize that vasculitis is a distinct complication of RA, when they described arteritis in the muscle biopsy in six out of 52 rheumatoid patients they studied [5].

The clinical features associated with rheumatoid vasculitis (RV) were established in several classic series in the 1960s [6–8]. The classic features were those of peripheral gangrene and mononeuritis multiplex [9, 10]. In addition to these features, there was a definite incidence of pericarditis, scleritis, nodules and of systemic disease. Renal, lung and cerebral

*The Vasculitides*. Edited by B.M. Ansell, P.A. Bacon, J.T. Lie and H. Yazici.
Published in 1996 by Chapman & Hall, London. ISBN 978-0-412-64140-4.

involvement was occasionally noted together with such rarities as aortic arch involvement or even temporal arteritis. As the spectrum extended, Bywaters made the provocative suggestion that all extra-articular disease in RA occurred on the basis of vasculitis [11].

Our review of 50 cases in 1981 suggested a shift away from the classical features to a broader spectrum of disease [12]. This recognized the common presence of nodules and skin lesions, both occurring in more than 80%. It is worth noting that 52% of the patients had minor digital lesions and 58% chronic leg ulcers, whereas only 17% had the classic peripheral gangrene. Non-specific systemic features were also seen in over four-fifths, particularly weight loss. We now suspect RV whenever unexplained weight loss occurs in RA. In contrast to those features, neuropathy was seen in less than half the patients, and only 14% had the classic mononeuritis multiplex. More patients had heart and lung involvement, which each occurred in about a third. Thus, the features historically bringing RV to attention were less common than other more non-specific features. This emphasizes the importance of a high index of suspicion for its occurrence and a willingness to obtain biopsy proof of involvement whenever necessary.

## THE CURRENT STATUS OF RV

It has been suggested that RV is getting both milder in its features and less frequent in its occurrence. Our data do not support either contention. Clinically, our cases seen since 1981 have had a remarkably similar spectrum of disease involvement to those reported earlier [12]. Analysis of cases seen in Birmingham between 1982 and 1986 showed virtually no difference from the previous picture [13]. A more recent analysis after 12 years (1981–93) confirms this except for a smaller incidence of cardiac and pulmonary involvement. However, the recent description of aortic root involvement [14, 15] shows that cardiac involvement does occur and there are still things to learn about RV.

One explanation for the lack of significant change in the spectrum of disease seen over the last three decades is that all the studies were from tertiary centers with an inherent referral bias. Thus the more seriously ill patients may have been referred to such centers even if they came from a very large population, that is the severity of vasculitis may indeed be declining. However, the recent description of a group of patients from Norwich seen between 1988 and 1993 suggest that this is not the case. These patients come from a stable mixed rural/urban area in eastern England (population 413 500 adults) served by a single district general hospital. Detailed analysis of the spectrum of disease (Table 18.1) shows almost the same pattern of disease as that seen in Birmingham during the same period.

**Table 18.1** Clinical features of patients with SRV

|  | Bath[1] (1977–81) % | Birmingham (1981–93) % | Norwich (1988–93) % |
|---|---|---|---|
| Nodules | 86 | 67 | 57 |
| Systemic | 86 | 72 | 49* |
| Cutaneous | 88 | 77 | 89 |
| Neurological | 42 | 36 | 38 |
| Lung | 34 | 13* | 28 |
| Gut | 10 | 9 | 4 |
| CVS | 34 | 6* | 19 |
| Renal | 24 | 6 | 25 |
| Eye | 16 | 9 | 9 |

[1] from Scott et al. (1981).
* $p < 0.05$ vs. group III.

The Norwich data also allowed a more accurate estimation of the incidence of systemic RV (SRV). An increased frequency of vasculitis was noted in the 1960s [16], probably related to the effect of the 'steroid epidemic' at that time [17]. This effect should be significantly lower in the 1980s and 1990s, so resulting in a reduced incidence of SRV. On the basis of our early studies (1977–81) we suggested an incidence of 0.6 per 100 000 per year (based on a poorly defined population). The Norwich study for 1988–93 found a higher incidence of 1.3 per 100 000 per year (1.6 per 100 000 in men and 0.9 per 100 000 in women). These figures may just reflect the fact that they are the first accurate data on incidence, but they certainly suggest that the incidence of SRV is not declining [18].

The Norwich/Birmingham comparison also afforded the opportunity to look at the outcome as another index of whether the disease is now milder. The mortality at one and two years was similar between the two centers and not significantly lower than the 1981 rate (Table 18.2). The difference between Norwich and Birmingham mortality rates do not reach statistical significance. The apparent increase in mortality in Norwich may simply reflect the older population seen in a district general hospital when compared with tertiary referral centers. Thus, a close look at the evidence does not suggest that RV is a disappearing

**Table 18.2** Mortality

|  | Bath % | Birmingham % | Norwich % |
|---|---|---|---|
| 1 year | 20 | 13 | 17 |
| 2 years | 46 | 16 | 27 |

disease with mild clinical symptoms. There is still a significant risk of mortality from serious systemic involvement. Early recognition and prompt institution of aggressive immunosuppressive therapy is still important. Early deaths are still seen in patients and there is also a significant morbidity in all three series, stressing the dangers in any delay in diagnosis and treatment.

## OVERLAP BETWEEN RV AND PAN

We have also used our clinical database to re-examine the concept proposed by Ball [4] that arteritis in RA is due to the concurrence of polyarteritis nodosa (PAN). This was based on the histological appearance which may be seen on biopsy of muscular arteries. However, the picture of RV is not always so acute. The subacute and healing phases may have few or no inflammatory cells [19], until the lesion has the appearance of the 'bland intimal proliferation' reported by Bywaters and Scott [7] in digital vessels with chronic obliterative arteritis. It is also important to emphasize that in RV any size of blood vessel may be involved, from capillaries through to the aortic root. PAN is classically the disease of small and large muscular arteries, and does not extend through into small-vessel disease [20, 21].

There are many similarities in the clinical spectrum of organ involvement between SRV and PAN, including heart, lung and peripheral nerve involvement [13, 22]. However, skin and musculoskeletal involvements were not only more common but also more serious in SRV. Gut and renal lesions were both more common and more serious in PAN. These important differences are not always apparent at presentation. We have seen five cases who presented with vasculitis, essentially identical to PAN, who subsequently developed progressive arthritis typical of RA [23]. The main difference between the two becomes obvious from investigations into etiopathogenesis. There is extensive evidence of circulating immunocomplexes containing rheumatoid factors in RV [24]. The close correlation with clinical activity suggests that they may be causally related (see below). This is not the case in PAN, where significant immunocomplexes or alterations in complement titers were not seen in our series (except in the very rare case associated with hepatitis B).

## THERAPY

The serious prognosis and extent of clinical involvement in SRV underlines the need for aggressive suppressive therapy. Steroids alone are usually ineffective and, indeed, the first introduction of steroids, or a sudden dose alteration, may precipitate vasculitis [17]. Cytotoxic drugs are presently the mainstay of therapy, with an emphasis on cyclophos-phamide, just as in the other systemic vasculitic diseases such as Wegener's granulomatosis and microscopic polyangiitis. We have been interested in the best route of delivery and have explored several regimes [25]. Analysis of 49 retrospective patients suggested that inter-mittent pulse cyclophosphamide together with methylprednisolone is more effective than continuous oral regimes [26]. We have been less impressed with azathioprine, which has a slower onset of action. Azathioprine may have a place in the treatment of severe SRV [27], but the number of patients in that study were small and the dose of steroid quite high. Azathioprine is probably less toxic than cyclopho-sphamide, as is another alternative, methotrexate, which is frequently used to treat RA. Our experience in vasculitis is small but, as with azathioprine, methotrexate appears less effective, although less toxic, than cyclophosphamide.

Alternative treatments have been used, sometimes together with cytotoxic drugs, in an attempt to improve the initial response. Plasma exchange may be effective in severely ill patients in the short term. Prostacyclin infusions may be helpful for peripheral vasospasm which exacerbates digital gangrene. Controlled studies have been undertaken to assess its role in cutaneous ulcers, but the results are not yet available. Although mortality has fallen from 46% (1970s) to 25% (1980s and 1990s) this by no means represents a cure. The most important effect of treat-ment appears to be a significant reduction in morbidity with a relapse rate falling from 48% to 17%. Despite this improvement relapses still occur. They may be seen at any stage, require treatment with cyclopho-sphamide, and stress the importance of long-term follow-up. New main-tenance regimes are required to keep patients in remission. There is currently no ideal candidate, but both thalidomide and pentoxyphyline have been useful in occasional patients in our hands. They deserve further study in the prevention of relapse.

ETIOPATHOGENESIS

The etiopathogenesis of RV is still not fully elucidated, but the common serological abnormalities suggests that this fits the classic model of immune complex-mediated vasculitis. The early descriptions empha-sized the high titers of IgM rheumatoid factors (RF). In our series, as in most others, they were present in the majority, but not all cases. We have been particularly interested in the role of IgG RF, which in a large cross-sectional survey was strongly associated with seropositive RA with extra-articular features, particularly vasculitis. Levels greater than twice normal were confined to the vasculitis group [28]. Serial studies and detailed observations in individual cases during therapy show a close association between IgG RF levels and clinical states of vasculitis, while IgM RF levels showed little change [29]. Another difference between these two rheumatoid factors is that the IgG RF is present in considerably higher levels in the serum than in synovial fluid [30]. A similar relationship between IgG RF and vasculitis has been noted in other centers [31]. The case for a direct pathogenetic role of IgG RFs is enhanced by the finding of hypocomplementemia, particularly depression of C4, concomitant with elevated IgG RF. Cryoglobulins, which contain significant immune complexes are also frequently detected. *In vitro* studies show that IgG RF participates in complement fixing immune complexes detected by an anti-C3b assay [32]. Finally, IgG RF is complexed to IgM RF in many cases, as demonstrated by column elution studies. However, immune complexes may not be the only factors, since we have also noted an increased percentage of circulating activated lymphocytes in SRV. *In vitro* studies of endothelial cell cytotoxicity suggests these may also have some relevance to the pathogenesis [33].

Other autoantibodies have also been noted in association with RV. Antinuclear antibodies are often elevated and precipitating antibodies to unidentified nuclear antigens have been reported in higher titer in RV than uncomplicated RA [34]. More recently, antineutrophil cytoplasmic antibodies have been detected by the Leiden group and related to antilactoferrin antibodies [35]. However, our own studies in collaboration with the Groningen group do not suggest that antilactoferrin antibodies are any commoner in RV than in active RA [Kallenberg and Bacon unpublished data]. Antiendothelial cell antibodies (AECA) have also been reported [36]. Such antibodies are in no sense specific to RV, having been reported in many forms of vasculitis [37]. It is unlikely, therefore, that they are related to the pathogenesis, but they may have an important use as a marker of disease activity in serial studies [38]. A marker to detect relapse in RV would be clinically very important. The roles of AECA and IgG RF in this respect need to be compared in prospective studies.

## SMALL-VESSEL INVOLVEMENT AND SUBCLINICAL DISEASE

In contrast to the serious nature of arterial inflammation in RA, small-vessel disease is widely accepted as a frequent and generally unimportant aspect of RA. Small nail edge or nail fold lesions are common (one-third in inpatients in one series [39]), but easily missed by both doctor and patient. The recorded incidence usually reflects the perseverance which has been put into observing it, as with other aspects of systemic RA [40, 41]. Although unimportant in themselves, such small-vessel lesions can both coexist with or herald major arteritis, as in Bywater's original case. The siting of such lesions bears a close relationship to local trauma. They are more common in patients using their hands actively, classically in the 'typus robustus' patient. They may even relate to simple local pressure, such as that produced by holding a pen [42]. The immunopathology of these lesions is not clear, but it is noteworthy that evidence of circulating immune complexes, RFs and complement consumption is far more widespread in active rheumatoid disease than simply in those few patients with active arteritis. Subclinical vasculitis has also been reported to occur in biopsies of apparently normal skin in about one-third of active RA patients in two series [43, 44].

The real significance of small-vessel and subclinical vasculitis remains to be determined, but is probably important. In the short term, it is likely to make a significant contribution to the systemic malaise which characterizes the active phase of RA and distinguishes it from non-inflammatory arthritis. Fever, weight loss and malaise are important features of RV but also can occur in RV, related to systemic cytokine production. The cytokines that drive this, such as IL-1, IL-6 and tumor necrosis factor (TNF), also have important roles in upregulating endothelial cell responses [45]. More accurate data on the frequency and distribution of blood vessel inflammation in RA would be more important. Recent data on salivary gland biopsies suggest that it is not simply confined to the skin [46].

The major significance of subclinical vasculitis may be in its potential to contribute to the overall morbidity and mortality of RA. The late consequences of vascular inflammation are increasingly being recognized as far from benign, even when healing has occurred. They present later as accelerated 'degenerative' vascular disease, and are well documented in late organ graft failure [47]. Both medial degeneration and vascular calcification are seen as manifestations of late RA [48]. As in SLE, it has been suggested these relate to steroids but the evidence from steroid doses used for SLE in those with and without vascular involvement does not support this [49].

We have recently proposed that the late consequences of vascular inflammation in RA, particularly the common subclinical vascular inflammation, could play a major role in the well documented but unexplained increased overall mortality of RA [50]. The excess mortality of RA does not relate principally to direct disease consequences, but to less specific factors including cardiovascular disease [51]. We hypothesize that this represents the late consequences of the vascular inflammation seen in the early active phases of RA.

The overall message from these studies is that RV remains an important aspect of rheumatoid disease in its acute manifestations, with serious consequences for the patients despite aggressive therapy. It is certainly not a disappearing disease and its incidence is higher than had previously been suggested. The importance of the less dramatic features, especially the extent of subclinical disease during active inflammation, remains an important undecided issue. If, as we hypothesize, it relates to the overall excess mortality in RA, then it requires much more serious investigation in the future.

## REFERENCES

1. Bannatyne, G.A. (1898) *Rheumatoid Arthritis*, 2nd edn, John Wright, Bristol.
2. Bywaters, E.G.L. (1957) Peripheral vascular obstruction in rheumatoid arthritis and its relationship to other vascular lesions. *Ann. Rheum. Dis.*, **16**, 84–103.
3. Ellman, P. and Ball, R.E. (1948) 'Rheumatoid disease' with joint and pulmonary manifestations. *Br. Med. J.*, **ii**, 816–21.
4. Ball, J. (1954) Rheumatoid arthritis and polyarteritis nodosa. *Ann. Rheum. Dis.*, **13**, 277–90.
5. Sokoloff, L., Wilens, S.L. and Bunim, J.J. (1951) Arteritis of striated muscle in rheumatoid arthritis. *Am. J. Pathol.*, **27**, 157–73.
6. Schmidt, F.R., Cooper, N.S., Ziff, M. and McEwan, C. (1961) Arteritis in rheumatoid arthritis. *Am. J. Med.*, **30**, 56–83.
7. Bywaters, E.G.L. and Scott, J.T. (1963) The natural history of vascular lesions in rheumatoid arthritis. *J. Chronic. Dis.*, **16**, 905–14.
8. Wilkinson, M. and Torrance, W.N. (1967) Clinical background of rheumatoid vascular disease. *Ann. Rheum. Dis.*, **26**, 475–9.
9. Hart, F.D., Golding, J.R. and Mackenzie, D.H. (1957) Neuropathy in rheumatoid disease. *Ann. Rheum. Dis.*, **16**, 471–80.
10. Pallis, C.A. and Scott, J.T. (1965) Peripheral neuropathy in rheumatoid arthritis. *Br. Med. J.*, **i**, 1141–7.
11. Bywaters, E.G.L. (1976) Vasculitis in rheumatoid arthritis, in *Non-articular Forms of Rheumatoid Arthritis, Proc. IV ISRA Symposium* (ed. T.E.W. Feltkamp), Stafleu's Scientific Publications, Leiden, pp. 82–4.
12. Scott, D.G.I., Bacon, P.A. and Tribe, C.R. (1981) Systemic rheumatoid vasculitis: a clinical and laboratory study of 50 cases. *Medicine (Baltimore)*, **34**, 843–50.
13. Bacon, P.A. and Scott, D.G.I. (1987) La vascularite rhumatoide, in *Polyarthrite Rhumatoide – Aspects Actuels et Perspectives* (ed. J. Sany), Médicine-Sciences Flammarion, Paris, pp. 27–43.
14. Scott, D.G.I. and Bacon, P.A. (1987) Cardiac involvement in immunology diseases, in *Clinics in Immunology and Allergy* (ed. W.A. Littler), Saunders, Baillière.
15. Townend, J.N., Emery, P., Davies, M.K. and Littler, W.A. (1991) Acute aortitis and aortic incompetence due to systemic rheumatological disorders. *Int. J. Cardiol.*, **33**, 253–8.
16. Hart, F.D. (1969) Rheumatoid arthritis: extra-articular manifestations. *Br. Med. J.*, **3**, 131–6.
17. Vollertson, R.S. *et al.* (1986) Rheumatoid vasculitis: survival and associated risk factors. *Medicine (Baltimore)*, **65**, 365–75.
18. Watts, R.A., Luqmani, R.A., Carruthers, D.M. *et al.* (1994) *Long-term Outcome in Systemic Rheumatoid Vasculitis (RV)*, Abstract to EULAR 1994 Symposium.
19. Tribe, C.R., Scott, D.G.I. and Bacon, P.A. (1981) The place of rectal biopsy in the diagnosis of systemic vasculitis. **34**, 843–50.
20. Kussmaul, A. and Maier, R. (1866) Über eine bisher nicht beschriebene

eigenthümliche Arterienerkrankung (Periarteritis nodosa), die mit Morbus Brightii und rapid fortschreitender allgemeiner Muskellähmung einhergeht. *Dtsch. Arch. Klin. Med.*, **1**, 484–518.

21. Jennette, J.C., Falk, R.J., Andrassy, K. *et al.* (1994) Nomenclature of systemic vasculitides – Proposal of an International Consensus Conference. *Arthritis Rheum.*, **37**(2), 187–92.

22. Bacon, P.A. (1993) Vasculitis from the rheumatologist's viewpoint, in *ANCA Associated Vasculitides* (ed. W. Gross), Plenum Press, London.

23. Farr, M., Tunn, E.J., Symmons, D. *et al.* (1986) Sulphasalazine in arthritis: Haematological problems associated with therapy. *Br. J. Rheum.* (Abstract), Spring Meeting, Dublin.

24. Scott, D.G.I. (1981) *Systemic Vasculitis in Bath–Bristol Area: Clinical, Pathological and Therapeutic Studies.* University of Bristol MD Thesis.

25. Hall, N.D., Bird, H.A., Ring, E.F.J. and Bacon, P.A. (1979) A combined clinical and immunological assessment of four cyclophosphamide regimes in rheumatoid arthritis. *Agents Actions*, **9**, 97–102.

26. Scott, D.G.I. and Bacon, P.A. (1984) Intravenous cyclophosphamide plus methylprednisolone treatment in systemic rheumatoid vasculitis. *Am. J. Med.*, **76**, 377–84.

27. Heurkens, A.H., Westedt, M.L. and Breedveld, F.C. (1991) Prednisolone plus azathioprine treatment in patients with rheumatoid arthritis complicated by vasculitis. *Arch. Intern. Med.*, **151**, 2249–54.

28. Allen, C., Elson, C.J., Scott, D.G.I. *et al.* (1981) IgG antiglobulins in rheumatoid arthritis and other arthritides: relationship with clinical features and other parameters. *Ann. Rheum. Dis.*, **40**, 127–31.

29. Scott, D.G.I., Bacon, P.A., Allen, C. *et al.* (1981) IgG rheumatoid factor, complement and immune complexes in rheumatoid synovitis and vasculitis: comparative and serial studies during cytotoxic therapy. *Clin. Exp. Immunol.*, **43**, 54–63.

30. Elson, C.J., Carter, S.D., Scott, D.G.I. *et al.* (1985) A new assay for IgG rheumatoid activity and its use to analyse rheumatoid factor reactivity with human IgG isotopes. *Rheumatol. Int.*, **5**, 175–9.

31. Pope, R.M. and McDuffy, S.J. (1980) IgG rheumatoid factor: analysis of various species of IgG for the detection by radioimmunoassay. *Arthritis Rheum.*, **23**, 733.

32. Elson, C.J., Carter, S.D., Cottrell, B.J. *et al.* (1985) Complement activating properties of complexes containing rheumatoid factor in synovial fluids and sera of patients with rheumatoid arthritis. *Clin. Exp. Immunol.*, **59**, 285–92.

33. Blann, A.D., Scott, D.G.I. and Bacon, P.A. (1991) Endothelial cell cytotoxicity in rheumatoid disease: a possible role for autoantibodies and immune complexes. *Int. J. Immunopathol. Pharmacol.*, **4**, 75–83.

34. Scott, D.G.I., Skinner, R.P., Bacon, P.A. and Maddison, P.J. (1984) Precipitating antibodies to nuclear antigens in systemic vasculitis. *Clin. Exp. Immunol.*, **56**, 601–6.

35. Covemans, I.E.M., Hagen, E.C., Daha, M.R. *et al.* (1992) Antilactoferrin antibodies in patients with rheumatoid arthritis are associated with vasculitis. *Arthritis Rheum.*, **35**, 1466–75.

36. Tan, E.M. and Pearson, C.M. (1972) Rheumatic disease sera reactive with capillaries in the mouse kidney. *Arthritis Rheum.*, **15**, 23–8.

37. Frampton, G., Jayne, D.R.W., Perry, G.J. *et al.* (1990) Autoantibodies to

endothelial cells and neutrophil cytoplasmic antigens in systemic vasculitis. *Clin. Exp. Immunol.*, **82**, 227–32.

38. Heurkens, A.H.M., Hiemstra, P.S., Lafeber, G.J.M. *et al.* (1989) Antiendothelial cell antibodies in patients with rheumatoid arthritis complicated by vasculitis. *Clin. Exp. Immunol.*, **78**, 7–12.
39. Dequeker, J. and Rosberg, G. (1967) Digital capillaritis in rheumatoid arthritis. *Acta Rheum. Scand.*, **13**, 299–307.
40. Gordon, D.A., Stein, J.L. and Broder, I. (1973) The extra-articular features of rheumatoid arthritis. A systematic analysis of 127 cases. *Am. J. Med.*, **54**, 445–52.
41. Kirk, J. and Cosh, J. (1969) The pericarditis of rheumatoid arthritis. *Q. J. Med.*, **152**, 397–423.
42. Edwards, J.C.W. (1980) Relationship between pressure and digital vasculitis in rheumatoid disease. *Ann. Rheum. Dis.*, **39**, 138–40.
43. Wested, M.L., Meijer, C.J.L., Vermeer, B.J. *et al.* (1984) Rheumatoid arthritis – the clinical significance of histo- and immunopathological abnormalities in normal skin. *J. Rheumatol.*, **11**, 448–53.
44. Fitzgerald, O.M., Barnes, L., Woods, R. *et al.* (1985) Direct immunofluorescence of normal skin in rheumatoid arthritis. *Br. J. Rheumatol.*, **24**, 340–5.
45. Pober, J.S. (1988) Cytokine mediated activation of vascular endothelium. Physiology and pathology. *Am. J. Pathol.*, **133**, 426–33.
46. Flipo, R., Janin, A., Hachulla, E. *et al.* (1994) Labial salivary gland biopsy assessment in rheumatoid vasculitis. *Ann. Rheum. Dis.*, in press.
47. Gao, S.Z., Schroeder, J.S., Hunt, S. and Stinson, E.B. (1988) Retransplantation for severe accelerated coronary disease in heart transplant recipients. *Am. J. Cardiol.*, **62**, 867.
48. Forsyth, C.C. (1960) Calcification of the digital vessels in a child with rheumatoid arthritis. *Arch. Dis. Child.*, **35**, 296.
49. Raynauld, J.P., Wolfe, F., Sibley, J.T. and Fries, J.F. (1993) Mortality by cardiovascular disease and use of corticosteroids in rheumatoid arthritis. *Arthritis Rheum.*, **36**, S193.
50. Bacon, P.A. and Kitas, G. (1994) The significance of vascular inflammation in rheumatoid arthritis. Editorial. *Ann. Rheum. Dis.*, In press.
51. Symmons, D.P.M. (1988) Mortality in rheumatoid arthritis. *Br. J. Rheum.*, **27**, 44–54.

# 19

# Vasculitis in primary Sjögren's syndrome

*A.G. Tzioufas, F.N. Skopouli, D. Boumba and H.M. Moutsopoulos*

## INTRODUCTION

Sjögren's syndrome (SS) is a chronic autoimmune disease, characterized by lymphocytic infiltration of the exocrine glands resulting in xerostomia and keratoconjunctivitis sicca. These manifestations can occur alone (primary SS) or in association with other autoimmune disorders, such as rheumatoid arthritis or systemic lupus erythematosus (secondary SS). Extraglandular manifestations are seen in approximately half of SS patients and include Raynaud's phenomenon, splenomegaly, lymphadenopathy, lung involvement, kidney involvement, liver involvement and vasculitis [1].

## CLINICAL PRESENTATION OF VASCULITIS

Inflammatory vascular disease is found in 5–10% of patients with SS. The most common manifestations are purpura, recurrent urticaria, skin ulcerations and mononeuritis multiplex. Raynaud's phenomenon is very frequently seen in SS patients with vasculitis.

Cases of systemic vasculitis with visceral involvement affecting kidneys, lungs, gastrointestinal tract, spleen, breast and reproductive tract have also been described. The most frequently involved site of the body is the skin. A palpable purpuric rash, localized in the lower extremeties and occasionally in the upper extremities and lower trunk, is seen in

*The Vasculitides.* Edited by B.M. Ansell, P.A. Bacon, J.T. Lie and H. Yazici.
Published in 1996 by Chapman & Hall, London. ISBN 978-0-412-64140-4.

many patients. In addition to the rash, some patients may have ulcerative lesions or violaceous discoloration of the fingers and toes. Peripheral nervous symptoms manifested as sensory or sensorimotor neuropathy are also observed. Sural nerve biopsy confirms the diagnosis of vasculitis. Gastrointestinal vasculitis resulting in infarcted bowel is rarely seen. Kidney involvement is manifested usually as glomerulonephritis, while acute necrotizing vasculitis is rarely observed [2].

## HISTOLOGIC CLASSIFICATION

The vasculitic lesions in primary SS involve small- and medium-sized vessels and express a variable histologic picture. All histologic patterns, however, may be categorized in four major histopathologic entities [2].

1. Acute necrotizing vasculitis. This may involve small- and medium-sized arteries. The entire vascular wall is infiltrated with acute and/or chronic inflammatory cells. Fibrinoid necrosis of the wall is characteristically present. These lesions resemble those observed in polyarteritis nodosa without, however, aneurysm formation.
2. Leukocytoclastic vasculitis. In this type, capillaries and post capillary venules are infiltrated with leukocytes. Fibrinoid necrosis of the wall, nuclear debris and extravasated red blood cells may be seen. This type of vasculitis is more frequently found in skin lesions.
3. Lymphocytic vasculitis. Capillaries and venules are also involved, but the inflammatory infiltrate consists of lymphocytes, plasma cells and histiocytes. Vascular cell necrosis is absent. This type of vasculitis is observed in the skin, but also in muscles and peripheral nerves.
4. Endarteritis obliterans. This is a non-inflammatory obstructive vasculitis involving medium-sized vessels. The lesion is characterized by fibrous thickening of the intima, resulting in compromise of the lumen, ranging from stenosis to complete obstruction and recanalization. This form may represent the healing process of pre-existing acute vasculitis, since it is usually seen in patients with a history of long-standing vasculitis.

## LABORATORY FINDINGS

Sera of patients with Sjögren's syndrome and vasculitis are characterized by polyclonal and oligomonoclonal B cell activation products. In fact, these patients present hypergammaglobulinemia, high titers of rheumatoid factor and antinuclear antibodies and antibodies to cellular autoantigens Ro/SSA and La/SSB [2, 3].

Antibodies to Ro/SSA are highly associated with purpuric skin lesions ($p = 0.0005$), vasculitis ($p = 0.0005$), rheumatoid factor ($p = 0.0005$),

**Table 19.1** Laboratory findings in patients with Sjögren's syndrome and vasculitis

|  | *Percentage positive* |
| --- | --- |
| *Rheumatoid factor* | 100% |
| Antinuclear antibodies | 100% |
| Cryoglobulinemia | 100% |
| Antibodies to Ro/SSA | |
|    neutrophilic type | 74% |
|    lymphocytic type | 11% |
| Antibodies to La/SSB | |
|    neutrophic type | 44% |
|    lymphocytic type | 11% |
| Low complement levels | 77% |

Data tabulated from [2] and [4].

cryoglobulins ($p = 0.001$) and hypocomplementemia ($p = 0.005$) [3]. Review of the histopathologic features of the biopsies from Sjögren's syndrome patients with and without precipitating antibodies to Ro/SSA showed that patients with antibodies to Ro/SSA have leukocytoclastic vasculitis characterized by predominance of polymorphonuclear cells within vascular infiltrates. In contrast, patients without antibodies to Ro/SSA had inflammatory vascular disease, which was characterized by the presence of mononuclear cells within the vessel walls [4].

Antibodies to neutrophil cytoplasm (ANCA) are not detected by immunofluorescence [unpublished observations]. A consistent serologic finding is cryoglobulinemia. Cryoglobulins are usually mixed monoclonal (type II) containing an IgMk monoclonal rheumatoid factor. The laboratory findings are summarized in Table 19.1.

## TYPE II CRYOGLOBULINEMIA AND MONOCLONAL RHEUMATOID FACTOR CROSS-REACTIVE IDIOTYPES IN SS VASCULITIS

Several cases of SS patients with extraglandular manifestation were shown to have circulating monoclonal immunoglobulins: a systemic study of serum and urine from unselected patients with primary SS demonstrated that 80% of patients with extraglandular disease had monoclonal light chains or immunoglobulins in their serum, while all patients excreted monoclonal light chains in the urine. In contrast, only 25% of the patients with disease limited to the exocrine glands had monoclonal light chains or immunoglobulins in their serum while 43% of these patients excreted light chains in the urine [5, 6]. One-third of SS patients have cryoglobulins in their serum. These are mixed cryoglobulins containing an IgMk monoclonal rheumatoid factor (mRF). The presence of cryoglobulins are associated with a higher prevalence of extraglandular disease and autoantibodies to Ro/SSA autoantigen as compared with patients without cryoglobulins [7]. Given that SS patients with vasculitis very often present cryoglobulinemia, we evaluated the prevalence, incidence and origin of cryoglobulinemia in 13 consecutive patients with primary SS in a five-year follow-up study and compared the results with the development of glomerulonephritis or peripheral neuropathy in the patients. Five patients with primary SS developed peripheral neuropathy and/or glomerulonephritis (peripheral neuropathy: prevalence 3.8, incidence 0.76; glomerulonephritis: prevalence 3.07, incidence 0.61). All five patients had mixed monoclonal cryoglobulinemia, preceding the development of vasculitis by three months to two years. Patients with peripheral neuropathy or glomerulonephritis had lower levels of the C4 complement component as compared with the rest of the patients (x = 8.4 mg/dl vs 36 mg/dl, $p < 0.001$). In contrast, mixed monoclonal cryoglobulinemia was observed in only 19 out of 125 patients without peripheral neuropathy or glomerulonephritis ($x^2 = 22.9$, $p < 0.001$) [8].

In order to delineate the origin and mechanisms of mRF production, several studies concentrated on their idiotypes. Monoclonal RFs have been shown to share extensively cross-reactive idiotypes (CRI) [9, 10]. Some CRI are common in SS and rheumatoid arthritis, while others, such as 17109 and G6, are found in RFs from SS patients [11, 12]. The 17109 idiotype is κ light chain specific, encoded by the hum Kv 325 germ line gene which belongs to the VKIIIb subgroup [13]. The G6 idiotype is encoded by the VH1 region of the heavy chain germ line gene [12]. Finally, a polyclonal anti-idiotypic antibody raised in a rabbit against an IgMk mRF from the cryoglobulin of one patient with SS (3rdSS idiotype) reacted with the sera of two-thirds of SS patients and one-quarter of RA patients. The CRI levels were significantly higher in SS patients with mixed monoclonal cryoglobulinemia and in patients with extraglandular manifestations [14].

In order to address the question of whether mRFs with particular CRIs are involved in the pathogenesis of vasculitis in SS, comparative studies of three CRIs were performed (Table 19.2). The third SS CRI was found in sera of all patients with peripheral neuropathy and/or glomerulonephritis and in 31 out of 125 SS patients without peripheral neuropathy or glomerulonephritis ($x^2 = 12.9$, $p < 0.005$) [unpublished observations].

These findings suggest that mixed monoclonal cryoglobulinemia plays an important role in the pathogenesis of peripheral neuropathy and glomerulonephritis in SS, possibly through the immune complex deposition. The mRFs of these cryoglobulins bear a particular idiotype (3rd SS).

**Table 19.2** Monoclonal rheumatoid factor cross-reactive idiotypes (CRI) in patients with Sjögren's syndrome (SS) with and without peripheral neuropathy (PN) and glomerulonephritis (GN)

| | SS patients | |
| CRIs | PN/GN(+ve) | PN/GN(−ve) |
| --- | --- | --- |
| 3rd SS | 5/5* | 31/125 |
| 17109 | 2/5 | 12/125 |
| G6 | 3/5 | 32/125 |

*$x^2 = 12.9$, $p < 0.005$.

Finally, SS patients should be evaluated for the presence of cryoglobulinemia, since its presence may predict vasculitis development in SS.

## THERAPY

The hypersensitivity forms of vasculitis, involving the skin, respond well to corticosteroids. Systemic vasculitis involving other organs should be treated with pulses of cyclophosphamide ($1 \text{ g/m}^2$/month) and prednisone (0.5–1 mg/kg/day) [15].

## REFERENCES

1. Moutsopoulos, H.M., Chused, T.M., Mann, D.L. *et al.* (1980) Sjögren's syndrome (sicca syndrome) current issues. *Ann. Int. Med.*, **92**, 212–26.
2. Tsokos, M., Lazarou, S.A. and Moutsopoulos, H.M. (1987) Vasculitis in primary Sjögren's syndrome. Histologic classification and clinical presentation. *Am. J. Clin. Pathol.*, **88**, 26–31.
3. Alexander, E.L., Arnett, F.C., Provost, T.T. and Stevens, M.B. (1983) Sjögren's syndrome: association of anti-Ro/SSA antibodies with vasculitis, hematologic abnormalities and serologic hyperreactivity. *Ann. Int. Med.*, **98**, 155–9.
4. Molina, R., Provost, T.T. and Alexander, E.L. (1985) Two histopathologic prototypes of inflammatory vascular disease in Sjögren's syndrome: differential association with seroreactivity to rheumatoid factor and antibodies to Ro/SSA and with hypocomplementemia. *Arthritis Rheum.*, **28**, 1251–8.
5. Moutsopoulos, H.M., Steinberg, A.D., Fauci, A.S. *et al.* (1983) High incidence of free monoclonal lambda light chains in the sera of patients with Sjögren's syndrome. *J. Immunol.*, **130**, 2263–5.
6. Moutsopoulos, H.M., Costello, R., Drosos, A.A. *et al.* (1985) Demonstration and identification of monoclonal proteins in the urine of patients with Sjögren's syndrome. *Ann. Rheum. Dis.*, **44**, 109–12.
7. Tzioufas, A.G., Manoussakis, M.N., Costello, R. *et al.* (1988) Cryoglobulinemia in autoimmune rheumatic diseases. Evidence of circulating monoclonal cryoglobulins in patients with primary Sjögren's syndrome. *Arthritis Rheum.*, **29**, 1098–104.
8. Tzioufas, A.G., Skopouli, F.N. and Moutsopoulos, H.M. (1994) Peripheral nerve involvement and glomerulonephritis in Sjögren's syndrome are asso-

ciated with mixed cryoglobulinemia. VII Mediterranean Congress of Rheumatology, June 16–18, Athens, Greece, p. 49.

9. Kunkel, H.G., Angello, V., Joslin, F.G. *et al.* (1973) Cross-idiotypic specificity among monoclonal IgM proteins with anti-gammaglobulin activity. *J. Exp. Med.*, **137**, 331–42.

10. Agnello, V., Arbetter, A., Ibanez, A. *et al.* (1980) Evidence for a subset of rheumatoid factors that cross react with DNA-histone and have a distinct cross-idiotype. *J. Exp. Med.*, **151**, 1514–27.

11. Fox, R.I., Chen, P., Carson, D.A. and Fong, S. (1986) Expression of a cross-reactive idiotype on rheumatoid factor in patients with Sjögren's syndrome. *J. Immunol.*, **136**, 477–83.

12. Mageed, R.A., Dearlove, M., Goodall, D.M. and Jefferis, R. (1986) Immunogenetic and antigenic epitopes of immunoglobulins, XVII-monoclonal antibodies reactive with common and restricted idiotypes to the heavy chain of rheumatoid factors. *Rheumatol. Int.*, **6**, 179–83.

13. Kipps, T.J., Tomhave, E., Chen, P.P. and Fox, R.I. (1989) Molecular characterization a major autoantibody associated cross-reactive idiotype in Sjögren's syndrome and rheumatoid factor associated cross reactive idiotype. *J. Immunol.*, **142**, 4261–8.

14. Katsikis, P.D., Youinou, R.Y., Galanopoulou, V. *et al.* (1990) Monoclonal process in primary Sjögren's syndrome and rheumatoid factor associated cross reactive idiotype. *Clin. Exp. Immunol.*, **82**, 509–14.

15. Lane, H.C., Moutsopoulos, H.M. and Fauci, A.S. (1982) Treatment of systemic vasculitis associated with Sjögren's syndrome (abstract). *Arthritis Rheum.*, **25** (Suppl.), 216.

# 20

# Mixed cryoglobulinemia: the role of HCV and organ-specific antibodies

*S. Bombardieri, A. Tavoni, M. Mosca, L. La Civita,*
*M.P. Dolcher, F. Lombardini, P. Migliorini and C. Ferri*

## INTRODUCTION

The term 'cryoglobulin' was introduced to define a group of proteins which have the common property of forming a reversible precipitate or gel in the cold [1]. Three groups of cryoglobulins have been identified.

Type I cryoglobulins are monoclonal, the most common one being immunoglobulin M (IgM), a protein associated with immunoproliferative disorders, such as Waldenstrom's macroglobulinemia, lymphoproliferative disorders and multiple myeloma.

Type II and type III are mixed cryoglobulins. Type II involves a group of polyclonal immunoglobulins of different classes and a monoclonal component, usually an IgM, that typically has rheumatoid factor (RF) activity. Type III cryoglobulins are exclusively polyclonal with an IgM component, again with RF activity [2]. Mixed cryoglobulins may either appear during the course of an infectious or autoimmune disease or be found as a distinct entity, the so-called essential mixed cryoglobulinemia (EMC) [3].

EMC, which was first described in 1966 by Meltzer and Franklin [3], is a systemic vasculitis affecting the small arteries and veins associated with the presence of large amounts of mixed IgM and IgG cryoglobulins with RF activity. The clinical hallmark of the disease is purpura, present

*The Vasculitides*. Edited by B.M. Ansell, P.A. Bacon, J.T. Lie and H. Yazici.
Published in 1996 by Chapman & Hall, London. ISBN 978-0-412-64140-4.

in over 90% of patients, which is due to a leukocytoclastic vasculitis and which is localized mainly in the lower extremities.

Liver involvement is another frequent feature of EMC and can seriously affect the overall prognosis of the disease. Cryoglobulinemic hepatitis, generally found in 70% of patients, shows a clinical course and histological findings comparable to other forms of chronic active hepatitis. In the majority of cases, the clinical signs of cryoglobulinemic hepatitis appear during the course of the disease, while in 27% it is one of the presenting symptoms, together with purpura, weakness, arthralgias and circulating mixed cryoglobulins. However, in about 20% of cases liver involvement may precede the typical clinicoserological features of EMC. One-third of patients may develop immune complex (IC) glomerulonephritis; other frequent clinical manifestations related to the IC vasculitis are purpura and polyneuritis (Table 20.1).

**Table 20.1** Demographic, clinical and serologic findings in 52 patients with mixed cryoglobulinemia*

| | |
|---|---|
| Age, mean – SD years – (range) | 59 – 9.4 (39–82) |
| No. of males/no. of females | 13/39 |
| Disease duration, mean – SD years – (range) | 10.3 – 5.5 (1–25) |
| Purpura | 95 |
| Weakness | 100 |
| Arthralgias | 90 |
| Raynaud's phenomenon | 31 |
| Sjögren's syndrome | 40 |
| Peripheral neuropathy | 69 |
| Renal involvement | 17 |
| Liver involvement | 62 |
| Cryocrit, mean, SD % | 5.8 – 11.9 |
| *Cryoglobulin composition* | |
| IgG/IgM | 47 |
| IgG/IgM/IgA | 53 |
| IgMk | 44 |
| CH50, mean – SD units (normal 160 – 220) | 81 – 80 |
| C3, mean – SD mg/dl (normal 60 – 130) | 76 – 40 |
| C4, mean – SD mg/dl (normal 20 – 55) | 13 – 14 |
| Antinuclear antibodies | 5 |
| Antimitochondrial antibodies | 3 |
| Antismooth muscle antibodies | 13 |
| Antiextractable nuclear antigen antibodies | 3 |

*Unless otherwise indicated, values are expressed as a percentage of the number of patients.

## EMC AS AN IMMUNE COMPLEX DISEASE AND AN IMMUNE PROLIFERATIVE DISORDER

Mixed cryoglobulinemia (MC) is commonly considered one of the most typical models of the human IC diseases [4]. It is, in fact, characterized by the presence of large amounts of circulating IC, low complement levels due to their consumption and to frequent reversible impairment of the mononuclear phagocytic system. Histologically, MC is characterized by inflammatory changes of the involved vessels, and immune deposits in target organs are demonstrable on immunochemical analysis.

On the other hand, it is also considered a lymphoproliferative disorder and an IgMk monoclonal RF is found in over 80% of cases. Moreover, patients with EMC are at higher risk of developing malignant B cell lymphoma [5]. As in IC diseases, the localization and severity of the lesions are linked to a number of predisposing factors, such as the nature of the antigen and of the immune response, the features of the IC (size, charge, antigen/antibody ratio, biological characteristics), the integrity of the clearance mechanisms, rheological factors and intrinsic characteristics of the target organ cells.

## ETIOLOGIC FACTORS

A number of observations have underlined the probable importance of environmental factors in the pathogenesis of EMC. While generally considered an exceedingly rare disorder, this condition is much more diffuse in certain geographical areas, such as Italy and southern Europe, and this peculiar distribution cannot be ascribed to genetic factors [6].

On the other hand, it appears that hepatotropic viruses may play a role as etiologic agents in this disease. First of all, these viruses tend to persist in the infected organism, triggering a chronic immune response [7]; and, secondly, in contrast with other well-known human models of systemic IC disease, chronic liver involvement is one of the major clinical features of EMC. It is, therefore, not surprising that over the past 20 years the putative role of the hepatotropic viruses has been thoroughly investigated in EMC.

### Viruses

Up until 1990 very little information was available on the possible etiologic factors responsible for EMC, and what there was, was often contradictory. Since liver involvement is frequently found during the course of the disease, various authors investigated the possible role of hepatotropic viruses in the etiopathogenesis of EMC [8].

Anecdotal cases reporting an association between hepatitis A virus and EMC appeared in the literature [9], while the association between Epstein–Barr virus and mixed cryoglobulinemia was suggested but not confirmed [10].

In 1974 the first report of an association between hepatitis B virus (HBV) and EMC was published [11]. Realdi *et al.* compared hepatitis B virus markers in the serum cryoprecipitates and supernatants of patients with MC secondary to chronic liver disease and in patients with EMC. In the group of patients with chronic liver disease, HBsAg was found in the serum cryoprecipitate and supernatant, but HBsAb was only found in the serum and cryoprecipitate. In 50% of the EMC patients, HBsAg was found in the serum and cryoprecipitate but not in the supernatant. Those autors concluded that HBV antigen–antibody complexes were formed in antigen excess in patients with secondary mixed cryoglobulinemia and in antibody excess in patients with EMC.

In contrast to the high prevalence shown in this and other studies [5], subsequent investigations using radioimunoassays and ELISA have demonstrated a low prevalence of HBsAg. The prevalence of markers for HBV in subsequent studies has varied from 0–74%. The variation in results may be due to differences in the patient populations. Although a high prevalence of serologic markers for HBV has been reported in patients with EMC, evidence for active HBV infection in these patients has rarely been found. It has not been established that hepatitis B IC play a role in EMC. None of the studies have demonstrated increased concentrations of HBV antigen and antibodies after dissociation of the high molecular complexes in the cryoglobulins. Therefore, the role of HBV in EMC certainly cannot be ruled out, but it may be confined to a minority of patients [12].

### Hepatitis C virus (HCV)

HCV has been demonstrated as the major cause of post-transfusion and sporadic non-A, non-B chronic hepatitis. The recent cloning and sequencing of the HCV genome has rekindled interest in the study of hepatotropic viruses in EMC. By recombinant techniques, a nucleocapsid peptide and a number of non-structural peptides have been produced. First generation assays (ELISA) and a recombinant immunoblotting assay (RIBA) can detect antibody to the c100–3 antigen located in the NS-3 region. Second generation ELISA and RIBA combine c100–3 with another non-structural protein, c33c, to form a new composite antigen, c200. These assays can also detect antibodies against c22–3, a nucleocapside antigen. The second generation assays have a greatly enhanced sensitivity and specificity [13].

In 1990, after the initial anecdotal observations, two different studies using a first generation ELISA found antibodies to HCV in 30% and in

**Table 20.2** Prevalence of HCV in mixed cryoglobulinemia and other rheumatic disorders

|  | *ELISA II +* | *Anti-HCV RIBA II+* | *HCV RNA (PCR+)* |
|---|---|---|---|
| MC patients (110) | 91 | 91 | 86 |
| Control diseases (110) | 9.1 | 6.4 | N.D. |
| Rheumatoid arthritis (30) | 6.7 | 6.7 | |
| Sjögren's syndrome (30) | 6.7 | 0 | |
| Systemic lupus erythematosus | 10 | 10 | |
| Systemic sclerosis (20) | 10 | 10 | |

55% of their patients, respectively [14, 15]. The significance of this association was confirmed by a number of subsequent reports, by the use of second generation ELISA and RIBA, and by the demonstration of HCV RNA using PCR techniques [16–23]. With these techniques the prevalence of serum antibodies found in EMC increased to 70–91%. The most recent results of our group are reported in Table 20.2 [24–25].

In sera without cryoglobulins and in cryoprecipitates, HCV antibodies were detected in the supernatants with the same frequency as that observed in the whole serum (91%), while they were less frequently found (30%) in isolated cryoglobulins. These data contrast with the observations of Agnello *et al.* who reported a significantly higher level of HCV antibodies in cryoprecipitates than in the supernatants [26].

While there is no doubt about the striking association between EMC and HCV, the exact role of HCV in relation to the signs and symptoms of EMC is not yet clear. Hypothetically, the virus could participate in the production of the CIC responsible for the lesions. In this respect it is intriguing that by quantitative PCR technology, HCV RNA was found to be concentrated up two thousand fold in the cryoglobulins, but a direct demonstration of HCV antigen in the vascular renal lesions characteristic of EMC is still lacking [22, 26].

As mentioned previously, the second hallmark of EMC is the presence of a chronic lymphoproliferative disorder [27]. There is increasing evidence suggesting that HCV infection could be at least partially responsible for this phenomenon.

To better define the role of HCV in the pathogenesis of the EMC, HCV infection of the peripheral lymphocytes was investigated in a series of patients. HCV RNA was present in the sera of eight out of 16 patients (50%), while its frequency markedly increased (13 out of 16, 81%) when genomic sequences were detected in the peripheral lymphocytes [28]. Thus, HCV infection may trigger and maintain EMC through different mechanisms. Among these, HCV infection of peripheral blood mononuclear cells could represent the primary step in the pathogenesis of the clinical manifestations; in particular, HCV lymphotropism could in part explain the relationship observed between EMC and other HCV-related diseases.

First of all, the infected lymphocytes may select HCV mutants with different clinical consequences and they could be the viral 'reservoir' responsible for the disease chronicity. Moreover, HCV may trigger the clonal B-cell expansion responsible for the production of various auto-antibodies, among these antinuclear, antismooth muscle, and antiliver–kidney microsomal antibodies are considered the serological hallmarks of different disease. When the prevalence of HCV infection is compared in two MC patient subsets, MC with and without a complicating B cell lymphoma, HCV related markers were detected in the majority of cases regardless of the presence or absence of complicating B cell lymphoma. These findings confirm the frequent association between MC and HCV and, more interestingly, suggest that in some MC patients chronic HCV infection could also be responsible for the evolution to malignant B cell lymphoma [26].

## PATHOGENETIC FACTORS

### Circulating immune complexes (CIC) and disease

The CIC and complement play a pivotal role in the pathogenesis of the clinical manifestations of EMC. The sites and the severity of the lesions found in immune complex diseases are closely related to a wide variety of factors favoring the deposition of CIC in target organs. The importance of these factors became immediately clear when it was established that in EMC there is no correlation between the amount of IC or cryoglobulins and disease manifestations or peculiar organ involvement [5]. This prompted us to study the role of a number of factors which are known to condition IC deposition and toxicity, among them the Ag/Ab ratio, reticuloendothelial system function, complement and rheological abnormalities.

### *Ag/Ab ratio*

Experimental evidence has demonstrated the importance of the CIC Ag/Ab ratio. IC with a large antigen excess are small in size and even if they remain in the circulation for long periods of time they do not localize in the tissues. In contrast, complexes with a slight antigen excess have a high inflammatory potential; they are large enough to activate complement fixation, but not so large as to be rapidly cleared. Complexes in Ab excess, especially those carrying multivalent antigens, are usually larger than the others and are rapidly removed without accumulating in the tissues.

Recently, our laboratory developed a method to calculate the Ag/Ab ratio in biological fluids [29]. EMC sera were tested by this method and generally appear to be in antigen excess. It was interesting to note that in several instances, aggressive therapy such as plasma exchange was capable of modifying these parameters [30].

### Reticuloendothelial system (RES)

Small quantities of IC are normally formed in the absence of disease. The antigens responsible for these normal antigen–antibody complexes are not all known, although at least some are antigens from ingested foods. Other environmental antigens and possibly autoantigens could be responsible for these normal CIC. Nevertheless, they are promptly eliminated through phagocytosis by cells of the mononuclear phagocyte system which contain receptors for the Fc portion of IgG and for C3. The functional state of the RES is one factor determining the levels of CIC.

Previous studies of RES function in IC diseases, particularly EMC, were done by measuring the half-life of heat-damaged red blood cells labeled with $^{99m}$Tc or $^{51}$Cr (T1/2 RES). The heat-damaged red blood cells are removed from the circulation, predominantly by the spleen [31]. In EMC patients, RES function studies showed a prolonged half-life of heat damaged red blood cells. This was corrected using both plasma-exchange and a low-antigen content diet [32].

### Complement

Complement is an enzymatic complex that is implicated in IC disease in two ways. Firstly, as an effector complex responsible for the signs and symptoms of inflammation. Secondly, complement plays a crucial role in modifying the composition and pathogenicity of IC, interfering with their solubilization and precipitation. In EMC low complement levels are usually found, particularly of the classical activation pathway componenents C1, C2 and C4. Similarly, alterations in complement function, inhibition of immunoprecipitation (ICPIC) and immune-complex solubilization capacity (ICSC) are present. However, a correlation between complement levels and function could not be established in follow-up studies of EMC patients [33–35].

*Rheological abnormalities*

Rheological factors may be of crucial importance in the localization of lesions in IC disorders. Given the large number of macromolecular complexes present in IC disorders, EMC patients could be considered at risk of developing rheological abnormalities. This prompted us to study blood viscosity and filtration at various temperatures in this condition. While a typical hyperviscosity syndrome was rare, a clear increase in blood viscosity was occasionally observed in mixed cryoglobulinemia, although significant differences were present in the plasma and serum viscosity of patients and controls. In contrast, blood filtration was severely impaired in a high percentage of cases and on average was significantly higher in patients than in controls. Indirect evidence suggests that blood viscosity is at least in part related to cryoglobulins. Thus, these data confirm that rheological factors may favor the IC localization in particular target organs [36].

## Role of organ-specific autoantibodies

A number of indications suggest that, in addition to CIC, other immune factors may play a role in certain types of organ damage. It is well known that EMC is characterized by spontaneous remissions and exacerbations. Moreover, in contrast to other features of the disease, certain manifestations, such as renal involvement and motor polyneuropathy [37], are present in only a minority of patients, analogous to what has been observed in other conditions.

Among the possible factors responsible for this behavior, we focused on the presence of organ-specific antibodies. Monoclonal antibodies against ANA in NZB-NZW and MRL-lpr/lpr mice bind glomerular antigens. In IgA nephropathy and Henoch–Schönlein nephritis, IgG antibodies recognize the 55 and 48 kD bands expressed in mesangial cells. Finally, antibodies to basement membrane and laminin have been detected in post-streptococcal glomerulonephritis [38]. The aim of our study was to explore the possibility that circulating antibodies can contribute to renal damage in EMC, binding renal antigens and thereby forming IC *in situ*. We investigated 33 patients, 11 with renal involvement and 22 without, for the presence of autoantibodies, using as our antigen source a glomerular extract which was run on an acrylamide gel, blotted on to a nitrocellulose sheet, and finally probed with the patient's sera. Out of the sera of 11 patients with active renal involvement, four did not react with the glomerular antigens, and seven reacted with a 50 kD antigen. In the group of 22 patients who did not show any renal involvement, 11 sera did not react with glomerular antigen, five reacted with one or two antigens and six reacted with three or more antigens.

These responses fluctuated during the course of the disease in four patients who were followed, with multiple sampling, over a longer period of time.

In conclusion, EMC sera without renal involvement either does not react with glomerular antigens or detects several bands over a wide range of molecular weights. EMC patients with renal involvement exhibit a very restricted response to a 50 kD band in the kidney and thymus, but not in the heart or liver. These antibodies fluctuate during follow-up. Further studies are in progress to characterize this antigen and to evaluate its pathogenetic role in EMC [39].

## REFERENCES

1. Lospalluto, J., Dorward, B., Miller, W. Jr. and Ziff, M. (1962) Cryoglobulinemia based on interaction between a gamma macroglobulin and 7S gamma-globulin. *Am. J. Med.*, **32**, 142.
2. Brouet, J.C., Clouvel, J.P., Danon, F. *et al.* (1974) Biologic and clinical significance of cryoglobulins. *Am. J. Med.*, **57**, 775.
3. Meltzer, M. and Franklin, E.C. (1966) Cryoglobulinemia: a study of twenty-nine patients. I. IgG and IgM cryoglobulins and factors affecting cryoprecipitability. *Am. J. Med.*, **40**, 828.
4. Gorevic, P.D., Kassab, H.J., Levo, Y. *et al.* (1980) Mixed cryoglobulinemia: clinical aspects and long term follow-up of 40 patients. *Am. J. Med.*, **69**, 287.
5. Gorevic, P.D. and Frangione, B. (1991) Mixed cryoglobulinemia cross reactive idiotypes: implication for relationships of MC to rheumatic and lymphopro-lipherative diseases. *Semin. Hematol.*, **28**, 79.
6. Migliorini, P., Bombardieri, S., Castellani, A. and Ferrara, G.B. (1981) HLA antigens in essential mixed cryoglobulinemia. *Arthritis Rheum.*, **24**, 932.
7. Zignego, A.L., Macchia, D., Monti, M. *et al.* (1992) Infection of peripheral mononuclear blood cells by hepatitis C virus. *J. Hepatol.*, **15**, 382.
8. Bombardieri, S., Ferri, C., Di Munno, O. and Pasero, G. (1979) Liver involvement in essential mixed cryoglobulinemia. *Ric. Clin. Lab.*, **9**, 361.
9. Inman, R.D., Hodge, M., Johnston, M.E. *et al.* (1986) Arthritis, vasculitis and cryoglobulinemia associated with relapsing hepatitis A virus infection. *Ann. Intern. Med.*, **105**, 700.
10. Fiorini, G.F., Sinico, R.A., Winearls, C. *et al.* (1988) EBV infection in patients with type 2 essential MC. *Clin. Immunol. Immunopathol.*, **47**, 262.
11. Realdi, G., Alberti, A., Rigoli, A. and Tremolada, F. (1974) Immune complexes and Australia antigen in cryoglobulinemic sera. *Z. Immunitatsfortsch Exp. Klin. Immunol.*, **147**, 114.
12. Galli, M. and the GISC members (1991) Cryoglobulinemia and serological markers of hepatitis viruses. *Lancet*, **758**.
13. Hosein, B., Fang, C.T., Popowskj, M.A. *et al.* (1991) Improved serodiagnosis of HCV infection with synthetic peptide antigen from capside protein. *Proc. Nat. Acad. Sci. USA*, **88**, 3647.
14. Pascual, M., Perrin, L., Giostra, E. and Schifer, J.A. (1990) Hepatitis C virus in patients with cryoglobulinemia type 2. *J. Infect. Dis.*, **162**, 569.
15. Ferri, C., Greco, F., Longombardo, G. *et al.* (1991) Antibodies to hepatitis C virus in patients with mixed cryoglobulinemia. *Arthritis Rheum.*, **34**, 1606–10.

16. Ferri, C., Greco, F., Longombardo, G. *et al.* (1991) Antibodies against hepatitis C virus in mixed cryoglobulinemia patients. *Infection*, **19**, 417–20.
17. Ferri, C., Greco, F., Longombardo, G., *et al.* (1991) Association between hepatitis C virus and mixed cryoglobulinemia. *Clin. Exp. Rheumatol.*, **9**, 621–4.
18. Bambara, L.M., Biasi, D., Caramaschi, P. *et al.* (1991) Cryoglobulinemia and hepatitis C virus infection. *Clin. Exp. Rheumatol.*, **9**, 95.
19. Harle, J.R., Disdier, P. and Durand, J.M. (1991) Cryoglobuliéme mixed au cours de l'infection par le virus de l'hepatite C. Dix cas. *Presse Méd.*, **20**, 1233.
20. Disdier, P., Harle, J.R. and Weiller, P.J. (1991) Cryoglobulinemia and hepatitic infection. *Lancet*, **338**, 1151.
21. Agnello, V., Chung, R.T., Dienstag, J.L. *et al.* (1991) Evidence for the association of hepatitic C virus with essential mixed cryoglobilinemia. *Arthritis Rheum.*, **34** (Suppl. 9), S45, Abstract 78.
22. Agnello, V,. Chung, R.T. and Kaplan, L.M. (1992) A role for hepatitis C virus infection in type 2 cryoglobulinemia. *N. Engl. J. Med.*, **327**, 1490.
23. Ferri, C., Longombardo, G., La Civita, L., Bombardieri, S. *et al.* (1992) Hepatitis C virus, autoimmune liver diseases and cryoglobulinemic hepatitis. *J. Hepatol.*, **16**, 242–3.
24. Ferri, C., Longombardo, G., La Civita, L. *et al.* (1992) Hepatitis C virus infection in mixed cryoglobulinemia. *Gastroenterology*, **103**, 1108–10.
25. Ferri, C., La Civita, L., Longombardo, G. *et al.* (1993) Hepatitis C virus and mixed cryoglobulinemia. *Eur. J. Clin. Invest.*, **23**, 399–405.
26. Ferri, C., Monti, M., La Civita, L. *et al.* (1993) Infection of peripheral mononuclear cells by hepatitis C virus in mixed cryoglobulinemia. *Blood*, **82**, 3701–4.
27. Abel, G., Zhang, Q. and Agnello, V. (1993) Hepatitis C virus infection in type II mixed cryoglobulinemia. *Arthritis Rheum.*, **36**, 1341–9.
28. Monteverde, A., Rivano, M.T., Allegra, G.C. *et al.* (1988) Essential mixed cryoglobulinemia, type II: a manifestation of a low grade malignant lymphoma? *Acta Haematol. (Basel)*, **79**, 20.
29. Bombardieri, S., Puccetti, A. and Pilo, A. (1989) A method for retracing the putative antigen/antibody ratio of immune complexes in biological fluids. *Clin. Exp. Rheumatol.*, **7/S–3** (Suppl.), 109.
30. Bombardieri, S., Caponi, L., Pilo, A. *et al.* (1989) Changes in the putative antigen/antibody ratio of circulating immune complexes in mixed cryoglobulinemia and systemic lupus erythematosus following selected removal of macromolecules. *Int. J. Art. Org.*, **12/S4**, 87.
31. Vitali, C., Tavoni, S. and Bombardieri, S. (1984) La fonction réticulo-endothéliale splénique dans les maladies à complexes immuns. *Médecine Hyg.*, **42**, 109.
32. Ferri, C., Pietrogrande, M., Cecchetti, R. *et al.* (1989) Low-antigen-content diet in the treatment of mixed cryoglobulinemia patients. *Am. J. Med.*, **87**, 519.
33. Schifferli, J.A., Morris, S.M., Dash, A. and Peters, D.K. (1981) Complement mediated solubilization in patients with systemic lupus erythematosus, nephritis or vasculitis. *Clin. Exp. Immunol.*, **46**, 557.
34. Schifferli, J.A., Wood, P. and Peters, D.K. (1982) Complement mediated inhibition of immune precipitation I. Role of the classical and alternative pathways. *Clin. Exp. Immunol.*, **47**, 555.
35. Balestrieri, G., Tincani, A., Migliorini, P. *et al.* (1984) Inhibitory effect of IgM rheumatoid factor on immune complex solubilization capacity and inhibition of immune precipitation. *Arthritis Rheum.*, **27**, 1130.

36. Ferri, C., Mannini, L., Bartoli, V. *et al.* (1990) Blood viscosity and filtration abnormalities in mixed cryoglobulinemia patients. *Clin. Exp. Rheumatol.*, **8**, 271.

37. Ferri, C., La Civita, L., Cirafisi, C. *et al.* (1992) Peripheral neuropathy in mixed cryoglobulinemia: clinical and electro-physical investigations. *J. Rheumatol.*, **19**, 889–95.

38. Kefalides, N.A., Pegg, M.T., Ohno, N. *et al.* (1986) Antibodies to basement membrane collagen and to laminin are present in sera from patients with post-streptococcal glomerulonephritis. *J. Exp. Med.*, **163**, 588.

39. Dolcher, M.P., Marchini, B., Sabbatini, A. *et al.* (in press) Autoantibodies from mixed cryoglobulinemia patients bind glomerular antigens. *Clin. Exp. Immunol.*, (In press).

# 21

# Cutaneous vasculitis

*J.J. Cream*

The vasculitides, cutaneous and other, encompass a diverse group of disorders mostly poorly understood, and in the absence of a satisfactory classification based on etiology and pathogenesis, most conveniently dealt with according to the size of vessel involved and the histology.

## SMALL VESSEL NECROTIZING VASCULITIS

Often labeled allergic or hypersensitivity vasculitis, histologically this group of disorders is characterized by endothelial cell swelling, neutrophil infiltration, leukocytoclasis, erythrocyte extravasation and fibrinoid necrosis affecting predominantly, in many instances, the dermal postcapillary venules [1]. Although indicative of an immune complex pathogenesis, this type of histology is not unique to, and diagnostic of, an immune complex vasculitis. It may also be found adjacent to cutaneous ulcers and occurs in sepsis, emboli and after repeated cold injury. The vessel damage leads to an increase in permeability and then hemorrhage (Plate 3). If the lumen is compromised ischemia follows.

The cutaneous manifestations include edema, urticaria, purpura and ulcers; pustules and bullae may develop reflecting the severity of the underlying vasculitis. Serum sickness-like features of arthralgia, arthritis, myalgias and fever may also be present.

A cause emerges in only a minority of patients and a less than comprehensive list of causes includes:

- Drugs:
  aspirin

*The Vasculitides.* Edited by B.M. Ansell, P.A. Bacon, J.T. Lie and H. Yazici. Published in 1996 by Chapman & Hall, London. ISBN 978-0-412-64140-4.

  penicillins
  thiazides
  sulfonamides.
- Infections:
  hepatitis B
  parvovirus B19
  hantavirus
  streptococcus
  TB.
- Immune complex:
  systemic lupus erythematosus (SLE)
  cryoglobulinemia
  inflammatory bowel disease.

Some types of small vessel vasculitis appear to be distinctive syndromes and a proportion can be more satisfactorily defined on the basis of the immunopathological findings.

### Henoch–Schönlein (or Schönlein–Henoch) purpura

One such distinctive syndrome is Henoch–Schönlein purpura which is notable because of activation of the IgA system with raised serum IgA levels, IgA deposits in the skin and IgA complexes in the blood as well as the risk of the development of an IgA mesangiopathy.

A petechial and purpuric rash may be associated with aching joints, as described by Schönlein, or with abdominal and renal complications as described by Henoch. The typical rash consists of crops of purpura, usually over the lower extremities and buttocks, and sometimes over the trunk, arms and face. The lesions can be urticarial initially. Vessel injury severe enough to give rise to necrosis may be followed by bullae and ulceration. In some patients, erythema may be more striking than the purpura. In others, areas of extensive confluent purpura and ecchymoses can be found. The purpura may appear at the sites of pressure beneath elasticated bands in socks and underclothing. Fever has been reported in between 45% and 75% of patients. Localized edema of face, hands, arms, feet or scrotum occurs in 45%. Transient arthralgia develops in about two-thirds of the patients, the knees and ankles being most frequently involved. More than half the patients develop abdominal symptoms or signs, such as colic, vomiting, occult or frank bleeding and intussusception due to vasculitis of the gastrointestinal vessels. In some 13% of patients, abdominal pain precedes the other features and it may be several days or longer before the tell-tale purpura appears. During this time there is a risk that the patient will come to appendicectomy or laparotomy. Intussusception occurs in about 3% and may not be revealed by barium enema. Perforation is rare. Torsion of the testis can occur and testicular swelling can mimic torsion. Protein-losing enteropathy has also been described. Extremely rare is pulmonary involvement which has been described in three cases, all fatal. Although Schönlein noted that cardiac involvement could occur, there are few reports in more recent times and in most instances the cardiac changes reflect secondary renal hypertension. Other rarities include acute pancreatitis, ocular involvement, intramuscular hemorrhage, and central nervous system involvement with fits, coma and paresis.

Exceptionally, attacks may recur over several weeks or months but, as a rule, the purpura and other manifestations settle over two to three weeks. Overall, the prognosis of Henoch–Schönlein purpura without clinical nephritis is excellent. At least half the patients, however, have renal involvement. In most this is trivial, being no more than transient microscopic hematuria. In a few, renal disease can be a serious complication of Henoch–Schönlein purpura and can lead to the development of proteinuria, nephrotic and acute nephritic syndromes, hypertension and acute renal failure. Hypertension occasionally develops in the absence of urinary abnormalities. The infrequency of serious renal problems in Henoch–Schönlein purpura is indicated in one series of 141 children of whom 39 (28%) had renal involvement amounting to clinical nephritis, and only three (2%) progressed to serious renal disease. There is always some degree of patient selection in any series and so the figures obtained from renal units might be expected to indicate the worst possible outcomes for those who develop chronic renal disease. In one series of 88 children with Henoch–Schönlein purpura nephritis followed in a renal unit over periods ranging from 6–21 years, there were, on review, 61 patients (69%) who were normal. The remainder (24%) had abnormal findings, six (6.8%) had minor urinary abnormalities, six were hypertensive with normal urine, four (4.5%) had heavy proteinuria, eight (9%) had chronic renal failure and four (4.5%) had died. There was no association with streptococcal infection or recurrence of the rash, and the prognosis was unaffected by treatment irrespective of whether this was steroids alone, immunosuppressives alone, or the two combined.

Deterioration in renal function can occur as long as two years after presentation. Renal disease in Henoch–Schönlein purpura accounts for a mortality of about 1–3% and about 7% of all childhood glomerular nephropathies are due to Henoch–Schönlein purpura. Adults with renal disease seem to fare slightly less well than the younger ones in that the mortality for adults with nephritis is 14% [2, 3].

Recurrence of the IgA mesangiopathy after renal transplant has been described in Henoch–Schönlein purpura patients. In at least one patient the IgA mesangiopathy recurred in the transplant, and at no time in the postoperative period did the patient develop any extrarenal manifestations of Henoch–Schönlein purpura. This and the fact that patients can have severe purpura without overt renal disease raises the possibility that somewhat different mechanisms may be at work in skin and glomeruli.

## Pathogenesis

Circulating complexes have been found in several studies and impaired clearance of complexes by the reticuloendothelial system has also been reported. Correlations between the level of IgA complexes and activity of the disease have been observed. It has been shown that IgA complexes are, unlike IgG ones, increased during the active phase of Henoch–Schönlein purpura and decline to normal levels during remission.

About 50% of Henoch–Schönlein purpura patients produce IgA rheumatoid factors, the highest concentrations of which are seen during the acute phase. It has been suggested that these IgA rheumatoid factors are the IgA components of the circulating IgA-containing complexes in Henoch–Schönlein purpura. As yet no foreign antigen has been identified in the complexes in Henoch–Schönlein purpura.

## Serum complement abnormalities

In Henoch–Schönlein purpura, although serum C3 levels are usually normal, high levels of the breakdown product of C3, C3d, are found. The properdin level may be low and in those with renal disease can remain low for a year or more. Depression of the serum properdin levels with normal levels of C1q and C4 implies that complement is being activated by the alternative pathway. In renal biopsies, C1q and C4 are invariably absent, but the deposition of properdin adjacent to C3 is again indicative of activation of the alternative pathway which could lead to the generation of factors chemotactic for inflammatory cells.

## Relationship of Henoch–Schönlein purpura to idiopathic IgA nephropathy

There are close histological resemblances between the most common primary glomerulonephritis, IgA nephropathy or Berger's disease and the nephritis of Henoch–Schönlein purpura to such an extent that it has been suggested that Berger's disease could be regarded as Henoch–Schönlein purpura without the rash. In both disorders, there is a male preponderance and commonly a preceding upper respiratory infection.

Patients with IgA nephropathy can present with or develop some of the clinical features of Henoch–Schönlein purpura, that is abdominal and joint pains, and, in a very few, urticaria and ecchymoses have been noted. Circulating IgA immune complexes are found in both diseases. As is the case in Henoch–Schönlein purpura, IgA rheumatoid factors have been found in approximately half the patients with IgA nephropathy. An antigen common to both diseases is suggested by observations that antibodies eluted from the glomeruli of Henoch–Schönlein purpura nephritis patients combined with the mesangial areas of biopsies from IgA nephropathy patients and vice versa. The eluates did not, however, attach to normal glomeruli or to glomeruli in biopsies from patients with other renal diseases.

There are some points of distinction. Familial cases of IgA nephropathy have been reported, but familial examples of Henoch–Schönlein purpura are very rare. Clinically, one obvious difference is that gastrointestinal bleeding or intussusception are never seen in Berger's disease. The nephropathies run different courses. Idiopathic IgA nephropathy is characterized by recurrent macroscopic hematuria and long-term studies provide evidence that in a number of patients the course is one of relentless progression to chronic renal failure. In contrast, Henoch–Schönlein purpura nephritis appears to be an acute disease the prognosis of which is related to the severity of the glomerular changes at the onset. In Henoch–Schönlein purpura, elevated serum IgA levels return to normal after the acute episode; in IgA nephropathy the serum IgA tends to remain high for months or years.

### Role of IgA

In both diseases IgA would appear to have a key role. In Henoch–Schönlein purpura, serum levels of IgA are often increased and this seems to be due to a marked increase in the rate of IgA synthesis. IgG and IgM synthesis rates are also increased but not to the same degree as IgA.

Much work has been done to characterize the IgA and thus determine its source. Ordinarily, in the serum 90% of the IgA occurs in monomeric form and is mostly IgA1. In contrast the IgA in the external secretions is almost exclusively polymeric and IgA1 and IgA2 are present in equal amounts. The IgA2 subclass and polymeric IgA have therefore been taken to indicate a mucosal origin.

Studies to determine the nature of the IgA in the mesangium in Henoch–Schönlein purpura and IgA nephropathy have shown that polymeric IgA is deposited. In both Henoch–Schönlein purpura and IgA nephropathy, IgA1 and IgA2 have been identified in the circulating complexes.

Patients with IgA nephropathy and Henoch–Schönlein purpura relapse during mucosal infections when they have been shown to have higher and more prolonged rises in polymeric IgA and circulating IgA complexes as compared with normal controls and patients with IgA-negative renal disease [4]. The vast majority of polymeric IgA producing cells are found in mucosal-associated lymphoid tissue and it is noteworthy that after experimental exposure to influenza virus hemagglutinin, the serum IgA antibodies produced were almost exclusively polymeric and of the IgA1 subclass.

It may be relevant that in IgA nephropathy *Hemophilus parainfluenzae* outer membrane antigens have been located in the mesangial deposits [5], and similar studies are awaited in Henoch–Schönlein purpura. Meanwhile, there is evidence that these patients have aberrant regulation of IgA production and mucosal factors are involved in this process.

*Treatment*

Symptomatic relief is afforded by bed rest and analgesics. For the severely affected it is tempting to resort to systemic steroids, despite the absence of controlled trials to demonstrate their value. There are anecdotal accounts that steroids will ameliorate arthralgia and abdominal pain. A retrospective comparison of steroid-treated and untreated children showed that steroids may bring about earlier resolution of the pain in the first 24 hours, but thereafter there was no difference between the two groups. A major objection to the use of steroids for the abdominal pain is that the signs of intussusception or perforation may be masked. Plasmapheresis has been resorted to on rare occasions. It has been claimed that renal complications are less likely to develop in those treated early with prednisolone. There are at present no trials which show that steroids or immunosuppressives affect established renal complications and the value of drugs, such as cyclophosphamide, remains to be assessed. For the very few with irreversible renal disease, renal transplant is an option. Recurrence of the nephritis can occur but is unusual. Fortunately, the large majority of patients will apparently have an uneventful recovery after Henoch–Schönlein purpura with or without nephritis, but it is wise to arrange for periodic blood pressure and urine checks.

## Urticarial vasculitis

Chronic urticaria is a common and, as a rule, self-limiting disorder in which investigations are generally fruitless. However, the association of urticaria and hypocomplementemia was first described as a distinct syndrome in 1973. Since then biopsy studies of chronic idiopathic urticaria indicate that as many as one in five patients has histological evidence of vasculitis and it has become evident that urticaria may be a prominent and only cutaneous manifestation of a vasculitis. Hypocomplementemia, which tends to be episodic, occurs in about 30%. The complement abnormalities are depression of CH50 and low serum levels of the early components of the classical pathway, C1q, C2 or C4. In some cases, an antibody against the CLR (collagen-like region) of C1q is present [6]. In two patients, the hypocomplementemia was associated with the IgG autoantibody–C3 nephritic factor (C3 NeF) [7].

Urticaria and angioneurotic edema are prominent cutaneous features and minimal residual staining – purpura or hemosiderin – is found in 35%. The urticaria can persist for an unusually long time, up to 72 hours. Erythema multiforme-like lesions have also been described. About 40% of patients have renal involvement giving rise to a glomerulonephritis with granular IgG deposits along the glomerular basement membrane. Gastrointestinal and neurological disturbances can occur and obstructive lung disease has been reported in 21%.

Patients with hypocomplementemic urticarial vasculitis may have a raised erythrocyte sedimentation rate (ESR), so this would appear to be a simple screening test, albeit an unreliable one.

Only a few patients will respond to antihistamines. Non-steroidal anti-inflammatory agents may be of more use. Systemic steroids may, however, have to be used, usually in low dose, 15–30 mg/day. Gold and colchicine have also been used.

Dapsone 100 mg/day may be effective in this condition and occasionally proves useful in other vasculitides. Dapsone has been shown to affect various neutrophil functions including chemotaxis, lysosomal activity and myeloperoxidase mediated iodination. It can also inhibit integrin-mediated neutrophil adherence and suppress cell attachment to IgA and IgG on basement membranes [8–10].

## Cryoglobulinemia

Single component cryoglobulins consist of a monoclonal immunoglobulin and are produced in benign monoclonal gammopathy and myeloma, Waldenström's macroglobulinemia and other lymphoproliferative diseases. They may give rise to Raynaud's phenomenon, cold urticaria, vasculitis, ulcers and hemorrhagic areas over the extremities or other exposed areas, such as the pinnae and cheeks. Nephritis has also been described in single component cryoglobulinemia.

More common are the mixed cryoglobulins which have been shown to be immune complexes consisting of a cold-reactive antiglobulin combined with IgG. Cold-reactive antiglobulins are usually of the IgM class, less often IgA and rarely IgG.

Two mechanisms account for the production of mixed cryoglobulins. First, there may be a clonal proliferation of immunoglobulin-secreting cells, as in lymphoma or Waldenström's macroglobulinemia. Secondly, they can occur in association with circulating antigens, as in an several infections including hepatitis B and C, shunt nephritis or in circumstances in which there is a source of continuously absorbed antigen, as after intestinal bypass surgery. They also occur in systemic lupus erythematosus (SLE), rheumatoid arthritis (RA) and liver disease.

### Clinical features of mixed cryoglobulinemia

Vasculitis develops in many but not all patients with immune complex cryoglobulins. In the skin the smaller upper dermal vessels are usually affected. Large vessel involvement with the formation of aneurysms is a rare event. Patients with mixed or immune-complex cryoglobulinemia have been reported in several series. The clinical features include purpura, urticaria, erythema multiforme-like lesions and leg ulcers. Other complications are arthralgia, weakness, nasal discomfort, hemoptysis, peripheral neuropathy, abdominal symptoms similar to those of Henoch–Schönlein purpura and, most serious of all, glomerulonephritis, which develops in at least 25% [11].

### Treatment

Many patients do not require treatment. Protection from the cold is an obvious measure. Chilling is certainly to be avoided, since there is at least one report of a patient who developed renal failure after sleeping in the open on a cold night in the winter. Where symptoms are severe or glomerulonephritis develops, the treatments available are cytotoxic drugs for those with an underlying tumor, or combinations of systemic steroids and immunosuppressants. Plasmapheresis may be invaluable for the severely ill patient in that it will temporarily lower the cryoglobulin level, improve reticuloendothelial clearance and reduce the temperature at which cryoprecitation occurs. About 60% of patients develop hypertension. The survival rate for patients with essential mixed cryoglobulinemia and renal disease is 75% at 10 years from the onset of symptoms. Recently, hepatitis C has been associated with mixed cryoglobulinemia and the point has been made that there may be a case for treating with interferon [12].

## Cryofibrinogenemia

This is a rare condition which can give rise to cold-intolerance, purpura, ecchymoses, eschars and hemorrhagic ulcers and bullae. Histologically, the vessels are occluded by thrombi and the disease comes into the differential diagnosis of an occlusive thrombopathy, some causes of which are thrombotic thrombocytopenic purpura, disseminated intravascular coagulation, Coumarin necrosis, protein C deficiency, cryoglobulinemia and the lupus anticoagulant. Often the cryofibrinogenemia is idiopathic. However, it can be associated with malignancy [13, 14].

## Benign hypergammaglobulinemic purpura of Waldenström

One distinctive group of patients with circulating complexes has been characterized. They were first described by Waldenström, who noted a polyclonal increase in gammaglobulins and named the disorder benign hypergammaglobulinemic purpura. Skin biopsies reveal necrotizing vasculitis. With the publication of two series of patients, it became apparent that another feature is the presence in the serum of intermediate size complexes, consisting of IgG reacted with IgG antiglobulins. Complexes are most easily detected by analytical centrifugation, when they are revealed as complexes intermediate in size between IgG and IgM, 13–14S. Treatment is rarely indicated since most patients have only minor discomfort and even though remission is rare the prognosis is good.

## Erythema elevatum diutinum (EED)

This is an exceptionally rare disorder as illustrated by the fact only 13 cases were recorded over a 60-year period at St John's Hospital for Diseases of the Skin, a major UK tertiary referral center for the London area and southern England.

Small purpuric lesions may occur as in many other vasculitides, but much more striking are the characteristic persistent red or orange-yellow plaques and violaceous nodules distributed acrally and symmetrically over the extensor surfaces, sometimes with blisters and ulcers. The nodules often have a juxta-articular distribution. Arthralgia is infrequent. There is one report of EED in association with RA. The disease tends to be very chronic and although resolution has been reported, this is exceptional and most patients continue to have lesions for many years.

The acute lesions show a leukocytoclastic vasculitis. The older lesions are distinctive with a complex histology consisting of granulation tissue, fibrosis and intracellular lipidosis – lipid droplets, myelin figures and rare cholesterol clefts within histiocytes and also within epidermal keratinocytes, mast cells, pericytes and lymphocytes.

Several observers have noted that the lesions can be induced at the site of injection of streptococcal antigen, but there are doubts about the specificity of these observations and therefore whether streptococci might have a pathogenetic role. For such an extraordinarily rare disorder it is perhaps notable that six patients with EED and human immunodeficiency virus (HIV) have been reported to date in the literature, and this has led to speculation that the disorder might represent an abnormal response to an infectious agent [15].

There is an undoubted association with paraproteinemia or hematological disease – five out of 17 in one series and six out of 13 in another; paraproteinemia and myeloma, polycythemia, hairy cell leukemia, IgG–IgM cryglobulinemia and the hypereosinophilic syndrome have all been reported.

Many patients will respond to dapsone or sulfonamides. Where there is a failure to respond to these drugs, intralesional and systemic steroids have been used. In those patients with myeloma, the use of alkylating agents does not in itself bring about any improvement in the skin lesions.

## LARGE VESSEL DISEASE

Systemic polyarteritis nodosa often does not present with skin lesions and is not ordinarily in the province of the dermatologist, being more usually dealt with by the general physician or rheumatologist. However, there is one type of polyarteritis nodosa (PAN) which will amost invariably come the way of the dermatologist. This is cutaneous polyarteritis nodosa also known as benign cutaneous polyarteritis nodosa.

### Cutaneous polyarteritis nodosa

This disease was first described in 1931. Subsequently, the description 'cutaneous polyarteritis nodosa' was reserved for a specific clinical and histological picture consisting of subcutaneous nodules due to an arteritis in the subcutis, usually accompanied by livedo reticularis in the affected area and most commonly involving the legs.

The histology is identical to that of systemic polyarteritis nodosa – a necrotizing vasculitis of small muscular arteries in subcutaneous tissue and muscle with, for the most part, sparing of visceral arteries, although occasionally there may be involvement of vessels in muscles or sensory nerves in the legs.

The cause is not known. There has been only one case reported with hepatitis B. Of considerable interest are three case reports of cutaneous polyarteritis nodosa in a mother and her neonatal infant [16]. There are occasional reports of cases with Crohn's disease and ulcerative colitis.

The clinical features in the largest series of 20 patients included subcutaneous nodules in 90%, livedo reticularis in 80%, leg ulcers in 35%, gangrene of the toes in 15%, and 45% had arthralgia and myalgia. Half the patients developed peripheral neuropathy. Two of the 20 patients subsequently progressed to systemic polyarteritis nodosa, but only after extraordinarily long relatively 'benign' phases of 18 and 19 years respectively [17]. An earlier view that the disorder is unrelated to classic systemic polyarteritis nodosa and is benign therefore requires qualification in that although in many patients the disease is not life-endangering and does not affect vital internal organs, there are occasional patients who may develop fatal vasculitis after a very protracted period.

### Treatment

In spite of its benign localized nature in many instances, it may result in considerable incapacity requiring steroids and imunosuppressive agents. Cyclosporin A may have a place in the management of difficult cases.

## Other disorders with livedo reticularis

The physical sign – livedo reticularis – a red-blue mottling of the skin in a netlike configuration indicates stasis in an inverted cone of vessels supplied by an arteriole. Rarely, it may arise from interference with the nerve supply to the subcutaneous vessels as in post-zoster reflex sympathetic dystrophy but, ordinarily, it is indicative of an intravascular disorder or a disease affecting the vessel wall. Intravascular diseases where stasis from hyperviscosity is the primary problem include: myeloproliferative disorders with very high cell counts, as in essential thrombocythemia, dysproteinemias, cryoglobinemia and cold agglutinin disease. The vasculitides responsible include both the localized and systemic forms of PAN, and SLE. It may occur as a result of vessel occlusion – in atheroembolic disease, the anticardiolipin syndrome and with an endarteritis obliterans.

One example of the rare endarteritis obliterans is Sneddon's syndrome in which widespread broken livedo reticularis is complicated by cerebrovascular episodes. In this disorder, skin biopsies may reveal endarteritis obliterans of deep dermal arteries. Cerebral arteriography has shown multiple occlusions in the medium-sized arteries. These patients do not necessarily have anticardiolipin antibodies [18].

A different disorder is livedo reticularis with ulceration or livedo vasculitis, and this occurs principally on the lower legs of women as painful, irregularly-shaped ulcers that are very slow to heal with scars resembling atrophie blanche. Skin biopsies show one or more vessels in the mid-dermis occluded by eosinophilic fibrinoid material.

As a cause of livedo, cholesterol emboli merit special mention. The disorder may masquerade as PAN with fever, myalgia, high ESR, hypertension, eosinophilia and a rash, mostly in patients over 50 years, but the disease is not unknown in younger individuals. The skin biopsy is diagnostic, and is one very good reason for always doing a skin biopsy even where the clinical picture could apparently be nothing other than a vasculitis. The biopsy reveals acicular needle-like clefts or birefringent cholesterol crystals within the vessels in 92% of cases. However, multiple sections may be needed and if positive will spare the patient anticoagulants and steroids. Older lesions show intimal fibrosis and fibrinous thrombosis. The emboli may arise spontaneously, after anticoagulation and angiography [19].

## REFERENCES

1. Jorizzo, J.L. (1993) Classification of vasculitis. [Review]. *J. Invest. Dermatol.*, **100**(1), 106S–10S.
2. Faull, R., Woodroffe, A., Aarons, I. and Clarkson, A. (1987) Adult Henoch–Schönlein nephritis. *Aus. N.Z. Med. J.*, **17**, 396–401.
3. Goldstein, A., White, R., Akuse, R. and Chantler, C. (1992) Long-term follow-up of childhood Henoch–Schönlein nephritis. *Lancet*, **339**, 280–2.
4. Jones, C., Powell, H., Kincaid-Smith, P. and Roberton, D.M. (1990) Polymeric IgA and immune complex concentrations in IgA-related nephropathy. *Kidney Int.*, **38**, 323–31.
5. Suzuki, S., Nakatomi, Y., Sato, H. *et al.* (1994) *Haemophilus parainfluenza* antigen and antibody in renal biopsy samples and serum of patients with IgA nephropathy. *Lancet*, **343**, 12–16.
6. Wisnieski, J.J. and Jones, S.M. (1992) Comparison of autoantibodies to the collagen-like region of C1q in hypocomplementemic urticarial vasculitis syndrome and systemic lupus erythematosus. *J. Immunol.*, **148**(5), 1396–403.
7. Carmichael, A. and Marsden, J. (1993) Urticarial vasculitis: a presentation of C3 nephritic factor. *Br. J. Dermatol.*, **128**, 589.
8. Harvath, L., Yancey, K. and Katz, S. (1986) Selective inhibition of human neutrophil chemotaxis to N-formyl-methionyl-leucyl-phenylalanine by sulfones. *J. Immunol.*, **137**, 1305–11.
9. Booth, S., Moody, C., Dahl, M. *et al.* (1992) Dapsone suppresses integrin-mediated neutrophil adherance function. *J. Invest. Dermatol.*, **98**, 135–40.
10. Thuong-Nguyen, V., Kadunce, D., Hendrix, J. *et al.* (1993) Inhibition of neutrophil adherence to antibody by dapsone: a possible therapeutic mechanism of dapsone treatment. *J. Invest. Dermatol.*, **100**, 349–55.
11. Frankel, A.H., Singer, D.R., Winearls, C.G. *et al.* (1992) Type II essential mixed cryoglobulinemia: presentation, treatment and outcome in 13 patients. *Q. J Med.*, **82**(298), 101–24.
12. Cacoub, P., Fabiani, L., Musset, L. *et al.* (1994) Mixed cryoglobulinemia and hepatitis C virus. *Am. J. Med.*, **96**, 124.
13. Beightler, E., Diven, D. G., Sanchez, R.L. and Solomon, A.R. (1991) Thrombotic vasculopathy associated with cryofibrinogenemia. *J. A. Acad. Dermatol.*, **24**(2 Pt 2), 342–5.
14. Jantunen, E., Soppi, E., Neittaanmaki, H. and Lahtinen, R. (1993) Essential cryofibrinogenemia, leukocytoclastic vasculitis and chronic purpura. *J. Intern. Med.*, **234**(3), 331–3.
15. LeBoit, P.E. and Cockerell, C.J. (1993) Nodular lesions of erythema elevatum diutinum in patients infected with the human immunodeficiency virus. *J. Am. Acad. Dermatol.*, **28**(6), 919–22.

16. Stone, M.S., Olson, R.R., Weismann, D.N. *et al.* (1993) Cutaneous vasculitis in the newborn of a mother with cutaneous polyarteritis nodosa. *J. Am. Acad. Dermatol.*, **28**(1), 101–5.
17. Chen, K.-R. (1989) Cutaneous polyarteritis nodosa: a clinical and histopathological study of 20 cases. *J. Dermatol.*, **16**, 429–42.
18. Burton, J. (1988) Livedo reticularis, porcelain white scars and cerebral thromboses. *Lancet*, **i**, 1263–4.
19. Falanga, V., Fine, M. and Kapoor, W. (1986) The cutaneous manifestations of cholesterol crystal embolization. *Arch. Dermatol.*, **122**, 1194–8.

# *Part Five*

## Management

# 22

# Treatment of severe vasculitis: general principles and the use of corticosteroids

*J.S. Cameron*

## CORTICOSTEROIDS IN THE TREATMENT OF THE VASCULITIDES

It is only very recently that any effective treatment has been available for vasculitis of any type. Prior to about 1930, divisions of the vasculitides were almost exclusively performed on the basis of clinical and macroscopic post-mortem observations, into periarteritis (later modified in 1903 to polyarteritis) as described by Kussmaul and Meier, small vessel vasculitides in the form of purpura described by Schönlein and Henoch, and the granulomatosis described by Klinger and Wegener, which bears the latter's name [1]. Then microscopic observations became common, with the introduction by Davson of the term 'microscopic polyarteritis' for those forms in whom neither necrotic lesions nor aneurysms were visible to the naked eye, which others such as Zeek called hypersensitivity angiitis [2], and yet others amalgamated as 'small vessel vasculitis'.

Thus, when corticosteroids were introduced into clinical practice in the early 1950s, descriptions of the vasculitides were incomplete and it is now sometimes difficult to identify exactly what form of disease was being treated: terms such as 'microscopic polyarteritis nodosa' and insufficient data are given to identify in contemporary terms what the patients may have suffered from. What is clear, however, is that prior to 1950 (apart from some cases of predominantly cutaneous hypersensitiv-

*The Vasculitides*. Edited by B.M. Ansell, P.A. Bacon, J.T. Lie and H. Yazici.
Published in 1996 by Chapman & Hall, London. ISBN 978-0-412-64140-4.

ity vasculitis affecting mainly capillaries and venules, such as Schönlein–Henoch purpura) vasculitis and especially polyarteritis was usually a post-mortem diagnosis, but even allowing for this the mortality was very high. Thus in 1957 Rose and Spencer [3] reported a five-year survival of only 4% in patients with polyarteritis from the time systemic symptoms were noted, and in the same year Walton [4] noted a mean survival of only five months in patients with Wegener's granulomatosis, with 82% of patients dead by the end of a year.

The introduction of corticosteroids in the form of cortisone or ACTH had a clear effect on most patients with vasculitis, suppressing symptoms but also appearing to prolong survival. This was evident in one of the early studies of the British Medical Research Council, published in 1960 [5], which showed a significant improvement in three-year survival from 37% in untreated patients to 62% in treated patients. In the Mayo Clinic series [6], five-year survival was 48% in treated patients, but in those with renal involvement it remained only 34%. However, neither study was a proper randomized prospective trial. Sack, Cassidy and Bole [7] reported a 64% survival of 40 patients with polyarteritis in 1978. Even as late as the 1970s, therefore, the exact role of corticosteroids in treating vasculitis remained undetermined. It is a cause for regret that since then no proper randomized controlled trial of any form of immunosuppression has been carried out in any form of vasculitis.

As early as 1954, nitrogen mustard had been used in the treatment of Wegener's granulomatosis, and by 1970 a limited number of patients were being treated with azathioprine, and later cyclophosphamide in addition [8, 9]. Since then almost all series of the different forms of vasculitis has included a high proportion of patients given both drugs, so that the effect of corticosteroids alone is now impossible to estimate. A few papers during the 1970s and early 1980s compared, in a non-controlled fashion, patients treated with corticosteroids alone, and those given cytotoxic agents in addition [10, 11]. Thus, in 1979, Leib, Restivo and Paulus [10] compared the survival of untreated patients with those given corticosteroids alone and those given cytotoxic agents in addition. Survival at five years was 12%, 50% and 80% respectively. In the study by Cohen, Conn and Ilstrup in 1980 [11], 36 patients treated with steroids alone and 14 treated with cytotoxic agents as well had survivals not significantly different from each other, those with double treatment faring rather worse; however, they had worse disease clinically, so no firm conclusions could be drawn. Many clinicians were influenced greatly by the seminal paper of Fauci *et al.* in 1979 [12] into using cyclophosphamide in addition to corticosteroids, or cyclophosphamide alone, on the basis that this study showed improvement even in cases that were already deteriorating, when cyclophosphamide was introduced into the therapeutic regime. Much of this lowering of mortality may have resulted from subsequent reduction in corticosteroid dosage, which is of course a legitimate advantage of double therapy.

The dose and route of administration of corticosteroids in vasculitis has never been studied adequately. Intravenous high dose 'boluses' of prednisolone or methylprednisolone [13] have become a popular component of induction regimes, followed by low maintenance doses, in contrast to the older high oral doses, tapering over the following weeks or months. The former regime has the advantage of fewer side-effects in the longer term, and it may be that the dramatic effect of the bolus dosage on lymphocytes has some real therapeutic advantage (see below). It should be remembered that the original justification for the introduction of i.v. methylprednisolone into rheumatology in 1974 for the treatment of severe lupus nephritis [14] was on the basis of the resemblance between the renal interstitial infiltrate of lupus and that of allograft rejection, for which this type of treatment was becoming popular. There is almost always a severe interstitial infiltrate in patients with small vessel vasculitis affecting the kidney, and in a few patients this may be seen in the absence of glomerular changes [15].

Corticosteroids have hitherto been used empirically: only *post hoc* has a rationale for the use of these drugs been evolved. Corticosteroids exert their many effects on the immune system, both physiological and pharmacological, by entering the cell and binding to an intracellular corticosteroid receptor which exists in association with the 90 kDa heat shock protein (HSP90). Activation of the receptor results in dissociation of the HSP90 and entry of the receptor–ligand complex into the nucleus. This is followed by binding of the complex directly to a promoter region of DNA, which influences transcription [16]. Thus corticosteroids are capable of inhibiting transcription of DNA coding for a number of molecules. The main effect is in inhibiting the synthesis and release from macrophages of inflammatory mediators, particularly TNF-$\alpha$ [17], IL-1/IL-6 [18], and eicosanoids through inhibition of phospholipase A2 [19]. In addition, expression of MHC class II antigens is inhibited [20], which might interfere with continued presentation of antigen to T helper (CD4 positive) lymphocytes.

A second level of effect is on these CD4 lymphocytes themselves. Injection of large doses of corticosteroids reduces the level of circulating lymphocytes dramatically for many hours, by redistribution into lymph nodes rather than by cytolysis as was originally thought. Transcription of the IL-2 gene is also inhibited [21], and consequently so is the generation of committed T cell clones. Thus corticosteroids powerfully inhibit the two components of delayed hypersensitivity reactions – the small number of activated IL-2 secreting CD4 positive lymphocytes and the much larger number of recruited monocytes/macrophages in various states of activation and maturation. The importance of this type of infiltrate in vasculitides of granulomatous type is obvious, and also applies to the more lymphocyte-rich infiltrate of microscopic polyangiitis.

Finally, older studies have shown that corticosteroids also inhibit inflammatory responses [22] through a decrease in exudation and inhibition of the rise in vascular permeability, inhibition of macrophage migration and division, a decrease in protease release, and stabilization of lysosomal membranes [23]; thus they have powerful anti-inflammatory effects on monocytes.

Thus, a personal summing up of the use of corticosteroids in vasculitic disease is that we have employed these drugs for 40 years without learning anything solid about their effects on the diseases in question. Indeed, it is possible today to ask the question of whether it is necessary to give corticosteroids at all, in view of their many side-effects, if cytotoxic agents are being employed [12]. Here there may be a trade off, since at least with oral cyclophosphamide, there are data from other disease states suggesting that it is possible to give larger dosages without marrow suppression and leukopenia in the presence of corticosteroids, probably because they stimulate the white cell precursors in the marrow.

## TREATMENT STRATEGIES IN VASCULITIS

It almost goes without saying that in beginning any therapeutic endeavor it is always useful to try and identify the goal of treatment. In the case of severe vasculitis (that is vasculitis affecting vital organs such as the lung, brain, kidney or gut), our goals are clear: in the initial induction phase, to limit directly destructive inflammatory processes and the secondary scarring and erosion of organ function they promote; and in the maintenance phase in chronic or relapsing types of disease, to achieve suppression of activity in the long term with minimum side-effects from the treatment. It is important, as in the treatment of severe lupus, to distinguish clearly between these phases of induction and maintenance treatment, because they may require different strategies and even different drugs.

Treatment of the various syndromes of vasculitis has presented, and still presents, major challenges to the physician. Progress has been made, almost entirely by empirical approach, and even in the severest forms of vasculitis survival and retention of organ function has improved steadily over the past two decades.

Nevertheless, with better survival, attention has inevitably turned to the destructive side-effects of these drugs when given in the medium and long term. Optimum management of the postacute phase is still controversial and poorly documented [24], and proper cost–benefit analysis has never been done, even for corticosteroids. Thus it is not clear whether early withdrawal of prednisolone or cytotoxic therapy, leading to larger numbers of relapses, with subsequent need for further higher dose induction treatment carries less of an overall burden than long-term, low-dose maintenance treatment, which in general avoids relapses but submits the patient to years of continuous treatment. In elderly patients especially, loss of skin protein, bone mineral and susceptibility to infections are heavy prices to pay. Concomitant cytotoxic therapy – although it has never been tested by controlled trial – appears to have benefit both on survival and on permitting lower doses of corticosteroids, but the type of drug, dose and route of administration remain the subject of debate, and this is enlarged upon elsewhere in this book.

## REFERENCES

1. Churg, J. (1991) Nomenclature of vasculitic syndromes: a historical perspective. *Am. J. Kidney Dis.*, **18**, 148–53.
2. Zeek, P.M. (1952) Periarteritis nodosa. A critical review. *Am. J. Clin. Pathol.*, **22**, 777–90.
3. Rose, G.A. and Spencer, H. (1957) Polyarteritis nodosa. *Q. J. Med.*, **26**, 43–81.
4. Walton, E.W. (1957) Giant cell granuloma of the respiratory tract (Wegener's granulomatosis). *Br. Med. J.*, **2**, 265–70.
5. Medical Research Council (1960) MRC study on treatment of polyarteritis nodosa with cortisone: results after three years. *Br. Med. J.*, **1**, 1399–400.
6. Frohnert, P.P. and Sheps, S.G. (1967) Long-term follow-up study of periarteritis nodosa. *Am. J. Med.*, **43**, 8–14.
7. Sack, M., Cassidy, J.T. and Bole, G.G. (1978) Prognostic factors in polyarteritis. *J. Rheumatol.*, **2**, 411–20.
8. Novack, S.N. and Pearson, C.M. (1971) Cyclophosphamide therapy in Wegener's granulomatosis. *New Engl. J. Med.*, **285**, 1493–6.
9. Fauci, A.S. and Wolff, S.M. (1973) Wegener's granuloma. A review of eighteen patients and a review of the literature. *Medicine (Baltimore)*, **52**, 535–61.
10. Lieb, E.S., Restivo, C. and Paulus, H.E. (1979) Immunosuppressive and corticosteroid therapy of polyarteritis nodosa. *Am. J. Med.*, **67**, 941–7.
11. Cohen, R.D., Conn, D.L. and Ilstrup, D.M. (1980) Clinical features, prognosis and response to treatment in polyarteritis. *Mayo Clin. Proc.*, **55**, 146–55.
12. Fauci, A.S., Katz, P., Haynes, B.F. and Wolff, S.M. (1979) Cyclophosphamide therapy of severe systemic necrotizing vasculitis. *New Engl. J. Med.*, **301**, 235–8.
13. Neild, G.H. and Lee, H.A. (1971) Methylprednisolone pulse therapy in the treatment of polyarteritis nodosa. *Postgrad. Med. J.*, **53**, 382–7.
14. Cathcart, E.S., Idelson, B.A., Scheinberg, M.A. and Couser, W.G. (1974) Beneficial effects of methylprednisolone 'pulse' therapy in diffuse proliferative lupus nephritis. *Lancet*, **i**, 163–6.
15. Cameron, J.S. (1991) New horizons in renal vasculitis. *Klin. Wschr.*, **69**, 536–51.
16. Hollenberg, S.M., Giguère, V., Segeui, P. and Evans, R.M. (1987) Colocalisation of DNA-binding and transcriptional activation functions in the human glucocorticoid receptor. *Cell*, **49**, 39–47.

17. Han, J., Thompson, P. and Beutler, B. (1990) Dexamethasone and pentaoxiphylline inhibit endotoxin-induced cachetin/tumor necrosis factor synthesis at separate points on the signalling pathway. *J. Exp. Med.*, **172**, 391–4.
18. Lee, S.W., Tsou, A.P., Chan, H. *et al.* (1988) Glucocorticosteroids selectively inhibit the transcription of the interleukin 1 beta gene and decrease the stability of interleukin 1 beta mRNA. *Proc. Nat. Acad. Sci. USA*, **85**, 1204–8.
19. Nakano, T., Ohara, O., Teraoka, H. and Arita, H. (1990) Glucocorticoids suppress group II phospholipase A2 production by blocking mRNA synthesis and post-transcriptional expression. *J. Biol. Chem.*, **265**, 12745–8.
20. Fertsch, D., Schoenberg, D.R., Germain, R.N. *et al.* (1987) Induction of macrophage Ia antigen expression by rIFNg and down-regulation by IFN-α-β and dexamethasone are mediated by changes in steady state levels of Ia mRNA. *J. Immunol.*, **139**, 244–9.
21. Northrop, J.P., Crabtree, G.R. and Mattila, P.S. (1992) Negative regulation of interleukin 2 gene transcription by the glucocorticoid receptor. *J. Exp. Med.*, **175**, 637–46.
22. Dannenberg, A.M. (1979) The antiinflammatory effects of glucocorticosteroids. A brief review of the literature. *Inflammation*, **3**, 329–43.
23. Wright, D.G. and Malawista, S.E. (1973) Mobilization and extracellular release of granular enzymes from human leukocytes during phagocytosis: inhibition by colchicine and cortisol but not by salicylate. *Arthritis Rheum.*, **16**, 749–58.
24. Gordon, M., Luqmani, R.A., Adu, D. *et al.* (1993) Relapses in patients with a systemic vasculitis. *Q. J. Med.*, **86**, 779–89.

# 23

# Therapy of vasculitis

*R.A. Luqmani, A. Pall, D. Adu, R.J. Moots,*
*N. Richards, A.J. Howie, J. Michael, P. Emery and*
*P.A. Bacon*

## INTRODUCTION

Vasculitis, first described by Kussmaul and Maier in 1866 [1], may affect any vessel, and may occur *de novo* or complicate pre-existing diseases. Vasculitis may be defined as inflammation of blood vessels leading to tissue damage or destruction. Vasculitis is an important component of the pathology of all connective tissue diseases; Bywaters [2] hypothesized that vascular inflammation was the basis of connective tissue disease. Vasculitis, which is confined to the skin, is discussed elsewhere in this volume.

The systemic vasculitides are rare multisystem diseases. The reported annual incidence of polyarteritis nodosa (PAN) varies from 2/million [3] to 4.6/million [4]. It has been suggested, however, that there has been an increase in the combined annual incidence of Wegener's granulomatosis and microscopic polyarteritis to 6.1/million [5].

In our experience, the most common forms of systemic vasculitis are rheumatoid vasculitis [6–8], Wegener's granulomatosis [9, 10], microscopic polyarteritis [11] and PAN [1, 12].

Classifications of vasculitis are based on vessel size, organ involvement, pathology and etiology when known [13, 14]. There is strong evidence to suggest an immunopathological etiology for at least some of these diseases. Vascular inflammation may be the end result of many different causes; conversely, any individual immunological abnormality

*The Vasculitides*. Edited by B.M. Ansell, P.A. Bacon, J.T. Lie and H. Yazici.
Published in 1996 by Chapman & Hall, London. ISBN 978-0-412-64140-4.

can produce a spectrum of clinical problems which often overlaps with that produced by other immunological events [15].

In practice, there is often considerable overlap of clinical features when using any system of classification. It is often more appropriate to keep the provisional diagnosis under review. Most clinicians managing vasculitis favor a more pragmatic approach to therapy, based on severity and importance of organ involvement, in order to avoid unnecessarily treating patients with toxic drugs. We have attempted to formally measure clinical disease activity, in order to rationalize the assessment and management of these patients [16]. Using a system based predominantly on clinical disease activity, we have demonstrated that the Birmingham Vasculitis Activity Score (BVAS) can differentiate between active and inactive disease, and may become an effective guide to therapy.

With such diversity it is not surprising to find a range of therapy. In 1950, Baggenstoss, Shick and Polley [17] first reported the successful use of steroids given to patients with vasculitis. Since then, there have been many studies using either steroids alone, or steroids in combination with cytotoxic agents, which have shown a reduction in the short-term mortality of these diseases [18–21]. Although in many studies different types of necrotizing vasculitis are grouped together, it is clear that their prognosis has been considerably improved by the use of cytotoxic agents.

## WEGENER'S GRANULOMATOSIS

Wegener [9] described this condition in three cases assumed to have an infectious disease with a poor prognosis. Godman and Churg [10] defined the characteristic triad of upper and/or lower respiratory tract necrotizing lesions in association with a focal segmental necrotizing glomerulonephritis and a systemic vasculitis. Common features at presentation include constitutional upset, pulmonary involvement, renal disease, evidence of an acute phase response, low hemoglobin, raised white cell count and platelets [22].

Initial reports of the efficacy of cyclophosphamide in Wegener's granulomatosis [23] were confirmed by the NIH group [24, 25]. Steroid therapy improved the average survival from five months to 12 months. The use of continuous low-dose oral cyclophosphamide achieved an 80% five-year survival [26]. In our experience of 28 patients, the relapse rate was 39% and the overall mortality was 21%, with a five-year survival rate of over 70% despite the aggressive use of cytotoxic agents, prednisolone and plasma exchange [22]. The prognosis in Wegener's remains serious; cyclophosphamide, or other cytotoxic agents, are not a panacea.

## LIMITED (NON-RENAL) WEGENER'S GRANULOMATOSIS

Classical Wegener's granulomatosis is a generalized disease but there are more limited forms [22, 27]. Although there may be no clinical evidence of renal involvement despite classic histological changes in the lungs, a careful post-mortem study [27] showed that some patients had microscopic evidence of granulomatous lesions in the kidney. In addition, the reverse situation has been described, with renal involvement (focal necrotizing glomerulonephritis in each case) predating the appearance of the respiratory tract lesions [28, 29]. We have suggested that these patients represent a spectrum of disease ranging from predominantly granulomatous to predominantly vasculitic pathology [22]. We have described patients shifting from non-renal to classic Wegener's granulomatosis despite immunosuppressive therapy [22]. Untreated, the disease carries a lower (but still significant) mortality risk than classic Wegener's [27]. In our experience, mortality in limited Wegener's granulomatosis is low with current therapy (no deaths in our series after 33 months median follow-up). Nevertheless, the disease can be locally destructive, and immunosuppressive therapy is required at this stage.

The antibiotic trimethoprim/sulfamethoxazole may be of value in the treatment of the classic form of the disease [30–32]. The mechanism of action may be truly anti-infective, possibly acting on an inhaled organism. If given at an early stage in the disease, this might obviate the need for more toxic immunosuppression. Alternative mechanisms of action, such as radicle scavenging, are also possible. A recent open study has suggested more benefit from this antibiotic in limited than in classic Wegener's granulomatosis [33].

## POLYARTERITIS NODOSA AND MICROSCOPIC POLYARTERITIS

This group of diseases has a similar clinical picture of severe multisystem disease to Wegener's granulomatosis, but without the consistent involvement of the upper and lower respiratory tract [34]. On the basis of the presence or absence of involvement of small vessels, it is divided into either microscopic polyarteritis (MPA) or polyarteritis nodosa (PAN) respectively [11, 15]. This distinction may be relatively arbitrary, since we have described patients who present with classical PAN, and who develop characteristic renal lesions of MPA on relapse [35]. Many previous series of cases with polyarteritis do not make this distinction clear. The untreated survival is reported to be poor, particularly in the elderly. The use of steroids on their own improved the one-year survival to 71% in an early study, but had much less effect on the long-term survival [36]. Leib, Restivo and Paulus [20] have demonstrated the important therapeutic role of cytotoxic drugs (using mainly azathioprine). In 64 cases, treatment was non-randomized, with cytotoxic agents being used in only the most severely affected group. This latter group had a longer mean survival (149 months) and higher five-year survival (80%) than either the group given steroids alone (63 months and 53% respectively) or the group on no therapy (three months and 12%). We have shown similar results [4]. Mortality was high in the untreated group, in some cases the diagnosis only becoming apparent at necropsy. Steroid therapy resulted in short-term improvement but mortality was > 60%. In the small group given cyclophosphamide together with steroids, the mortality was only 25%. Comparison of these cases from 1974–80 [4] with our more recent data in our renal and rheumatological units [35] gives a clear view of the overall change in prognosis as well as some factors relating to prognosis. The earlier series was largely in a precyclophosphamide era, and the majority received steroids alone. In the latter series almost all patients were treated with cytotoxic agents (chiefly cyclophosphamide). The overall mortality in the first series was 53%; in the second series, the 10-year mortality was between 25% and 45% for vasculitis with renal involvement as compared with nil for vasculitis without renal involvement. While the main cause of death in the earlier series was active vasculitis, supplemented by complications of vasculitis, such as hypertension, sepsis was the most common single cause of death in the later renal series. Thus while mortality due to vasculitis has diminished, the problem of sepsis has become more prominent and is related to therapy [37].

Previous studies of MPA have demonstrated similar mortality figures from 20–50% despite aggressive therapy [38–41]. Risk factors for poor outcome include age, significantly elevated serum creatinine (greater than 500 µmol/l) at onset, oliguria and the presence of crescents on renal histology.

## RHEUMATOID VASCULITIS

The term systemic rheumatoid vasculitis (SRV) refers to patients with rheumatoid arthritis (RA) who have clinical and/or histological evidence of vasculitis, or who have nailfold infarcts in the presence of significant extra-articular features. Cutaneous lesions are one of the commonest lesions of SRV [42]. It has been suggested that the necrotizing vasculitis complicating rheumatoid disease is the exact equivalent of PAN [43]. The histology is identical and the clinical picture bears many similarities [6, 7]. However, there are significant differences in the serological abnormalities found in SRV as compared with PAN. In SRV, high levels of IgG rheumatoid factor and complement-fixing, rheumatoid factor-containing immune complexes are found which correlate well with clinical features [44]. These are not present in PAN [45]. Brief improvement is seen following removal of circulating complexes by plasma exchange [46]. Cytotoxic drugs appear to delay the relapse seen after plasma exchange, with azathioprine being less effective than cyclophosphamide.

SRV is a serious condition which is often relapsing and has a significant and cumulative mortality despite cytotoxic therapy [42, 47]. Steroids may be responsible for inducing the vasculitis of SRV [48, 49]. Chlorambucil [50] and azathioprine [51, 52] have been used with variable success. However, cyclophosphamide remains the drug of choice [53, 47].

## CHURG–STRAUSS SYNDROME

In the original description of this disorder [54], it was shown to be a separate disease from PAN. Lanham *et al.* [55] define it as a triad of asthma, eosinophilia and a systemic vasculitis involving two or more extra-pulmonary organs. The pathological lesions closely resemble those seen in PAN. There is dispute as to whether or not granulomatous inflammation could be used to distinguish this disease from PAN [56].

Steroids alone are beneficial in most cases, although cytotoxic agents and additional plasma exchange are occasionally used [57].

## LARGE-VESSEL VASCULITIS

Takayasu's arteritis and giant cell arteritis (GCA) share similar pathological appearances and are often responsive to corticosteroids alone. Both conditions may cause considerable morbidity, as may the effects of using long-term steroids. In addition, the relapse rate in GCA may be high when steroids are tapered rapidly. In view of these difficulties, there is increasing interest in the use of additional therapy with methotrexate, and also cyclosporin A either in the initial management, or where steroid related side-effects or relapse of disease are present.

## CYCLOPHOSPHAMIDE REGIMES

### Pulse vs continuous therapy

We have experimented with several regimes of cyclophosphamide in order to deliver a large dose of the drug at intervals for maximal efficiency while trying to minimize side-effects [58, 59]. We have devised a regime of intermittent bolus doses of cyclophosphamide combined with methylprednisolone in our patients which, in SRV, appears to be more successful than continuous oral regimes [21]. Pulse cyclophosphamide has been shown to be effective in treating lupus nephritis [60]. It is uncertain, however, whether or not pulse cyclophosphamide is effective therapy for systemic necrotizing vasculitides other than SRV [61].

We have used both oral as well as intravenous pulse therapy. The same total dose of cyclophosphamide was used, 15 mg/kg, but spread over three consecutive days (i.e. at 5 mg/kg/day). Each pulse of cyclophosphamide is accompanied by a pulse of prednisolone given at a total dose of 10 mg/kg, again spread over three consecutive days (i.e. 3.3 mg/kg/day). In our retrospective review of systemic necrotizing vasculitis [35], we were unable to demonstrate any differences in outcome between different regimens of cyclophosphamide.

We therefore attempted to compare the efficacy and toxicity of pulse vs continuous cyclophosphamide (and prednisolone) in treatment of

**Table 23.1** Dose schedules for high-dose intermittent cyclophosphamide and prednisolone with alterations for age, renal function and marrow toxicity

| | Cyclophosphamide | | Prednisolone |
|---|---|---|---|
| Usual dose* | 15 mg/kg (up to 1 000 mg/bolus) | 10 mg/kg | (up to 1 000 mg/bolus) |
| Renal impairment For serum creatinine | | | |
| < 150 µmol/l | 15 mg/kg | 10 mg/kg | |
| 150–250 µmol/l | 10 mg/kg | 10 mg/kg | |
| 251–500 µmol/l | 7.5 mg/kg | 10 mg/kg | |
| > 500 µmol/l | 5 mg/kg | 7 mg/kg | |
| Marrow toxicity | Stop until blood count returns to within normal range Reduce dose by 25% | 10 mg/kg | |
| Previous cytotoxic agents | Stop previous drugs Wait two weeks before commencing bolus | 10 mg/kg | |
| Age > 70 years | 10 mg/kg | 10 mg/kg | |

* Each bolus dose is given either as an intravenous infusion in one day (prednisolone followed by cyclophosphamide) or orally as the same total dose as the i.v. regime, divided over a three-day period. Bolus doses are given at increasing intervals with clinical response.

systemic vasculitis in a randomized prospective trial. The end points included: survival, drug toxicity, primary treatment failure, relapse and renal function. All patients presenting with newly diagnosed classic polyarteritis, microscopic polyarteritis, Wegener's granulomatosis or limited Wegener's granulomatosis who were between the ages of 15 and 70 at entry were asked to provide informed consent to the study. Details of the two therapeutic regimens used are shown in Table 23.1. Treatment consisted of either pulses of cyclophosphamide (15 mg/kg) and prednisolone (10 mg/kg) at 0, 2, 4, 7, 10, 13, 17, 21, 25, 30, 35, 40, 46 and 52 weeks (n = 20) or continuous cyclophosphamide (2.0 mg/kg) for three to six months followed by azathioprine (1.5 mg/kg) and prednisolone (initial dose 0.85 mg/kg) (n = 27). Escalation therapy consisting of plasma exchange, IVIG or further pulses of steroids and cyclophosphamide were given if treatment was judged ineffective at the discretion of the attending physician. Analysis of the results of this study (which is

**Table 23.2** Diagnosis of patients treated with either pulse or continuous cyclophosphamide and prednisolone (figures represent number of patients in each group)

| Patients | Pulse | Continuous |
|---|---|---|
| Wegener's granulomatosis (WG) | 10 | 9 |
| Limited Wegener's granulomatosis (LWG) | 4 | 1 |
| Polyarteritis nodosa (PAN) | 2 | 5 |
| Microscopic polyarteritis (MPA) | 4 | 12 |
| Total | 20 | 27 |

**Table 23.3** Demographic details, initial renal function and initial renal histology of patients treated with either pulse or continuous cyclophosphamide and prednisolone (see Table 23.2 for abbreviations)

| Patients | Pulse | Continuous |
|---|---|---|
| All patients | | |
| n | 20 | 27 |
| Median age | 48.5 (22–70) | 62 (15–70) |
| Male:female ratio | 15:5 | 16:11 |
| MPA and WG groups | | |
| Serum creatinine | 180.5 (72–938) | 370 (60–1082) |
| % glomeruli | | |
| Active | 22 (3–79) | 45 (9–68) |
| Chronic | 15 (2–73) | 21 (3–76) |

still ongoing) are shown in Tables 23.2 to 23.5. The two groups were of similar age, sex and disease duration. Renal function and renal histological abnormalities were worse (but not significantly so) in the group treated with continuous cyclophosphamide. Forty-six patients were randomized to the study, of whom 38 have been followed-up for more than one year. The diagnoses were as follows: Wegener's granulomatosis (n = 24), microscopic polyarteritis (n = 16), and PAN (n = 7). After one year follow-up, there were no significant differences in mortality (pulse three vs continuous three). Kaplan–Meier survival curves were similar in both groups (Figure 23.1). Requirement for escalation therapy was similar in both groups (pulse: plasma exchange in five, IVIG in four, IV prednisolone in one; continuous: plasma exchange in 12, IVIG in five, i.v. prednisolone in seven). Primary treatment failure (three vs two) and relapses (five vs six) were similar in both groups; chronic dialysis (one vs two); and hemorrhagic cystitis (zero vs zero). There were more frequent episodes of clinically important infection in the continuous group (17 vs eight). These were often of the lower respiratory tract or in the form of septicemias. There was no significant difference in the systemic vasculitis activity scores (16) between the treatment groups.

**Table 23.4** Follow-up and outcome data of patients treated with either pulse or continuous cyclophosphamide and prednisolone

|  | *Pulse* | *Continuous* |
| --- | --- | --- |
| Follow-up (months) | 17.5 (< 1–40) | 24 (2.5–41) |
| Deaths | 3 | 3 |
| Treatment failure | 3 | 2 |
| Relapse | 5 | 6 |
| Chronic dialysis | 1 | 2 |
| *Use of escalation therapy (figures represent number of patients in each group)* | | |
| *n* | 20 | 27 |
| Plasma exchange | 5 | 12 |
| i.v. Prednisolone | 1 | 7 |
| IVIG | 4 | 5 |

**Table 23.5** Toxic effects in patients treated with either pulse or continuous cyclophosphamide and prednisolone (figures represent number of patients in each group)

|  | *Pulse* | *Continuous* |
| --- | --- | --- |
| Marrow suppression | 4 | 10 |
| Infection | 8 | 17 |
| Hemorrhagic cystitis | 0 | 0 |
| Neoplasia | 0 | 0 |

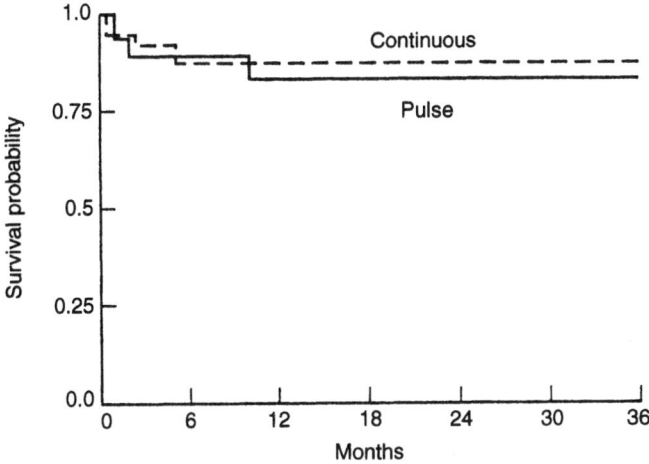

**Figure 23.1** Kaplan–Meier survival curves for mortality in necrotizing vasculitis for patients treated with either pulse or continuous cyclophosphamide and prednisolone. Note the significant difference between the two curves.

The use of intravenous bolus cyclophosphamide combined with a single dose of methylprednisolone appears to have definite benefit in a wide variety of types of necrotizing vasculitis. It produces rapid improvement and is still the simplest way to initiate therapy in severely ill patients. However, the use of intermittent oral pulses is much easier to manage on a long-term basis from patients' point of view. This appears to diminish the side-effects except the gastrointestinal ones, which can be controlled with an anti-emetic plus an $H_2$-antagonist. Our study suggests that a combination of i.v. and oral pulses is as effective as continuous low-dose oral cyclophosphamide and predniso- lone at inducing remission of vasculitis. The duration between pulses (starting at two weeks) as well as the concomitant use of pulse predni- solone are significantly different from the regimen employed by Hoff- man et al. [61]. These differences may well explain why our results with pulse therapy are better than those of the NIH group.

## Mechanisms of action of cyclophosphamide

Cyclophosphamide is effective in patients with a variety of necrotizing vasculitides with widely different pathogeneses. It is likely that the cyclophosphamide is affecting the cells mediating the vascular damage in these diseases rather than the underlying trigger factors. Potential targets would include polymorphonuclear leukocytes, natural killer cells, activated cytotoxic CD8+ cells, as well as B cells. Cyclophospha- mide may have a direct effect on endothelial cells both to diminish surface receptor expression of adhesion molecules and to diminish cytokine release.

## Side-effects

Toxic effects of cyclophosphamide are well recognized. The bladder toxicity is well known, due chiefly to the effect of acrolein, one of its urinary metabolites. As well as causing hemorrhagic cystitis and bladder fibrosis, it has been associated with bladder carcinoma [62, 63].

Suppression of the immune system inevitably exposes patients to potentially serious infections. In 15 patients with various types of vasculitis treated with daily oral cyclophosphamide and steroids, 10 developed significant infections, of whom two died [37]. There did not appear to be any clear relationship between dose or degree of leukopenia and the infective episodes. Although this highlights the potential for severe infections, the proportion of patients with problems was considerably higher than our own experience and those of others [26].

*In vitro* studies of peripheral blood mononuclear cells and T cell clones derived from patients treated with cyclophosphamide demonstrate an increased frequency of hypoxanthine-guanine phosphoribosyl transferase mutants [64] and increased rates of sister chromatid exchange [65]. The implication of these observed abnormalities of DNA has not been established at a clinical level.

## OTHER IMMUNOSUPPRESSIVE THERAPIES IN VASCULITIS

### Azathioprine

The usual dose of azathioprine in vasculitis is 1.5–2 mg/kg/day often combined with prednisolone. Marrow suppression may result and requires dosage adjustment. We temporarily stop therapy and then reintroduce it at a 25% reduction of the previous dose when the total white cell count and platelet count reach the normal range. Other effects include gastrointestinal intolerance and, rarely, hepatotoxicity. Although there may be long-term effects including the induction of lymphoid tumors [66], the drug has been used safely in pregnancy [67].

The treatment of both polymyalgia rheumatica and giant cell arteritis is chiefly with prednisolone, although this probably has little effect on the underlying pathogenetic mechanisms, which may involve lymphocyte subset changes (CD4+ cells have been found in affected vessels [68], thus explaining the relapse rate on stopping steroid therapy [69]). Azathioprine has been used with considerable success in steroid-resistant cases [70]. Although it can induce remission of polyarteritis [20], azathioprine has been superseded by cyclophosphamide in the necrotizing vasculitides. Many groups are using it as a maintenance therapy, after cyclophosphamide, and this is currently the subject of our controlled, randomized prospective study.

## Cyclosporin A

Cyclosporin A originates from *Tolypocladium inflatum Gams* and in recent years has been widely used as an immunosuppressant agent [71]. Experience of its effects in patients with vasculitis remain limited to a handful of cases where other therapy had failed [72–74] apart from its more extensive use in Behçet's syndrome, especially in the presence of active uveitis [75, 76] where it has been shown to be more effective than pulse i.v. cyclophosphamide. The reluctance to use it more widely is largely as a result of its recognized toxic effects on the kidney. The frequent combination of cyclosporin A with other immunosuppressive agents make it difficult to delineate its toxic effects, but since it does not damage polymorphonuclear leukocytes or pulmonary alveolar macrophages [77], it is less likely to predispose to bacterial and fungal infections.

## Methotrexate

Methotrexate is widely used for the treatment of rheumatoid arthritis [78]. Its role in the treatment of systemic vasculitis is increasing with the search for less toxic, long-term maintenance therapies in vasculitis, given the initial good response to more aggressive cytotoxic regimens in the early stages of disease. It has been used to treat rheumatoid leg ulcers, the etiology of which may include vasculitis [79], although there is controversy over whether or not, like steroids, the drug may be involved in the development of SRV [80]. In Wegener's granulomatosis [81] and Takayasu's arteritis [82], methotrexate has successfully been introduced, largely in an attempt to reduce the dose of steroid necessary to control the above diseases. Most of these studies were on patients without life-threatening disease, for example those with severe renal involvement. It may, therefore, be useful in either moderately severe disease, or as a maintenance therapy following more intensive immunosuppression. Studies are needed to determine its effectiveness in this role.

## Therapeutic apheresis

Cytotoxic agents have wide-ranging, non-selective effects on the immune system. More selective therapy might diminish the side-effects and aid an understanding of disease mechanisms. Plasmapheresis has the attraction of immediately removing soluble mediators of inflammation from the circulation. Other potential benefits are less clear [83], and its efficacy in treating rheumatoid vasculitis [46] and non-virus associated PAN [84] has been challenged.

Hind *et al.* [85] showed that the recovery of renal function in patients with rapidly progressive glomerulonephritis who were dialysis dependent (but who did not have antiglomerular basement membrane disease) was improved if steroids and immunosuppressive agents were combined with plasma exchange. Unfortunately, the morbidity and mortality (chiefly due to sepsis) of this combination is well recognized [85–87].

CONCLUSIONS

In a prospective randomized parallel group trial comparing intermittent bolus with continuous oral cyclophosphamide (both with prednisolone) as initial therapy, we found no difference in mortality between the two treatment groups after one-year follow-up. Requirement for escalation therapy, primary treatment failure, relapse and requirement for chronic dialysis were similar in both groups. There were more frequent episodes of clinically important infection in the continuous group, often of the lower respiratory tract or in the form of septicemia. Although further follow-up will elucidate the long-term outcome of these regimens, these data strongly support the use of either regimen of cyclophosphamide as effective therapies for obtaining remission of disease. Further work should be aimed at testing the effectiveness of less toxic regimens in the maintenance phase of treatment, after the latter has been induced by cyclophosphamide.

Cyclophosphamide represents a real advance in the therapy of a wide range of vasculitides (particularly the necrotizing vasculitides which previously had a poor prognosis). Cyclophosphamide is usually combined with steroids and may be used in a variety of ways. Intermittent regimes can be given to increase the dose of cyclophosphamide at any one time, which may produce more rapid benefit. The increased interval between the doses may account for the diminished side-effects seen with the intermittent regimes. An intermittent intravenous cyclophosphamide and methylprednisolone combination can be used for seriously ill patients, but the dose needs to be tailored for diminished renal function. An oral intermittent regime forms a useful and easy-to-manage maintenance therapy. It should be stressed that the low-dose oral continuous regimens are equally successful in treating vasculitis. In cases responding well, the duration of total therapy is still undetermined. The risk of relapse must be balanced against the risks of therapy and this will vary for different diseases. Despite all of these advantages, the blanket immunosuppressive effect of cyclophosphamide is clearly a disadvantage. The future may lie in exploring combinations of cytotoxic drugs with more selective immunomodulating actions.

## REFERENCES

1. Kussmaul, A. and Maier, R. (1866) Über eine bisher nicht beschreibene eigenthümliche Arterienerkrankung (Periarteritis nodosa), die mit Morbus Brightii und rapid fortschreitender allgemeiner Muskellähmung einhergeht. *Dtsch. Arch. Klin. Med.*, **1**, 484–518.
2. Bywaters, E.G.L. (1976) Vasculitis in rheumatoid arthritis, in *Non-articular Forms of Rheumatoid Arthritis* (ed. T.E.W. Feltkamp), Proc IV ISRA Symposium, Stafleu's Scientific Publications, Leiden, pp. 82–4.
3. Sack, M., Cassidy, J.T. and Bole, G. (1975) Prognostic factors in polyarteritis. *J. Rheumatol.*, **2**, 411–20.
4. Scott, D.G.I., Bacon, P.A., Elliott, P.J. et al. (1982) Systemic vasculitis in a district general hospital 1972–1980: clinical and laboratory features, classification and prognosis of 80 cases. *Q. J. Med.*, **51**, 292–311.
5. Andrews, M., Edmunds, M. and Campbell, A. (1990) Systemic vasculitis in the 1980s – is there an increasing incidence of Wegener's granulomatosis and microscopic polyarteritis. *J. Royal Coll. Phys.*, **24**, 284–8.
6. Ball, J. (1954) Rheumatoid arthritis and polyarteritis nodosa. *Ann. Rheum. Dis.*, **13**, 277–90.
7. Sokoloff, L. and Bunim, J.J. (1957) Vascular lesions in rheumatoid arteritis. *J. Chronic Dis.*, **5**, 668–87.
8. Kulka, J.P. (1959) The vascular lesions associated with rheumatoid arthritis. *Bull. Rheum. Dis.*, **10**, 201–2.
9. Wegener, F. (1936) Über generalisierte septische Gefässerkrankungen. *Verh. Dtsch. Path. Gesellsch.*, **29**, 202–9.
10. Godman, G.C. and Churg, J. (1954) Wegener's granulomatosis: pathology and review of the literature. *A. M. A. Arch. Path.*, **58**, 533–53.
11. Davson, J., Ball, J. and Platt, R. (1948) The kidney in periarteritis nodosa. *Q. J. Med.*, **17**, 175–202.
12. Frohnert, P.P. and Sheps, S.G. (1967) Long term follow up study of periarteritis nodosa. *Am. J. Med.*, **43**, 8–14.
13. Zeek, P.M. (1952) Periarteritis nodosa: a critical review. *Am. J. Clin. Pathol.*, **22**, 777–90.
14. Jennette, J.C., Falk, R.J., Andrassy, K. et al. (1994) Nomenclature of systemic vasculitides. *Arthritis Rheum.*, **37**, 187–92.
15. Fan, P.T., Davis, J.A. and Somer, T. (1980) A clinical approach to systemic vasculitis. *Semin. Arthritis Rheum.*, **9**, 248–304.
16. Luqmani, R.A., Bacon, P.A. and Moots, R.J. (1994) Birmingham vasculitis activity score (BVAS) in systemic necrotizing vasculitis *Q. J. Med.*, **87**, 161.
17. Baggenstoss, A.H., Shick, R. M. and Polley, H.F. (1950) The effect of cortisone on the lesions of periarteritis nodosa. *Am. J. Pathol.*, **27**, 537–9.
18. Fauci, A.S. (1978) The spectrum of vasculitis. *Ann. Intern. Med.*, **89**, 660–76.
19. Fauci, A.S., Katz, P., Haynes, B.F. and Wolff, S.M. (1979) Cyclophosphamide therapy of severe systemic necrotizing vasculitis. *N. Engl. J. Med.*, **301**, 235–8.
20. Leib, E.S., Restivo, C. and Paulus, H.E. (1979) Immunosuppressive and corticosteroid therapy of periarteritis nodosa. *Am. J. Med.*, **67**, 941–7.
21. Scott, D.G.I. and Bacon, P.A. (1984) Intravenous cyclophosphamide plus

methylprednisolone in the treatment of systemic rheumatoid vasculitis. *Am. J. Med.*, **76**, 377–84.

22. Luqmani, R.A., Bacon, P.A., Beaman, M. *et al.* (1994) Classical versus non-renal Wegener's granulomatosis. *Q. J. Med.*, **87**, 161–7.

23. Novack, S.N. and Pearson, C.M. (1971) Cyclophosphamide therapy in Wegener's granulomatosis. *N. Engl. J. Med.*, **284**, 938–42.

24. Fauci, A.S. and Wolff, S.M. (1973) Wegener's granulomatosis: studies in eighteen patients and a review of the literature. *Medicine (Baltimore)*, **52**, 535–61.

25. Wolff, S.M., Fauci, A.S., Horn, R.G. and Dale, D.C. (1974) Wegener's granulomatosis. *Ann. Intern. Med.*, **81**, 513–25.

26. Fauci, A.S., Haynes, B.F., Katz, P. and Wolff, S.M. (1983) Wegener's granulomatosis: prospective clinical and therapeutic experience with 85 patients for 21 years. *Ann. Intern. Med.*, **98**, 76–85.

27. Carrington, C.B. and Liebow, A.A. (1966) Limited forms of angiitis and granulomatosis of Wegener's type. *Am. J. Med.*, **41**, 497–527.

28. Anderson, C.L. and Stravides, A. (1978) Rapidly progressive renal failure as the primary manifestation of Wegener's granulomatosis. *Am. J. Med. Sci.*, **275**, 109–12.

29. Woodworth, T.G., Abuelo, J.G., Austin H.A. III and Esparza, A. (1987) Severe glomerulonephritis with late emergence of classic Wegener's granulomatosis. *Medicine (Baltimore)*, **66**, 181–91.

30. DeRemee, R.A., McDonald, T.J. and Weiland, L.H. (1985) Wegener's granulomatosis: observations on treatment with antimicrobial agents. *Mayo Clin. Proc.*, **60**, 27–32.

31. West, B.C., Todd, J.R. and King, J.W. (1987) Wegener's granulomatosis and trimethoprim-sulphamethoxazole: complete remission after a twenty year course. *Ann. Intern. Med.*, **106**, 840–2.

32. Axelson, J.A., Clark, R.H. and Ancerewicz, S. (1987) Wegener's granulomatosis and trimethoprim-sulphamethoxazole. *Ann. Intern. Med.*, **107**, 600.

33. Reinhold-Keller, E., Beigel, A., Duncker, G. *et al.* (1993) *Clin. Exp. Immunol.*, **93** (Suppl. 1), 38.

34. Rose, G.A. and Spencer, H. (1957) Polyarteritis nodosa. *Q. J. Med.*, **26**, 43–81.

35. Gordon, M., Luqmani, R.A., Adu, D. *et al.* (1993) Relapses in patients with a systemic vasculitis. *Q. J. Med.*, **86**, 779–89.

36. Pickering, G., Bywaters, E.G.L., Danielli, J.F. *et al.* (1960) Treatment of polyarteritis nodosa. Results after three years: report to the Medical Research Council by the Collagen Disease and hypersensitivity panel. *Br. Med. J.*, **i**, 1399–400.

37. Bradley, J.D., Brandt, K.D. and Katz, B.P. (1989) Infectious complications of cyclophosphamide treatment for vasculitis. *Arthritis Rheum.*, **32**, 45–53.

38. Droz, D., Noel, L.H., Leibowitch, M. and Barbanel, C. (1979) Glomerulonephritis and necrotizing angiitis, in *Advanced Nephrology*, Vol. 8, Year Book Medical Publishers, Chicago, pp. 343–63.

39. Serra, A., Cameron, J.S., Turner, D.R. *et al.* (1984) Vasculitis affecting the kidney: presentation, histopathology and long term outcome. *Q. J. Med.*, **53**, 181–207.

40. Savage, C.O., Winearls, C.G., Evans, D.J. *et al.* (1985) Microscopic polyarteritis: presentation, pathology and prognosis. *Q. J. Med.*, **56**, 467–83.

41. Adu, D., Howie, A.J., Scott, D.G.I. *et al.* (1987) Polyarteritis and the kidney. *Q. J. Med.*, **62**, 221–37.

42. Ellman, P. and Ball, R.E. (1948) Rheumatoid disease with joint and pulmonary manifestations. *Br. Med. J.*, ii, 816–20.

43. Scott, D.G.I., Bacon, P.A. and Tribe, C.R. (1981) Systemic rheumatoid vasculitis: a clinical and laboratory study of 50 cases. *Medicine (Baltimore)*, **60**, 288–97.

44. Scott, D.G.I., Bacon, P.A., Allen, C. *et al.* (1981) IgG rheumatoid factor, complement and immune complexes in rheumatoid synovitis and vasculitis: comparative and serial studies during cytotoxic therapy. *Clin. Exp. Immunol.*, **43**, 54–6.

45. Allen, C., Elson, C.J., Scott, D.G.I. *et al.* (1981) IgG antiglobulins in rheumatoid arthritis and other arthritides: relationship with clinical features and other parameters. *Ann. Rheum. Dis.*, **40**, 127–31.

46. Scott, D.G.I., Bacon, P.A. and Bothamley, J.E. (1981) Plasma exchange in rheumatoid vasculitis. *J. Rheumatol.*, **8**, 433–9.

47. Luqmani, R.A., Moots, R.J. and Scott, D.G.I. (1994) Long term outcome in systemic rheumatoid vasculitis (SRV) *Br. J. Rheumatol.*, **33** (Suppl. 1), 138.

48. Kemper, J.W., Baggenstoss, A.H. and Slocumb, C.H. (1957) The relationship of therapy with cortisone to the incidence of vascular lesions in rheumatoid arthritis. *Ann. Intern. Med.*, **46**, 831–51.

49. Vollertsen, R.S., Conn, D.L. and Ballard, D.J. (1986) Rheumatoid vasculitis: survival and associated risk factors. *Medicine (Baltimore)*, **65**, 365–75.

50. Kahn, M.F., Bedoiseau, M. and deSeze, S. (1967) Immunosuppressive drugs in the management of malignant and severe rheumatoid arthritis. *Proc. R. Soc. Med.*, **60**, 130–3.

51. Nicholls, A., Snaith, M.L., Maini, R.N. and Scott, J.T. (1973) Controlled trial of azathioprine in rheumatoid vasculitis. *Ann. Rheum. Dis.*, **32**, 589–91.

52. Heurkens, A.H., Westedt, M.L. and Breedveld, F.C. (1991) Prednisone plus azathioprine treatment in patients with rheumatoid arthritis complicated by vasculitis. *Arch. Intern. Med.*, **151**, 2249–54.

53. Weisman, M. and Zvaifler, N.J. (1975) Cryoglobulinaemia in rheumatoid arthritis. Significance in serum of patients with rheumatoid vasculitis. *J. Clin. Invest.*, **56**, 725–39.

54. Churg, J. and Strauss, L. (1957) Allergic granulomatosis, allergic angiitis, and periarteritis nodosa. *Am. J. Pathol.*, **27**, 277–301.

55. Lanham, J.G., Elkon, K.B., Pusey, C.D. and Hughes, G.R.V. (1984) Systemic vasculitis in asthma and eosinophilia: a clinical approach to the Churg–Strauss syndrome. *Medicine (Baltimore)*, **63**, 65–81.

56. Finan, M.C. and Winkelmann, R.K. (1983) The cutaneous extravascular necrotizing granuloma (Churg–Strauss granuloma) and systemic disease: a review of 27 cases. *Medicine (Baltimore)*, **62**, 142–58.

57. Adu, D., Luqmani, R.A. and Bacon, P.A. (1993) Polyarteritis, Wegener's granulomatosis and Churg–Strauss syndrome, in *Oxford Textbook of Rheumatology* (eds P.J. Maddison, D. Isenberg, P. Woo and D.N. Glass), Oxford University Press, Oxford, pp. 846–59.

58. Hall, N.D., Bird, H.A., Ring, E.F.J. and Bacon, P.A. (1979) A combined clinical and immunological assessment of four cyclophosphamide regimes in rheumatoid arthritis. *Agents Actions*, **9**, 97–102.

59. Bacon, P.A. (1979) Circulating immune complexes in systemic rheumatoid disease. *Rheum. and Rehab.*, (Suppl.), 11–15.

60. Austin, H.A., Klippel, J.H., Balow, J.E. *et al.* (1986) Therapy of lupus nephritis: controlled trial of prednisolone and cytotoxic drugs. *N. Engl. J. Med.*, **314**, 614–19.

61. Hoffman, G.S., Leavitt, R.Y., Fleisher, T.A. *et al.* (1990) Treatment of Wegener's granulomatosis with intermittent high-dose intravenous cyclophosphamide. *Am. J. Med.*, **89**, 403–10.

62. Stillwell, T.J., Benson, R.C. Jr., deRemee, R.A. *et al.* (1988) Cyclophosphamide-induced bladder toxicity in Wegener's granulomatosis. *Arthritis Rheum.*, **31**, 465–70.

63. Hoffman, G.S., Kerr, G.S., Leavitt, R.Y. *et al.* (1992) Wegener's granulomatosis: an analysis of 158 patients. *Ann. Intern. Med.*, **116**, 488–98.

64. Palmer, R.G., Dore, S. and Denman, A.M. (1986) Cyclophosphamide induces more chromosome damage than chlorambucil in patients with connective tissue diseases. *Q. J. Med.*, **59**, 395–400.

65. Palmer, R.G., Smith-Burchnell, C.A. and Pelton, B.K. (1988) Use of T cell cloning to detect *in vivo* mutations induced by cyclophosphamide. *Arthritis Rheum.*, **31**, 757–61.

66. Rosman, M. and Bertino, J.R. (1973) Azathioprine. *Ann. Intern. Med.*, **79**, 694–700.

67. Schein, P.S. and Winokur S.T. (1975) Immunosuppressive and cytotoxic chemotherapy: long term complications. *Ann. Intern. Med.*, **82**, 84–95.

68. Andersson, R., Jonsson, R., Tarkowski, A. *et al.* (1987) T cell subsets and expression of immunological activation markers in the arterial walls of patients with giant cell arteritis. *Ann. Rheum. Dis.*, **46**, 915–23.

69. Dasgupta, B., Duke, O. and Timms, A. (1989) Selective depletion and activation of CD 8+ lymphocytes from peripheral blood of patients with polymyalgia rheumatica and giant cell arteritis. *Ann. Rheum. Dis.*, **48**, 307–11.

70. De Silva, M. and Hazelman, B.L. (1986) Azathioprine in giant cell arteritis/polymyalgia rheumatica in a double blind study. *Ann. Rheum. Dis.*, **45**, 136–8.

71. Kahan, B.D. (1989) Cyclosporine. *N. Engl. J. Med.*, **321**, 1725–38.

72. Wendling, D., Hory, B. and Blanc, D. (1985) Cyclosporine: a new adjuvant therapy for giant cell arteritis? *Arthritis Rheum.*, **28**, 1078–9.

73. Borleffs, J.C., Derksen, R.H. and Hene, R.J. (1987) Treatment of Wegener's granulomatosis with cyclosporin (letter). *Ann. Rheum. Dis.*, **46**, 175.

74. Gremmel, F., Druml, W., Schmidt, P. and Graninger, W. (1988) Cyclosporin in Wegener's granulomatosis. *Ann. Intern. Med.*, **108**, 491.

75. Özyazgan, Y., Yurdakul, S., Yazici, H. *et al.* (1991) Low dose cyclosporin A versus pulsed cyclophosphamide in Behçet's syndrome: a single masked trial, in *Behçet's Disease: Basic and Clinical Aspects* (eds J.D. O'Duffy and E. Kökmen), Marcel Dekker, New York, pp. 591–5.

76. Assaad-Khalil, S.H. (1991) Low dose cyclosporin in Behçet's disease, in *Behçet's Disease: Basic and Clinical Aspects* (eds J.D. O'Duffy and E. Kökmen), Marcel Dekker, New York, pp. 603–12.

77. Drath, D.B. and Kahan, B.D. (1984) Phagocytic cell function in response to immunosuppressive therapy. *Arch. Surg.*, **119**, 156–60.

78. Songsiridej, N. and Furst, D.E. (1990) Methotrexate – the rapidly acting drug. *Baillière's Clin. Rheumatol.*, **4**, 575–93.

79. Espinoza, L.R., Espinoza, C.G., Vasey, F.B. and Germain, B.F. (1986) Oral methotrexate therapy for chronic rheumatoid arthritis ulcerations. *J. Am. Acad. Dermatol.*, **15**, 508–12.

80. Scully, C.J., Anderson, C.J. and Cannon, G.W. (1991) Long-term methotrexate therapy for rheumatoid arthritis. *Semin. Arthritis Rheum.*, **20**, 317–31.

81. Hoffman, G.S., Leavitt, R.Y., Kerr, G.S. and Fauci, A.S. (1992) The treatment of Wegener's granulomatosis with glucocorticoids and methotrexate. *Arthritis Rheum.*, **35**, 1322–9.

82. Hoffman, G.S., Leavitt, R.Y., Kerr, G.S. *et al.* (1994) Treatment of glucocorticoid resistant relapsing Takayasu arteritis with methotrexate. *Arthritis Rheum.*, **37**, 578–82.

83. Lockwood, C.M., Woolledge, S. and Nicholas, A. (1979) Reversal of impaired spenic function of patients with nephritis or vasculitis (or both). *N. Engl. J. Med.*, **300**, 524–30.

84. Guillevin, L., Le Clerc, P., Cohen, P. *et al.* (1993) Treatment of severe polyarteritis nodosa without HBV infection and Churg–Strauss syndrome: a prospective trial in 57 patients comparing prednisone, pulse cyclophosphamide with and without plasma exchange. *Arthritis Rheum.*, **36** (Suppl. 1), S96.

85. Hind, C.R.K., Paraskevakou, H., Lockwood, C.M. *et al.* (1983) Prognosis after immunosuppression of patients with crescentic nephritis requiring dialysis. *Lancet*, **1**, 263–5.

86. Wing, E.J., Bruns, F.J., Fraley, D.S. *et al.* (1980) Infectious complications with plasmapharesis in rapidly progressive glomerulonephritis. *J.A.M.A.*, **244**, 2423–6.

87. Lhote, F., Guillevin, L., Leon, A. *et al.* (1988) Complications of plasma exchange in the treatment of polyarteritis nodosa and Churg–Strauss angiitis and the contribution of adjuvant immunosuppressive therapy: a randomised trial in 72 patients. *Artif. Organs.*, **12**, 27–33.

# 24

# Treatment of polyarteritis nodosa and Churg–Strauss syndrome

## Indications of plasma exchanges: meta-analysis of four prospective controlled trials

*L. Guillevin and F. Lhote*

SUMMARY

To define the most effective treatment for polyarteritis nodosa (PAN) and Churg–Strauss syndrome (CSS), we undertook four consecutive prospective therapeutic trials including 244 patients and tried to answer several important questions: should cyclophosphamide (CYC) be given as the first-line treatment? What is the place of plasma exchanges (PE) in the treatment of systemic vasculitis? Does hepatitis B virus (HBV) related PAN require specific treatment?

Our first randomized trial in 71 patients (1981–83) compared the association of CYC with corticosteroids (CS) and PE to CS and PE, in order to evaluate the efficacy of CYC given as the first-line treatment to control disease activity and subsequent survival of PAN and CSS patients. Between December 1983 and December 1988, we conducted two trials simultaneously: one aimed at patients without HBV markers and the second at patients with HBV markers. In 78 patients without HBV markers, we compared prednisone and PE to prednisone alone as the initial therapeutic regimen. In 33 patients with PAN related to HBV, a

*The Vasculitides.* Edited by B.M. Ansell, P.A. Bacon, J.T. Lie and H. Yazici.
Published in 1996 by Chapman & Hall, London. ISBN 978-0-412-64140-4.

new therapeutic strategy was applied as an alternative to long-term steroid and immunosuppressive therapy: short-term steroid therapy and PE were used to control the evolution of PAN and antiviral therapy was administered to suppress the etiological agent of the vasculitis. In the last protocol, which included 62 patients and addressed severe PAN without HBV markers or CSS, we showed that PE did not improve the prognosis and control of the disease.

Twelve years after the beginning of the trials on PAN and CSS patients, we think that the therapeutic strategy should be as follows:

- In PAN without HBV and CSS: Prednisone in association with CYC improves the control of the disease despite infectious side-effects which may be reduced by better CYC dose adaptation.
- In PAN related to HBV: the first-line treatment should be the association of antiviral agents and PE. This treatment was effective and cured a majority of patients within two to three months; half of them seroconverted. The length of HBV infection before its diagnosis, delay before initiation of treatment and previous immunosuppressive therapy led to a poor seroconversion rate.
- The role of PE in the treatment of systemic necrotizing vasculitis: PE are obviously useful in PAN related to HBV where immune complex deposition has been demonstrated. When PAN is not related to HBV and in CSS, even in severe cases, there is presently no argument supporting systematic administration of PE at the time of diagnosis.

## INTRODUCTION

Thirty years ago, treatment of PAN, a disease which was first described by Kussmaul and Maier [1], consisted primarily of CS [2], and the five-year survival rate with steroid treatment alone was 48%. Since 1979, retrospective studies have shown that adding immunosuppressive agents, especially CYC [3, 4], to the treatment was effective and improved the prognosis. Other authors [5, 6] did not report similar results with azathioprine used systematically at the onset of the disease. Several studies have also indicated that PE are effective in the treatment of PAN [6–8], although no data from a large series of patients are available. All these studies [2–9] were non-controlled and retrospective.

When we decided, between August 1980 and June 1993, to conduct prospective therapeutic trials in PAN and CSS patients, several questions were important: should CYC be given as the first-line treatment? What is the place of PE in the treatment of systemic vasculitis? Does PAN related to HBV require specific treatment including PE?

The association of PAN with HBV was described by Gocke *et al.* [10]

and Trepo and Thivolet [11]. The frequency of this association was confirmed by many reports, and PAN related to HBV accounts for one-third (36%) of all cases of systemic PAN in France [9]. In chronic hepatitis B without vasculitis, steroids had deleterious effects and were shown to enhance viral replication [12]. PAN related to HBV is characterized by the high rate of HBV replication, and HBV infection is usually chronic. CS that have proved effective against the symptoms of vasculitis may perpetuate chronic HBV infection. The evolutions of PAN and HBV infection are usually dissociated with recovery from vasculitis despite chronic HBV infection. Steroid-induced immunosuppression therefore facilitates progression to cirrhosis, which may be complicated in the future by hepatocellular carcinoma. In our experience, the persistence of chronic HBV infection can be associated with relapses of PAN as long as the HBV infection has not been cleared, and patients who recover from PAN can die of liver cirrhosis [13, 14]. In a retrospective study [13], we found that the one-year survival rates were significantly lower in patients with HBs antigenemia (70%) than in the absence of this infection (85%) ($p < 0.05$). This heightened mortality was related to gastrointestinal complications, including digestive bleeding and perforations which were observed more frequently in HBV-associated polyarteritis.

A new therapeutic strategy was therefore proposed [12, 15, 16] as an alternative to long-term steroid and immunosuppressive therapy: short-term steroid therapy and PE were used to control the evolution of PAN and antiviral therapy was administered in order to suppress the etiological agent of the vasculitis.

Thirteen years after the beginning of these trials, we can now define a therapeutic strategy for PAN and CSS and discuss the indications of CYC, PE and antiviral treatments.

PATIENTS AND METHODS

**Patients**

Criteria for entry into the studies were as follows:

1. age > 15 years and < 75 years;
2. systemic PAN with recent symptoms and diagnosed clinically by the presence of multiple system involvement;
3. histologic evidence of vascular lesions indicative of a diagnosis of vasculitis (focal or segmental vascular lesions, fibrinoid necrosis and/or pleomorphic inflammatory cell infiltration of the arterial wall) [17];
4. in the absence of histologic criteria, arteriographic evidence of vasculitis (microaneurysms, multiple stenoses and occlusions of medium-

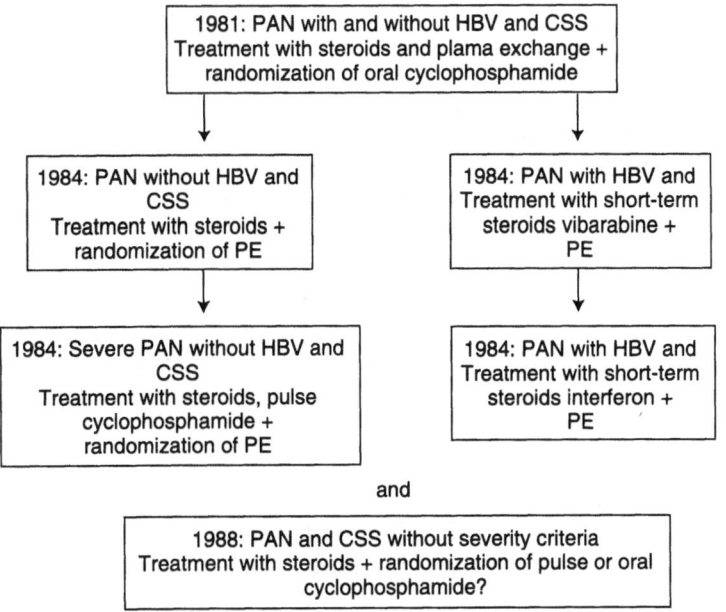

**Figure 24.1** General design of the protocols organized by the French Vasculitis Network.

sized arteries) [18], or adequation with the criteria defined by the American College of Rheumatology for PAN or CSS;
5. vasculitis sufficiently active to justify the use of corticosteroids. Since December 1983 patients' sera have been systematically screened to determine the existence of HBV replication [19]. The general design of the protocols is given in Figure 24.1

### Principles of treatment common to the protocols

*Long-term prednisone therapy*

In the trials in which it was indicated, prednisone was given at an initial dosage of 1 mg/kg/day for one month. In those patients who had clinical and laboratory evidence of improvement, the dosage was tapered according to the following system. The daily dose was decreased by 2.5 mg every 10 days for one month, and then by 2.5 mg every week until a level equivalent to half the initial dosage was reached. This dosage was maintained for three weeks and then further decreased by 2.5 mg every week, to approximately 20 mg/day. A more careful tapering schedule was followed for doses < 20 mg/day.

### Cyclophosphamide

Oral CYC was given at the dose of 2 mg/kg bw/day for one year. White blood counts were evaluated every two weeks for two months, then at least every month during the year of treatment so that the CYC dose could be adjusted to avoid severe neutropenia (defined as a neutrophil count < 1500 cells/mm$^3$). If neutropenia occurred, the CYC dose was reduced or temporarily stopped until the neutrophil count returned to normal. At this time, half the CYC dose was reintroduced. The dose was increased to the initial dose a fortnight later if neutrophils remained above 1500/mm$^3$. CYC was stopped after one year in order to avoid or reduce the incidence of long-term side-effects, such as leukemia or lymphoma, which can result from the use of immunosuppressive agents.

### Plasma exchanges

The amount of plasma scheduled for each session was 60 ml/kg of body weight. The replacement fluid consisted of 500 ml of fluid gelatin and 4% albumin. Fresh-frozen plasma was used in some situations, mainly for patients with decreased levels of coagulation factors. When PE side-effects occurred, or when venous access was not possible, exchanges were stopped or withdrawn entirely.

### Evaluation of disease activity

Disease activity was defined by clinical and laboratory criteria. The disease was considered controlled when the patient's general condition improved, no new clinical manifestations related to PAN appeared and the erythrocyte sedimentation rate (ESR) returned to normal. Stabilization or improvement (partial or total) in peripheral neuropathy and in renal and cardiac function (if abnormalities existed previously) were also necessary.

The patient was considered to be cured when the required criteria for control of the disease were met and maintained for at least 18 months after discontinuation of treatment: HBeAg/anti-HBeAb sero-conversion was not considered a criteria for complete recovery from PAN. The patient was considered to be in clinical remission when clinical symptoms were stable or improved and laboratory abnormalities regressed or disappeared under constant treatment. If there was no evidence of control of the disease activity under the assigned treatment or if relapse occurred, the trial medication was stopped and the patient was withdrawn from the study. Relapses were defined as new systemic manifestations of PAN or worsening of the initial symptoms of the disease.

## Statistical analysis

A chi-square test with Yates' correction for small numbers, when appropriate, was used for comparison between qualitative variables (for example, data on remission, relapses, withdrawal from the study) and Student's *t* test was used for comparison between quantitative variables expressed as means $\pm$ SD [20]. The actuarial method was used to construct life-table plots. The log rank statistical analysis was used to compare the survival curves. For the treatment groups, survival time was evaluated relative to the time of diagnosis.

## Protocols

### Protocol 1 (August 1980 to December 1983)

The trial evaluated the systematic use of CYC as the first-line treatment of PAN and CSS. Patients were consecutively enrolled and randomly assigned to one of the following treatment groups: prednisone and PE (group 1A), or prednisone, CYC and PE (group 1B). In this study, patients were included without consideration of HBV status.

Treatment protocol:

- prednisone: every patient took prednisone (according to the schedule described above);
- plasma exchanges: PE were performed in both groups (1A and 1B). For each patient, 13 PE were planned during the first six months of treatment (three during the first week, two during the second week, then one session 10, 15, 21 and 30 days later and one each month during the last four months);
- cyclophosphamide: oral CYC (2 mg/kg/day) was prescribed after randomization in group 1B and stopped after one year.

### Protocols 2 and 3 (December 1983 to December 1988)

Considering the results obtained in the interim analyses of protocol 1 at three years, we decided to stop that trial and to organize two new therapeutic trials. Patients enrolled in the studies were treated according to HBV status and CYC prescribed as the second-line treatment.

### Protocol 2

Patients without HBV infection were enrolled in the study to evaluate the systematic use of PE and the effectiveness of CYC as the second-line treatment of PAN and CSS. The study patients were randomly assigned to one of two treatment groups: prednisone plus PE (group 2A) or prednisone alone (group 2B).

Treatment protocol:

- prednisone: every patient took prednisone (according to the schedule described above);
- plasma exchanges: patients assigned to group 2A underwent 12 PE during the first two months of treatment (three during each of the first two weeks, two during the third week, then one session 10, 15, 21 and 30 days later);
- cyclophosphamide: oral CYC (2 mg/kg/day) was prescribed only in the case of failure of the assigned treatment.

*Protocol 3*

HBV-infected patients were enrolled in the study to evaluate the effectiveness and the tolerance of antiviral agents associated with short-term immunosuppression for the treatment of PAN related to HBV. The presence of HBs and HBe antigens with active replication of the virus was required for inclusion.

Treatment protocol:

- prednisone: every patient took prednisone at a dose of 1 mg/kg/day during the first week of treatment. Steroids were stopped at the end of the second week;
- vidarabine (Vira A): after stopping prednisone, every patient received Vira A for three weeks. Vira A was administered by continuous intravenous infusion at a dose of 15 mg/kg/day during the first week and then 7.5 mg/kg/day during the second and the third weeks. When HBeAg/anti-HBeAb seroconversion was not obtained within four months, the treating physician in charge could prescribe a second cycle of Vira A without PE, independent of the evolution of the PAN;
- plasma exchanges: every patient underwent PE. PE were begun just after the patient's inclusion in the study. During each of the first two weeks, three PE were performed. After stopping prednisone, every patient underwent 14 PE during the three weeks of treatment with Vira A (four during the first week, five during the second and the third weeks). After stopping Vira A, PE were performed three times a week for three weeks and then twice a week for two weeks. After this period, PE were performed once or twice a week depending upon the clinical results observed and then stopped;
- interferon-α 2b: since January 1987, interferon-α 2b has been administered as the second-line antiviral agent when HBeAg/anti-HBeAb seroconversion was not obtained within four months and when a second cycle of Vira A treatment was contraindicated (i.e. intolerance to Vira A, essentially severe myalgias). Interferon-α 2b was initiated at

a dose of 3 million units, three times a week for a maximum period of one year.

## Protocol 4

The purpose of this protocol was to evaluate the interest of PE in severe PAN without HBV infection and in CSS. Severity was assessed by at least one of the following criteria which were previously shown to be associated with a poor prognosis: age > 50 years, gastrointestinal (GI) tract involvement, cardiomyopathy, renal failure and/or glomerulonephritis, central nervous system (CNS) involvement, white blood cell count (WBC) > 12 000/mm$^3$. After inclusion, randomization assigned half the patients to receive PE: three sessions/week for three weeks; 60 ml/kg of plasma were removed and replaced with 4% albumin. On days 1, 2 and 3, every patient received methylprednisolone (15 mg/kg), then oral prednisone (1 mg/kg bw/d) for one month which was then tapered. Patients also received monthly pulse CYC (0.6 mg/m$^2$) for one year. End points were: death, recovery, remission and side-effects.

## RESULTS

### Clinical findings

Between August 1980 and December 1993, 244 patients were eligible for the prospective studies as follows:

- protocol 1: 71 patients (39 in group 1A and 32 in group 1B) between August 1980 and December 1983;
- protocol 2: 78 patients (36 in group 2A and 42 in group 2B) between January 1984 and December 1988;
- protocol 3: 33 patients between January 1984 and December 1988;
- protocol 4: 62 patients between January 1989 and June 1993.

### Clinical manifestations

The clinical manifestations of PAN and CSS are summarized in Table 24.1. These are symptoms that are classically encountered in PAN and CSS. While there were some differences between groups, they were not statistically significant for the majority of symptoms. However, we did notice that PAN related to HBV was never associated with asthma or other respiratory signs. Orchitis was observed in six of the 17 male patients with HBV markers included in protocol 3, in none of the 39 men included in trial 1 and none of 30 men in trial 2.

**Table 24.1** Main clinical and biological manifestations observed in polyarteritis nodosa and Churg–Strauss syndrome

| | Protocol 1 | Protocol 1 | Protocol 2 | Protocol 2 | Protocol 3 | Protocol 4 | Protocol 4 |
|---|---|---|---|---|---|---|---|
| Reference | 14 | | 20 | | 15 | Not published | |
| Time of inclusion | 1980–83 | 1980–83 | 1983–88 | 1983–88 | 1983–88 | 1989–1993 | 1989–1993 |
| n | n = 39 | n = 32 | n = 36 | n = 42 | n = 33 | n = 28 | n = 34 |
| Treatment group | CS + PE | CS + PE + CY | CS | CS + PE | Vira A + PE | CS + pCY | CS + pulse CY + PE |
| *Clinical* | | | | | | | |
| Fever | 51.2 | 71.8 | 66.6 | 61.9 | 79 | 50 | 61.7 |
| Weight loss | 59 | 65.6 | 61.1 | 64.2 | 52 | 71 | 79.4 |
| Peripheral neuropathy | 57.9 | 68.8 | 69.4 | 80.9 | 76 | 64 | 55.9 |
| Myalgia | 53.8 | 65.6 | 55.5 | 59.5 | 30 | 46 | 44 |
| Skin involvement | 46.2 | 59.4 | 50 | 52.3 | 21 | 46 | 35.3 |
| Arthralgia | 43.5 | 46.8 | 33.3 | 47.6 | 45 | 25 | 29.4 |
| Kidney involvement | 28.9 | 31.2 | 33.3 | 26.1 | 30 | 43 | 41.2 |
| Hypertension | 28.9 | 31.2 | 30.5 | 30.9 | 45 | 36 | 29.4 |
| Abdominal pain | 25.6 | 43.8 | 30.5 | 16.6 | 24 | 53.6 | 38.2 |
| Asthma | 17.9 | 25 | 19.4 | 23.8 | 0 | 29 | 17.6 |
| Lung involvement (except asthma) | | | | | 0 | | |
| Cardiac involvement | 12.8 | | 13.8 | 11.9 | 3 | 7.1 | 8.8 |
| Cardiac failure | 7.7 | 3.1 | | | 3 | | |
| Pericarditis | | 15.6 | | | 3 | | |
| *Biological* | | | | | | | |
| ESR | 61.5 | 68.7 | 66.6 | 73.8 | nd | 80 | 84 |
| ANCA (by immunofluorescence) | nd | nd | nd | nd | | 25 | 17.6 |
| HBV markers (HBe antigenemia) | 25.6 | 9.3 | 0 | 0 | 100 | 0 | 0 |

## Laboratory findings

### Eosinophilia

Eosinophilia was greater than $500/mm^3$ in 29 patients included in protocol 1, in 31 in protocol 2, in 10 in protocol 3 and in 11 in protocol 4. It was associated with asthma in 14 out of 28 cases in protocol 1, 16 out of 31 cases in protocol 2, in no cases out of 33 in protocol 3 and in 13 out of 56 in protocol 4.

### HBV serology

HBsAg and other evidence of HBV replication were found in 13 out of 71 patients in protocol 1, in every patient in protocol 3 and in no patients in protocols 3 and 4.

### Liver function at inclusion

Transaminases were twice the normal range in six cases. Hepatic cytolysis was not observed in protocol 2 and 4. In protocol 3 transaminases were elevated in 24 patients.

### Liver biopsies

In protocol 1, 20 out of 71 patients underwent liver biopsies for two reasons: diagnosis or evaluation of hepatic lesions; in two cases, PAN vascular lesions were observed; histologic evidence of acute hepatitis was noted in two cases, chronic hepatitis in four, steatosis in four and liver cirrhosis in one. The liver biopsy was normal in seven cases. No liver biopsy was performed in protocols 2 and 4. In protocol 3, at the time of diagnosis, a liver biopsy was performed in 18 out of 33 patients but only one provided evidence of vasculitis; histological evidence of acute or subacute hepatitis was found in six cases, persistent chronic hepatitis in five, chronic aggressive hepatitis in three, chronic lobular hepatitis in three and granulomatous hepatitis in one. They were not performed in the other patients because liver function tests were normal or numerous hepatic artery aneurysms were present.

### Renal function

Proteinuria was found in 28 out of 71 patients included in protocol 1: 20 patients were in renal failure and two were undergoing chronic hemodialysis. In protocol 2, 19 patients were in renal failure with serum creatinine levels > 1.6 mg/l; nine patients (seven in group 2A and two in group 2B) had severe renal disease with serum creatinine levels > 3 mg/dl, and seven needed hemodialysis; one of these nine patients (group 2A) died of a cerebral hemorrhage, two were started on long-term hemodialysis, five recovered normal or subnormal renal function and one successfully underwent renal transplantation. In protocol 3, six patients suffered from renal insufficiency with serum creatinine levels > 1.6 mg/dl; renal failure was always associated with multiple renal infarctions; two patients had glomerulonephritis, one of these with proteinuria of 7 g/24 h; a renal biopsy was contraindicated in all six because of multiple aneurysms of both kidneys; one of these patients died three months after inclusion in the study despite hemodialysis. In protocol 4, 26 patients presented kidney involvement including 17 with renal failure. Mean creatininemia was 126 μmol/l.

## Results of plasma exchanges

A total of 2160 PE were performed in 170 patients.

*Protocol 1*

A total of 813 PE were performed in 71 patients (11.4 ± 2.4 patients). The mean exchanged volume was comparable in both groups (3042 ± 619 ml in group 1A, 3236 ± 807 ml in group 1B). Centrifugation was used in 678 PE sessions (83.4%) and filtration in 128 (15.7%) (no data were available on the technique used in seven cases). The replacement fluid was 4% albumin in 745 sessions and fresh-frozen plasma in 115; eight patients received both during 47 sessions.

*Protocol 2*

A total of 412 PE were performed in 36 patients. The mean volume exchanged was 3362 ml (range: 2000–5000 ml).

*Protocol 3*

A total of 662 PE were performed in 33 patients. The mean number of PE was 23 (range: 6–34). The mean volume exchanged was 3212 ml. Only one patient needed to supplement replacement fluid with fresh-frozen plasma because of a sharp decrease in clotting factors.

*Protocol 4*

A total of 256 PE were performed in 30 patients. PE were stopped after the fourth session in one patient for hemodynamic intolerance and in one patient for hematoma at the puncture site.

## Results of follow-up

### Protocol 1

The absence of CYC's superior efficacy on survival and the superiority of CYC, prednisone and PE to control the disease were taken into account by the Coordinating Committee in its decision to stop the trial before the calculated adequate sample size was reached. Recruitment was stopped after 71 patients had entered the study. The censoring date was 1 May 1987, 42 months after the last patient started treatment. The mean plus SD follow-up period was $39 \pm 19$ months (range: 1–78 months) for group 1A and $42 \pm 18$ months (range: 1–72 months) for group 1B.

### Protocol 2

The high relapse rate in both groups (10 in group 2A and eight in group 2B) and the absence of prednisone plus PE's superior efficacy on survival was taken into account to stop the trial before the calculated adequate sample size was reached. Recruitment was stopped after 78 patients had entered the study, since the interim analyses at three years demonstrated no greater efficacy of CS plus PE on disease control than CS alone. The censoring date for the trial was 31 December 1988, one month after the last patient started treatment. Patients were followed until the censoring date even after withdrawal for treatment failure or side-effects, or until death occurred. Mean follow-up was $42.2 \pm 23$ months (range: 1–82 months) for group 2A and $45.8 \pm 23.7$ months (range: 1–83 months) for group 2B.

### Protocol 3

Recruitment was stopped after 33 patients had entered the study. The censoring date for the trial was 31 December 1988, which was one month after the last patient started treatment. Follow-up continued until the censoring date, even after withdrawal from the study because of treatment failure or side-effects, or until the patient's death. The mean plus SD follow-up period was $58.35 + 28$ months (range: 2–93 months).

### Protocol 4

Recruitment in this trial was closed in December 1993. At present the preliminary results are available in 62 patients. The mean follow-up period is $28 \pm 16$ months.

**Outcome**

*Protocol 1*

In 10 patients (nine in group 1A, one in group 1B; $p < 0.05$), the study therapy failed to control the disease activity, and the patients worsened despite therapy. Among the nine group 1A patients, four received CYC and CS were increased as for the group 1B patient. Relapses were significantly higher in group 1A (15 patients, 38.5%) than in group 1B (three patients, 9.4%) ($p < 0.001$). In patients from group 1A, higher doses of steroids were given to 13 patients. CYC was introduced in four cases and PE were reinstigated in three cases. Among the 15 patients from group 1A who relapsed, a second remission was obtained in 10 patients, but three of them developed end-stage renal disease which required chronic hemodialysis; three patients died, two of them from acute manifestations of systemic vasculitis; the other two patients were still in relapse at the time of the study analysis. In group 1B, relapses were rare: two patients underwent remission with CS, one of them died after a third relapse despite high doses of CS, PE and CYC; the other one died from hepatic encephalopathy due to liver cirrhosis of unknown origin 48 months after the PAN onset. A third patient developed a middle-stage renal disease and was still alive at the time of analysis. Among the 52 (73.2%) patients still alive at the end of the study, 27 (38%) recovered completely with no clinical and biologic symptoms of systemic vasculitis after having stopped treatment for at least 18 months. Twenty-five patients (35.2%) no longer had manifestations of PAN or CSS and were considered to be in remission, but they remained under low-dose steroid treatment (approximately 10 mg/day).

*Protocol 2*

One year after the start of the treatment, the average steroid dose was 10 mg/day in group 2A and 13.7 mg/day in group 2B ($p$: not significant). A high relapse rate was observed in both groups (10 in group 2A, eight in group 2B). In 26 patients (14 in group 2A, 12 in group 2B), the randomized treatment failed to control disease activity and PAN or CSS deteriorated or relapsed despite treatment. The steroid dosage was increased for 10 patients (four in group 2A, six in group 2B). Sixteen patients (10 in group 2A, six in group 2B) received another treatment: in group 2A, oral CYC (2 mg/kg/day) was added for nine patients who had not improved and dapsone (100 mg/day) for one; in group 2B, CYC was added for five patients and dapsone for one. Nine of these 26 patients (five in group 2A, four in group 2B) continued to deteriorate and died. Two of them had received CYC in association with CS, and one had simultaneously been treated with PE. Among the 63 patients still alive at the end of the study, 56 (71.8%) recovered completely with no remaining clinical and biologic symptoms of systemic vasculitis after having stopped treatment for at least 18 months. Seven (9%) no longer experienced manifestations of PAN or CSS and were considered to be in remission, that is they remained under low-dose steroids which were necessary to control the disease and to avoid relapses. Among the 26 patients for whom the randomized treatment failed, nine of them died, 17 are alive, including 15 who have recovered and two who are in remission. Two patients are chronically hemodialyzed. In most cases, failure and death related to the vasculitis occurred during the first six months after the start of the assigned trial, except for two patients from group 2B who died respectively, 38 and 60 months after the beginning of the treatment. In both cases, cause of death was due to myocarditis with cardiac failure.

### Protocol 3

Among the 25 patients still alive at the end of the study, 24 (96%) recovered completely, with no clinical or laboratory evidence of systemic vasculitis after at least 18 months without treatment. One (4%) no longer had manifestations of PAN and was considered to be in remission; that is he remained on low-dose steroids (< 15 mg/day), which were necessary to control the disease and to avoid relapse. Among the eight patients who had previously been treated with CS and CYC, six recovered, one was in remission and one died. Among the 25 patients who had not undergone prior treatment with CS, 18 recovered and seven died. No relapses were observed during this prolonged follow-up. None of the seroconverters experienced HBV reactivation. Among the patients considered to be in remission, none developed new manifestations of PAN. Some patients presented sequelae of PAN: eight cases of hypertension still require treatment by beta-blockers or converting-enzyme inhibitors, three others suffer from moderate chronic renal insufficiency (serum creatinine < 2.3 mg/dl), and two patients are handicapped by peripheral neuropathy and one by optic neuropathy.

### Protocol 4

During the study a remission was obtained in 46 patients (19 in group A and 27 in group B). A relapse after an initial remission was observed in seven other patients (four in group A and three in group B). Relapses occurred three, nine, 12 and 18 months after the onset of treatment in group A, and four, 12 and 40 months in group B. A second remission was obtained in every case: under steroids alone in two patients, pulse CYC and prednisone in one, oral CYC and CS in three, CS plus oral weekly methotrexate in one patient.

### Virologic results and liver function follow-up

Fifteen patients included in protocol 1 and all protocol 3 patients were considered.

*Protocol 1*

Among the patients who presented PAN associated with HBV, only one patient seroconverted after three months of treatment with CS and PE.

*Protocol 3*

Within two weeks following the end of the Vira A and PE cycle, transaminases increased in all nine patients who underwent early ser-oconversion to anti-HBeAb. In most cases, cytolysis was of minor importance, but in three patients it increased by more than 10-fold over the normal range. These three patients developed jaundice: two of them recovered within one month but one patient died of fulminant hepatitis. After the first cycle of Vira A, 12 patients (36.4%) seroconverted to anti-HBeAb and five (15.2%) to anti-HBsAb. Three of the eight patients who had received steroids and eventually CYC before inclusion in this trial seroconverted to anti-HBeAb, as did nine of the 25 patients who had not been treated previously. Seroconversion was delayed and occurred in most cases a few months after the end of active therapy. Eight patients underwent a second cycle of Vira A administration four months after the end of the first one; two of them seroconverted 24 months after the entry into the study. Interferon-$\alpha$ 2b was given to four patients: one of them seroconverted to anti-HBe and anti-HBs Abs during the 12th month of interferon-$\alpha$ 2b treatment. Two other patients who had undergone a second cycle of Vira A administration lost HBeAg and no longer expressed serologic evidence of replication as assessed by HBV-DNA spot-test hybridization, but they have not yet produced anti-HBeAb. At the end of the study, 15 (45.5%) patients had seroconverted to anti-HBeAb and six (18.2%) had also seroconverted to anti-HBsAb. In all, 17 patients (51.5%) no longer expressed serologic evidence of HBV replication.

### Causes of death

*Protocol 1*

A total of 19 patients (26.8%) died during the study period: 11 group 1A patients (28.2%) and eight group 1B patients (25%) (*p*: not significant). The causes of death are given in Table 24.2. The median survival time was 80 months from the time of diagnosis. Survival curves showed that at five years, 72% of the patients in group 1A were alive versus 75% in group 1B (*p*: not significant); survival rates at seven years were the same.

## Protocol 2

A total of 15 patients (19.2%) died during the study period: six in group 2A (16.7%) and nine from group 2B (21.4%). The causes of death are given in Table 24.2. Survival curves show that at five years 83% of patients in group 2A were alive versus 79% in group 2B (*p*: not significant). Survival rates at seven years were the same.

## Protocol 3

A total of eight patients (24.2%) died during the study period. The causes of death are listed in Table 24.2. The survival curve shows that at seven years 76% of the patients were alive.

## Protocol 4

Nine patients died during the study period: five in group A (17.8%) and four in group B (11.8%). The causes of death are given in Table 24.2. Kidney involvement was not associated with a mortality increase (0 out of 12 in group A and two out of 14 in group B). Conversely, three of the four patients who presented abnormal surgery for GI tract involvement related to PAN died.

## Treatment side-effects

### Protocol 1

Side-effects of the steroid treatment were diffuse osteoporosis in four patients, aseptic necrosis of the femoral head in one patient, duodenal ulcers in one patient, gastric bleeding in five patients, pneumonia in nine patients (four in group 1A and five in group 1B), pulmonary miliary tuberculosis in one patient (group 1B) and septicemia in two patients. CYC side-effects included: neutropenia (*n* = 17) with septicemia (*n* = 1), toxic hepatitis (*n* = 1) and hemorrhagic cystitis (*n* = 4). Two hundred and fifty-one complications were reported in 60 patients during 206 (25.3%) of the 813 completed PE: 47 sessions (5.8%) were temporarily stopped as a result of complications. The most common problems were technical difficulties (in 90 sessions), moderate or severe hypotension (in 52 sessions) and allergy to the replacement fluid (in 51 sessions). Hepatitis B antigen was iatrogenically transferred by PE into one patient. In four

**Table 24.2** Cause and time of deaths during the study period

| Protocol | Protocol 1 | Protocol 1 | Protocol 2 | Protocol 2 | Protocol 3 | Protocol 4 | Protocol 4 | Total |
|---|---|---|---|---|---|---|---|---|
| Reference | 14 | | 20 | | 15 | not published | | |
| Time of inclusion | 1980–83 | 1980–83 | 1983–88 | 1983–88 | 1983–88 | 1989–93 | 1989–93 | |
| Number (%) | 11 (28.2) | 8 (25) | 6 (16.6) | 9 (21.4) | 8 (24.2) | 5 (17.8) | 4 (11.8) | 51 (20.9) |
| *Systemic vasculitis* | 3 | 2 | 4 | 5 | 3 | 1 | 2 | 20 |
| Bowel infarction/ perforation | 2 | 0 | 1 | 1 | 1 | 1 | 2 | 8 |
| Hemoperitoneum | 0 | 0 | 0 | 1 | 0 | 0 | 0 | 1 |
| Cardiac insufficiency | 1 | 0 | 1 | 2 | 0 | 0 | 0 | 4 |
| Multivisceral involvement | 0 | 2 | 0 | 0 | 1 | 0 | 0 | 3 |
| Respiratory failure | 0 | 0 | 0 | 1 | 0 | 0 | 0 | 1 |
| Stroke | 0 | 0 | 2 | 0 | 0 | 0 | 0 | 2 |
| Renal failure | 0 | 0 | 0 | 0 | 1 | 0 | 0 | 1 |
| *Infections as side-effects* | 3 | 2 | 0 | 1 | 0 | 1 | 0 | 7 |
| Septicemia | 3 | 1 | 0 | 1 | 0 | 1 | 0 | 6 |
| Tuberculosis | 0 | 1 | 0 | 0 | 0 | 0 | 0 | 1 |

| *Other causes* | | | | | | | |
|---|---|---|---|---|---|---|---|
| Myocardial infarct | 5 | 4 | 2 | 3 | 5 | 3 | 2 | 24 |
| Sudden death (unknown) | 0 | 0 | 0 | 0 | 0 | 1 | 0 | 1 |
| Shock (unknown) | 2 | 0 | 0 | 0 | 3 | 1 | 0 | 6 |
| Lymphoma | 0 | 0 | 0 | 0 | 0 | 1 | 1 | 2 |
| Cancer | 0 | 0 | 0 | 0 | 0 | 0 | 1 | 1 |
| Liver cirrhosis (post-HBV) | 1 | 1 | 1 | 1 | 0 | 0 | 0 | 4 |
| Fulminant hepatitis | 0 | 2 | 0 | 0 | 0 | 0 | 0 | 2 |
| Pulmonary embolism | 0 | 0 | 0 | 0 | 1 | 0 | 0 | 1 |
| Traffic accident | 1 | 0 | 0 | 1 | 0 | 0 | 0 | 2 |
| Suicide | 1 | 0 | 0 | 0 | 0 | 1 | 0 | 1 |
| Unknown | 0 | 1 | 0 | 0 | 0 | 0 | 0 | 1 |
| Cerebral bleeding | 0 | 0 | 1 | 1 | 0 | 0 | 1 | 2 |
| | 0 | 0 | 0 | 0 | 1 | 0 | 0 | 1 |

patients, PE were stopped permanently because of the severe side-effects (thrombosis: one; circulatory collapse: one; extensive hematoma: one; pulmonary embolism due to peripheral thrombosis: one). No patient died during a session. The side-effects and complications due to PE were usually mild and transient.

### Protocol 2

Side-effects of the steroid treatment were severe diffuse osteoporosis in two patients, aseptic necrosis of the femoral head in two patients, aseptic necrosis of the humeral head in one patient, duodenal ulcers in two patients and pneumonia in one patient. During PE, eight complications were reported in seven patients. No serious side-effects were observed in patients treated with dapsone.

### Protocol 3

No side-effects of the steroid treatment were noted. Complications of PE were transient: hemolysis occurred in two patients; one arteriovenous fistula thrombosis was recorded. One patient who developed angina during a session later had a myocardial infarction; the relationship between PE and myocardial infarction was not clearly established. During Vira A administration, three patients experienced transient thrombocytopenia (50 000 < platelets < 100 000/mm$^3$) at the end of the second week of treatment. The drug was withdrawn and readministered later, when platelet levels had returned to normal.

### Protocol 4

In addition to the PE side-effects reported above, two patients from the group without PE died from tuberculosis, and in the group with PE, one patient presented a pneumopathy due to *Klebsiella pneumoniae*.

### DISCUSSION

For several decades, an overall improvement in the prognosis for PAN and CSS has been observed as a consequence of the systematic use of steroids and extensive use of CYC [2–5]. Results obtained with CS and CYC are controversial. Fauci *et al.* [4] demonstrated that CYC was able to improve survival in PAN when steroids and other immunosuppressive treatments failed. Analysis of a large series of patients using immunosuppressive therapy showed that survival rates were similar to those obtained with steroids alone: 55% in the study by Cohen, Conn and Ilstrup [5] and 58% in our previous study [9]. In contrast, Leib, Restivo and Paulus [3] demonstrated the superiority of the association of cyto-toxic agents with steroids to control PAN: they obtained a five-year survival rate of 53% with steroids and 80% with the association of cytotoxic agents and CS.

In the early 1980s, PE were widely used as a treatment for PAN [6–8, 19]. It was demonstrated that PE were effective and that they were able to successfully treat patients with PAN related to HBV [19] and to improve the course of PAN after failure of steroids and CYC [7]. PE are able to remove immune complexes and to improve the capacities of the reticuloendothelial system to clear these complexes [8].

The results of the first protocol [14] showed that there was no differ-ence in the 10-year survival rate between the patients treated with CS and PE and those treated with CS, CYC and PE, despite better control of the disease activity in group 1B (CYC as the first-line treatment) and a higher number of relapses in group 1A.

In protocol 2 [20], our purpose was to test the indications of PE and to evaluate if this treatment was able to improve the control of the disease and the survival rate. We demonstrated that PE did not improve the results in the group that underwent this treatment. The results were not significantly different from those who took CS alone. The failure of therapy to control the disease activity was observed more often than in our previous study in which CYC was given initially in combination with CS ± PE. In contrast, lethal infections related to the therapy were rare in this study (one case), and the survival rate was higher than that seen in other studies including our earlier one: 83% vs 75% at seven years.

In protocol 4, we tested indications of PE in a subgroup of patients with severe PAN without HBV infection or with CSS. Severity has been assessed on criteria described in detail elsewhere [9] and summarized above. In this group of patients PE has not improved the prognosis and has not facilitated the control of the disease. We think that PE should be prescribed only in vasculitis failing to respond to steroids and cyclophos-phamide, and should not be prescribed as the initial treatment of this subgroup of vasculitis.

In PAN related to HBV, we proposed a regimen of vidarabine and PE after a short prednisone treatment. The overall therapeutic results obtained in this group of patients were excellent: 77% of the patients no longer had any symptoms of vasculitis following therapy and none relapsed during the prolonged follow-up period. In all, 17 patients (51.5%) no longer expressed serologic evidence of HBV replication. In most cases, seroconversion was obtained within a few weeks after stopping the treatment, but in five cases it occurred more than one year after the end of the Vira A plus PE cycle. In two patients, HBeAg/anti-HBeAb seroconversion was obtained 24 months after a second course of Vira A and, in one other, 65 months after stopping treatment; one patient seroconverted during interferon-α 2b therapy. We can argue in some cases that late HBeAg/anti-HBeAb seroconversion may not be the direct effect of the antiviral agent, but rather the natural outcome and/or the consequence of stopping CS, which have been demonstrated to delay HBeAg/anti-HBeAb seroconversion but whose withdrawal triggers it. These results are much better than those obtained with CS which rarely [13], if ever, allow seroconversion but, on the contrary, favor high levels of virus replication and thereby the development of chronic liver disease.

Twelve years of therapeutic trials in PAN and CSS enable us to propose the following strategies according to disease characteristics.

### In PAN without HBV and CSS

Prednisone in association with CYC improves the control of the disease despite infectious side-effects, which could be reduced with a better dose adaptation as a function of neutrophil and lymphocyte counts. We can expect that other modalities of treatment with CYC could favor better clinical results and a lower rate of infectious side-effects: we are presently conducting a trial to evaluate CYC pulses which might be more effective than oral administration, as was demonstrated in the treatment of systemic lupus erythematosus. No long-term side-effects were observed and no leukemia or lymphoma was attributed to the CYC treatment during long-term follow-up. Such complications did not occur in the group of patients who received CYC for only 12 months. Conversely, leukemia and carcinoma were observed in Wegener's granulomatosis patients [21] and in rheumatoid arthritis patients [22] who had received long-term treatments. Twelve months of CYC are enough in PAN. It is also possible that CYC could be more effective in some PAN subgroups, for instance those with clinical symptoms of poor prognosis. We are presently attempting to optimize CYC prescription in PAN and CSS, and thereby improve prognosis.

### In PAN related to HBV

The first-line treatment should be the association of antiviral agents and PE. This treatment is effective and cures a majority of patients within two to three months. Half of them seroconverted and are not exposed to future complications of HBV replication, such as liver cirrhosis. Unfortunately, it is not possible to clear the virus in every patient as was observed in the treatment of chronic hepatitis. The length of infection before the diagnosis, the lapse of time before the initiation of treatment, and previous immunosuppression lead to a poor seroconversion rate.

## The role of PE in the treatment of systemic necrotizing angiitis

PE are obviously useful in PAN related to HBV. Side-effects of PE were minor and transient. No patient developed long-term side-effects of PE, such as HBV or hepatitis C virus infections, because albumin was systematically used as the replacement fluid during PE. For PAN not related to HBV and CSS, there is presently no argument to support systematic prescription of PE at the time of diagnosis.

## APPENDIX

The following institutions and principal investigators comprise the Cooperative Study Group for Polyarteritis Nodosa and contributed to this study: M. Alcalay (Poitiers), P. Amarenco (Paris), B. Amor (Paris), G. Andreu (Bobigny), K.S. Ang (St-Brieuc), Ph. Arlet (Toulouse), P. Babinet (St-Denis), J. Baillet (Amiens), G. Baralis (Arles), J. Barrier (Nantes), J.M. Beaufils (Mayenne), M. Bentata-Pessayre (St-Germain), E. Bercoff (Rouen), J.F. Besancenot (Dijon), O. Blétry (Paris), M. Bourel (Rennes), Ph. Brissaud (Rennes), J. Brun (Caen), A. Bussel (Paris), J. Cabane (Paris), Ph. Cassan (Vichy), J.P. Cassuto (Nice), J.M. Chalopin (Dijon), B. Christoforov (Paris), F. Cordier (Laval), F. Coulomb (Dreux), D. Daupleix (Bobigny), J. Debray (Paris), J.F. Delfraissy (Clamart), F. Delrieu (Paris), O. Deshayes (Rouen), R. Desproges-Gotteron (Limoges), B. Desrumeaux (Rouen), G. Dorsit (Moulins), P. Dournovo (Eaubonne), P. Dreyfus (Bobigny), J.P. Ducroix (Amiens), P. Dujardin (Nice), Y. Echard (Montfermeil), E. Eliazewicz (Créteil), A. Faradji (Strasbourg), J. Fermanian (Necker, Paris), A. Fournier (Amiens), A. Franco (Grenoble), J.G. Fuzibet (Nice), S. Gié (Rennes), P. Godeau (Paris), P. Grosbois (Rennes), C. Guillemot (La Roche-sur-Yon), L. Guillevin (Bobigny), R. Himler (Strasbourg), M. Huart (La Roche-sur-Yon), G. Huchon (Paris), J.N. Hugues (Bobigny), B. Hurault de Ligny (Caen), J.M. Idatte (Paris), J.C. Imbert (Paris), J. Ph. Jais (Paris), G. Janin (Macon), V. Jeantils (Bondy), J. Jouquan (Brest), M.F. Kahn (Paris), D. Laplane (Paris), C. Laroche (Paris), P. Lebon (Le Mans), J. Lecoz (Paris), V. Lemaître (Valenciennes), A. Leon (Bobigny), A. Léonard (Elbeuf), Le Thi Huong Du (Paris), F. Liozon (Limoges), R. Makdassi (Amiens), J. Mallecourt (Dreux), H. Mallet (Thonon-les-Bains), J.C. Marche (Clermont-Ferrand), A. Marsac (Paris), R. Marteau (Paris), D. Molle (Paris), M. Mougeot-Martin (Creil), G. Osterman (Reims), G. Pagniez (Creil), J. Pasquier (Lyon), P. Pasquier (Paris), Y. Pennec (Brest), B. Pépin (Paris), A.M. Piette (Suresnes), J.C. Piette (Paris), B. Pinel (Rouen), T. Ponge (Nantes), J. Pourrat (Toulouse), O. Pourrat (Poitiers), J.F. Quaranta (Nice), F. Rossi (Vendôme), B. Roualdes (Créteil), J.C. Roujeau (Créteil), H. Rousset (St-Étienne), M. Ruel (Senlis), L. Sary (Chartres), A. Schaeffer

(Créteil), A. Schannen (Paris), J. Sebaoun (Bobigny), D. Sereni (Paris), J.M. Simon (St-Brieuc), S. Smail (Amiens), A. Sobel (Créteil), Y. Tanter (Dijon), B. de Toffol (Tours), C. Trepo (Lyons), R. Treves (Limoges), M. Vantderstigel (Créteil), P. Veyssier (Compiègne), E. Vidal (Limoges), D. Vital-Durand (Lyons), B. Wechsler (Paris), B. You (Nancy), J.M. Ziza (Paris).

ACKNOWLEDGEMENTS

This work was based on prospective trials which were supported by grants from the Institut National pour la Santé et la Recherche Médicale (INSERM), the Caisse Nationale d'Assurance-Maladie des Travailleurs Salariés (CNAMTS) and the Association pour la Recherche sur les Angéites Nécrosantes (ARAN). Research was carried out with the help of the Cooperative Study Group for Polyarteritis Nodosa (see Appendix) and the Sociéte Nationale Française de Médecine Interne (SNFMI).

REFERENCES

1. Kussmaul, A. and Maier, K. (1866) Über eine bisher nicht beschreibene eigenthümliche Arterienerkrankung (Periarteritis nodosa), die mit Morbus Brightii und rapid fortschreitender allgemeiner Muskelhämung einhergeht. *Dtsch. Arch. Klin. Med.*, **1**, 484–518.
2. Frohnert, P. and Sheps, S. (1967) Long term follow-up study of polyarteritis nodosa. *Am. J. Med.*, **48**, 8–14.
3. Leib, E., Restivo, C. and Paulus, H. (1979) Immunosuppressive and corticosteroid therapy of polyarteritis nodosa. *Am. J. Med.*, **67**, 941–7.
4. Fauci, A., Katz, P., Haynes, B. and Wolff, S. (1979) Cyclophosphamide therapy of severe necrotizing vasculitis. *N. Engl. J. Med.*, **301**, 235–8.
5. Cohen, R., Conn, D. and Ilstrup, D. (1980) Clinical features, prognosis and response to treatment in polyarteritis. *Mayo Clin. Proc.*, **55**, 146–5.
6. Guillevin, L., Tanter, Y., Bletry, O. *et al.* (1983) Treatment of severe polyarteritis nodosa with plasma exchange. *Prog. Artif. Organs*, **204**, 723–6.
7. Bletry, O., Bussel, A., Badelon, I. *et al.* (1982) Intérêt des échanges plasmatiques au cours des angéites nécrosantes. Onze cas. *Nouv. Presse Méd.*, **11**, 2827–32.
8. Lockwood, C., Rees, A. and Pinching, A. (1977) Plasma exchanges and immunosuppression in the treatment of fulminating immune complex crescentic nephritis. *Lancet*, **1**, 63–7.
9. Guillevin, L., Le, T.H.D., Godeau, P. *et al.* (1988) Clinical findings and prognosis of polyarteritis nodosa and Churg–Strauss angiitis: a study in 165 patients. *Br. J. Rheumatol.*, **27**, 258–64.
10. Gocke, D., Hsu, K., Morgan, C. *et al.* (1970) Association between polyarteritis and Australia antigen. *Lancet*, **2**, 1149–53.
11. Trepo, C. and Thivolet, J. (1970) Antigen Australia, hépatite à virus et périartérite noueuse. *Presse Méd.*, **78**, 1575.
12. Trepo, C., Ouzan, D., Delmont, J. and Tremisi, P. (1988) Supériorité d'un nouveau traitement curateur des périartérites noueueses induites par le

virus de l'hépatite B grâce à l'association corticothérapie brève, vidarabine et échanges plasmatiques. *Presse Méd.*, **17**, 1527–31.

13. Guillevin, L., Le, T.H.D. and Gayraud, M. (1989) Systemic vasculitis of the polyarteritis nodosa group and infection with hepatitis B virus: a study in 98 patients. *Eur. J. Intern. Med.*, **1**, 97–105.

14. Guillevin, L., Jarrousse, B., Lok, C. *et al.* (1991) Longterm follow-up after treatment of polyarteritis nodosa and Churg–Strauss angiitis with comparison of steroids, plasma exchange and cyclophosphamide to steroids and plasma exchange. A prospective randomized trial of 71 patients. The Cooperative Study Group for Polyarteritis Nodosa. *J. Rheumatol.*, **18**(4), 567–74.

15. Guillevin, L., Lhote, F., Leon, A. *et al.* (1993) Treatment of polyarteritis nodosa related to hepatitis B virus with short-term steroid therapy associated with antiviral agents and plasma exchanges. A prospective trial in 33 patients. *J. Rheumatol.*, **20**, 289–98.

16. Wood, J., Czaja, A., Taswell, H. *et al.* (1987) Hepatitis B virus desoxyribonucleic acid in serum during hepatitis Be antigen clearance in corticosteroid-treated severe chronic active hepatitis B. *Gastroenterology*, **93**, 1225–30.

17. Arkin, A. (1930) A clinical and pathological study of polyarteritis nodosa. *Am. J. Pathol.*, **6**, 401–2.

18. Bron, K., Strott, C. and Shapiro, A. (1983) The diagnostic value of angiographic observations in polyarteritis nodosa. A case of multiple aneurysms in the visceral organs. *Arch. Intern. Med.*, **116**, 450–4.

19. Chalopin, J., Rifle, G., Turc, J. *et al.* (1980) Immunological findings during successful treatment of HBs Ag-associated polyarteritis nodosa by plasmapheresis alone. *Br. Med. J.*, **1**, 368.

20. Guillevin, L., Fain, O., Lhote, F. *et al.* (1992) Lack of superiority of steroids plus plasma exchange to steroids alone in the treatment of polyarteritis nodosa and Churg–Strauss syndrome. A prospective, randomized trial in 78 patients. *Arthritis Rheum.*, **35**, 208–15.

21. Hoffman, G.S., Kerr, G.S., Leavitt, R.Y. *et al.* (1992) Wegener's granulomatosis: an analysis of 158 patients. *Ann. Intern. Med.*, **116**, 488–98.

22. Kahn, M., Arlet, J., Bloch-Michel, H. *et al.* (1979) Leucémies aiguës après traitement par agents cytotoxiques en rhumatologie. 19 observations chez 2006 patients. *Presse Méd.*, **8**, 1393–7.

# 25

# Relapses of Wegener's granulomatosis: role of micro-organisms and prophylaxis with trimethoprim/sulfamethoxazole

*C.G.M. Kallenberg*

Wegener's granulomatosis (WG) is characterized by granulomatous necrotizing inflammation of the upper and lower respiratory tract in conjunction with systemic vasculitis and necrotizing crescentic glomerulonephritis (NCGN) [1]. In its extended presentation, all three elements are present (extended WG). In non-renal WG the kidney is not involved in the disease process, whereas limited WG is generally characterized by granulomatous and vasculitic lesions in the respiratory tract only without involvement of other major organ systems. Treatment with cyclophosphamide in combination with corticosteroids is very effective in most cases, although side-effects may be severe and sometimes lethal [2]. After remission is achieved, the course of the disease is highly variable and unpredictable. Most patients have relapses at variable intervals requiring reinstitution of immunosuppressive therapy.

The etiopathogenesis of the disease as well as the factors that underly the occurrence of relapses are presently unknown. It has been noticed that the disease frequently follows an indolent course for weeks to months or even years before the full-blown manifestations are apparent [3]. This prodomal indolent period is characterized, in many cases, by ongoing inflammation of the upper airways. These findings may suggest that infectious micro-organisms are involved in the pathogenesis of the

*The Vasculitides*. Edited by B.M. Ansell, P.A. Bacon, J.T. Lie and H. Yazici.
Published in 1996 by Chapman & Hall, London. ISBN 978-0-412-64140-4.

disease. In this chapter the role of micro-organisms in WG as well as the possible therapeutic potential of antibiotics, in particular trimethoprim/ sulfamethoxazole (TS), in the treatment of WG will be discussed.

INFECTION AND WG

It has been noticed by many physicians that new cases of WG do not present evenly distributed throughout the year. Falk *et al.* [4] analyzed the time points of presentation of their cases of WG, and found a predilection for the winter months during which 38% of their patients had the onset of symptoms, whereas the onset of symptoms during summer occurred in 11% of their patients only.

This may suggest that seasonal factors, one of which is possibly of microbial origin, are involved in the induction of WG. No particular micro-organism from either bacterial or viral origin has, however, been incriminated. It is of interest to note that the first description of anti-neutrophil cytoplasmic antibodies (ANCA) came from an Australian study reporting cases of necrotizing crescentic glomerulonephritis possibly in conjunction with an arbovirus infection [5].

Infections have been described in association with relapses of WG. Pinching *et al.* [6] reported that nine out of 20 relapses in patients with WG were provoked by bacterial or viral infection. Seven of these 20 relapses occurred following reduction in immunosuppressive drugs, whereas in four relapses no relation with either infection or reduction in immunosuppression was found. Infections were caused by different common micro-organisms both from bacterial (*Hemophilus influenzae, Staphylococcus aureus, Klebsiella aerogenes,* etc.) and viral (influenza B, respiratory syncytial virus) origin. In addition, infections were not restricted to one particular locality, but occurred in the respiratory tract as well as in other organ systems. Ronco *et al.* [7] noticed the occurrence of cytomegalovirus in the blood in a case of WG before treatment was started.

Stegeman *et al.* [8] recently observed that chronic nasal carriage of *Staph. aureus* is a substantial risk factor for the occurrence of relapses in WG. It proved that 36 out of 57 patients with WG (63%) were chronic nasal carriers of *Staph. aureus* defined as the presence of $\geq$ 75% of nasal cultures positive for *Staph. aureus* (cultures were taken at each out-patient visit every four to six weeks). Relapses occurred predominantly in carriers of *Staph. aureus* (relative risk of carriage for an ensuing relapse, 7.16) (Figure 25.1). During this three-and-a-half year prospective study relapses were not related to diagnosed infections. Otherwise, an association was observed between the persistence of ANCA and the occurrence of relapses. Twenty-two of the 33 patients who were intermittently or persistently positive for ANCA during follow-up developed

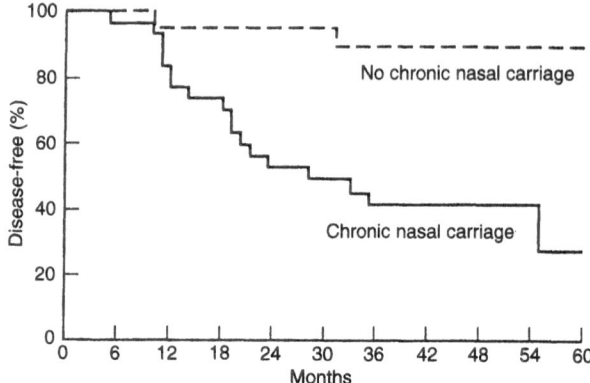

**Figure 25.1** Disease-free interval of 57 patients with Wegener's granulomatosis grouped according to *Staphylococcus aureus* carrier status. The time of the disease-free interval was counted from the beginning of the most recent period of disease activity (either initial diagnosis or relapse; $p < 0.001$). (With permission from [8].)

a relapse as opposed to only one of 21 patients who were persistently negative for ANCA during follow-up.

What do we have to conclude from the foregoing data? At present, WG cannot be considered to be an infectious disease. Micro-organisms do, however, seem to play a role in the expression of the disease. In this respect, particularly the role of *Staph. aureus* needs further study. Several toxins of *Staph. aureus* may function as a super-antigen and stimulate B cells and probably T cells in an unrestricted and T-helper cell-independent way [9, 10]. Cell wall components of *Staph. aureus* are effective B cell mitogens that may, among others, induce autoreactive B cells to produce ANCA [11, 12]. *Staph. aureus* can also directly stimulate neutrophils *in vitro* [13], and proteinases from *Staph. aureus* are able to inactivate human proteinase inhibitors like $\alpha_1$-antitrypsin [14]. All these effects may play a role in the induction of relapses of WG. Otherwise, it has been found that ANCA are able to activate primed neutrophils to the production of reactive oxygen species and the release of lytic enzymes [15]. As stated, neutrophils have to be primed before ANCA-induced activation can occur as priming results in the expression of the targets of ANCA, such as proteinase 3 and myeloperoxidase, at the cell membrane [16]. Priming of neutrophils occurs in the presence of low concentrations of substances such as TNF, which are produced at the site of infections. In this way, infection can create the conditions in which neutrophil activation by ANCA can occur. As such, infections can play a role in ANCA-associated vasculitis [17].

## TREATMENT AND PROPHYLAXIS OF WG WITH TRIMETHOPRIM/ SULFAMETHOXAZOLE (TS)

Although controlled prospective studies are still lacking, several observations suggest that TS may have some beneficial effect in the treatment or prophylaxis of WG. In 1985 DeRemee, McDonald and Weiland [18] described their results in treating indolent patients with WG who did not improve with cyclophosphamide with or without steroids, with 160/ 800 mg TS daily. During a period in which approximately 100 new patients with WG were examined, they treated 12 patients with TS. Eleven patients improved, four of them without concurrent or previous immunosuppressive treatment, five of them when TS was added to the immunosuppressive regimen, and two cases improved with TS alone after immunosuppressives had been given unsuccessfully. It should be noted that in the series treatment was particularly effective for manifestations of WG in the upper and lower airways (including one case with dacryocystitis). The effect of TS on renal manifestations is not directly apparent from the data. West, Todd and King [19] described the case history of a patient who, during a disease period of almost 20 years, repeatedly relapsed when cyclophosphamide was withdrawn. After addition of TS (160/800 mg twice daily, later on 160/800 mg daily) they were able to discontinue cyclophosphamide without the development of relapse. Reinhold-Keller *et al.* [20] recently reported their experience with TS (160/800 mg twice daily) in the long-term treatment of WG. In patients with limited WG, 11 out of 17 (65%) showed sustained improvement after a mean of 27 months (range 6–82). The remaining six patients had to stop, either because of progress of respiratory disease (five patients) or because of the development of generalized disease (one patient). In generalized WG the use of TS was accompanied by maintenance of remission in 14 out of 32 patients (44%) during a three-year period.

Some years ago we started a prospective placebo-controlled study on 81 patients with WG in order to investigate whether prophylactic treatment with TS (160/800 mg twice daily) can prevent the occurrence of relapse. Relapses occurred significantly less frequently in patients assigned to co-trimoxazol (seven out of 41) compared with those assigned to placebo (16 out of 40). Also, patients on co-trimoxazol less frequently suffered from infections than those on placebo. Thus, prolonged treatment with co-trimoxazol leads to a reduced incidence of relapses [21].

How can the beneficial effect of TS in WG be explained? First, the drug may exert its effect via its antimicrobial activity suggesting, as originally postulated by Wegener [22, 23], that WG is in some way related to a microbial infection. Secondly, TS may act as an immunosuppressive agent [24].

REFERENCES

1. Fauci, A.S., Haynes, B.F. and Katz, P. (1978) The spectrum of vasculitis. Clinical, pathologic, immunologic and therapeutic considerations. *Ann. Intern. Med.*, **89**, 660–76.
2. Fauci, A.S., Haynes, B.F., Katz, P. and Wolff, S.M. (1983) Wegener's granulomatosis: prospective clinical and therapeutic experience with 85 patients for 21 years. *Ann. Intern. Med.*, **98**, 76–85.
3. Cohen Tervaert, J.W., van der Woude, F.J. and Kallenberg, C.G.M. (1987) De ziekte van Wegener: een ernstige aandoening met een sluipend begin. *Ned. Tijdschr. Geneeskd.*, **131**, 1391–4.
4. Falk, R.J., Hogan, S., Carey, T.S. and Jennette, J.C. (1990) Clinical course of antineutrophil cytoplasmic autoantibodies-associated glomerulonephritis and systemic vasculitis. *Ann. Intern. Med.*, **113**, 656–63.
5. Davies, D.J., Moran, J.E., Niall, J.F. and Ryan, G.B. (1982) Segmental necrotizing glomerulonephritis with antineutrophil antibody: possible arbovirus aetiology? *Br. Med. J.*, **285**, 606.
6. Pinching, A.J., Rees, A.J., Pussell, B.A. *et al.* (1980) Relapses in Wegener's Granulomatosis: the role of infection. *Br. Med. J.*, **281**, 836–8.
7. Ronco, P., Verroust, P., Mignon, F. *et al.* (1983) Immunopathological studies of polyarteritis nodosa and Wegener's granulomatosis: a report of 43 patients with 51 renal biopsies. *Q. J. Med.*, **206**, 212–23.
8. Stegeman, C.A., Cohen Tervaert, J.W., Sluiter, W.J. *et al.* (1994) Association of chronic nasal carriage of *Staphylococcus aureus* and higher relapse rates in Wegener's granulomatosis. *Ann. Intern. Med.*, **120**, 12–17.
9. Fleischer, B. and Schrezenmeier, H. (1988) T cell stimulation by staphylococcal enterotoxins. *J. Exp. Med.*, **167**, 1697–707.
10. Fraser, J.D. (1989) High-affinity binding of staphylococcal enterotoxin A and B to HLA-Dr. *Nature*, **339**, 221–3.
11. Dziarski, R. (1982) Preferential induction of autoantibody secretion in polyclonal activation by peptidoglycan and lipopolysaccharide. I. *In vitro* studies. *J. Immunol.*, **128**, 1018–25.
12. Dziarski, R. (1982) Preferential induction of autoantibody secretion in polyclonal activation by peptidoglycan and lipopolysaccharide. II. *In vivo* studies. *J. Immunol.*, **128**, 1026–30.
13. Bates, E.J., Ferrante, A. and Beard, L.J. (1991) Characterization of the major neutrophil stimulating activity present in culture medium conditioned by *Staphylococcus aureus*-stimulated mononuclear leucocytes. *Immunology*, **72**, 448–50.
14. Potempa, J., Watorek, W. and Travis, J. (1986) The inactivation of human plasma $\alpha_1$-proteinase inhibitor by proteinases from *Staphylococcus aureus*. *J. Biol. Chem.*, **261**, 14330–4.
15. Falk, R.J., Terrell, R.S., Charles, L.A. and Jennette, J.C. (1990) Anti-neutrophil cytoplasmic autoantibodies induce neutrophils to degranulate and produce oxygen radicals in vitro. *Proc. Nat. Acad. Sci. USA*, **87**, 4115–19.
16. Charles, L.A., Caldas, M.L.R., Falk, R.J. *et al.* (1991) Antibodies against granule proteins activate neutrophils *in vitro*. *J. Leukocyte Biol.*, **50**, 539–46.
17. Kallenberg, C.G.M., Brouwer, E., Weening, J.J. *et al.* (1994) Anti-neutrophil

cytoplasmic antibodies: current diagnostic and pathophysiological potential. *Kidney Int.*, **46**, 1–15.

18. DeRemee, R.A., McDonald, T.J. and Weiland, L.H. (1985) Wegener's granulomatosis: observations on treatment with antimicrobial agents. *Mayo Clin. Proc.*, **60**, 27–32.

19. West, B.C., Todd, J.R. and King, J.W. (1987) Wegener's granulomatosis and trimethoprim-sulfamethoxazole. Complete remission after a twenty-year course. *Ann. Intern. Med.*, **106**, 840–2.

20. Reinhold-Keller, E., Beigel, A., Duncker, G. *et al.* (1993) Trimethoprim/sulfamethoxazole in the long-term treatment of Wegener's granulomatosis. *Clin. Exp. Immunol.*, **93**(S1), 38.

21. Stegeman, C.A., Cohen Tervaert, J.W., de Jong, P.E. and Kallenberg, C.G.M. (1995) Prevention of relapses of Wegener's Granulomatosis by treatment with trimethoprim-sulfamethoxazole (abstract). *Kidney Int.*, (in press).

22. Wegener, F. (1936) Über generalisierte septische Gefässerkrankungen. *Verh. Dtsch. Pathol. Ges.*, **29**, 202–9.

23. Wegener, F. (1939) Über eine eigenartige rhinogene Granulomatose mit besonderer Beteiligung des Artieriensystem und der Nieren. *Beitr. Pathol. Anat.*, **102**, 37–68.

24. DeRemee, R.A. (1988) The treatment of Wegener's Granulomatosis with trimethoprim/sulfamethoxazole: illusion or vision? *Arthritis Rheum.*, **31**, 1068–72.

# 26

# Alternative treatments for systemic vasculitis

*C.M. Lockwood*

INTRODUCTION

The progress towards specific immunotherapy for autoimmune disease described within this chapter begins with the introduction of intensive plasma exchange as a means by which removal of circulating nephrotoxic autoantibodies in patients with Goodpasture's syndrome could be effected immediately, thus tiding the patient over whilst waiting for immunosuppressive therapy to inhibit their further production [1]. The success of this approach and the ease with which it could be carried out meant that many patients with other autoimmune diseases could also benefit. The strategy has now developed into a sophisticated combined regimen, in which both the autoantibodies and the cells orchestrating their production can be targeted with a level of specificity not thought possible when plasma exchange was first introduced 17 years ago.

So what criteria need to be fulfilled before embarking on specific immunotherapy for autoimmune disease? On the one hand, as far as the disease is concerned, first, it is important to demonstrate that autoimmune mechanisms are operating in the condition and that they might contribute to its pathogenesis; secondly, the ability to monitor treatment is essential and so suitable indices of the autoimmune response, as well as parameters of clinical disease activity, need to be evaluable in the management programme; thirdly, it should be evident that the side-effects of conventional therapy are unacceptable. On the

*The Vasculitides.* Edited by B.M. Ansell, P.A. Bacon, J.T. Lie and H. Yazici.
Published in 1996 by Chapman & Hall, London. ISBN 978-0-412-64140-4.

other hand, as far as the specific immunotherapy is concerned, this should be safe, specific, efficient and economical at least in the long term.

What diseases are suitable candidates for this approach? For the purpose of this chapter I shall use as an example the systemic vasculitides. These form a group of closely related disorders, defined pathologically by the presence of inflammation and necrosis in blood vessel walls of different sizes and at different sites in the body, thereby producing a wide variety of clinical syndromes [2]. The vasculitides can be subdivided into primary forms, affecting blood vessels exclusively, and secondary forms, which arise in association with pathology elsewhere, for example close to a tumor, or downstream from an abscess cavity. The primary forms of vasculitis are the ones under consideration here. Ideally, they should be classified by pathogenetic mechanisms but, because so little was known about this until recently, the classification has been a morphological one, dependent on the size of vessel involved and the presence or absence of granulomata. Thus there are large vessel vasculitides, such as Takayasu's disease and giant cell arteritis, as well as small vessel Wegener's granulomatosis (WG), in which granulomata may be found; by contrast, in large vessel vasculitides such as polyarteritis nodosa (PAN), or small vessel vasculitides such as microscopic polyangiitis (MP), Kawasaki disease and Henoch-Schönlein purpura, granulomata are typically absent.

## AUTOIMMUNITY IN SYSTEMIC VASCULITIS

The primary forms of small vessel vasculitis are now emerging as autoimmune diseases in which both humoral and cellular components can be identified [3]. The presence of circulating autoantibodies to neutrophil cytoplasm antigens (ANCA) seems to be an integral feature of untreated small vessel vasculitis. Variations in their isotype [4–7] subclass [8, 9], affinity [9, 10] and idiotype [11] can influence clinical expression of the vasculitis. Furthermore, as detailed below, there is growing evidence that ANCA contribute directly to the pathogenesis of these disorders [12] and so regulation of their production is a suitable goal for immunotherapy. There is less evidence for the role of cellular mechanisms in the generation of vasculitis. However, T cells are prominent in vasculitic lesions, and in certain patients they may be directly vasculotoxic; here, the clinical efficacy of therapeutic agents with specificity for lymphocytes, particularly T cells, has indicated that this is a suitable route along which to develop specific immunotherapy [13].

It is notable that although ANCA with specificity for proteinase 3 (Pr3) are closely associated with WG, such a close association is not found between ANCA of other specificities and other vasculitic syndromes [12]. Thus, although antibodies to myeloperoxidase (MPO) are frequently found in patients with MP [14], antibodies to elastase [15], lysozyme [16], cathepsin G [17] and lactoferrin [18], which are also neutrophil cytoplasm antigens, have been reported in diseases indistinguishable clinically from MP. Furthermore, the indirect immunofluorescence appearances which anti-MPO antibodies produce in immobilized human neutrophils, the 'p' ANCA pattern, can also be produced by circulating autoantibodies (of as yet undertermined specificity) in patients with chronic inflammatory gastrointestinal diseases [19] such as chronic active hepatitis [20, 21], ulcerative colitis [22–24] and Crohn's disease [25], in which multisystem vasculitis is not a dominant feature. No final concensus has been reached as to how ANCA produce tissue injury. Three mechanisms have been proposed. Most emphasis has been placed on the finding that they may directly activate polymorphonuclear neutrophil leukocytes (PMN) [26]. This can promote PMN adhesion by up regulation of the cell surface adhesion molecule CD18 [27], and local endothelial cell injury is consequent on PMN release of toxic oxygen species during activation [28]. A second mechanism follows a more indirect pathway. There is now evidence that certain cytokines can orchestrate the incorporation of proteinases such as Pr3 into the cell membranes of PMN and endothelial cells [29]. At sites of local cytokine release, ANCA will bind to membrane incorporated Pr3 and so produce injury. Finally, a third mechanism involves disturbance of the physiological mechanism whereby proteins are inactivated locally after their release from PMN. ANCA compete with α1-antitrypsin for Pr3, leaving the enzymic site unblocked [30]. Further tissue destruction is a consequence of the failure of the inactivation of proteinases, the latter causing injury to neighbouring endothelial cells.

## MANAGEMENT OF SYSTEMIC VASCULITIS

### Contribution of ANCA

The ability to detect ANCA has provided the first laboratory test capable of diagnosing the systemic vasculitides. As well as allowing an assessment of the prevalance of these disorders (as frequent as systemic lupus in the experience of most centers), their recognition has greatly improved management by sustantiating what otherwise is usually a diagnosis based on clinical suspicion, since biopsy evidence of vasculitis is notoriously hard to obtain. Thus treatment can be introduced early and, hopefully, morbidity reduced. However, although they are valuable for diagnosis, there is no evidence that any of the features they possess at presentation have usefulness in predicting prognosis. In a two-center, prospective study of 60 patients [31], neither level of ANCA at presentation, nor specificity, were any guide to the outcome of treatment, either in terms of organ involvement, time to remission or likelihood of relapse. Nevertheless, ANCA levels could be helpful in management when measured serially during treatment. In the same study it was found that 23 (38%) of the patients entering remission relapsed on treatment during 12 months of follow-up. Rising ANCA titers predicted relapse a mean of 7.8 weeks before it was clinically detectable in 13 out of 23 (57%), while positive ANCA titers were present at the time of relapse in 19 out of 23 (83%). This led to the recommendation that monthly serology, without necessarily the need for patient–doctor contact, could be helpful during follow-up; if found, rising ANCA titers would indicate the need for extra clinical vigilance. However, it is necessary to remember that rarely, in certain patients, high titers of ANCA persist despite little or no evidence of disease activity (perhaps due to variation in ANCA affinity or specificity), and in others vasculitis appears to progress without ANCA being detectable.

### Other laboratory tests

Other less specific indices of active autoimmune disease activity should also be incorporated in the management strategy. The most useful are the time honored measurements of erythrocyte sedimentation rate (ESR) and C reactive protein (CRP). The latter is the most valuable [32], having a shorter half-life, hours rather than days, as is the case for ESR, which is also affected by the fibrinogen level and, therefore, routinely elevated in any nephrotic patient with moderate to heavy proteinuria. However, CRP levels are also elevated in any non-specific inflammatory process, for example during intercurrent infection, and so their fluctuation cannot be taken as a true reflection of the autoimmune involvement. Nevertheless, in our study a rise of one of these, ANCA, CRP or ESR, occurred in all but one of the instances of relapse we observed. More recently other parameters of immune activity have been assessed, for example, ILR2 [33], cytokine [34] and adhesion molecule [35] levels, but, just as for ESR and CRP, they too have the disadvantage that their fluctuation may have a multifactorial basis, again infection being the most likely responsible factor.

Standard simple tests of organ function should also be used to monitor progress, for example examination of the urine deposit for blood, protein or red cell casts and plasma creatinine, for patients with renal vasculitis, or measurement of KCO (the diffusing capacity of carbon monoxide) for patients with pulmonary hemorrhage due to lung vasculitis. On a daily basis these can provide a sensitive index of treatment response.

Finally, for reasons not understood, the platelet count, almost always elevated in active vasculitis, can provide further corroborative evidence for fluctuation in underlying disease [36].

### Non-invasive assessment of vasculitis

[111]Indium-labeled, autologous polymorph scans, routinely used in most diagnostic imaging departments to localize cryptic abscess cavities associated, for example, with diverticular disease, can be usefully adapted for objective, non-invasive assessment of degree and extent of vasculitic inflammation [37]. The adaptation involves quantitative scanning in the region of interest and the performance of two scans at 24 hour intervals (to estimate degree) as well as whole body imaging (to estimate extent). The particular findings were that of 32 patients with WG whose diagnosis was made by accepted clinical, histological and serological criteria, there was abnormal uptake of the labeled white cells in the upper respiratory airways, nasopharynx, sinuses, larynx or trachea in 29. By contrast, in 18 patients with MP, abnormal uptake in these areas was found in only four ($p < 0.01$). These scans were more sensitive that CT or routine radiology. Similar abnormalities of uptake were observed at other sites of suspected WG vasculitis, for example in the lung although not in the kidney, probably because granulomata of sustantial size generally are not found there and because the degree of PMN infiltration is limited anyway. As well as being an objective assessment of the vasculitis, often revealing sites not previously suspected clinically, scans performed serially have also proved valuable for monitoring treatment; see for example Figure 26.1.

### CONVENTIONAL TREATMENT OF SYSTEMIC VASCULITIS

Cytotoxic drugs and steroids are the mainstays of treatment for the systemic vasculitides. Usually, these are combined at high dose in an induction regimen to gain effective control of the disease activity at presentation followed after a short while, usually two to three months, by a lower dose, maintenance regimen, in which other immunosuppressive agents may substitute for the cyclophosphamide.

Induction therapy: empirically it was determined that a dose of cyclophosphamide at 3 mg/kg body weight (rounded down to the nearest 50 mg) was the most suitable to induce remission. This dose was lowered or discontinued temporarily if the total white cell count fell to less than $4.0 \times 10^9/l$ or the PMN count fell to less than $2.0 \times 10^9/l$, or if severe infection occurred. In patients aged 55 years or over, a lower induction dose of 2 mg/kg was used because of the greater susceptibility of the elderly to bone marrow suppression and infection. Steroids were given at a high dose, initially prednisolone 60 mg/day, tapering at

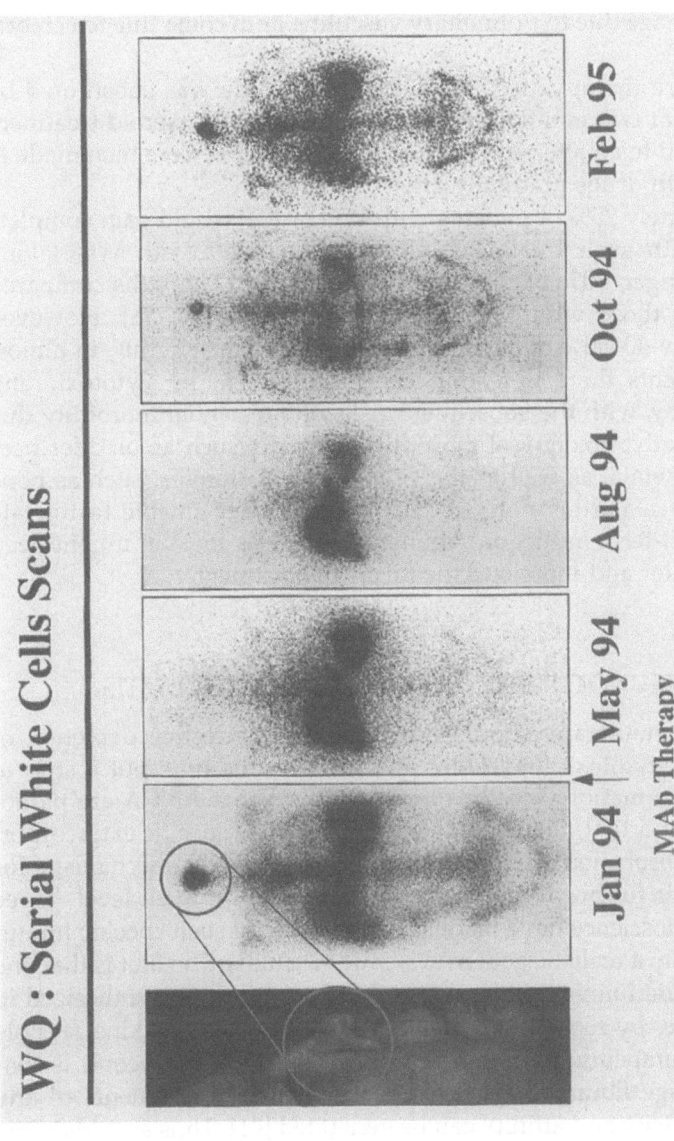

**Figure 26.1** Serial leukocyte scintiscanning. Autologous [111]Indium-labeled neutrophils scanned after reinjection show abnormal localization in the paranasal sinuses in the symptomatic patient (normal clearance occurs to liver, spleen and axial skeleton) pretreatment. Nine months later, although asymptomatic, abnormal localization again becomes detectable. Nasal stuffiness, sinus pain and epistaxis only become symptomatic four months after the scan first shows reappearance of an abnormality.

weekly intervals, until at two months the patient was receiving 10 mg/day. In patients with fulminating vasculitis threatening vital organ function, intensive plasma exchange was found to be useful. Thus evidence from several studies shows that such an approach might benefit patients with renal vasculitis (requiring dialysis support), with lung hemorrhage due to pulmonary vasculitis or in coma due to cerebral vasculitis.

Maintenance therapy, usually cyclophosphamide was substituted by azathioprine at two months, and at the same time the steroid treatment was converted to an alternate day regimen. Attempts were then made to withdraw both drugs gradually after 12 months.

Approximately 75% of patients with WG or MP would gain complete remission with such treatment, albeit the patients with WG taking somewhat longer, 50% of those with WG taking 12 months compared with 66% of those with MP taking two months [36, 38]. However, approximately 40–50% of such patients would relapse. Thus, in almost half the patients there is a long-term requirement for cytotoxic and steroid therapy, with the subsequent risks of substantial morbidity due to the cumulative toxicity of cyclophosphamide, such as bladder neoplasm or leukemia, as well as the side-effects of steroids, such as bone loss or cataract. Furthermore, certain patients seem unable to tolerate even the short-term induction regimen. There was thus an urgent need to develop safer and more specific forms of treatment.

## SPECIFIC IMMUNOTHERAPY FOR SYSTEMIC VASCULITIS

Specific immunotherapy should aim to achieve selective depletion of circulating vasculotoxic molecules and at the same time put a stop to their further formation. Ideally, therefore, given that ANCA are pathogenetically important, the strategy should combine specific extracorporeal immunoabsorption of the autoantibodies, with a mechanism for inhibiting their further formation, either at the B or T cell level. Three advances in bioscience have made the possibility of such specific immunotherapy seem a realistic goal in systemic vasculitis. The first is that one of the important human autoantigens, MPO, can now be synthesized in large quantities by recombinant DNA technology, thus making feasible its use for therapeutic immunoabsorption [39, 40]. The second is that using the phage library display approach, antibody fragments of any isotype, specificity and affinity can be generated [41]. Thus anti-idiotypic antibody therapy is a real possibility. The third is that 'humanized' monoclonal antibodies with specificity for lymphocytes, particularly T lymphocytes, are now available for use in humans, allowing immunomodulation of T cell function to be a treatment option [42].

## SPECIFIC IMMUNOABSORPTION

Already semispecific immunoabsorption can now be effected with extra-corporeal circuits containing devices that deplete ANCA on a charge basis [43]. In pilot experience with two patients, substantial depletion of ANCA was achieved without unwanted lowering of other immunoglobulin populations, as judged by levels of antibodies to tetanus toxoid or levels of overall IgG. The advantage of this approach compared with total plasma exchange is that the balance of the immunoglobulins, composing the repertoire of the patient's own humoral response, is retained. Introduction of recombinant antigen for absorption should make the specific depletion as efficient as possible and, by the nature of the technology, which allows production of the recombinant antigen in large quantities, it will eventually be cheap as well. Moreover, the fact that the patient receives no plasma products in exchange will contribute to reducing the risk of infectious transmission, as well as to the economy of the treatment. All of these features therefore satisfy the aims of specific immunotherapy – that it be safe, specific, efficient and economical.

## REGULATION OF B CELL FUNCTION

We have explored the use of pooled normal human immunoglobulins, commercially prepared for therapeutic use (IVIg), in the treatment of systemic vasculitis [44, 45]. Although IVIg therapy may produce its effects by a variety of mechanisms, including control of T cell function and interference with cytokine action, as well as Fc receptor blockade, it is the regulation of autoantibody production by B cells, through idiotypic anti-idiotypic reactions, that make this treatment attractive. Rossi *et al.* [11] and others [46] have shown the importance of ANCA idiotypic anti-idiotypic reactions *in vitro* and that these could be influenced by anti-idiotypic determinants present in IVIg. Thus $F(ab)_2$ fragments of IVIg could, in a dose dependent fashion, block ANCA binding to antigen. The degree of inhibition was variable, ranging up to 100% for the ANCA-containing sera from certain patients. Similar inhibitory activity was found in the sera of patients in remission after treatment and in the occasional patient whose disease remitted spontaneously, without drugs being used. Furthermore, when serial samples were studied, the anti-idiotypic reactivity was found to reciprocate with the autoantibody levels. In open studies we found that treatment with IVIg, given as Sandoglobulin 2 g/kg divided into five daily doses of 0.4 g/kg, was beneficial in patients with ANCA-associated vasculitis, many of whom had failed to respond to conventional immunosuppressive therapy. Of 14 patients given IVIg, 13 entered remission, and 11 were still in remission at 10 months. ANCA titers generally fell to 50% of pretreatment values, remaining reduced well after the half-life of the administered IVIg. There were no major side-effects, although a small number of patients showed a transient exacerbation of their symptoms between one and two weeks after commencing treatment. However, we did not feel that IVIg was justified for the treatment of patients with renal vasculitis, who had developed crescentic change, the histological hallmark of so-called rapidly progressive glomerulonephritis (RPGN). This was because, at that time, combined cyclophosphamide and steroid treatment was documented as particularly efficacious for the crescentic nephritis [47] and other reports suggested that IVIg might exacerbate progressive nephritis [48, 49]. Patients with nephritis whom we have treated more recently have not shown any ill effects from their IVIg therapy. IVIg would appear to be particularly valuable in the management of vasculitis in the elderly, the very young and in pregnancy, as well as for those at any age who are vulnerable to infection. We have more recently used IVIg as sole treatment for systemic vasculitis in six patients for whom benefit was seen in four (D.R.W. Jayne, unpublished). Thus now that the approach to production of synthetic anti-idiotypic antibody fragments is feasible, treatment of vasculitis by human antibodies seems a promising goal for specific immunotherapy.

## HUMANIZED ANTI-T CELL ANTIBODY TREATMENT

A major advance in the treatment of autoimmune disease has been the development of humanized monoclonal antibodies with specificity for lymphocytes, notably T lymphocytes. Their recent availability for therapeutic use proved particularly valuable for certain patients referred with intractable vasulitis, in whom biopsy showed a preponderance of T lymphocytes in vasculitic lesions. We have now treated five such patients all with severe disease refractory to steroids, cyclophosphamide, azathioprine and cyclosporin A, used alone or in combination [13]. None of the patients have circulating ANCA and all had biopsy evidence of abnormal T cell infiltrates in their vasculitic lesions.

The monoclonal antibodies used were anti-CD52 (CD52 is a glycoprotein complex carried on most lymphocytes, but not on NK cells) and anti-CD4. The anti-CD52 is lytic whereas the anti-CD4 is a blocking antibody; both are isotype IgG1. Similar antibodies to these act synergistically in animal models for tolerance induction and the two together, but not individually, can arrest the progression of an experimental autoimmune arthritis, whereas either can prevent its induction. Thus combination therapy might well be the optimal strategy for human disease, which is usually well established by the time the patient is referred for treatment.

The monoclonal antibodies were administered intravenously on a daily basis using doses of up to 40 mg/day for up to 10 days. The effect of treatment was monitored by the CD4 counts and standard tests for organ function. In patients with T cell associated vasculitis the monoclonal antibody treatment was given as a last resort to try to control what appeared to be a directly vasculotoxic effect of the infiltrating T cell population.

The significance of the response, within 72 hours in most instances, emphasized the impact of the novel therapeutic effect. The nature of the immunomodulation was not entirely clear, although a major reprogramming of the immune system was evident, as judged by the long term depletion of the patients' CD4 cells.

The results in this group of patients led to consideration of the same treatment for patients with ANCA associated vasculitis. Initially, patients who were resistant to conventional treatment with cytotoxic drugs and steroids were given additional humanized monoclonal antibody therapy. Control of the autoimmune response was temporally associated with the treatment given, but a delayed response to the routine drugs could not be ruled out. Monoclonal antibody therapy was then given to patients in whom conventional therapy was contraindicated. In those patients, autoantibody levels fell promptly after treatment was started and symptoms abated. However, the measurement of circulating ANCA is only a surrogate. Direct evidence of the regression of vasculitis with treatment comes with the radiological evidence of biopsy proven vasculitic lesions disappearing after starting monoclonal antibody therapy.

Thus far we have treated 11 patients with humanized monoclonal antibodies, five with T cell-associated vasculitis, and six with ANCA-associated disease. The monoclonal antibody therapy has not had any untoward side-effects nor has it been complicated by any systemic opportunistic infections. It thus promises also to be safe, specific, efficient and economical, at least in the long term.

## CONCLUSION

The progress towards specific immunotherapy for vasculitis offers exciting prospects for the treatment of this and other similar autoimmune diseases in the not too distant future.

## REFERENCES

1. Lockwood, C.M., Boulton-Jones, J.M., Lowenthal, R.M. *et al.* (1975) Recovery from Goodpasture's syndrome after immunosuppressive treatment and plasmapheresis. *Br. Med. J.* **2**, 252–4.
2. Savage, C.O.S. and Lockwood, C.M. (1993) Systemic vasculitides, in *Clinical*

*Aspects of Immunology* (eds D.K. Peters and P.J. Lachmann), Blackwell, Vol. II, pp. 1205–16.

3. Gross, W.L., Schmitt, W.H. and Csernok, E. (1993) ANCA and associated diseases: immunodiagnostic and pathogenetic aspects. *Clin. Exp. Immunol.*, **91**, 1–12,

4. Jayne, D.R.W., Jones, S.J., Severn, A. *et al.* (1989) Severe pulmonary haemorrhage and systemic vasculitis in association with circulating antineutrophil antibodies of IgM class only. *Clin. Nephrol*, **32**, 101–6.

5. Esnault, V.L.M., Soleimani, B., Keogan, M.T. *et al.* (1992) Association of IgM with IgG ANCA in patients presenting with pulmonary haemorrhage. *Kidney Int.*, **41**(5), 1304–10.

6. Thomas, D.M., Moore, R., Donovan, K. *et al.* (1992) Pulmonary-renal syndrome in association with anti-GBM and IgM ANCA antibodies. *Lancet*, **339**, 1304.

7. Ronda, N., Esnault, V.L.M., Layward, L. *et al.* (1993) Antineutrophil cytoplasm antibodies (ANCA) of IgA isotype in adult Henoch-Schonlein purpura. *Clin. Exp. Immunol.*, **95**, 49–55.

8. Jayne, D.R.W., Weetman, A.P. and Lockwood C.M. (1991) The IgG subclass of autoantibodies to neutrophil cytoplasmic antigens in systemic vasculitis. *Clin. Exp. Immunol.*, **84**, 476–82.

9. Esnault, V.L.M., Jayne, D.R.W., Weetman, A.P. and Lockwood, C.M. (1991) IgG sublass distribution and relative functional affinity of anti myeloperoxidase antibodies in systemic vasculitis at presentation and follow-up. *Immunology*, **74**, 714–18.

10. Esnault, V.L.M., Ronda, N., Jayne, D.R.W. and Lockwood, C.M. (1993) Association of ANCA isotype and affinity with disease expression. *J. Autoimmun.*, **6**, 197–205.

11. Rossi, F., Jayne D.R.W., Lockwood, C.M. and Kazatchkine, M.D. (1991) Anti-idiotypes against anti-neutophil cytoplasmic antigen autoantibodies in normal human polyspecific IgG for therapeutic use and in the remission sera of patients with systemic vasculitis. *Clin. Exp. Immunol.*, **83**, 298–303.

12. Lockwood, C.M. (1993) Specificity and pathogenicity of antineutrophil cytoplasm antibodies. *Exp. Nephrol.*, **1**, 13–18.

13. Lockwood, C.M., Thiru, J.D., Hale, G. and Waldmann, H. (1993) Humanised monoclonal antibody therapy for intractable systemic vasculitis. *Lancet*, **341**, 1620–2.

14. Falk, R.J. and Jennette, J.C. (1988) Antineutrophil cytoplasmic antibodies with specificity for myeloperoxidase in patients with systemic vasculitis and idiopathic necrotizing and crescentic glomerulonephritis. *N. Engl. J. Med.*, **318**, 1651–7.

15. Cohen Tervaert, J.W., Huitema, M.G., Dolman, K.M. *et al.* The clinical significance of autoantibodies to human leucocyte elastase. *Br. J. Rheumatol*, in press.

16. Schmitt, W.H., Csernok, E., Flesch, B.K. *et al.* (1993) Autoantibodies against lysozyme: A new target antigen for antineutrophil cytoplasm antibodies (ANCA). Proceedings of the 4th ANCA International Workshop, Lübeck, May 1992. *Adv. Exp. Med. Biol.*, **336**, 267–72.

17. Lesaure, P., Nusbaum, P. and Halbacks-Mecarelli, L. (1993) Methods of Detection of Anticathepsin G Autoantibodies in Humans. Proceedings of the 4th ANCA International Workshop, Lübeck, May 1992. *Adv. Exp. Med. Biol.*, **336**, 257–62.

18. Lesavre, P., Chen, N., Nusbaum, P. *et al.* (1990) Antineutrophil cytoplasmic

antibodies (ANCA) with antilactoferrin activity in vasculitis. *Kidney Int.*, **37**, 442.

19. Peter, H.H., Metzger, D., Rump, A. and Rother, E. (1993) ANCA in diseases other than systemic vasculitis. *Clin. Exp. Immunol.*, **93**(Suppl. 1), 12–13.

20. Klein, R., Eisenberg, J., Weber, P. *et al.* (1991) Significance and specificity of antibodies to neutrophils detected by Western blotting for serological diagnosis of primary sclerosing cholangitis. *Hepatology*, **14**, 1147–52.

21. Mulder, L., Horst, G., Haagsma, E. *et al.* (1993) Prevalence and characterisation of neutrophil cytoplasmic antibodies in autoimmune liver diseases. *Hepatology*, **17**, 411–17.

22. Saxon, A., Shanahan, F., Landers, C. *et al.* (1990) A distinct subset of antineutrophil cytoplasmic antibodies is associated with inflammatory bowel disease. *J. Allergy Clin. Immunol.*, **86**, 202–10.

23. Rump, J.A., Schölmerick, J., Gross, V. *et al.* (1990) A new type of perinuclear antineutrophil cytoplasmic antibody (pANCA) in active ulverative colitis but not in Crohn's disease. *Immunology*, **181**, 406–13.

24. Duerr, R.H., Targan, S.R., Landers, C.J. *et al.* (1991) Antineutrophil cytoplasmic antibodies ulcerative colitis: a link between primary sclerosing cholangitis and ulcerative colitis. *Gastroenterology*, **100**, 1385–91.

25. Halbacks-Mecarelli, L., Nusbaum, P., Noel, L.H. *et al.* (1992) Antineutrophil cytoplasmic antibodies (ANCA) directed against cathepsin G in ulcerative colitis Crohn's disease and primary sclerosing cholangitis. *Clin. Exp. Immunol.*, **90**, 79–84.

26. Falk, R.J., Terrell, R.S., Charles, L.A. and Jennette, J.C. (1990) Antineutrophil cytoplasm antibodies induce neutrophils to degranulate and produce toxic oxygen radicals in vitro. *Proc. Nat. Acad. Sci. USA*, **87**, 4115–19.

27. Mayet, W.J., Meyer, Zm. and Buschenfelde, K.H. (1993) Antibodies to proteinese 3 increase adhesion of neutrophils to human endothelial cells. *Clin. Exp. Immunol.*, **94**, 440–6.

28. Ewert, B.H., Jennette, J.C. and Falk, R.J. (1992) Antimyeloperoxidase antibodies stimulate neutrophils to damage human endothelial cells. *Kidney Int.*, **41**, 375–83.

29. Maget, W.J., Csernok, E., Szmkowick, C. *et al.* (1993) Human endothelial cells express proteinase 3 the target antigen of anticytoplasmic antibodies in Wegener's granulomatosis. *Blood*, **82**, 1221–9.

30. Dolman, K.M., Stegeman, C.A., Van den Weil, B.A. *et al.* (1993) Relevance of classic antineutrophil cytoplasmic autoantibody (C-ANCA)-mediated inhibition of proteinase 3-α1-antitrypsin complexation to disease activity in Wegener's granulomatosis. *Clin. Exp. Immunol.*, **93**, 405–10.

31. Jayne, D.R.W., Gaskin, G., Pusey, C.D. and Lockwood, C.M. (1995) ANCA and the prediction of relapse in systemic vasculitis. *Q. J. Med..*, **88**, 127–33.

32. Hind, C.R.K., Winearls, C.G., Lockwood, C.M., Rees, A.J. and Pepys, M.B. (1984) Objective monitoring of Wegener's granulomatosis by measurement of C-reactive protein concentration. *Clin. Nephrol.*, **21**, 341–5.

33. Schmitt, W.H., Heesen, C., Csernok, E. *et al.* (1992) Elevated serum levels of soluble interleukin 2 receptor in patients with Wegener's granulomatosis. *Arth. Rheum.*, **35**, 1088–96.

34. Grau, G.E., Roux Lombard, P., Gysler, C. *et al.* (1989) Serum cytokine changes in systemic vasculitis. *Immunology*, **68**, 196–8.

35. Pall, A., Adu, D., Drayson M. *et al.* (1993) Circulating adhesion molecules in systemic vasculitis. *Clin. Exp. Immunol.*, **93**(Suppl. 1), 32.

36. Hoffman, G.S., Kerr, G.S., Leavitt, R.Y. *et al.* (1992) Wegener's granulomatosis: An analysis of 158 patients. *Ann. Int. Med.*, **116**, 488–98.
37. Reuter, H., Wraight, P., Zhao, M.H. *et al.* (1995) Clinical significance of radiological imaging and determing autoantibody specificity in systemic vasculitis. *Q. J. Med.*, **88**, 509–17.
38. Savage, C.O.S., Winearls, C.G., Evans, D.J. *et al.* (1985) Microscopic polyarteritis: presentation, pathology and prognosis. *Q. J. Med.*, **220**, 467–83.
39. Moguilevsky, N., Garcia-Quintana, L., Jacquet, A. *et al.* (1991) Structural and biological properties of human recombinant myeloperoxidase produced by Chinese hamster ovary cell lines. *Eur. J. Biochem.*, **197**, 605-14.
40. Short, A.K., Lockwood, C.M., Bollen, A. and Moguilevsky, N. (1991) Recombinant myeloperoxidase as an antigen in ANCA positive vasulitis. *Clin. Exp. Immunol.*, **93**(Suppl. 1), 21.
41. Finnern, R., Lockwood, C.M. and Ouwehand, W. (1993) Anti-neutrophil cytoplasm antibody (ANCA) fragments from a human v-gene library. *Clin. Exp. Immunol.*, **93**(Suppl. 1), 21.
42. Mathieson, P.W., Cobbold, S.P., Hales, G. *et al.* (1990) Monoclonal antibody therapy in systemic vasculitis. *N. Engl. J. Med.*, **323**, 250–4.
43. Brownlee, A., Chapman, P., Mobey, L. and Lockwood, C.M. (1993) The effect of preincubation with ion exchange resins on the binding activity of anti-neutrophil antibodies (ANCA). *Clin. Exp. Immunol.*, **93**(Suppl. 1), 23.
44. Jayne, D.R.W., Davies, MJ., Fox, C.J.V. *et al.* (1991) Treatment of systemic vasculitis with pooled intravenous immunoglobulin. *Lancet*, **337**, 1137–40.
45. Jayne, D.R.W., Esnault, V.L.M. and Lockwood, C.M. (1993) ANCA anti-idiotype antibodies and the treatment of systemic vasculitis with intravenuous immunoglobulin. *J. Autoimmun.*, **6**, 207-19.
46. Pall, A.A., Varagunam, M., Adu, D. *et al.* (1994) Antiidiotypic antibodies against myeloperoxidase antibodies in pooled human immunoglobulin. *Clin. Exp. Immunol.*, **95**, 257-62.
47. Pusey, C.D., Rees, A.J., Evans, D.J. *et al.* (1991) Plasma exchange in focal necrotizing glomerulonephritis without anti-GBM antibodies. *Kidney Int.*, **40**, 757–63.
48. Jordan, S.C. (1989) Intravenous gammaglobulin therapy in systemic lupus erythematosus and immune complex disease. *Clin. Immunol. Immunopathol.*, **53**, S164–9.
49. Schifferli, J., Leski, M., Favre, H. *et al.* (1991) High dose intravenous IgG treatment and renal function. *Lancet*, **i**, 457–8.

# Part Six

# Vasculitis in Childhood

Many of the disorders already described, including Behçet's syndrome, macrosopic polyarteritis nodosa, Wegener's granulomatosis, Churg–Strauss syndrome, Takayasu's arteritis, and leukocytoclastic angiitis can all be seen, albeit rarely, in children. Seropositive juvenile rheumatoid arthritis can be complicated by vasculitis, while clinical manifestations, including vasculitis in systemic lupus erythematosus, are similar in the young as in the adult. Systemic sclerosis, although much rarer than localized disease in childhood, does not differ clinically from that of the adult. The disease itself, for example Takayasu's arteritis, may cause alteration in the growth of limbs, while chronic illness and corticosteroid therapy impair overall growth.

The purpose of this section is to highlight certain syndromes which appear to be either specific to childhood, such as Kawasaki syndrome, or more common in childhood, such as Henoch–Schönlein purpura and the vasculitis of familial Mediterranean fever, and cutaneous vasculitis induced by streptococcal infections, while the hallmarks of juvenile dermatomyositis are vasculitis early and calcinosis late.

---

Vasculitis in childhood

---

Polyarteritis
Cutaneous (streptococcal)
Kawasaki disease
Granulomatous vasculitis
   Wegener's
   Churg–Strauss
Leukocytoclastic vasculitis
   Hypersensitivity
   Henoch–Schönlein purpura
Giant cell arteritis (Takayasu)
Behçet's disease
Familial Mediterranean fever

---

# 27

# Juvenile dermatomyositis

*B.M. Ansell*

## INTRODUCTION

Juvenile dermatomyositis is a relatively uncommon rheumatic disease which is characterized by non-suppurative inflammation of muscle and skin; early in its course, vasculitis of varying severity occurs, and later calcinosis. The peak age of onset in childhood is between five and 14 years; boys are slightly less frequently affected than girls.

The etiology of juvenile dermatomyositis is unknown. It has been described in a few children with hypogammaglobulinemia in assoication with echovirus infections [1]. It has also been noted in children with IgA deficiency [2], and has been reported in C2 complement component deficiency [3]. The increased frequency of HLA B8 and DR3 and the recent suggestion of HLA DQA1*0501 as a risk factor, support an immunogenetic predisposition [4]. There are some tantalizing suggestions of a viral background with numerous infectious agents considered [5]. There have been a number of relatively recent reviews of juvenile dermatomyositis [6–7, 8], so this chapter will deal particularly with vascular problems and their management.

## VASCULITIS IN JUVENILE DERMATOMYOSITIS

The concept of juvenile dermatomyositis being a systemic angiopathy was introduced in 1966 by Banker and Victor [9] who identified the necrotizing vasculitis in arterioles, capillaries and venules of the striated muscles, skin and subcutaneous tissue and the gastrointestinal tract; they also noted there might not be a prominent inflammatory compo-

*The Vasculitides*. Edited by B.M. Ansell, P.A. Bacon, J.T. Lie and H. Yazici.
Published in 1996 by Chapman & Hall, London. ISBN 978-0-412-64140-4.

nent. Subsequently, Banker showed ultrastructural alterations of muscles and intramuscular blood vessels [10].

Deposits of immunoglobulin IgG and IgM and complement C3 in the vessel wall of nine out of 11 children suggested that it might be immunologically mediated [11]. Later the deposition of IgM, $C_3D$ fibrin and C5–9 membrane attack complex was noted in arterioles and capillaries in 10 out of 12 muscle biopsies studied [12]. Endothelial cells contain reticulotubular inclusion bodies suggesting local production of interferon-$\alpha$ [13]. This necrotizing vasculitis is identifiable in arterioles, capillaries and venules of striated muscle, the gastrointestinal tract, skin and subcutaneous tissue. The muscle pathology reflects the vascular component with perivascular atrophy, equal involvement of type I and type II fibers, as well as an inflammatory infiltrate, areas of infarction and loss of the muscle capillary network. In the gastrointestinal tract, the vasculitis can lead to infarction, ulceration, diffuse bleeding and even perforation, as well as being responsible for impaired absorption. The nail-fold capillary loop abnormalities have been described as distinctive with dilatation of isolated loops, the drop out of surrounding loops together with arborized clusters of other capillary loops [14]. In their long-term studies, Crow, Boye and Levinson [15] felt they had identified a group of children where there was muscle infarction, lymphocytic vasculitis and a non-inflammatory vasculopathy in children who also had extensive erythematous and ulcerative cutaneous disease and who tended to develop calcinosis.

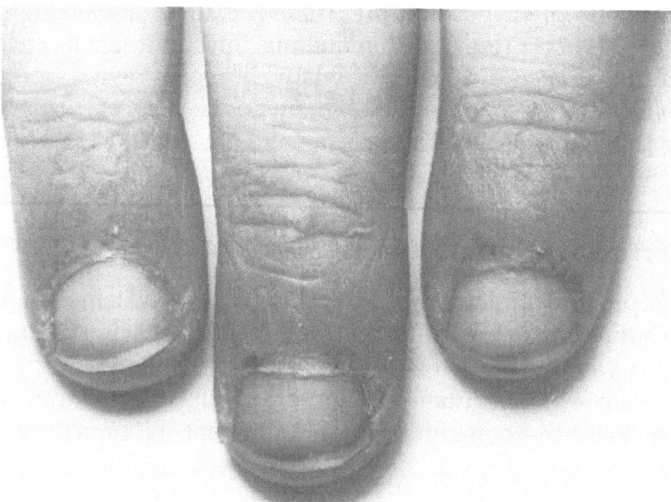

**Figure 27.1** Erythema around the nail edge with telangiectasia.

Clinically, periungual erythema and telangiectasia of the nail-beds are characteristic (Figures 27.1 and 27.2), as is the telangiectasia along the eyelids, which may well be accompanied by edema. Occlusive endarteropathy is probably responsible for the cutaneous lesions with ulceration (Figure 27.3). At times this cutaneous ulceration may occur in a livedo reticular pattern. Characteristically, there is ulceration around the eyes and in the skin folds, particularly the axillae (Figure 27.4) and groins, and also on the buttocks (Figure 27.5). Infarction of the palate can occur; it is often associated with rapidly weakening palatal movement. The transient retinal exudates, often referred to as cytoid bodies and looking like cotton wool spots, are thought to result from the occlusion of small vessels. Rarely there is involvement of the myocardium, causing various degrees of atrioventricular block and occasionally complete heart block [16].

Death can occur from an acute gastrointestinal ulceration with bleeding or perforation, myocarditis and also from infection, particularly septicemia associated with infected skin ulcers.

## ASSESSMENT

Levels of factor VIII-related antigen and fibrinopeptide A are increased in the plasma of children with active dermatomyositis [17]. Unfortunately, factor VIII-related antigen is not specific for the vasculitis of

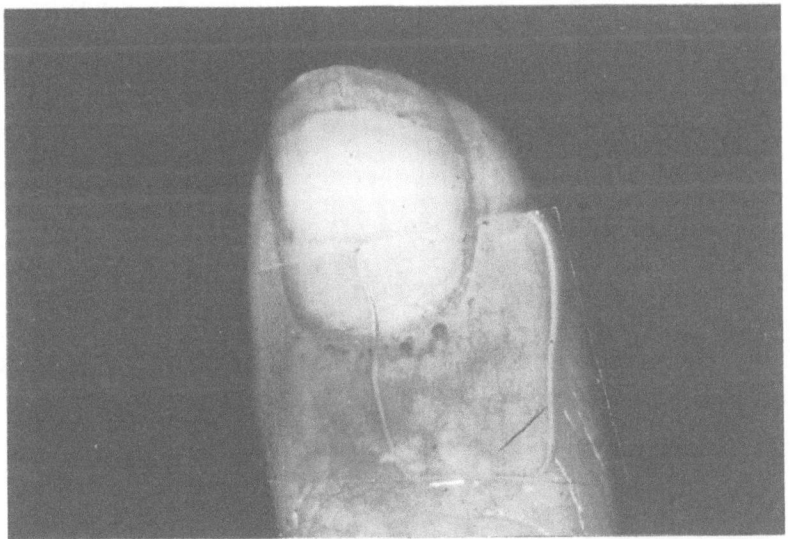

**Figure 27.2** Capillary nail fold lesions enhanced by oil under sellotape on a finger tip.

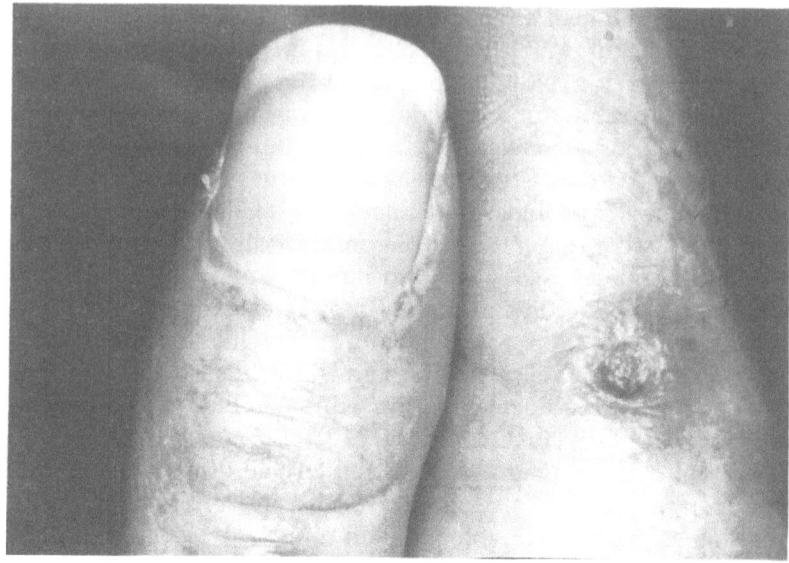

**Figure 27.3** Erythema with telangiectasia around the nail fold, and ulceration of the skin over the next MCP.

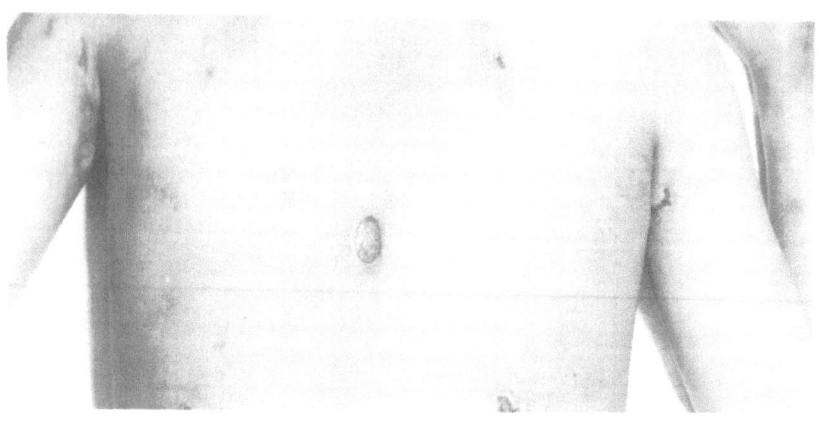

**Figure 27.4** Multiple ulcerations of varying ages in acute dermatomyositis. Note the severe axillary lesions of varying age.

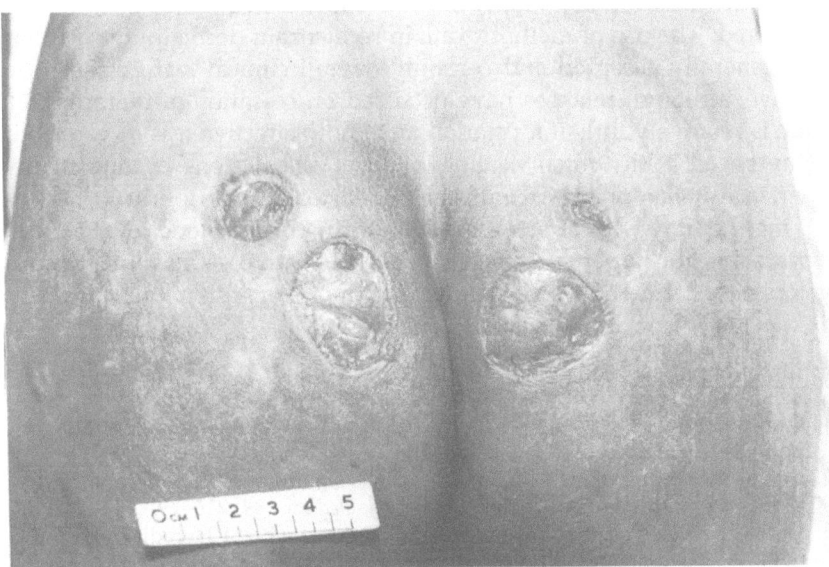

**Figure 27.5** Scarring and new ulceration on the buttocks during a relapse.

juvenile dermatomyositis, and its overall role in monitoring the severity and persistence of the vasculitis in dermatomyositis has not been established [18]. The possible use of neopterin warrants further study [19].

Using serial *in vivo* nail-fold microscopy Silver and Marciq [20] were able to assess the frequency and degree of vasculopathy in nine children, with the degree of morphologic changes correlating with their clinical course of dermatomyositis. This may be one of the more effective ways of monitoring patients from early in their disease.

## MANAGEMENT

The diagnosis of dermatomyositis calls for the immediate use of corticosteroids. With severe gastrointestinal involvement or vasculitic ulcers, it is probably wise to start with intravenous bolus methylprednisone, going to oral administration of prednisone or deflazacort as the disease comes under control. If serious vasculitic ulcerative skin lesions are present, the question of intravenous gammaglobulin should be considered. Intravenous gammaglobulin at 1 gm/kg given daily for two days and repeated monthly for six months has been found to be effective in dermatomyositis, and the one case with a vasculitic rash responded [21]. Other references on the overall use of intravenous gammaglobulin have suggested initially a four-day course of treatment. Certainly, if there is little response and persistence of widespread ulceration and severe changes in the capillary loops, immunosuppressive therapy should be considered. The use of methotrexate in recalcitrant dermatomyositis has been generally accepted and certainly overall clinical features seem to improve, although relapses have occurred on cessation of therapy [22]. There is relatively little information as to how effective it is in vasculitis or the speed with which healing of ulcerative lesions occurs; in my experience it has not been entirely satisfactory. Cyclosporin at a dose of 5 mg/kg/day with appropriate monitoring of the blood level as well as watching for hypertension, has proved effective, as has intravenous cyclophosphamide 10 mg/kg given monthly over a six-month period.

## REFERENCES

1. Webster, A.D.M. (1984) Echovirus disease in hypogammaglobulinaemia patients. *Clin. Rheum. Dis.*, **10**, 189–203.
2. Carroll, J.E., Silverman, A., Isobe, Y. *et al.* (1976) Inflammatory myopathy, IgA deficiency and intestinal malabsorption. *J. Pediatr.*, **89**, 216–19.
3. Leddy, J.P., Griggs, R.C., Klemperer, M.R. *et al.* (1975) Hereditary complement (C2) deficiency with dermatomyositis. *Am. J. Med.*, **58**, 83–91.
4. Reed, A.M., Ober, C. and Pachman, L.M. (1991) Molecular genetic studies of MHC gene in children with juvenile dermatomyositis: increased risk associated with HLA DQA1*0501. *Human Immunol.*, **32**, 235–40.
5. Christensen, M.L., Pachman, L.M., Schniederman, R. *et al.* (1986) Prevalence of Coxsackie B virus antibodies in patients with juvenile dermatomyositis. *Arthritis Rheum.*, **29**, 1365–70.
6. Roberts, L.J. and Fink, C.W. (1988) Childhood polymyositis/dermatomyositis. *Clin. Dermatol.*, **6**, 36–48.
7. Pachman, L.M. (1990) Juvenile dermatomyositis, in *Paediatric Rheumatology Update* (eds. P. Woo, P. White and B.M. Ansell), Oxford University Press, Oxford, pp. 171–81.
8. Ansell, B.M. (1991) Juvenile dermatomyositis. *Rheum. Dis. Clin. Nth Am.*, **17**, 931–42.
9. Banker, B.Q. and Victor, M. (1966) Dermatomyositis (systemic angiopathy) of childhood. *Medicine*, **45**, 261–89.
10. Banker, B. (1975) Dermatomyositis in childhood: ultra-structural alterations of muscle and extra-muscular blood vessels. *J. Neuropath. Exp. Neurol.*, **34**, 46–75.
11. Whitaker, J.M. and Engel, W.K. (1972) Vascular deposits of immunoglobulin and complement in idiopathic inflammatory myopathy. *N. Engl. J. Med.*, **286**, 333–8.
12. Kissell, J.T., Mendell, J.R. and Rammohan, K.W. (1986) Microvascular deposition of complement membrane attack complex in dermatomyositis. *N. Engl. J. Med.*, **314**, 329–34.
13. Fidzianska, A. and Goebel, H.H. (1989) Tubuloreticular structures and cylindric confronting cisterine in childhood dermatomyositis. *Acta Neuropath.*, **79**, 310–16.
14. Spencer-Green, G., Crowe, W.E. and Levinson, J.E. (1982) Nail fold capillary abnormalities and clinic outcome in childhood dermatomyositis. *Arthritis Rheum.*, **25**, 954–8.
15. Crowe, W.E., Boye, K.E. and Levinson, J.E. (1982) Clinical and pathogenetic

implication of histopathology in childhood polydermatomyositis. *Arthritis Rheum.*, **25**, 126–39.

16. Askari, A.M. (1984) Cardiac abnormalities in inflammatory disorders of muscle. *Clin. Rheum. Dis.*, **10**, 131–50.
17. Bowyer, S.L., Ragsdale, C. and Sullivan, D.B. (1989) Factor VIII related antigen and childhood rheumatic disease. *J. Rheumatol.*, **16**, 1093–7.
18. Guzman, J., Ross, P. and Malleson, P.N. (1994) Monitoring disease activity in juvenile dermatomyositis: the role of Von Willebrand factor and muscle enzymes. *J. Rheumatol.*, **21**, 739–43.
19. Benedetti, F., De Amici, D., De Aramini, L. *et al.* (1993) Correlation of serum neopterin concentrations with disease activity in juvenile dermatomyositis. *Arch. Dis. Child.*, **69**, 232–5.
20. Silver, R.M. and Marciq, H.R. (1989) Childhood dermatomyositis. Serial microvascular studies. *Pediatr.*, **83**, 278–83.
21. Lang, B.A., Laxer, R.M., Murphy, G. *et al.* (1991) Gammaglobulin for dermatomyositis. *Am. J. Med.*, **91**, 169–72.
22. Miller, L.C., Sisson, B.A., Tucker, L.B. *et al.* (1992) Methotrexate treatment of recalcitrant childhood dermatomyositis. *Arthritis Rheum.*, **35**, 1143–9.

# 28

# Kawasaki syndrome

*M.J. Dillon*

Vasculitis is a feature of many different diseases and syndromes of childhood [1]. In some it is the predominant manifestation of the condition; in others, it may be one aspect of a multisystem disease. Of the various vasculitic syndromes seen in children, Kawasaki disease (mucocutaneous lymph node syndrome) is comparatively common and is of some importance since, unlike many vasculitides, there is good evidence pointing to an infective initiating agent. It was first described in Japan in 1967 [2]. Since then over 80 000 cases have been reported from that country alone [3], although it is of worldwide distribution, affecting predominantly infants and young children under five years of age [4]. There is an ethnic bias towards Oriental or Afro-Caribbean children, a male preponderance, some seasonality and occasional epidemics [5, 6]. In Japanese populations, the incidence is 150 per 100 000 children aged under five years per year [3]. Elsewhere, the incidence is substantially less: 10.3 in the USA, and approximately 3.0 per 100 000 children under five years of age in several European countries including the UK [7–9].

## CLINICAL FEATURES

The principal manifestations are outlined in Table 28.1, based on the diagnostic guidelines prepared by the Japan Kawasaki Disease Research Committee [10]. At least five of the six items in Table 28.1 should be present for the diagnosis to be established. Patients with four items can be diagnosed as having Kawasaki disease if coronary artery aneurysms are also present on two-dimensional echocardiography or coronary angiography. However, there may be a need to review the diagnostic

*The Vasculitides.* Edited by B.M. Ansell, P.A. Bacon, J.T. Lie and H. Yazici.
Published in 1996 by Chapman & Hall, London. ISBN 978-0-412-64140-4.

**Table 28.1** Principal symptoms of Kawasaki disease

- Fever persisting for five days or more
- Changes in the peripheral extremities (reddening of palms and soles, indurative edema in initial stage and membranous desquamation of the finger tips in the convalescent phase)
- Polymorphous exanthema
- Bilateral conjunctival injection
- Changes of lips and oral cavity (reddening of lips, stawberry tongue, diffuse injection of oral and pharyngeal mucosa)
- Acute non-purulent cervical lymphadenopathy

**Table 28.2** Other significant symptoms and findings in Kawasaki disease

- Cardiovascular system: heart murmurs, gallop rhythm, ECG changes, cardiomegaly, two-dimensional echo findings of pericardial effusion, coronary artery aneurysms, aneurysms of peripheral arteries, angina pectoris and myocardial infarction
- Gastrointestinal tract: diarrhea, vomiting, abdominal pain, hydrops of the gallbladder, ileus and jaundice
- Blood: leukocytosis, thrombocytosis, increased erythrocyte sedimentation rate, increased C-reactive protein, hypoalbuminemia and anemia
- Urine: proteinuria, increased leukocytes in sediment
- Skin: transverse furrows of fingernails
- Respiratory tract: cough and rhinorrhea
- Joints: pain and swelling
- Neurological system: pleocytosis in cerebrospinal fluid, convulsions and facial palsy

criteria in the future in view of the increasing number of cases of incomplete Kawasaki disease that are being recognized [11]. Other significant symptoms or findings seen in Kawasaki disease are listed in Table 28.2.

Cardiovascular complications are variable but may be up to 35% if transient coronary artery dilatation, pericardial effusion, ECG abnormalities, pericarditis, myocardial infarction, ventricular aneurysm, mitral incompetence and cardiac failure are included [4, 6, 12, 13]. The incidence of coronary artery aneurysms ranges from 20–30% depending on which series is being considered.

## LABORATORY INVESTIGATIONS

Investigations reveal a polymorphonuclear leukocytosis, thrombocythemia, circulating immune complexes [14] and both antineutrophil cytoplasmic antibodies (ANCA) [15, 16] and antiendothelial cell antibodies

(AECA) [17–19], which may have diagnostic as well as etiopathological roles. The ANCA findings on immunofluorescence are very characteristic, with a diffuse cytoplasmic staining that is distinct from the granular cytoplasmic pattern seen in Wegener's granulomatosis or the perinuclear pattern in renal associated disease [20]. These findings are discussed further below (see etiopathogenesis).

## TREATMENT

Aspirin and high-dose intravenous gammaglobulin either as five daily doses of 400 mg/kg/day [21], or one dose of 2 g/kg [22], are currently recommended. Dipyridamole has also been used in addition to aspirin by some groups. In the presence of very large coronary artery aneurysms, especially if there is myocardial ischemia, intravenous prostacyclin [4] or plasma exchange or exchange transfusions [23] may also have a place. Intra-arterial or intravenous urokinase has been used when a coronary artery becomes occluded with thrombus [24]. Coronary revascularization surgery may be indicated for critical stenotic lesions in the convalescent phase of the illness [13]. Steroids remain controversial and are not advocated, but some argue that they might have a role if used in conjunction with aspirin therapy. In support of this is the observation that steroid treatment has been of value in children with aggressive disease who appear resistant to gammaglobulin and aspirin therapy at the Hospital for Sick Children, London.

## PROGNOSIS

The general outlook for children with Kawasaki disease is good, although there is an acute mortality rate of 0.1–0.3% due to myocardial infarction. This may be reduced by alertness of clinicians to the diagnosis, and the early use of gammaglobulin and antiplatelet therapy. After the acute phase there is morbidity and occasionally mortality due to coronary artery stenotic lesions in later life [25] with ischemic myocardial sequelae. It has interestingly been postulated that adult atheromatous coronary heart disease may have its origins in childhood and be due to previous covert or overt Kawasaki disease [26].

## ETIOPATHOGENESIS

It seems likely that Kawasaki disease has an infective basis, although the nature of the organism and the mechanism involved remains in doubt. A number of organisms have been considered including *Streptococcus sanguis*, *Propionibacterium acnes*, EB virus, human herpes virus-6, *Chlamydia*, *Rickettsia* and retroviruses, but none have clearly been implicated

[7, 8]. A retroviral role was considered because of the profound immunoregulatory disturbance in the acute phase of the illness, suggesting the involvement of a lymphotropic agent [27, 28], but this has not been confirmed [29]. At present there is interest in the role of a toxin-producing organism in view of the finding of selective expansion of T cells expressing T cell receptor variable regions Vβ2 and Vβ8 in Kawasaki disease patients [30]. These changes might be caused by new clones of toxic shock syndrome toxin producing *Staphylococcus aureus* or pyogenic exotoxin producing streptococci since a substantial proportion of Kawasaki disease patients have been shown to be infected with such bacteria [31].

A number of cellular and humoral immunological abnormalities have been documented during the acute phase of Kawasaki disease. These include deficient T8-positive suppressor/cytotoxic T cells, increased T4-positive activated helper T cells bearing HLA-DR surface antigens and grossly increased numbers of B cells spontaneously secreting IgG and IgM [32]. High levels of the plasma cytokines interleukin-1 (IL-1) and tumor necrosis factor (TNF) and spontaneous secretion of cytokines have also been documented [33–35]. Immune complexes have been detected in the circulation two to four weeks after onset of the disease in several studies [14, 36] and have been shown to induce platelet activation and release of vasoactive mediators [14]. Furthermore, there appeared to be an association between the presence of immune complexes and the characteristic thrombocytosis occurring in the second and third weeks of the illness.

In addition, Leung *et al.* [17, 37] demonstrated complement dependent endothelial cell toxicity of Kawasaki serum for cultured endothelial cells treated with TNF, IL-1 and gamma interferon, and argued that immune activation results in elaboration of cytokines that activate endothelium with the induction of new surface antigens and that the resulting antibody response caused vascular injury. However, antibodies that did not depend on prior cytokine stimulation of target endothelium have also been detected by ELISA [18, 19]. Cytotoxicity can also occur without prestimulation of endothelial cells with cytokine, although it is enhanced by such pretreatment [19]. Cytokine stimulation may therefore be necessary to promote maximal cytotoxic effects of antiendothelial cell antibodies but may not be necessary for their formation. Leung *et al.* [37] also demonstrated that endothelial activation antigens were expressed *de novo* (for example, ELAM-1) or showed increased expression (for example, ICAM-1) in skin biopsies of children in the acute stages of the disease and this disappeared in the majority of cases following treatment with gammaglobulin.

There is evidence to support a contribution from neutrophils as well in inducing endothelial injury in Kawasaki disease. IgG and IgM antineutrophil cytoplasmic antibodies are elevated in acute Kawasaki disease [4, 15, 16, 20, 38]. The indirect immunofluorescence appearances are unusual [15, 20] in that there is a diffuse cytoplasmic staining distinct from the granular pattern seen in Wegener's granulomatosis and the perinuclear staining seen in renal associated disease thought to be due to antibodies directed against proteinase (PR3) and myeloperoxidase (MPO) respectively [39]. Kaneko *et al.* [20] showed on studying antigenic specificities of these antibodies that a proportion of patients had antibodies to the α fraction of neutrophil primary granules but not to MPO, leaving a number of patients who had positive ANCA for acid extract of neutrophils in whom the precise epitope was unknown. More recently, Gilbert *et al.* [40] showed that a substantial proportion of acute Kawasaki disease patients' sera contained antibodies directed against cathepsin G. It is, however, unclear whether the specificity is in anyway meaningful in terms of the pathophysiology of the disease or whether it just reflects non-specific B cell activation.

However, it is known that ANCA can induce neutrophil activation, and contribute to endothelial injury in vasculitis [41, 42]. There is also evidence that endothelial cells may also have the ability to contribute to neutrophil activation and trigger the release of reactive oxygen species [43, 44]. What is more, endothelial bound MPO and PR3 can be recognized by ANCA [45], and PR3 is expressed under the influence of cytokines by endothelial cells and can bind anti-PR3 antibodies [46]. These findings suggest a relationship between ANCA, neutrophils and endothelial cells that may have implications in terms of the vasculitic process.

On the other hand, a recent paper has cast some doubt on the pathogenetic roles of ANCA and AECA in Kawasaki disease since their incidence was not significantly different when comparing patients with the condition and those with febrile infections [47].

CONCLUSION

The pathogenesis of the vascular injury in Kawasaki disease is complex and involves a range of mechanisms as outlined above. However, in view of the circumscribed nature of the condition and the possibility of an infective etiology, it might be a very suitable model for identifying the types of infective agents that may be involved in precipitating vasculitic processes that could be applied in other systemic vasculitides in which the etiology is generally unknown.

REFERENCES

1. Hicks, R.V. (ed.) (1988) *Vasculopathies of childhood*, PSG Publishing, Littleton.
2. Kawasaki, T. (1967) Acute febrile mucocutaneous syndrome with lymphoid

involvement with specific desquamation of the fingers and toes in children. *Jap. J. Allergy*, **16**, 178–222.

3. Yanagawa, H. and Nakamura, Y. (1986) Nationwide epidemic of Kawasaki disease in Japan during winter of 1985–1986. *Lancet*, **ii**, 1138–40.

4. Tizard, E.J., Suzuki, A., Levin, M. and Dillon, M.J. (1991) Clinical aspects of 100 patients with Kawasaki disease. *Arch. Dis. Child.*, **66**, 185–8.

5. Hicks, R.V. and Melish, M.E. (1986) Kawasaki syndrome. *Pediatr. Clin. N. Am.*, **33**, 1151–75.

6. Rowley, A.H., Gonzalez-Crussi, F. and Shulman, S.T. (1988) Kawasaki syndrome. *Rev. Infect. Dis.*, **10**, 1–15.

7. Shulman, S.T. (ed.) (1987) Kawasaki disease: Proceedings of the Second International Kawasaki Disease Symposium. *Progress in Clinical Biological Research*, Vol. 250, Alan R. Liss, New York.

8. Kawasaki, T. (ed.) (1988) *Proceedings of the Third International Kawasaki Symposium*, Japan Heart Foundation, Tokyo.

9. Hall, S. and Newton, L. (1992) Kawasaki disease, in *British Paediatric Surveillance Unit Seventh Annual Report*, pp. 8–9.

10. Japan Kawasaki Disease Research Committee (1984) *Diagnostic guidelines of Kawasaki disease*, 4th revised edition, Japan Red Cross Medical Center, Tokyo, Japan.

11. Rowley, A.H., Gonzalez-Crussi, F., Giddings, S.S. *et al.* (1987) Incomplete Kawasaki disease with coronary artery involvement. *J. Pediatr.*, **110**, 409–13.

12. Kato, H., Koike, S., Yamamoto, M. *et al.* (1975) Coronary aneurysms in infants and young children with acute febrile mucocutaneous lymph node syndrome. *J. Pediatr.*, **86**, 892–8.

13. Suzuki, A., Kamiya, T., Ono, Y. *et al.* (1985) Indication of aortocoronary bypass for coronary arterial obstruction due to Kawasaki disease. *Heart Vessels*, **1**, 94–100.

14. Levin, M., Holland, P.C., Nokes T.J. *et al.* (1985) Platelet immune complex interaction in pathogenesis of Kawasaki disease and childhood polyarteritis. *Br. Med. J.*, **290**, 1456–60.

15. Savage, C.O.S., Tizard, E.J., Jayne, D. *et al.* (1989) Antineutrophil cytoplasm antibodies in Kawasaki disease. *Arch. Dis. Child.*, **64**, 360–3.

16. Tizard, E.J. and Dillon, M.J. (1990) Laboratory investigation in the diagnosis and management of childhood vasculitis. *Prosp. Pediatr.*, **20**, 35–9.

17. Leung, D.Y.M., Collins, T., Lapierre, L.A. *et al.* (1986) Immunglobulin M antibodies present in the acute phase of Kawasaki syndrome lyse cultured vascular endothelial cells stimulated by gamma interferon. *J. Clin. Invest.*, **77**, 1428–35

18. Tizard, E.J., Baguley, E., Hughes, G.R.V. and Dillon, M.J. (1991) Anti-endothelial cell antibodies detected by a cellular based ELISA in Kawasaki disease. *Arch. Dis. Child.*, **66**, 189–92.

19. Kaneko, K., Savage, C.O.S., Pottinger, B.E. *et al.* (1993) Cytotoxic autoantibodies to endothelial cells in Kawasaki disease, in *Proceedings of 4th International Symposium on Kawasaki Disease* (eds. M. Takahashi and K. Taubert), American Heart Association, pp. 185–91.

20. Kaneko, K., Shah, V., Gaskin, G. *et al.* (1993) Kawasaki disease has distinct anti neutrophil cytoplasmic antibodies, in *Proceedings of 4th International Symposium on Kawasaki Disease* (eds, M. Takahashi and K. Taubert), American Heart Association, pp. 192–7.

21. Newburger, J.W., Takahashi, M., Burns, J.C. *et al.* (1986) The treatment of

Kawasaki syndrome with intravenous gamma globulin. *N. Engl. J. Med.*, **315**, 341–7.

22. Newburger, J.W., Takahashi, M., Beiser, A.S. *et al.* (1991) A single intravenous infusion of gammaglobulin as compared with four infusions in the treatment of acute Kawasaki syndrome. *N. Engl. J. Med.*, **324**, 1633–9.

23. Tizard, E.J. and Dillon, M.J. (1991) Plasmapheresis in childhood. *Care Crit. Ill*, **7**, 51–5.

24. Terai, M., Ogata, M., Sugimoto, K. *et al.* (1985) Coronary arterial thrombi in Kawasaki disease. *J. Pediatr.*, **106**, 76–8.

25. Tatara, K., Kusakawa, S., Itoh, K. *et al.* (1989) Long term prognosis of Kawasaki disease in patients with coronary artery obstruction. *Heart Vessels*, **5**, 47–51.

26. Brecker, S.J.D., Gray, H.H. and Obedershaw, P.J. (1988) Coronary artery aneurysms and myocardial infarction: adult sequelae of Kawasaki disease. *Br. Heart J.*, **59**, 509–12.

27. Shulman, S.T. and Rowley, A.H. (1986) Does Kawasaki disease have a retroviral aetiology? *Lancet*, **2**, 545–6.

28. Burns, J.C., Geha, R.S., Schneeberger, E.E. *et al.* (1986) Polymerase activity in lymphocyte culture supernatants from patients with Kawasaki disease. *Nature*, **323**, 814–16.

29. Melish, M.E., Marchette, N.J., Kaplan, J.C. *et al.* (1989) Absence of significant RNA dependent DNA polymerase in lymphocytes from patients with Kawasaki syndrome. *Nature*, **337**, 288–90.

30. Abe, J., Kotzin, B.L., Jujo, K. *et al.* (1992) Selective expansion of T cells expressing T-cell receptor variable regions Vβ2 and Vβ8 in Kawasaki disease. *Proc. Nat. Acad. Sci. USA*, **89**, 4066–70.

31. Leung, D.Y.M., Meissner, H.C., Fulton, D.R. *et al.* (1993) Toxic shock syndrome toxin-secreting *Staphylococcus aureus* in Kawasaki syndrome. *Lancet*, **342**, 1385–8.

32. Leung, D.Y.M., Chu, E.T. and Wood, N. (1983) Immunoregulatory T cell abnormalities in mucocutaneous lymph node syndrome. *J. Immunol.*, **130**, 2002–4.

33. Furukawa, S., Matsubara, T., Jojoh, K. *et al.* (1988) Peripheral blood monocyte/macrophages and serum tumour necrosis factor in Kawasaki disease. *Clin. Immunol. Immunopathol.*, **42**, 247–51.

34. Maury, C.P.J., Salo, E. and Pelkonen, P. (1988) Circulating interleukin-1b in patients with Kawasaki disease. *N. Engl. J. Med.*, **319**, 1670–1.

35. Rowley, A.H., Shulman, S.T., Preble, O.T. *et al.* (1988) Serum interferon concentrations and retroviral serology in Kawasaki syndrome. *Pediatr. Infect. Dis. J.*, **7**, 663–5.

36. Kohsaka, T., Abe, J., Nakayama, M. *et al.* (1988) The significance of IgM-immune complex and complement breakdown products in Kawasaki syndrome, in *Proceedings of the 3rd International Kawasaki Disease Symposium* (ed. T. Kawasaki), Japan Heart Foundation, Tokyo, pp. 138–40.

37. Leung, D.Y.M., Cotran, R.S. and Kurt-Jones, E. (1989) Endothelial cell activation and high interleukin 1 secretion pathogenesis of acute Kawasaki disease. *Lancet*, **ii**, 1298–302.

38. Dillon, M.J. and Tizard, E.J. (1991) Anti-neutrophil cytoplasmic antibodies and anti-endothelial cell antibodies. *Pediatr. Nephrol.*, **5**, 256–9.

39. Jennette, J.C. and Falk, R.J. (1990) Antineutrophil cytoplasmic autoantibodies and associated diseases: a review. *Am. J. Kidney Dis.*, **15**, 517–29.

40. Gilbert, R.D., Shah, V., Reader, J. *et al.* (1993) Cathepsin G specific IgM anti

neutrophil cytoplasmic antibodies (ANCA) and anti endothelial cell antibodies (AECA) in acute Kawasaki disease. *Clin. Exp. Immunol.*, **93**(Suppl. 1), 30.

41. Savage, C.O.S., Pottinger, B.E., Gaskin, G. *et al.* (1992) Autoantibodies developing to myeloperoxidase and proteinase 3 in systemic vasculitis stimulate neutrophil cytotoxicity towards cultural endothelial cells. *Am. J. Pathol.*, **141**, 335–42.

42. Ewert, B.H., Jennette, J.C. and Falk, R.J. (1992) Anti-myeloperoxidase antibodies stimulate neutrophils to damage human endothelial cells. *Kidney Int.*, **41**, 375–83.

43. Von Asmuth, E.J.U., Van der Linden, C.J., Leeuwenberg, J.F.M. *et al.* (1991) Involvement of the CD 11b/CD18 integrin, but not the endothelial cell adhesion molecules ELAM-1 and ICAM-1 in tumour necrosis factor induced neutrophil toxicity. *J. Immunol.*, **147**, 3869–75.

44. Savage, C.O.S. (1993) The endothelial cell: active participant or innocent bystander in primary vasculitis? *Clin. Exp. Immunol.*, **93**(Suppl. 1), 6–7.

45. Savage, C.O.S., Gaskin, G., Pusey, C.D. *et al.* (1993) Anti-neutrophil cytoplasm antibodies (ANCA) can recognize vascular endothelial cell-bound ANCA-associated autoantigens. *J. Exp. Nephrol.*, **1**, 190–5.

46. Mayet, W.J., Hermann, E.H., Csernok, E. *et al.* (1993) *In vitro* interactions of C-ANCA (antibodies to proteinase 3) with human endothelial cells, in *ANCA-Associated Vasculitides* (ed. W.K. Gross), Plenum Press, New York, pp. 109–13.

47. Guzman, J., Fung, M. and Petty, R.E. (1994) Diagnostic value of anti-neutrophil cytoplasmic and anti-endothelial cell antibodies in early Kawasaki disease. *J. Pediat.*, **124**, 917–20.

# 29

# Kawasaki syndrome: the Florentine experience

*F. Falcini, M. Ermini, S. Trapani, G. Taccetti,*
*G. Bartolozzi, L. De Simone, A. Manetti, S. Turchini,*
*A. Farsi, A. Lombardi, M. Matucci-Cerinic, G. Keser,*
*M. Khamashta, H. Direskeneli and G.R.V. Hughes*

As already described [1], Kawasaki syndrome is an acute mucocutaneous, self-limited febrile illness characterized by a systemic vasculitis of small-and medium-sized vessels involving the coronary arteries [2]. Untreated it may lead to coronary aneurysms and myocardial infarction which are responsible for death either in the acute phase of the disease or subsequently. Intravenous gammaglobulin has been shown to reduce the risk of coronary artery abnormalities when administered during the first 10 days of the disease [3].

The first case in Italy was reported in 1977 [4], and until 1980 only eight further cases had been described [5–7]. In 1980, two boys with typical Kawasaki syndrome were admitted to the Rheumatology Unit of the Paediatric Department of Florence University. They were the first cases seen in Tuscany and the 10th and 11th reported in Italy [8]. Since then 52 further cases, 34 males and 18 females (M:F = 2:1), have been admitted to this department; all except one child came from Tuscany. The highest number of cases, 26 children, was observed in 1991 and 1992 with 11 cases clustered between February and July 1991. Seventeen cases occurred in winter, 15 in spring, 13 in summer and nine in autumn. Two cases occurred in children attending the same kindergarten and two patients were first-degree cousins sharing the same house. The age at

*The Vasculitides.* Edited by B.M. Ansell, P.A. Bacon, J.T. Lie and H. Yazici.
Published in 1996 by Chapman & Hall, London. ISBN 978-0-412-64140-4.

**Table 29.1** Signs and symptoms of 54 Kawasaki syndrome children

| *Major manifestations* | |
|---|---|
| Fever duration of five days or more | 98% |
| Changes of lips or mucosa | 91% |
| Rash | 87% |
| Conjunctivitis | 77% |
| Changes of extremities | 68% |
| Lymph node enlargement | 46% |
| | |
| *Minor Manifestations* | |
| Diarrhea and/or abdominal pain | 36% |
| Cardiac complications | 27.7% |
|    coronary dilatation | 16.6% |
|    aneurysm | 3.7% |
|    pericarditis | 5.5% |
|    mitral regurgitation | 1.8% |
| Pneumonia | 16% |
| Hydrops of gallbladder | 12.5% |
| Neurological complications | 10% |
| Urinary tract infection | 6% |
| Jaundice | 4% |
| Otitis | 4% |
| Epistaxis | 2% |

onset ranged from two months to 14 years with an average of 33 months. Eighty per cent of the patients were younger than four years. The interval between disease onset and diagnosis ranged from 3–28 days (mean 13 days). In 50 of the 54 children (92.5%) the presentation was typical, but in four (7.5%) the Kawasaki syndrome was incomplete [9]. The frequency of major and minor clinical manifestations is shown in Table 29.1. The duration of the fever ranged from 5–28 days (mean 10 days). In five patients the fever lasted less than five days.

The basic laboratory data (Table 29.2), showed thrombocytopenia in

**Table 29.2** Blood tests laboratory data

| | *Range* | *Mean value* |
|---|---|---|
| ESR (mm/h) | 43–100 | 85 |
| WBC ($\times 10^3$/mmc) | 7400–31 300 | 8540 |
| Hb (g/dl) | 7–12 | 10.4 |
| PLTS ($\times 10^3$/mmc) | 508 000–1 000 000 | 620 000 |
| ALT (U/l) | 11–450 | 102 |
| AST (U/l) | 17–590 | 93 |

two cases early in the disease. The blood parameters returned to normal in all patients during the fourth week. Immunological studies were performed in 30 of the 54 patients and showed an increase in CD4+ and CD19+ lymphocyte subsets, reduced CD8+ and increased CD4/CD8 ratio. Circulating immune complexes were detected in 40 of the 54 children [10, 11]. ANCA testing performed in 30 of the 54 patients was positive (c-ANCA pattern) in six patients; all sera were negative using ELISA specific for anti-Pr3 antibodies. Anticardiolipin antibodies (solid phase assay) was performed in 29 of the 54 patients and was positive in nine patients (seven IgG, two IgM). AECA (ELISA, antiendothelial cell antibodies), were detected in five out of 20 patients tested in the acute phase of the disease [12, 13]. HLA A,B,C,DR tissue typing was performed in 40 of the 54 children; no genetic predisposition to Kawasaki syndrome was found. Plasma angiotensin converting anzyme (ACE) levels (fluorimetric method) were measured in 40 of the 54 patients to try to evaluate endothelial injury during the course of the disease. This evaluation was made in the acute phase of the disease, at four weeks and at one year from disease onset. ACE values were significantly lower $(2.0 \pm 1.5 \text{ pmol/ml/min})$ than in age and sex-matched controls $(8.9 \pm 3.1 \text{ pmol/ml/min})$ in the acute phase and progressively increased $(4.3 \pm 2.1 \text{ pmol/ml/min})$ one month after the disease onset [14].

A physical examination, ECG, M-mode, 2D ECHO and, since 1990, color Doppler echocardiography was performed in all children to detect cardiac involvement. Echocardiography was performed twice a week during the first month, once a month during the first three months and then every six months for two years. Two to three weeks after disease onset 15 of the 54 children (27.7%) had evidence of cardiac problems. Ten children (18.5%) had coronary dilatation (ranging from 3.2 to 4.5 mm) and two children (3.7%) developed a coronary aneurysm of the left coronary artery (diameter > 5 mm); three children had a mild epipericardial separation. In one patient with coronary dilatation, mild mitral regurgitation lasting three months was detected by color Doppler echocardiography.

High serum cholesterol was detected in 50% of children, 3–18 months after their disease onset [15]; levels normalized within the third year of the disease. There were no deaths among these patients and the cardiac abnormalities resolved during the second month of illness in all except five children who had coronary dilatation lasting more than two years. In the two children with coronary aneurysms, remodelling was seen on 2D ECHO, 6–12 months after diagnosis. These patients are still being checked by color Doppler echocardiography twice a year.

Thallium 201 dipyridamole myocardial scintigraphy was undertaken in five patients with long-standing coronary dilatation and/or aneurysm.

No patient showed any defect of myocardial perfusion during pharmacological manipulation, exercise and stress [16]. At the last follow-up none of the patients had cardiac symptoms and the ECG was normal in all.

One boy aged five years required a cholecystectomy during the third month of the disease due to the persistence of severe gallbladder hydrops [17].

All children were treated with aspirin 30 mg/kg/day during the active phase of the disease and 5 mg/kg/day up to the third month of illness in the absence of cardiac abnormalities. In the presence of coronary dilatation, dipyridamole 2 mg/kg/day was added. Forty of the 54 children, all in the group diagnosed after 1986, were given intravenous gammaglobulin 10–14 days after disease onset; 25 received 400 mg/kg/day for five days and 15 a single infusion of 2 gm/kg. In this latter group we observed a more rapid regression of fever [18–21]. Three patients were given a second cycle of gammaglobulin, 2 gm/kg, as the fever persisted despite the treatment [22]. An immediate resolution of fever was noted after the completion of the second infusion. No significant differences were noted between the group treated with gammaglobulin and the group treated with only aspirin, as far as the incidence of cardiac complications. Coronary damage risk factors consisted of fever for more than 12 days, male gender (14 boys and one girl had cardiac involvement), age > two years and initial elevation of ESR to >100 mm/h. There was no significant relationship between the presence of aCL, ANCA or AECA and coronary abnormalities. ACE levels might be considered a useful marker of endothelial injury showing a clear restoration of enzyme activity with the disease resolution [15].

The incidence of coronary aneurysms in these children was lower than that reported, with the outcome of the disease favorable in most patients [23]. Early diagnosis and an early therapeutic approach may well have had a role in reducing the risk of coronary involvement in our patients.

## REFERENCES

1. Dillon, M.J. (1996) Kawasaki syndrome, in *The Vasculitides* (eds B.M. Ansell, P.A. Bacon, J.T. Lie and H. Yazici), Chapman & Hall, London, pp. 384–91.
2. Kawasaki, T., Kosaki, F., Okawa, S. *et al.* (1974) A new infantile acute mucotaneous lymph node syndrome (MLNS) prevailing in Japan. *Pediatrics*, **54**, 271.
3. Suzuki, A., Tizard, E.J., Gooch, V. *et al.* (1990) Kawasaki disease: Ecocardiographic features in 91 cases presenting in the United Kingdom. *Arch. Dis. Child.*, **65**, 1142–6.
4. Della Porta, G. and Alberti, A. (1977) La sindrome linfomucocutanea di Kawasaki. *Minerva Ped.*, **28**, 1745.
5. Pecorari, R., Saltari, P. and Furlani, M. (1979) La malattia di Kawasaki. *Acta. Paed. Lat.*, **32**, 218.

6. Giovannini, M., Riva, E. and Besana, R. (1979) La sindrome di Kawasaki. *Minerva Ped.*, **31**, 537.

7. Scherini, A., Vegni, M. and Elli, P. (1980) La malattia di Kawasaki in Italia, tre nuovi casi. *Minerva Ped.*, **32**, 1111.

8. Giani, I., Lapi, E., Falcini, F. *et al.* (1982) La sindrome linfomucocutanea. *Ped. Med. Chir.*, **4**, 297.

9. Centers for Disease Control (1980) Kawasaki disease - New York. *M.M.W.R.*, **29**, 61.

10. Falcini, F., Trapani, S., Pampaloni, A. *et al.* (1993) Hydrops of gallbladder requiring cholecystectomy. *Clin. Exp. Rheum.*, **11**, 99.

11. Barron, K.S. (1991) Immune abnormalities in Kawasaki Disease: prognostic implications and insight into pathogenesis. *Cardiol. Young.*, **1**, 206.

12. Savage, C.O.S., Tizard, J., Jayne, D. *et al.* (1989) Antineutrophil cytoplasm antibodies in Kawasaki disease. *Arch. Dis. Child.*, **64**, 360.

13. Guzman, J., Fung, M., Thomas, E. *et al.* (1992) Diagnostic value of serum antineutrophil cytoplasmic antibodies (ANCA) in Kawasaki Disease. *Arthr. Rheum.*, **35**, S229.

14. Rider, L.G., Wener, M.H., French, J. *et al.* (1993) Autoantibody production in Kawasaki syndrome. *Clin. Exp. Rheum.*, **11**, 445.

15. Falcini, F., Lombardi, A., Leoncini, G. *et al.* (1993) Angiotensin Converting Enzyme in Paediatric Vasculitides. *Arth. Rheum.*, **36**, D170.

16. Salo, E., Pesonen, E. and Jorma, V. (1991) Serum cholesterol levels during and after Kawasaki disease. *J. Pediatr.*, **119**, 557.

17. Fukazawa, M., Fukushige, J., Taeuchit *et al.* (1993) Discordance between Tallium 201 scintigraphy and coronary angiography in patients with Kawasaki disease: myocardial ischemia with normal coronary angiogram. *Pediatr. Cardiol.*, **14**, 67.

18. Newburger, J.W., Takahashi, M., Burns, J.C. *et al.* (1986) The treatment of Kawasaki Syndrome with intravenous gammglobulin. *New Engl. J. Med.*, **315**, 341.

19. Rowley, A.H. and Shulman, S.T. (1991) Current therapy for acute Kawasaki syndrome. *J. Pediatr.*, **118**, 987.

20. Barron, K.S., Murphy, D.J., Silverman, E.D. *et al.* (1990) Treatment of Kawasaki syndrome: A comparison of two dosage regimens of intravenously administered immune globulin. *J. Pediatr.*, **117**, 638.

21. Engle, A.M., Fatica, N.S., Bussel, J.B. *et al.* (1989) Clinical trial of single-dose intravenous gammaglobulin in acute Kawasaki Disease. *A.J.D.C..*, **143**, 1300.

22. Sundel, R.P., Burns, J.C., Baker, A. *et al.* (1993) Gammaglobulin re-treatment in Kawasaki Disease. *J. Pediatr.*, **123**, 657.

23. Akagi, T., Rose, V., Benson, L.N. *et al.* (1992) Outcome of coronary artery aneurysms after Kawasaki disease. *J. Pediatr.*, **121**, 689.

# 30

# Vasculitis and its relation to the streptococcus

*J. David*

The term scarlet fever has been used for centuries. To the layman it conveys a sense of drama, unlike measles, which is accepted as a childhood complaint. In the days of the militia, scarlet fever was used as signifying a female passion for the red coats. This usage comes from the British national proneness to belittle something that is serious, rather than to suggest that the epidemic disease was a frivolous condition. In the twentieth century, scarlet fever has lost most of its virulence, but the name retains much of the dramatic and the doctors tend to talk of scarlatina to their patients, implying that it is a milder form of the disease. The disease is certainly mild in Europe, but not universally so. Its epidemic future cannot be forecast. If the present trend continues, the term scarlet fever may become of historical interest only.

The relationship between the group A β-hemolytic streptococcus and acute rheumatic fever and one type of acute glomerulonephritis has been recognized for many years. More recently, two additional clinical syndromes have been reported in children, also caused by prior infection with this organism. These are poststreptococcal reactive arthritis and polyarteritis nodosa (PAN). There are rapidly changing views about these conditions and the etiopathogenic role of the streptococcus.

*Streptococcus pyogenes* is one species of the large family of streptococci. It its distinguished by its ability to form β-hemolysis on blood agar and to produce soluble hemolysins in fluid media. The coccus is 0.5–0.75 μm in diameter; they are arranged in chains of 10 or more. The organism is Gram positive and non-motile. Capsules may be seen in the young

*The Vasculitides.* Edited by B.M. Ansell, P.A. Bacon, J.T. Lie and H. Yazici.
Published in 1996 by Chapman & Hall, London. ISBN 978-0-412-64140-4.

cultures, but are not prominent features. Their presence is related to the relative amounts of hyaluronic acid and hyaluronidase. The organism possesses an extractable polysaccharide antigen, which can be demonstrated by a precipitation test. This polysaccharide antigen is situated deep within the cell wall. It enables β-hemolytic streptococci to be separated into the Lancefield groups – labeled A to R. This group classification corresponds to the pathogenicity and the hemolytic properties of the organism. Thus group A streptococci are mainly pathogenic for humans, group B are the most frequent streptococci in cattle. As well as the polysaccharide group antigens, group A streptococci possess surface protein antigens of two varieties, M and T; their identification enables the group to be subdivided into at least 57 types [1].

Streptokinase is a toxic product of *S. pyogenes*. It can dissolve human fibrin, possibly by activating a lytic factor present in human serum. It is antigenic, provoking the formation of antistreptokinase, which inhibits the lytic property of streptokinase. A somewhat similar product is deoxyribonuclease or streptodornase. Proteinase is also liberated by the organism. It is these surface M proteins which are the major virulence factors and to which one acquires immunity to homologous-type organisms. M proteins 1, 5, 6 and 19 have been associated with heart cross-reactivity and are known to be important in rheumatic heart disease.

The possible relationship between childhood PAN and group A streptococcal infection was mentioned by Fink [2] at the first Park City meeting on pediatric rheumatology. Six out of seven children with PAN had a preceding sore throat or otitis and five had elevated ASO titers. Shortly thereafter, Blau, Morris and Yunis [3] also pointed out the high incidence of significant antibody titers to group A streptococci in this disease. Fink described the recurrent, painful skin nodules and livedo reticularis in children with cutaneous polyarteritis [4]. Diaz-Perez and Winklemann [5] have suggested that cutaneous PAN should be separated from the systemic form of the disease.

At Northwick Park Hospital, London, 15 children with a vasculitic illness that occurred in association with streptococcal infection were followed over a 20-year period. Twelve of these are described elsewhere [6]. Eight of the 12 children were boys and four were girls; their mean age at onset was eight years (range 4–11 years).

The duration of their disease from onset to diagnosis was one to 12 weeks. All had a preceding sore throat and there was an associated upper respiratory tract infection in nine (60%). All had malaise with marked irritability, a recurrent fever of 38–39°C and weight loss of up to 10% body weight, and 10 (66%) had non-specific central abdominal pain. Table 30.1 outlines the clinical signs present.

In all patients, painful erythematous nodules occurred on the medial

**Table 30.1** Clinical signs on presentation

| Case/sex/age at onset | Edema | | Cutaneous vasculitis | | | | Arthritis |
|---|---|---|---|---|---|---|---|
| | Muscle | Periorbital | Instep | Limbs | Trunk | Face | |
| 1. FA/F/6 | + | + | + | + | − | + | + |
| 2. KN/M/5 | − | − | + | + | + | − | + |
| 3. ID/M/5 | − | − | + | + | + | − | + |
| 4. YQ/M/11 | + | + | + | + | + | − | + |
| 5. GH/M/8 | − | − | + | + | + | − | + |
| 6. TP/F/10 | + | + | + | + | − | + | + |
| 7. PF/M/9 | + | − | + | + | + | − | + |
| 8. ND/M/7 | + | + | + | + | − | − | + |
| 9. DH/M/11 | − | − | + | + | − | − | + |
| 10. MR/F/9 | + | + | + | + | + | − | + |
| 11. LB/F/4 | + | − | + | + | − | − | + |
| 12. RG/M/8 | + | + | + | + | + | + | + |

aspect of the foot and sole. All patients had livedo reticularis on their arms and shins. Seven patients had lesions on their trunk and three on their face. The rash on the limbs was associated with brawny edema of muscles in seven out of 12; six patients had periorbital edema. An evanescent arthritis, most commonly affecting the ankles and knees, was present in all the patients. In six, muscle pain was severe enough to cause immobilization.

Three patients gave a history of rheumatic fever, although none had documented carditis and all had normal echocardiograms. Digital desquamation and stomatitis, which are often seen in Kawasaki disease, were not present. One child had developed gangrene of a terminal digit and one an acute anterior iritis. Neither Raynaud's phenomenon, nor hypertension were seen and there were no deaths.

An acute phase response was present with an elevated ESR of more than 50 mm/h and a CRP of more than 55 mg/l (range 1–112 mg/l). Table 30.2 lists the presenting hematological parameters. Peripheral eosinophilia was not present. The mean hemoglobin concentration was 102 g/l and was of a normochromic/normocytic type. The platelet count mean was $698 \times 10^9$/l. There were two patients who had platelet counts of more than $1000 \times 10^9$/l. Hypergammaglobulinemia was present with all having an IgG more than 15 g/l. The ASO titer was raised at 1060 units in all patients (normal range less than 200). Antihyaluronidase titer was raised (256 units) in all six patients in whom it was measured. Three patients had β-hemolytic streptococcus grown on throat swab. Routine autoantibody screens, ANCA and antiglomerular basement membrane antibodies were all negative, as was hepatitis B surface antibody.

**Table 30.2** Hematology parameters on presentation

|        | Hb (g/dl) | WBC ($\times 10^9$/l) | Neutrophil ($\times 10^9$/l) | Platelets ($\times 10^9$/l) |
|--------|-----------|-----------|------------|-----------|
| 1. FA  | 7.7  | 23.2 | 17.4 | 1188 |
| 2. KN  | 9.9  | 21.0 | 14.7 | 1030 |
| 3. ID  | 12.0 | 12.2 | NK   | 455  |
| 4. YQ  | 11.6 | 90.0 | 72.1 | 505  |
| 5. GH  | 10.0 | 25.6 | 19.9 | 924  |
| 6. TP  | 12.3 | 11.5 | 4.8  | 680  |
| 7. PF  | 10.1 | 24.5 | NK   | 570  |
| 8. ND  | 8.5  | 65.0 | 60.0 | 800  |
| 9. DH  | 9.3  | 11.0 | NK   | 499  |
| 10. MR | 11.2 | 15.9 | 14.0 | 635  |
| 11. LB | 10.5 | 15.5 | 14.0 | 500  |
| 12. RG | 10.0 | 40.1 | 34.0 | NK   |

NK=not known.

Six patients had skin lesion biopsies and a necrotizing vasculitis was present involving the vessels of the deep dermis. No giant cells or granulomata were noted.

The patients were treated with oral penicillin and nine with an NSAID. Ten patients required corticosteroids with a mean starting dose of 2 mg/kg/day, which was reduced within 14 days to 1 mg/kg/day. The fever and acute phase reactants all regressed rapidly. Prophylactic penicillin was continued long term.

Seven of the 12 patients had a relapsing course with vasculitic lesions on steroid reduction. Five fully recovered. The mean follow-up time was eight years (1–19 years). One patient was free of disease for eight years, then within six months of discontinuing penicillin developed tonsillitis and a severe vasculitic illness ensued. He had a fever, widespread skin lesions and bilateral peroneal nerve palsies. His neutrophil count was 65 $\times$ $10^9$/l and hemoglobin 78 g/l. CRP and ESR were both markedly elevated. Visceral arterial angiography demonstrated microaneurysms of the hepatic and renal arteries. This patient was treated with methyl-prednisolone given intravenously at 20 mg/kg on three consecutive days. Cyclophosphamide was added in at 0.5 gm/m$^2$ every three weeks for six doses.

A further patient, who also had severe recurrent disease, was shown to have renal microaneurysms. His renal function was entirely normal. He too was treated with parenteral cyclophosphamide. A third patient with relapsing disease had normal arteriography. She had recurrent episodes which followed non-specific infections and occurred when the steroid dose was lowered below 12.5 mg/day. She was subsequently treated with oral azathioprine at 2.5 mg/day. Recurrence of disease in these three patients was associated with an increase in ASO titer of more than 1060 units. Betahemolytic streptococcus was not isolated from their throat swabs.

Rose and Spencer [7] reviewed 111 patients with polyarteritis and noted that concomitant rheumatic fever was found in 12.5% of the patients. This association has also been made in a necropsy series of Freidberg and Gross [8]. Fordham *et al.* described polyarteritis in post-streptococcal glomerulonephritis in three patients [9].

The three patients described by David, Ansell and Woo [6] who developed recurrent vasculitic disease with microaneurysms suggest that cutaneous PAN is part of a continuum of PAN and can occur after streptococcal infection. Most cases of cutaneous PAN have a benign course [10, 11]. In contrast, Magilav reported nine patients with a chronic course with evidence of systemic vasculitis on angiography, but without renal deterioration [12], although Fisher and Orkin suggest that cutaneous involvement is inversely related to the degree of systemic involvement [13].

It is recommended that arteriography be performed in systemically ill patients or those with relapsing cutaneous lesions. If arteritis is demonstrated, more aggressive cytotoxic treatment will be required for effective control of disease.

With regard to the immunological aspect of streptococcal-related disease, Martini and colleagues described a patient with a vasculitic illness much like the cutaneous PAN [14]. This patient also had myositis. There was a relapsing course from childhood with recurrences associated with an elevation in ASOT. Their studies demonstrated regions of homology between streptococcus type 5M protein and skeletal myosin. They felt that there were two possible mechanisms which might explain the association between streptococcal infection and the flare of disease. The first being superantigen T cell stimulation and the second molecular mimicry. Tomai and colleagues [15] reported that streptococcal type 5M protein may act as a superantigen binding to the major histocompatibility complex, thus activating T cells bearing Vβ8 chain receptors. This may lead to a non-specific activation of autoreactive T cell clones. Molecular mimicry is considered one of the possible mechanisms by which infectious agents induce autoimmune disorders. Group A streptococci contain certain antigens that are immunologically cross-reactive with human tissues, as described with type 5M protein and the human myocardium. Martini performed a computer search for the presence of shared sequences between 5M streptococcal antigens and human skeletal myosin and found four homologous regions shared by the streptococcal type 5M protein. They were long enough to represent a putative shared epitope. Murine monoclonal antibodies to *S. pyogenes* are able to cross-react with skeletal muscle myosin [16].

In selected individuals, streptococcal infections, possibly through the mechanisms discussed above, may be responsible for vasculitic disease.

## REFERENCES

1. Fox, E.N., Wittner, M.K. and Dorman, A. (1961) Antigenicity of the M proteins of group A hemolytic streptococci. *J. Exp. Med.*, **124**, 1135.
2. Fink, C.W. (1977) Polyarteritis and other diseases with necrotizing vasculitis in childhood. *Arthritis Rheum.*, **20**, 378–84.
3. Blau, E.B., Morris, R.F. and Yunis, E.J. (1977) Polyarteris nodosa in older children. *Pediatrics*, **60**, 227–34.
4. Fink, C.W. (1991) The role of the streptococcus in poststreptococcal reactive arthritis and childhood polyarteritis nodosa. *J. Rheumatol*, **18** (Suppl. 29), 14–20.
5. Diaz-Perez, J.L. and Winklemann, R.K. (1980) Cutaneous periateritis nodosa: a study of 33 cases. *Vasculitis* (eds. R.K. Winklemann and K. Woolf), Lloyd Luke, London, pp. 273–84.
6. David, J., Ansell, B.M. and Woo, P. (1993) Polyarteritis nodosa asscociated with streptococcus. *Arch. Dis. Child.*, **69**, 685–8.
7. Rose, G.A. and Spencer, H. (1957) Polyarteritis nodosa. *Q. J. Med.*, **26**, 43–81.
8. Freidberg, C.K. and Gross, L. (1934) Periateritis nodosa (necrotising arteritis) associated with rheumatic heart disease, with a note on abdominal rheumatism. *Arch. Intern. Med.*, **54**, 170.
9. Fordham, C.C. III, Epstein, F.H., Huffines, W.D. and Harrington, J.T. (1964) Polyarteritis and acute poststreptococcal glomerular nephritis. *Ann. Intern. Med.*, **61**, 89–97.
10. Golding, D.N. (1970) Polyarteritis presenting with leg pains. *Br. Med. J.*, i, 277–8.
11. Sack, M., Cassidy, J.T. and Boyle, G.C. (1975) Prognostic factors in polyarteritis. *J. Rheumatol*, **2**, 411–20.
12. Magilavy, D.B., Petty, R.E., Cassidy, J.T. and Sullivan, D.B. (1977) A syndrome of childhood polyarteritis. *J. Pediatr.*, **91**, 25–30.
13. Fisher, I. and Orkin, M. (1964) Cutaneous form of polyarteritis nodosa – an entity? *Arch. Dermatol.*, **89**, 180–9.
14. Martini, A., Ravelli, A., Albani, S. *et al.* (1992) Recurrent juvenile dermatomyositis and cutaneous necrotizing arteritis with molecular mimicry between streptococcal type 5/M/protein and human skeletal myosin. *J. Pediatr.*, **121**(5), 739–42.
15. Tomai, M., Kotb, M., Majumdar, G. and Beachey E.H. (1990) Superantigenecity of streptococcal M protein. *J. Exp. Med.*, **172**, 359–62.
16. Krisher, K. and Cunningham, M.W. (1985) Myosin: a link between streptococi and heart. *Science*, **227**, 413–5.

# 31

# Schönlein–Henoch purpura nephritis

*P. Niaudet*

## INTRODUCTION

Schönlein–Henoch purpura (SHP) is a clinical syndrome characterized by the association of skin, joint and gastrointestinal symptoms which may occur in successive attacks. SHP is characterized by widespread vasculitis. However, although the clinical symptoms of this disease are characteristic, the diagnosis is not always easy to establish because other forms of systemic vasculitis, mainly the microscopic form of periateritis nodosa, may mimic the disease. In addition, in contrast to systemic lupus erythematosus, there are no biological tests that can identify the disease with certainty. Immunofluorescence microscopy has made an important contribution to both the diagnosis and the study of the pathogenesis of the disease, particularly as it has demonstrated the presence of IgA deposits in the glomeruli and in the vessel walls. These findings brought confirmation of the immunologic nature of the pathologic lesions. There are no biological tests to identify the disease with certainty but it may now be considered that IgA deposits in the glomeruli and/or vessel walls is diagnostic.

## CLINICAL MANIFESTATIONS

SHP occurs mainly in children between the ages of three and 15 years and is uncommon in adults. Male patients are more commonly affected than female patients, with a male:female ratio ranging from 1.5:1 to 2:1.

*The Vasculitides*. Edited by B.M. Ansell, P.A. Bacon, J.T. Lie and H. Yazici.
Published in 1996 by Chapman & Hall, London. ISBN 978-0-412-64140-4.

The extrarenal manifestations of SHP may occur in any order and at any time over a period of several days or weeks. They principally involve the skin, joints and gastrointestinal tract.

The purpura is the initial manifestation in at least half the children. The rash is usually symmetrical, purpuric and affects areas of pressure in particular. The rash primarily involves the legs especially around the malleoli and buttocks, but may also be found on the elbows. Occasionally, purpura may occur almost anywhere on the body. Successive episodes of purpura are common during the first week of the disease, and occasionally are associated with recurrence of abdominal and joint symptoms. In a few patients, attacks of purpura may be observed for up to seven years [1, 2].

Arthralgia is present in about two-thirds of children and may be the initial manifestation. The arthralgia most commonly affects the knees and ankles, less commonly the wrist and small joints of the hands. Occasionally, other joints may be affected. The joints are often normal on examination, but there may be periarticular swelling. The arthralgia is always transient and may migrate from joint to joint, but no permanent damage to the joint occurs.

The abdominal symptoms associated with SHP may be mild, with colicky pain being the only manifestation. However, the pain may be severe enough to warrant a laparotomy. Intussusception secondary to severe purpura of the bowel is common in older children. Vomiting is frequently associated with any of the abdominal manifestations. Occasionally, perforation or necrosis of the bowel occurs. Ileal stricture also has been reported as a late complication of SHP.

Fever and weight loss are common prodromes. Hemorrhage into the calf, or subcutaneously, is common. The testis is also a site for painful bleeding. Periorbital and subconjunctival bleeding may occur, along with epistaxes. Acute hemorrhagic pancreatitis or parotiditis is seen occasionally, and lung purpura may be severe, simulating Goodpasture's syndrome. The vasculitic process can affect both the central nervous system and, more rarely, the peripheral nervous system. Occasionally, myocardial involvement has been reported in these patients, as has muscle involvement. Among the other complications, stenosing ureteritis has been reported several times, and renal infarction is possible.

## RENAL MANIFESTATIONS

In unselected groups of children with SHP, the frequency of renal involvement varies. These variations may depend on the methods of detection of nephritis. The proportion of patients with renal involvement has been reported to be 10–60%. Of 1987 unselected patients reported in different series, 531 (27%) had renal involvement. Glomerular involvement is usually evident within the first weeks from the onset of purpura but it may be the first manifestation of the disease, preceding the purpura [1, 2]. In other cases, it may develop with one of the later relapse of SHP. The severity of the dermal, gastrointestinal and articular manifestations bears no constant relationship to the severity of renal involvement and there is agreement that, even with the mildest extrarenal symptoms, the kidney may be affected. Symptomless hematuria, gross or microscopic, is by the far the most frequent sign of renal involvement. In some cases proteinuria in variable amounts is also present. A nephrotic syndrome is frequent in referral series (up to 50%) and exceptional in the general population with SHP nephritis. Early renal insufficiency may be associated with proteinuria and/or nephrotic syndrome but is usually moderate and rarely requires dialysis. Hypertension is in some instances observed during the initial period of the nephropathy and has been reported in children with minimal urinary abnormalities. Of the 151 patients with SHP nephritis whom we have studied and followed for more than one year, two had symptomless hematuria, 58 proteinuria with hematuria, and 89 a nephrotic syndrome, with early renal insufficiency in 26 [3].

## PATHOLOGY

The disease is characterized by the diffuse presence of granular deposits that always contain predominant IgA, localized primarily within the mesangium. These deposits may encroach on the capillary walls and in a few cases the IgA may be predominantly peripheral in an endomembranous location. The presence of small granules of IgA suggesting subepithelial deposits is not exceptional. The presence of peripheral deposits is usually associated with more severe forms of the disease. IgG and IgM deposits are found in the same location as IgA in 40% of cases and are most often less abundant. Fibrinogen is present in 80% and scattered granules of C3 in 85% of cases. C1q and C4 are almost invariably absent. The secretory component is not detectable in the deposits. IgA subclasses have also been studied with the use of monoclonal antibodies. It has been found that IgA deposits are mainly composed of IgA1 and there are discrepancies concerning the presence of IgA2.

By light microscopy the glomerular involvement is that of a mesangiopathy characterized by the presence of mesangial deposits with varying degrees of hypercellularity and of superimposed crescent formation. Tubules and interstitium show changes that are usually proportionate to the degree of glomerular involvement and arterioles may show some endarteritic changes in severe cases. Four patterns may be described:

- *Mesangiopathic glomerulonephritis* is characterized by the presence of normal or nearly normal glomeruli with an enlargement of mesangial stalks due to the presence of deposits which are exceptionally seen by light microscopy.
- *Focal and segmental glomerulonephritis* is characterized by the presence of a varying number of crescents or synechiae to Bowman's capsule superimposed on a mesangiopathic glomerulonephritis. The size of the crescents is usually small.
- *Diffuse proliferative endocapillary glomerulonephritis* is characterized by a combination of cellular proliferation mainly composed of mesangial cells with an increase in mesangial matrix and more or less numerous double contours. A pattern mimicking membranoproliferative glomerulonephritis with diffuse interposition of mesangial cells and matrix is, however, exceptional.
- *Endo- and extracapillary glomerulonephritis* is characterized by the presence of a variable proportion of crescents superimposed on a diffuse proliferative glomerulonephritis. In this proliferative variety of glomerular involvement, the segments affected (and therefore the crescents) tend to be larger. The latter may occasionally be circumferential.

CLINICOPATHOLOGIC CORRELATIONS

The clinicopathologic correlations may be illustrated by our experience concerning 151 children having rather severe forms of SHP nephritis [3]. Overall, the higher the percentage of glomeruli affected by segmental lesions and crescents, particularly when combined with diffuse mesangial hypercellularity, the more severe the clinical presentation (Table 31.1). Patients who have hematuria only usually show a mesangiopathic glomerulonephritis or a focal and segmental glomerulonephritis with few segmental lesions. Of the patients who had proteinuria associated with hematuria, 14% had no crescents at all, 62% had fewer than 50% of their glomeruli affected by crescents, and only 23% had 50% or more of the glomeruli involved with crescents. At the other extreme, of the patients who had heavy proteinuria or the nephrotic syndrome with renal impairment at onset of their nephritis, 78% had 50% or more of crescentic glomeruli and, in most, this was combined with diffuse endocapillary proliferation. However, it is difficult to predict the histopathology in patients presenting with the nephrotic syndrome. Therefore, a renal biopsy is justified in such patients.

PROGNOSIS

Nephritis is the major cause of morbidity and mortality in patients with SHP. In Europe, SHP accounts for approximately 3% of children progres-

**Table 31.1** Clinicopathologic correlations at the time of initial renal biopsy

| | n | Isolated hematuria | Proteinuria < 1g/d | Proteinuria ≥ 1 g/d | Nephrotic syndrome | Proteinuria/ nephrotic syndrome + renal failure |
|---|---|---|---|---|---|---|
| Mesangiopathic glomerulonephritis | 2 | 1 | 1 | | | |
| Focal segmental glomerulonephritis | 47 | 1 | 11 | 18 | 15 | 2 |
| Diffuse proliferative endocapillary glomerulonephritis | 13 | | 1 | 6 | 4 | 2 |
| Endo- and extracapillary glomerulonephritis | | | | | | |
| crescents < 50% | 21 | | 3 | 4 | 12 | 2 |
| crescents ≥ 50% | 68 | | | 14 | 33 | 21 |
| Total | 151 | 2 | 16 | 42 | 64 | 27 |

sing to terminal renal failure, which makes the disease an important cause of renal failure in childhood. This might also be true in the USA [4]. Of the 151 patients followed by Habib, Niaudet and Levy [3] for one to 18 years, 29 progressed to end-stage renal failure, whereas 24 had persistent nephropathy at latest examination. However, the prognosis for unselected children is relatively good. In the series of Kobayashi *et al.* [5], only three out of 123 children with nephritis died, whereas 86% achieved complete remission. In the series reported by Koskimies *et al.* [6], one out of 39 children who had urinary abnormalities progressed to end-stage renal disease, and two had persistent nephropathy.

With long-term follow-up, however, the prognosis is not always as benign as it appears. In the experience of Counahan *et al.* [7], half of the children were in complete remission after two years, whereas a third of the patients exhibited persistent urinary abnormalities with normal renal function. Of the 88 patients who underwent follow-up for a mean duration of 10 years, most of the children who had urinary abnormalities with normal renal function at two years had gone into remission, whereas a proportion of those with decreased renal function had progressed to renal failure. However, with a mean follow-up of 23.4 years after onset, of 78 patients from the initial cohort of 88 subjects [8], 17 patients deteriorated clinically. It is of interest that seven of these patients had apparently completely recovered after a follow-up of 10 years. Moreover, in women, 16 (36%) of the 44 successful pregnancies were complicated by hypertension, persistent proteinuria or both abnormalities. The development of severe hypertension 10–20 years after onset had already been reported by Counahan *et al.* [7] in some patients with apparently healed glomerulonephritis.

Among the 151 children followed by Habib, Niaudet and Levy [3] for one to 18 years, of the 53 patients who had a poor outcome (defined by the persistence of a proteinuria ≥1 g/d and/or the progression to end-stage renal failure), 35 had a massive proteinuria with or without nephrotic syndrome associated or not with impaired renal function at time of biopsy. However, if massive proteinuria at presentation was a feature of poor prognosis, it is of interest that 56 of the patients who presented with these symptoms recovered or had minimal urinary abnormalities with normal renal function at latest examination. Therefore, although the renal illness may appear most severe in the early stages of the disease, remarkable recovery is possible. Exacerbation of renal symptoms and biopsy-confirmed worsening of glomerular lesions may be observed in patients who have repeated attacks of purpura or recurrent macroscopic hematuria.

In conclusion, patients with nephritis secondary to SHP who present with minimal urinary abnormalities have an excellent prognosis and only those patients who present with marked proteinuria with or without nephrotic syndrome and/or impaired renal function are at risk of developing chronic renal failure. In such patients a renal biopsy should be performed before initiating any form of therapy since the best prognostic indicator is the pathology (Table 31.2).

TREATMENT

There is no definite evidence, at present, that any form of treatment alters the course of nephritis associated with SHP [9]. However, some uncontrolled studies have reported an improvement in renal symptoms in patients treated with immunosuppressive agents, often in association with corticosteroids. Other studies have failed to demonstrate any benefit from such treatments. However, the results of therapy are difficult to compare in the various studies because the treated patients are not homogeneous in terms of severity of their nephropathy [10].

A serious problem arises with patients in whom adverse features are found on clinical and renal biopsy examinations, including the nephrotic syndrome (with or without renal insufficiency at presentation), extensive crescent formation, or both. In our opinion, treatments should be restricted to such patients. In view of the fact that each center has only a small number of patients presenting with severe renal disease, there are no controlled therapeutic trials available in the literature.

Our own experience is based on a retrospective study on the outcome,

**Table 31.2** Correlations between biopsy findings and late outcome in patients followed for more than one year

|  | n | Clinical recovery | Proteinuria < 1g/d | Proteinuria ≥ 1 g/d without renal failure | Renal failure |
|---|---|---|---|---|---|
| Mesangiopathic glomerulonephritis | 2 | 2 |  |  |  |
| Focal segmental glomerulonephritis | 55 | 29 | 15 | 6 | 5 |
| Diffuse proliferative endocapillary glomerulonephritis | 13 | 8 | 3 | 1 | 1 |
| Endo- and extracapillary glomerulonephritis |  |  |  |  |  |
| crescents < 50% | 21 | 11 | 3 | 6 | 1 |
| crescents ≥ 50% | 60 | 16 | 11 | 11 | 22 |
| Total | 151 | 66 | 32 | 24 | 29 |

as it relates to treatment, in 74 patients with severe forms of nephritis secondary to SHP who were evaluated between 1957 and 1980 [3, 11]. These patients had presented with nephrotic syndrome and/or 50% or more crescentic glomeruli on initial biopsy. Forty-five patients received various combinations of immunosupressive agents, corticosteroids and anticoagulants and 12 of these (27%) progressed to end-stage renal failure. Of the 29 untreated patients, 11 (38%) progressed to end-stage renal failure. These figures, although non-significant, suggested that therapy might have had a beneficial effect.

Following the initial reports of Cole *et al.* [12] and Rose, Cole and Robson [13] describing the beneficial role of steroid pulse therapy in the treatment of severe glomerulopathies, including SHP nephritis, we undertook, in 1980, a prospective study of the treatment of severe forms of SHP nephritis with methylprednisolone pulse therapy [10]. Among the 34 patients who presented, as in our retrospective study, with nephrotic syndrome and/or crescentic glomerulonephritis involving 50% or more of the glomeruli, five progressed to end-stage renal failure (Table 31.3). It is of interest that, of these five patients, three had been treated late in the course of their disease. We therefore recommend that children with severe forms of SHP nephritis should be treated as early as possible with methylprednisolone pulse therapy followed by a three-month course of oral prednisone.

**Table 31.3** Correlations between biopsy findings and outcome in patients treated with methylprednisolone pulses

| | n | Clinical recovery | Proteinuria < 1g/d | Proteinuria ≥ 1 g/d without renal failure | Renal failure |
|---|---|---|---|---|---|
| Focal segmental glomerulonephritis | 4 | 2 | 2 | | |
| Diffuse proliferative endocapillary glomerulonephritis | 2 | 2 | | | |
| Endo- and extracapillary glomerulonephritis | | | | | |
| crescents <50% | 7 | 5 | | 2(1)* | |
| crescents ≥50% | 21 | 9 | 2 | 5(1)* | 5 |
| Total | 34 | 18 | 4 | 7 | 5 |

*( )=recent relapse.

Finally, in patients with very extensive crescent formation, plasma exchange has been proposed, usually in combination with immunosuppressive agents, corticosteroids, or both [14, 15]. However, the specific role of plasma exchange is difficult to ascertain in protocols consisting of combined therapy. Moreover, plasma exchange is not without risk in such patients.

REFERENCES

1. Hurley, R.M. and Drummond, K.N. (1972) Anaphylactoid purpura nephritis: clinocopathological correlations. *J. Pediatr.*, **81**, 904.
2. Meadow, S.R., Glasgow, E.F., White, R.H.R. *et al.* (1972) Schönlein-Henoch nephritis. *Q. J. Med.*, **41**, 241.
3. Habib, R., Niaudet, P. and Levy, M. (1994) Schölein-Henoch purpura nephritis and IgA nephropathy, in *Renal Pathology with Clinical and Functional Correlations* (eds. C.C. Tisher and B.M. Brenner), JB Lippincott Company, Philadelphia, pp, 472–523.
4. Bunchman, T.E., Mauer, S.M., Sibley, R.K. and Vernier, R.L. (1988) Anaphylactoid purpura: charateristics of 16 patients who progressed to renal failure. *Pediatr. Nephrol.*, **2**, 393.
5. Kobayashi, O., Wada, H., Okawa, K. and Takeyama I. (1977) Schönlein-Henoch's syndrome in children. *Contrib. Nephrol.*, **4**., 48.
6. Koskimies, O., Rapola, J., Savilahti, E. and Vilska, J. (1974) Renal involvement in SHP. *Acta. Paediatr. Scand.*, **63**, 357.
7. Counahan, R., Winterborn, M.H., White, R.H.R *et al.* (1977) Prognosis of Henoch-Schönlein nephritis in children. *Br. Med. J.*, **2**, 11.

8. Goldstein, A.R., White, R.H.R, Akuse, R. and Chantler, C. (1992) Long-term follow-up of childhood Henoch-Schönlein nephritis. *Lancet*, **339**, 280.
9. Austin, H.A. and Balow, J.E. (1983) Henoch-Schönlein nephritis: prognostic features and the challenge of therapy. *Am. J. Kidney. Dis.*, **2**, 515.
10. Niaudet, P., Murcia, I., Beaufils, H. *et al.* (1993) Primary IgA nephropathies in children: prognosis and treatment, in *Advances in Nephrology* (eds J.P. Grunfeld, J.F. Bach, H. Kreis and M.H. Maxwell), Mosby Year Book, pp. 121–40.
11. Levy, M., Broyer, M., Arsan, A. *et al.* (1976) Anaphylactoid purpura nephritis in childhood. Natural history and immunopathology. *Adv. Nephrol.*, **6**, 183.
12. Cole, B.R., Brockleband, J.T., Kienstra, R.A. *et al.* (1976) 'Pulse' methylprednisolone therapy in the treatment of severe glomerulonephritis. *J. Pediatr.*, **88**, 307.
13. Rose, G.M., Cole, B.R. and Robson, A.M. (1981) The treatment of severe glomerulopathies in children using high dose intravenous methylprednisolone pulses. *Am. J. Kidney Dis.*, **1**, 148.
14. McKenzie, P.E., Taylor, A.E., Woodroffe, A.J. *et al.* (1979) Plasmapheresis in glomerulonephritis. *Clin. Nephrol.*, **12**, 97.
15. Camerone, G., Garelli, S., Valbonesi, M. and Mosconi, L. (1982) Plasma exchange treatment in a patient with severe SHP. *Minerva. Med.*, **18**, 1185.

# 32

# Vasculitis in familial Mediterranean fever

*M. Pras, P. Langevitz, A. Livneh and D. Zemer*

Familial Mediterranean fever (FMF) is an ancient disease – there is evidence that it has existed for thousands of years. The disease was recognized only 50 years ago, when the affected population who lived in geographic isolation came into contact with modern medicine [1–3].

Febrile painful attacks form the hallmark of FMF and are characterized by a marked elevation of body temperature accompanied by inflammation of the membranes of the peritoneum, pleura and synovia [4, 5]. Most febrile episodes are of short duration but sometimes synovial attacks may be prolonged and resume a protracted course that may last for weeks and months [6]. The inflammatory processes of FMF are characterized by the accumulation of sterile fluid containing a large number of polymorphonuclear cells, which is indicative of the existence of chemotactic activity during the febrile attacks. This is related most probably to the activity of the FMF gene product, the nature of which is unknown at present.

Amyloidosis is a very frequent event in FMF patients [7–9]. Amyloid deposition of the AA type occurs in various tissues and organs but mainly affects the kidneys [7, 8]. The amyloid fills the glomeruli gradually, causing a nephrotic syndrome at an early stage which progresses to renal failure. Five to seven years after the first symptom of the proteinuria, the patients reach end-stage renal disease.

Amyloidosis of the AA type is very frequent in Jewish patients of Sephardic origin and in Anatolian Turks suffering from FMF. It occurs

*The Vasculitides*. Edited by B.M. Ansell, P.A. Bacon, J.T. Lie and H. Yazici.
Published in 1996 by Chapman & Hall, London. ISBN 978-0-412-64140-4.

also in Arabs, Ashkenazi Jews and Armenians in Armenia, but amyloidosis is quite rare in Armenian patients living in California [9].

Daily prophylactic treatment with colchicine was suggested by Goldfinger [10] in 1973 and causes complete remission in 65% of patients, and a partial remission in which the rate and severity of attacks decrease markedly in 30% of patients. In 5% of patients, the rate of attacks remains unchanged, despite continuous colchicine therapy [11].

Continuous colchicine treatment prevents the development of kidney amyloidosis. Only four out of 906 compliant FMF patients developed a mild nephropathy following 4–11 years of colchicine treatment, while 16 out of 54 non-compliant patients developed a nephrotic syndrome during this period [12].

Two years ago, as part of a systematic genome-wide search, we found evidence of linkage between FMF and chromosome 16 DNA markers. Two loci from the subtelomeric region of the short arm of chromosome 16 had high lod scores of 9 and 14, sufficient to establish linkage [13].

All symptoms and signs of FMF could be explained by the metabolic defect which is caused by a product of the abnormal gene of chromosome 16. We do not have any reason to assume that the inflammatory episodes of FMF are caused by immunological reactions. We do not believe that the deposition of AA amyloidosis is related to an immune process. The complement levels are normal in this disease, no autoantibodies could be demonstrated, the elevated ESR is related to the high fibrinogen level rather than to hyperglobulinemia, and corticosteroid therapy is not effective in the arthritis and other symptoms of the disease.

About 3% of FMF patients develop manifestations of vasculitis. Two forms of vasculitis which are found in FMF patients are known to be caused by immunological reactions. We may add a third kind of vasculitis, perhaps more specific to FMF, which we also believe is mediated by immunological processes.

These three forms of vasculitis affect only a small group of FMF patients, but in a high frequency compared to the general population:

- *Henoch–Schönlein purpura (H–S)* [14–20] occurred in over 40 of our patients admitted to our hospital and in a number of other FMF patients that were treated in other hospitals in Israel. Most of the patients were children and young adults (Plate 4). Although the disease demonstrated the usual manifestations of H–S, in FMF the disease was characterized by a prolonged and severe course that required steroid therapy in most cases.
- *Polyarteritis nodosa (PAN)* has been reported in 15 cases of FMF [21–28]. All 15 cases involved young patients whereas PAN generally occurs in the fifth or sixth decade of life. PAN is a rare disease with

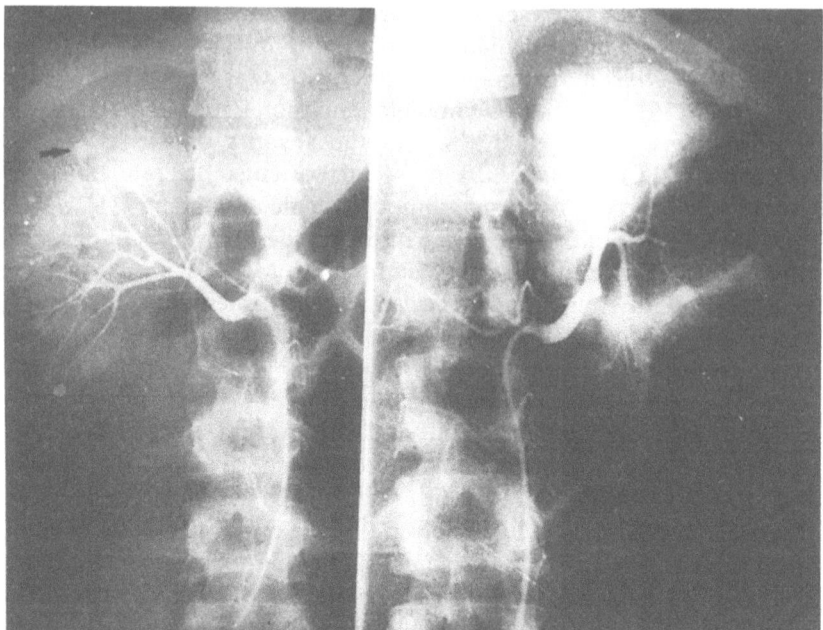

**Figure 32.1** Renal aneurysms in a child with FMF.

an incidence of five to six per million and the 15 cases of PAN recorded among approximately 10 000 cases of FMF are more than would be expected by chance. We cannot say that FMF patients are prone to develop PAN but there is no doubt that the disease is more frequent in FMF patients. The course in PAN and FMF is severe despite the relatively younger age of the patients. A peculiar feature of PAN associated with FMF is the occurrence of perirenal hematomas due to ruptured aneurysms [28] (Figure 32.1).

Before 1980 most patients with this complication of FMF died despite steroid therapy. Since cyclophosphamide therapy was introduced all PAN patients have survived [27–28]. A newly described syndrome designated by us as:

- *Protracted febrile myalgia* (PFM) occurred in 14 of our FMF patients [29]. All 14 patients had suffered from typical febrile attacks of FMF since early childhood and most of them were treated successfully with colchicine. All 14 patients had been admitted to hospitals in Israel with fever and disabling myalgia. Eleven out of 14 had suffered from long-standing abdominal pain without signs of peritonitis. In all patients the ESR was greatly elevated (97–145 mm/h). Moderate leukocytosis was found and polyclonal hyperglobulinemia (mean 5.5 g%) was observed in all patients. In one patient a skin biopsy was performed, and revealed a granulocytic infiltration in the walls of arterioles with disposition of IgA on immunofluorescent staining. While NSAIDs did not reduce the severe myalgic pain, prednisone 1 mg/kg promptly relieved the pain and other manifestations. The histological findings in the skin biopsy of this patient, coupled with the hyperglobulinemia in all patients, make us think that PFM should, perhaps, be classified in the spectrum of vasculitides associated with FMF.

# REFERENCES

1. Sohar, E., Gafni, J., Pras, M. *et al.* (1967) Familial Mediterranean fever. A survey of 470 cases and review of the literature. *Am. J. Med.*, **43**, 227–53.
2. Siegal, S. (1945) Benign paroxysmal peritonitis. *Am. J. Med.*, **22**, 1–21.
3. Mamou, H. and Cattan, R. (1952) La maladie periodique (sur 14 cas personnels dont 8 compliqués de nephropathies). *Semaine Hip. Paris*, **28**, 1062–70.
4. Heller, H., Gafni, J., Michaeli, D. *et al.* (1966) The arthritis of familial Mediterranean fever (FMF). *Arthritis Rheum.*, **9**, 1–17.
5. Sohar, E., Pras, M. and Gafni, J. (1975) Familial Mediterranean fever and its articular manifestations. *Clin. Rheum. Dis.*, **1**, 195–209.
6. Sneh, E., Pras, M., Michaeli, D. *et al.* (1977) Protracted arthritis in familial Mediterranean fever. *Rheum. Rehabil.*, **16**, 102–6.
7. Heller, H., Sohar, E., Gafni, J. *et al.* (1961) Amyloidosis in familial Mediterranean fever. *Arch. Intern. Med.*, **107**, 539–50.
8. Gafni, J., Ravid, M. and Shoar, E. (1968) The role of amyloidosis in familial Mediterranean fever, a population study. *Isr. J. Med. Sci.*, **4**, 995–9.
9. Pras, M., Bronshpigel, N., Zemer, D. and Gafni, J. (1982) Variable incidence of amyloidosis in familial Mediterranean fever among different ethnic groups. *Johns Hopkins Med. J.*, **150**, 22–6.
10. Goldfinger, S.E. (1972) Colchicine for familial Mediterranean fever. *N. Engl. J. Med.*, **287**, 1032.
11. Zemper, D., Revach, M., Pras M. *et al.* (1974) A controlled trial of colchicine in preventing attacks of familial Mediterranean fever. *N. Engl. J. Med.*, **291**, 932–4.
12. Zemer, D., Pras, M., Sohar, E. *et al.* (1986) Colchicine in the prevention and treatment of the amyloidosis of familial Mediterranean fever. *N. Engl J. Med.*, **314**, 1001–5.
13. Pras, E., Aksentijevich, I., Gruberg, L. *et al.* (1992) Mapping of a gene causing familial Mediterranean fever to the short arm of chromosome 16. *N. Engl. J. Med.*, **326**, 1509–13.
14. Niaudet, P. (1996) See chapter 32 of this volume.
15. Allen, D.M., Diamond, L.K. and Howell, D.A. (1960) Anaphylactoid purpura in children (Henoch–Schönlein syndrome). *Am. J. Dis. Child.*, **49**, 833–4.
16. Sibber, D.L. (1972) Henoch–Schönlein syndrome. *Pediatr. Clin. N. Am.*, **19**, 1061–70.
17. Rotem, Y. and Federgruen, A. (1962) Schönlein–Henoch syndrome in familial Mediterranean fever. *Harefuah*, **62**, 1–5.
18. Flatau, E., Kohn, D., Schiller, D. *et al.* (1982) Schönlein–Henoch syndrome in patients with familial Mediterranean fever. *Arthritis Rheum.*, **25**, 42–7.

19. Levinsky, R.J. and Barratt, T.M. (1979) IgA immune complexes in Henoch-Schönlein purpura. *Lancet*, ii, 1100–3.
20. Schlesinger, M., Vardy, P.A. and Rubinow, A. (1985) Henoch-Schönleing purpura and familial Mediterranean fever. *Isr. J. Med. Sci.*, **21**, 83–5.
21. Dollberg, L., Gavty, R., Grifman, M. and Rosenfeld, J. (1966) Periarteritis nodosa in a patient with familial Mediterranean fever. *Dapim. Refuim*, **25**, 408–15.
22. Padovani, J., Kasbarian, M., Pollini, J. *et al.* (1974) Intérêt de l'arteriographie rénale dans la périarterite noueuse. *Ann. Radiol.*, **17**, 135–40.
23. Mongin, M., Imbert, P., Dor, J.F. and Kasbarian, M. (1972) Hématome souscapsulaire rénale spontane et maladie periodique (A propos d'un cas). *Marseille Médicale*, **109**, 483–9.
24. Dor, J.F., Clauvel, J.P., Degos, L. and Mongin, F. (1979) Hématome peri-renale spontane au cours d'une maladie periodique. *Nouv. Presse Med.*, **8**, 1927–9.
25. Benhamou, E., Albou, A., Destaing F. *et al.* (1954) Périarterite noueuse et maladie periodique. *Bull. Soc. Med. Hop. Paris*, **70**, 247–54.
26. Bonan, H. (1955) Périarterite noueuse et maladie periodique. *La Tunisie Médicale*, **43**, 79–80.
27. Sachs, D., Langevitz, P., Morag, B. and Pras, M. (1987) Polyarteritis nodosa and familial Mediterranean fever. *Br. J. Rheumatol.*, **26**, 139–41.
28. Glikson, M., Galun, E., Schlezinger, M. *et al.* (1989) Polyarteritis nodosa and familial Mediterranean fever. A report of two cases and review of the literature. *J. Rheum.*, **16**, 536–9.
29. Langevitz, P., Zemer, D., Livneh, A. *et al.* (1994) Protracted febrile myalgia in patients with familial Mediterranean fever. *J. Rheumatol.*, **21**, 1708–9.

# 33

# Behçet's syndrome in children

*H. Özdoğan*

## INTRODUCTION

Behçet's syndrome (BS) is a multisystem disease of unknown etiology, but viral and streptococcal infections and autoimmunity have been implicated in its pathogenesis. The syndrome is characterized by oral and/or genital ulcerations, skin lesions and inflammatory eye lesions. A positive 'pathergy' reaction is one of the possible diagnostic features of the syndrome, especially in patients originating from the Mediterranean basin and Japan [1]. This geographical variation is also observed in the association of BS with the HLA-B51 antigen [2]. A summary of the clinical findings is shown in Table 33.1 [3[.

The onset of the disease occurs infrequently before the age of 16 or after 40 years. The onset is usually in the second and third decade of life, and the worst prognosis is in young males with an age of onset before 25 years [4]. Even though it is a topic of intensive investigation, little is reported on BS in children. Lang *et al.* reviewed the literature from 1965 to 1990 and found 37 cases [5]. Recently, data from four different centers made it possible to get an overview of the clinical expression of juvenile Behçet's syndrome (JBS), although the criteria preferred and the patient selection were not uniform [6–9].

## CLINICAL FEATURES

Tables 33.2 and 33.3 compare the data of the four centers. The French data is different because it gives the results of a nationwide survey, whereas the other three reports are retrospective reviews of their large,

*The Vasculitides*. Edited by B.M. Ansell, P.A. Bacon, J.T. Lie and H. Yazici.
Published in 1996 by Chapman & Hall, London. ISBN 978-0-412-64140-4.

**Table 33.1** Clinical findings in Behçet's syndrome*

| Lesion | Approximate prevalence (%) | Features |
|---|---|---|
| Oral aphthous ulceration | 97–100 | Minor or major ulceration or herpetiform lesions |
| Genital lesions | 80–90 | Usually scrotal in men |
| Skin lesions | 80–90 | Erythema nodosum-like lesions, acne, other skin vasculitides |
| Eye lesions | 50 | Anterior and/or posterior uveitis and retinal vasculitis which may lead to blindness |
| Arthritis | 40–50 | A non-deforming mono- or oligoarthritis |
| Arteritis | 1–3 | Occulsion, aneurysm formation, hemorrhage |
| Thrombophlebitis | 25 | May involve the venae cavae |
| Neurological involvement | 1–10 | Benign intracranial hypertension, multiple sclerosis-like picture, pyramidal involvement |
| Gastrointestinal involvement | 0–25 | Intestinal ulceration |

* From Yazici and Moutsopoulos [3].

**Table 33.2** Comparison of the clinical characteristics of patients from four centers

| | Tunisia [6] | Iran [7] | France [8] | Turkey [9] |
|---|---|---|---|---|
| Juvenile/total | 14/582 (2%) | 67/2175 (3%) | 17 | 52/2154 (2%) |
| M:F | 9:5 | 33:34 | 11:6 | 24:28 |
| Mean age at onset | 12 | -** | 11 (2 mo.–14 y) | 11.6 ± 3 (3–16 y) |
| Age at diagnosis | 15 ± 5 (10–32 y) | – | 11 (3–18 y) | 14.9 ± 1.3 (12–16 y) |
| Diagnostic criteria | ISGC* | Japanese | Mason Barnes ISGC | ISCG |
| Mean follow-up | 6 ± 4 (1–15 y) | – | (1–2.2 y) | 2.4 ± 2 (1 mo.–10 y) |
| Prepubertal onset | – | – | – | 14 |
| Family history | 2 (14%) | – | 3 (18%) | 4 (8%) |

*ISGC: Interational Study Group Criteria.
**(-): Not given in the original paper.

**Table 33.3** Comparison of clinical features of patients with JBS, from four centers

|                       | Tunisia (%) | Iran (%) | France (%) | Turkey (%) |
|-----------------------|-------------|----------|------------|------------|
| Oral ulcers           | 100         | 77       | 94         | 100        |
| Genital ulcers        | 60          | 26       | 53         | 70         |
| Skin lesions          | 86          | 61       | 92         | –          |
| ostiofolliculitis     | 71          | –**      | –          | 77         |
| erythema nodosum      | 14          | –        | –          | 59*        |
| Eye involvement       | 14          | 31       | 47         | 48         |
| Arthritis             | 21          | 13       | 69         | 34         |
| CNS involvement       | 7           | –        | 18         | 6          |
| Vascular involvement  | 33          | –        | 6          | 12         |
| Pathergy              | 65          | –        | 60         | 68         |
| HLA-B5                | 30          | –        | 54         | 52         |
| Pharyngeal stenosis   | 1           | –        | –          | 1          |

\* EN: 100% in children with prepubertal onset.
\*\* (–): Not given in the original paper.

hospital-based BS populations. According to these reports, approximately 2–3% of the patients fulfil the diagnostic criteria for BS before the age of 16. It is concluded that the prevalence of JBS in France is one out of 600 000, which is considerably lower than the rates given for BS in northern Europe (one out of 300 000), Japan (one out of 10 000) and Turkey (10/10 000, 40/10 000) [10]. The sex ratio is almost equal to one in the Turkish and Iranian series; however, males predominated in the other two. The mean age of onset is about 12, with a wide range starting from two months to 16 years. The diagnosis is usually delayed for an average of three years. The role of puberty is reported only in the Turkish group, and the disease started before puberty in approximately one-quarter of these cases.

### Mode of onset

The initial symptom is usually recurrent oral ulcers in two-thirds of the patients. Rarely, uveitis, genital ulcers or arthritis can be the first manifestation of the disease.

### Skin and mucosal involvement

Virtually all patients with JBS experience oral ulceration, the majority developing usually multiple, minor aphthous ulcers. Major or herpetiform ones are observed less frequently. Genital ulcerations, which often scar (especially on the male scrotum), are more prevalent in girls. The frequency is less in the prepubertal children with BS but increases after puberty. The presence of perianal aphtosis has been emphasized by Hamza [6]. He observed these lesions in 14% of the children compared to 1% in adult BS.

The frequency of ostiofolliculitis was similar in the adult and post-pubertal group, and acne-like lesions were present only in children with an onset after puberty. However, erythema nodosum was observed in every prepubertal child with BS compared to 44% in the postpubertal group of the Turkish patient population.

## Eye involvement

Eye involvement is the most serious manifestation of the syndrome with regard to severe morbidity. It is most prevalent among young male patients in whom there is up to 70% prevalence compared to 5% in postmenopausal females [4]. It has been suggested that ocular involvement was uncommon in JBS [5, 6]. The frequency of uveitis is 14% in the Tunisian JBS population; however, it is 47% and 48% in the French and Turkish patients, respectively. Eye involvement is especially prevalent in boys (71%).

## Joint involvement

Arthritis of BS is usually mono- or oligoarticular, may be episodic or continuous but rarely causes joint damage. The ankles, knees, wrists and elbows are the most commonly affected joints. There is no sex or age predilection, but a geographical variation is observed in the frequency of arthritis between French (69%) and Iranian (13%) patients.

## Central nervous system involvement

Although rare, CNS involvement may occur in children with BS. Seizures, raised intracranial pressure with associated headache and cerebral stem syndrome are characteristic. There appears to be a geographical variation in the reported prevalence of CNS involvement, being highest in France (18%), but in only 6–7% of Turkish and Tunisian patients. This discrepency is also true for the adult BS populations of the USA and northern Europe (10–25%) compared to that of Turkey (5%) [11].

## Vascular involvement

Both venous and arterial involvement may occur in patients with JBS and signal a poor outcome in terms of mortality. Various forms of thrombophlebitis, affecting peripheral calf veins, superior and inferior venae cavae and hepatic veins, are observed. Some patients develop thrombosis of more than one site as reported in the Turkish and Tunisian children. Aneurysm of the pulmonary artery was the cause of death in two patients. Vascular involvement is observed only in boys with JBS.

## Pharyngeal stenosis

Although very rare, this has been reported in a single patient in 3 of the 4 groups. Aspiration pneumonia secondary to pharyngeal stenosis was the cause of death in a Turkish patient.

## Pathergy reaction

The frequency of pathergy positivity is similar among the above-mentioned patient populations and does not seem to be affected by either sex or age.

## HLA-B5

Approximately half the children carry this antigen. Recently, it has been suggested that this association is linked with more severe manifestations of the syndrome [10].

## THERAPY

Azathioprine at a dose of 2.5 mg/kg/day has been shown to control the progression of existing, and the development of new, eye disease [12]. Cyclosporin-A at 5 mg/kg/day is also effective for severe eye involvement [13]. To control the inflammation in severe cases of eye disease and vasculitis, systemic corticosteroids alone or in combination with immunosuppressives are administered. Colchicine is effective in treating erythema nodosum and joint manifestations [14]. In our Turkish JBS population, 27% of the children received immunosuppressive treatment, azathioprine alone or in combination with cyclosporin-A, mainly for severe ocular disease.

## CONCLUSION

Although rare, BS may be seen in children. Two to 3% of patients with BS fulfil the diagnostic criteria before the age of 16 years.

Early onset and male sex are poor prognostic signs with regard to eye involvement. All of the patients with vascular involvement are male. These findings are in concordance with the previous reports stating that

**Table 33.4** Vascular involvement in Tunusian, French and Turkish children with JBS

|                                 | Tunusia | France | Turkey |
|---------------------------------|---------|--------|--------|
| Total                           | 4       | 1      | 7      |
| Intracranial                    |         |        |        |
|   hypertension        | 3       | –      | –      |
|   SSS* thrombosis     | 3       | –      | –      |
| Budd-Chiari                     | 1       | –      | 1      |
| Superior + inferior             |         |        |        |
| venae cavae th.**               | 2       | 1      | 2      |
| Calf vein th.**                 | –       | –      | 7      |
| Pulmonary aneurysm              | 1       | –      | 1      |
| Died                            | 2       | –      | 2      |

*SSS: superior sagittal sinus.
**th.: thrombophlebitis.

BS runs a distinctly more severe course among males and young patients compared to female and older patients [4].

All of the patients with prepubertal onset develop erythema nodosum. This observation may prove useful in differential diagnosis.

In general, in spite of some minor variations, the clinical picture of childhood BS resembles that of adult disease.

## REFERENCES

1. Yazici, H., Chamberlain, M.A., Tüzün, Y. *et al.* (1984) A comparative study of the pathergy reaction among Turkish and British patients with Behçet's disease. *Ann. Rheum. Dis.*, **43**, 74–5.
2. Zouboulis, C.C., Büttner, P., Djawari, D. *et al.* (1993) HLA-Class I antigens in German patients with Adamantiades-Behçet's disease and correlation with clinical manifestations, in *Behçet's Disease* (eds. B. Wechsler and P Godeau), Excerpta Medica, Amsterdam, pp. 175–80.
3. Yazici, H. and Moutsopoulos, H. (1985) Behçet's disease, in *Current Therapeutics, Immunology and Rheumatology* (eds. L.M. Lichtenstein and A.S. Fauci), Decker, Philadephia, pp. 194–7.
4. Yazici, H, Tüzün, Y. Pazarli, H. *et al.* (1984) Influence of age of onset and patient's sex on the prevalence and severity of manifestations of Behçet's syndrome. *Ann. Rheum. Dis.*, **43**, 783–9.
5. Lang, B.A., Laxer, B.M., Thorner, P. *et al.* (1990) Pediatric onset of Behçet's syndrome with myositis: Case report and literature review illustrating unusual features. *Arthritis Rheum.*, **33**, 418–25.
6. Hamza, M. (1993) Juvenile Behçet's disease, in *Behçet's Disease* (eds. B. Wechsler and P. Godeau), Excerpta Medica, Amsterdam, pp. 377–80.
7. Shafaie, N., Shahram, F., Davatchi, F. *et al.* (1993) Behçet's disease in children, in *Behçet's Disease* (eds. B. Wechsler and P. Godeau), Excerpta Medica, Amsterdam, pp. 381–3
8. Koné-Paut, I. and Bernard, J.L. (1993) Behçet's disease in children: a French

nationwide survey, in *Behçet's Disease* (eds. B. Wechsler and P. Godeau), Excerpta Medica, Amsterdam, pp. 385–9.

9. Yurdakul, S., Özdoğan, H., Kasapçopur, Ö. *et al.* (1993) Behçet's syndrome with juvenile onset: Report of 44 patients. *Clin. Exp. Rheum.*, **11**, S9, 71.

10. Yurdakul, S., Günaydin, İ, Tüzün, Y. *et al.* (1988) The prevalence of Behçet's syndrome in a rural area in northern Turkey. *J. Rheumatol.*, **15**, 820–2.

11. Serdaroğlu, P., Yazici, H., Özdemir, C. *et al.* (1989) Neurological involvement in Behçet's syndrome: a prospective study. *Arch. Neurol.*, **46**, 265–9.

12. Yazici, H., Pazarli, H., Barnes, C.G. *et al.* (1990) A controlled trial of azathioprine in Behçet's syndrome. *N. Engl. J. Med.*, **322**, 281–5.

13. Özyazgan, Y., Yurdakul, S., Yazici, H. *et al.* (1992) Low dose cyclosporin A versus pulsed cyclophosphomide in Behçet's syndrome: a single masked trial. *Br. J. Ophthalmol.*, **76**, 241–3.

14. Aktulga, A., Altac, M., Müftüoğlu, A. *et al.* (1980) A double blind study of colchicine in Behçet's disease. *Haematologica*, **65**, 399–402.

# Index

Page numbers appearing in *italics* refer to tables; page numbers appearing in **bold** refer to figures; differential diagnosis is indicated by *vs*.

ACTH 310
Acute necrotizing vasculitis 278
Adhesion molecules 14
  blood levels of circulating *90*
  counter-receptors *216*
  detection in body fluids 88
  cDNA identification 83
  endothelial cells *43*
  leukocytes *43*
  soluble 88
  systemic vasculitis 365
  tissue infiltration 83
Ag/Ab ratio 288–9
AIDS-related complex 249
Allergic angiitis 26
  granulomatous 8, 24
  *see also* Eosinophilic angiitis
Allergic granuloma 9
Allergic vasculitis 24
  complement cascade activation 40
Allergy, vascular manifestations 135
Alveolar hemorrhage 222
  classification *224*
  lung vasculitides 224–5
American College of Rheumatology
  classification (1990) 26, *32*, 48–9

Diagnostic and Therapeutic Criteria
  Committee 152
  Takayasu arteritis classification 184
  Wegener's granulomatosis
    classification 147, *148*, *149*
ANCA 10, 16–17, 48–52
  antigen binding block 369
  autoantibodies 223
    to myeloid granular proteins 52–4
  binding to target antigens 59
  cell surface binding 44
  Churg–Strauss syndrome 123, 223,
    230, 233
  classical PAN 122, 123
  cross-recognition 42
  depletion by specific
    immunoabsorption 369
  endothelial cell target 161
  immunological marker 13
  Kawasaki syndrome 385–6, 394
  microscopic polyangiitis 123
  microscopic polyarteritis 141
  neutrophil activation 358, 388
  pathophysiological role 55, 59
  PMN activation 364
  polyarteritis nodosa (PAN) 130

ANCA *contd*
  proteinase 3 specificity 363
  pulmonary vasculitides 223
  relapse prediction of systemic
    vasculitis 365
  signal transduction pathway effects
    161
  small-vessel vasculitis 363
  systemic vasculitis management
    364–5
  testing in diagnostic approach 55
  tissue injury mechanisms 364
  Wegener's granulomatosis 156, 159,
    **160**, 161, 223
    marker 49–50
    relapse 357–8
ANCA-associated vasculitis 162
ANCA-cytokine-sequence theory 159,
  **160**
c-ANCA 44, 50
  diagnostic use 55
  Wegener's granulomatosis 50–1, 147
    markers 150, 225
p-ANCA 44, 50, 51, 151
  diagnostic use 55
Aneurysm 5, 6
  Behçet's disease 116, 207
  classical polyarteritis nodosa (PAN)
    109, 110, 111
  familial Mediterranean fever **414**
  pulmonary 237, **238**, 239
  Takayasu arteritis 113, 185, **186**, **187**
Angiitis of the central nervous system
  246, 247
  antiphospholipid syndrome 259
  Behçet's disease 256
  Churg–Strauss syndrome 255
  Cogan's syndrome 256–7
  fibromuscular dysplasia 259
  giant cell arteritis 255
  Hodgkin's disease-associated 252
  hypersensitivity 255
  inflammatory bowel disease 257
  lymphoid granulomatosis 257
  malignant angioendotheliomatosis
    **258**, 259
  Moyamoya disease 259
  non-granulomatous **250**, 252, **253–4**

  polyarteritis nodosa 255
  primary 246–8, **247**
  primary granulomatous 246, 248–9,
    **250–1**, 252, **253**
    clinical course 252
    clinical features 248–9
    diagnosis 248–9
    etiology 249, 252
    fibrosis 249
    immunosuppressive therapy 252
    incidence 248
    lymphoma/leukemia association
    249
    mycoplasmal infection 249, 252
    pathology 248, 249
    spinal cord 252, **253–4**
  sarcoidosis 257
  secondary 246, **247**, 252, 255–7, 257,
    **258**, 259
  systemic lupus erythematosus (SLE)
    256
  systemic necrotizing vasculitis 255
  Takayasu arteritis 255
  temporal arteritis 255
  Wegener's granulomatosis 255
Angiography 11, 109
  Behçet's disease 116
  cardiac myxoma embolism 259
  cerebral vasculitis 249
  classical polyarteritis nodosa (PAN)
    109–12
  hepatic 110, **111**
  intravenous digital subtraction 114
  large-vessel vasculitis 112–14, **115**,
    116
  magnetic resonance 114
  polyarteritis nodosa (PAN) 130
  renal 110–11
  small-vessel vasculitis 112
  systemic necrotizing arteritis 109–12
  Takayasu's arteritis 113–14, **115**
  temporal arteritis 112–13
  treatment 117
  vasculitis look-alikes 116–17
Angiotensin converting enzyme 394
Ankylosing spondylitis 202
Anti-CD4 antibody 165, 166
Anti-CD25 antibody 165

Anti-CD52 monoclonal antibodies 370
Anti-Clq antibodies 43
Anti-DNA antibodies 70, 71
Anti-elastase antibodies 54
Anti-idiotypic antibody therapy 368
Anti-MPO 10, 51, 52, 53–4, 123, 233,
    363–4, 388
  assays 55
  positivity **56, 58**
Anti-PR3 52, *53*
  antibodies 388
  assays 55
  inhibition 59
  positivity **57, 58**
Anti-VLA-4 monoclonal antibody 92
Antibodies, vascular damage 42
Anticardiolipin antibodies 217, 259
Antiendothelial cell antibodies
    (AECA) 42, 65, 66, 217
  antigen characterization 77
  antigen specificity 76–7
  binding characteristics 66–7
  CNS disease 73–4
  complement levels 69
  cytotoxic 75–6
  detection 66–7
  endothelial cytotoxicity 75–6
  idiopathic inflammatory myopathies
    75
  immunological marker 13
  Kawasaki syndrome 385–6, 394
  nephritis association 69–71
  pathogenetic mechanisms in
    systemic vasculitis 74–7
  pathogenic role 42
  rheumatoid vasculitis 272
  systemic lupus erythematosus (SLE)
    67–74
  thrombosis 72
  vascular damage 77
Antiglomerular basement membrane
    antibodies 224, 225
Antilactoferrin antibodies 54, 272
Antimyeloperoxidase antibodies, *see*
    Anti-MPO
Antineutrophil cytoplasmic
    antibodies, *see* ANCA
Antinuclear antibodies (ANA) 10

cross-recognition 42
  rheumatoid vasculitis 272
Antiphospholipid antibodies 72, 72–3
  cross-recognition 42
Antiphospholipid syndrome 71, 259
Antiproteinase III 10
Antistreptokinase 398
Antithrombin III impairment 72
$\alpha_1$-antitrypsin 59
Aortic insufficiency in Takayasu
    arteritis 182
Arterial wall inflammation 124
Arteriosclerosis obliterans 194
Artery
  dissection in Takayasu arteritis 185
  thickening 4
Arthralgia
  Churg–Strauss syndrome 231
  polyarteritis nodosa (PAN) 126
Aspirin 386
  Kawasaki syndrome 395
Asthma
  Churg–Strauss syndrome 123, 230
  corticodependent 230
  polyarteritis nodosa (PAN)
    association 339
Azathioprine 16, 324–5
  Behçet's syndrome 205, 422
  maintenance therapy 325
  microscopic polyarteritis treatment
    318
  polyarteritis nodosa treatment 318
  polymyalgia rheumatica treatment
    324
  prednisolone combination 324
  rheumatoid vasculitis 271
  systemic rheumatoid vasculitis
    treatment 319
  systemic vasculitis treatment 321,
    368
  treatment use 310

B cell lymphoma, malignant 285
B cells 17
  IgG/IgM secretion in Kawasaki
    syndrome 387
Behçet's disease/syndrome 199, *200*,
    **200, 201,** 202–5

428 *Index*

Behçet's disease/syndrome *contd*
  adhesion molecules 216–17
  aneurysm 207
  angiography 116
  antiendothelial cell antibodies
    (AECA) 75, 217
  aphthae 203, 237, **418**, 420
  arterial occlusion 207
  arthritis *418*, 421
  azathioprine 422
  Budd–Chiari syndrome 207
  CD11b/CD18 expression 216
  CD44 expression 216
  central nervous system involvement
    *418*, 421
  childhood 417, *418–19*, 420–3
  clinical features **200**, 417, *418–19*,
    420–2
  colchicine 422
  cyclosporin A 325, 422
  cytokines in lesions 217
  diagnostic criteria 237
  endothelial cell
    antithrombotic/antiaggregant
    activity 211, **212**, 213
    damage mechanism 218
    dysfunction 207–9
  endothelin levels 214–15
  endothelin-1, 2 218
  erythema nodosum 423
  etiology 207–8
  eye involvement *418*, 421
  familial clustering 202, 203
  genetic transmission 203, 204
  genital ulceration *418*, 420–1
  HLA B5 antigen association 203,
    207, 422
  HLA B51 antigen association 202,
    417
  inferior vena cava thrombosis *201*
  interleukin-1 (IL-1) 217, 218
  interleukin-6 (IL-6) 217
  mortality 204–5
  onset 417, 420
  oral aphthosis 203, 237, **418**, 420
  pathergy *201*, 202, **203**, 422
  pharyngeal stenosis 422
  plasma fibrinolytic activity 210

  plasminogen activator-inhibitor-1
    (PA-1) levels 214
  prognosis 205, 421
  prostacyclin (PGI$_2$) levels 211, **212**,
    213, 218
  pulmonary arterial aneurysms **200**,
    239
  pulmonary vasculitides 222, *223*, 237
  L-selectin expression 216
  skin involvement *418*, 420–1
  thrombomodulin 213, 218
  thrombosis 207, 237
  TNF-α 217, 218
  treatment 205, 422
  triad 207
  urate crystal test *201*, 202, 203
  vascular endothelium 209
  vascular involvement 207, 421–2,
    422–3
  vascular occlusion mechanisms *219*
  vascular tree 199
  vasculitis 208
  venous lesions 207
  von Willebrand factor (vWF) 210
  von Willebrand factor-antigen 218
Benign hypergammaglobulinemic
    purpura of Waldenström 302
Berger's disease, *see* IgA nephropathy
Biopsy 21
Birmingham Vasculitis Activity Score
    (BVAS) 316
Brain in Wegener's granulomatosis
    **101**, *102*, 103, 108
Budd–Chiari syndrome 207
Buerger's disease, *see* Thromboangiitis
    obliterans

C1q receptor 45
  endothelial surface 66
C3
  complement factor fixation 66
  deposits in juvenile
    dermatomyositis 378
C3b 41
  deposition 44
  receptor 66
C5–9 membrane attack complex 378
c-reactive protein

systemic vasculitis 365
Wegener's granulomatosis 156
Cadherins 215
Cardiac involvement in Churg–Strauss
  syndrome 232
Cardiac myxoma embolism 259
Cardiomegaly 127
Cathepsin G antibodies 54
Kawasaki syndrome 388
CD3+-CD4+ T cells, Behçet's disease
  208
CD4+ cells in Wegener's
  granulomatosis 158, *159*
CD4 cells
  corticosteroid effects 311
  depletion monitoring 371
CD11 adhesion molecules 43
CD11a 216
CD25+ cells 158, *159*
CD31 expression 215
CD34 86
CD62 86
Cell adhesion molecules 83
  circulating in disease 87–91
  classification 83, **84**
  immunoglobulin superfamily
    receptors 85
  integrin binding motif 96
  integrins 84–5
  lymphocyte circulation 87
  role 87
  selectins 85–6
Central nervous system
  Behçet's syndrome *418*, 421
  disease 73–4
  polyarteritis nodosa (PAN) 126
  Wegener's granulomatosis **101**, *102*,
    103
Cerebral sarcoidosis 252
Cerebral vasculitis
  angiography 249
  drug use/abuse 257
  hypersensitivity angiitis 255
Chemotactic factor release 40
*Chlamydia* 386
Chlorambucil 319
Cholesterol 394
  emboli 304–5

Churg–Strauss syndrome 22, 121, 123
  American College of Rheumatology
    classification 32
  ANCA 223, 230, 233
  aneurysms 111
  angiitis of the central nervous
    system 252–9
  anti-MPO 53, 55
    specificity 59
  antiviral treatment 334
  asthma 230
  cardiovascular manifestations 232
  classification *148*
  clinical manifestations 339, *340*
  cranial nerve palsy 232
  cyclophosphamide treatment 336,
    337
  gastrointestinal involvement 233
  glomerulonephritis 232
  investigations 233
  laboratory findings 341–2
  leukocyte margination 112
  lung vasculitides 230–3
  markers 32–3, **34**
  mortality 346–7, *348–9*
  myeloperoxidase antibodies 223
  neurological symptoms 231–2
  pathological characteristics *30–1*
  plasma exchange 319, 336, 337, 342,
    351
  prednisolone long-term therapy 335,
    337, 338
  prednisone treatment 339
  prognosis 131
  pulmonary vasculitides 222, 223
  renal involvement 232
  rheumatic symptoms 231
  skin manifestations 232
  steroids 319
  systemic necrotizing arteritis with
    granulomatosis 112
  therapy trial results 339–50, *348–9*
  treatment 319, 332–4
    side effects 347, *350*
Circulating immune complexes
  Ag/Ab ratio 288–9
  essential mixed cryoglobulinemia
    287, 288–90

Circulating immune complexes *contd*
  reticuloendothelial system 289
  rheological abnormalities 290
Clotting cascade, endothelium 72
Clotting factors 210
Co-trimoxazole 359
Cogan's syndrome 256–7
Colchicine
  Behçet's syndrome 422
  familial Mediterranean fever 413
Cold-reactive antiglobulins 300
Complement
  binding 41
  cascade 40, 44
  essential mixed cryoglobulinemia
    289
  function alteration 289
  membrane-bound regulator
    expression of activation 45
  system activation 41
Computed tomography (CT)
  Q-P study 106–7, 108
  Wegener's granulomatosis 99, 103,
    *104*, **104–5**, 106–7, 108, 227, **228–9**
Connective tissue disease, pulmonary
  vasculitis 240
Coronary artery vasculitis 127
Corticosteroids 316
  CD4 cell effects 311
  cost-benefit analysis need 312
  cyclophosphamide combination
    310–11
  cytotoxic agent combination 316
  dosage 311
  immune system effects 311
  inflammatory response inhibition
    312
  juvenile dermatomyositis 381
  polyarteritis nodosa (PAN) 332, 400
  route 311
  treatment of vasculitides 309–13
  *see also* Methylprednisolone;
    Prednisolone; Prednisone;
    Steroids
Cortisone 310
Cotrimoxazole 158
CR1 receptor 41, 42
Cranial nerve palsy 126

Cryofibrinogenemia 301–2
Cryoglobulin
  mixed
    immune complex 300
    production 300
  Sjögren's syndrome 279
  types I-III 283
Cryoglobulinemia
  alveolar hemorrhage 225
  mixed 283–4, 301
    blood viscosity 290
    glomerulonephritis 301
    hepatitis C infection 288
    monoclonal 180
    treatment 301
    *see also* Essential mixed
      cryoglobulinemia
  pulmonary vasculitis 236–41
  small-vessel necrotizing vasculitis
    300–1
  type II 279–80
Cryoglobulinemic vasculitis 10
  *vs.* microscopic polyarteritis 139
CS1 peptide 95
Cutaneous vasculitis 294
  childhood 376
  large vessel disease 303–4
  small-vessel necrotizing vasculitis
    294–303
Cyclophosphamide 13, 14, 15
  Churg–Strauss syndrome treatment
    336, 337
  continuous therapy 319–24, 326
  familial Mediterranean fever 414
  intermittent pulse 271
  mechanisms of action 323–4
  methylprednisolone combination
    271
  microscopic polyarteritis 142, 318
  necrotizing vasculitis 323–4
  polyarteritis nodosa treatment 318,
    332, 336, 337
    duration 350
  prednisolone combination 320–3,
    326, 350
  pulse therapy 319–23, 326
    route 320
  regimes 319–24

rheumatoid vasculitis 271
second-line treatment 336–7
side effects 324, 347–8, 368
steroid combination 15, 142
systemic rheumatoid vasculitis
 treatment 319
systemic vasculitis treatment 366,
 368
toxic effects 322, 323
treatment use 310
Wegener's granulomatosis 162, 163,
 *164*, 226, 316
Cyclosporin, juvenile dermatomyositis
 382
Cyclosporin A 325
 Behçet's syndrome 325, 422
 cutaneous polyarteritis nodosa
  treatment 304
 giant cell arteritis 319
 large-vessel vasculitis 319
 Takayasu's arteritis 319
 Wegener's granulomatosis 163, 165
Cytoadhesins 85
Cytokines 43
 neutrophil activation 44
 systemic vasculitis 365
Cytomegalovirus 10, 357
 polyarteritis nodosa (PAN) 124
Cytoplasmic antigens 44
Cytotoxic therapy 313


Dapsone
 erythema elevatum diutinum (EED)
  303
 urticarial vasculitis 300
Decay accelerating factor (DAF) 45
Dermatomyositis 75
Dipyridamole 386
 Kawasaki syndrome 395
Disease activity
 indices 12
  damage 12–13
  functional 13
 measurement 12
DPB1*1401 alleles 74
Drug use/abuse, cerebral vasculitis
 257

Eicosanoids 311
Endarteritis obliterans 278, 304
Endothelial cell layer permeability 41
Endothelial cells
 adhesion molecules *43*, 86–7
 anti-inflammatory deregulation 40,
  44–5
 antibody 42, 65, 217–19
  binding 39
 anticoagulant activity deregulation
  44–5
 antigen presentation 74–5
 antithrombotic/antiaggregant
  activity 211, **212**, 213
 Behçet's disease 209–11, **212**, 213–14
 cyclophosphamide action 323–4
 cytotoxicity 75–6
 damage 39
 dysfunction in Behçet's disease 218–
  19
 fibrinolytic activities 213–14
 functions *209*
 glomerular 71
 hemostatic function 209–11, **212**,
  213–14
 juvenile dermatomyositis 378
 lesion prevention 39
 leukocyte interaction 215–19
 myeloperoxidase deposition 161
 phagocyte interaction 39–40
 PR3 deposition 161
 procoagulant activities 209–11
 prostacyclin release 73
 surface binding of molecules 210
 T cell interaction 39–40, 75
 tissue specific antigens 74
 vascular disease pathogenesis 65
 vascular tone 214–15
 vasculitis 68–9
Endothelial derived relaxing factor, *see*
 Nitric oxide
Endothelial leukocyte adhesion
 molecule (ELAM–1) 74, 215
Endothelial-selectin, *see* E-selectin
Endothelin 214–15
 vascular endothelial damage marker
  89
Endothelin-1 214

Endothelin-1,2 218
Endothelium, *see* Vascular
    endothelium
Eosinophilic angiitis 26
    histopathology *29*
Epitheloid cells 9
Epstein-Barr virus 286, 386
Erythema elevatum diutinum (EED)
    302–3
Erythrocyte sedimentation rate
    systemic vasculitis 365–6
    Takayasu's arteritis 183–4
    Wegener's granulomatosis 156
Essential mixed cryoglobulinemia 236,
    283–4
    circulating immune complexes 287,
        288–90
    clinical signs 283–4
    complement 289
    etiology 285
    exacerbation 290
    hepatitis B virus 286
    hepatitis C virus 286–8
    IgM rheumatoid factor 285
    immune complex disease 285
    immune proliferative disorder 285
    lymphoproliferative disorders 287
    malignant B cell lymphoma risk 285
    organ-specific autoantibodies 290–1
    pathogenetic factors 288–91
    spontaneous remission 290
    viral etiology 285–8
Ethmoidal cells **100**, *102*, **102**

F(ab')₂ fraction binding 76
Factor V 210
Factor VIII 379
Familial Mediterranean fever 202,
    412–15
    amyloidosis 412–13
    childhood 376
    chromosome 16 DNA markers 413
    clinical signs 412
    colchicine 413
    nephrotic syndrome 412, 413
    prednisone 415
    protracted febrile myalgia 414–15
    renal aneurysm **414**

vasculitis incidence 413–14
Felty's syndrome 54
Fibrin deposition 72
Fibrinogen 405
Fibrinolytic activity, clotting cascade
    72
Fibrinopeptide A 379
Fibromuscular dysplasia 259
fibronectin binding site on VLA–4 95

Gammaglobulin, intravenous 381
    Kawasaki syndrome 386, 395
Gangrene 267
Gastrointestinal vasculitis 128, 233
Giant cell angiitis 26
    *see also* Granulomatous angiitis
Giant cell arteritis 171–2
    American College of Rheumatology
        classification 32
    angiitis of the central nervous
        system 252, 255–9
    clinical presentation 175–6
    corticosteroid therapy 177
    cyclosporin A 319
    diagnosis 176–7
    disseminated 236
    elderly people 171
    epidemiology 172–3
    etiology 171–2, 173–4
    foreign-body giant cell reaction
        174–5
    HLA-DRB1 alleles 173
    inflammatory response 174
    interdigitating reticulum cells 174
    internal elastic membrane
        calcification 175
    lung vasculitides 233–6
    macrophage infiltration 174, 175
    methotrexate 319
    morphology 174–5
    ophthalmic complications 176
    pathological characteristics *30–1*
    polymyalgia rheumatica association
        173
    prednisolone 177, 178
    prognosis 178
    racial incidence 172
    rheumatoid arthritis association 173

sex differences 172–3
steroid therapy 176
T cell HLA-DR antigen expression 174
T cell infiltration 174
temporal artery biopsy 176
treatment 177, 319
    adverse effects 177–8
    vascular damage 173–4
Giant cell vasculitis 26
Giant cells 9
    foreign body 174, 175, 186, 249
    Langhans type 175, 186, 249
    myogenic 186
    Takayasu's arteritis 186, **190**
    Wegener's granulomatosis 155
Glomerular nephropathy 127
Glomerulonephritis
    rapidly progressive 122, 370
    in Schönlein–Henoch purpura 406
β-glucuronidase antibodies 54
GlyCAM–1 86
β2 glycoprotein 1 72
Glycosamino-glycans 211
Goodpasture's syndrome 122, 224–5
    alveolar hemorrhage 225
    plasma exchange 362
    Schönlein–Henoch purpura simulation 404
Granule proteases 43
Granulocytes
    karyorrhexis 33, **35**
    margination 112
Granuloma
    systemic vasculitis 363
    Wegener's granulomatosis *102*, *103*
Granulomatous angiitis 26, *28*
    allergic overlapping **28**
    histopathology *29*
    primary of the central nervous system 252
Granulomatous aortitis 26
Granulomatous disease, overlapping **28**
Granulomatous vasculitis 9
Group A β-hemolytic streptococcus 397
    Lancefield groups 398

90 kDa heat shock protein 311
HBeAg/anti-HBeAb seroconversion 351
Hemolytic streptococci 10
*Hemophilus influenzae* 357
*Hemophilus parainfluenzae* 299
Henoch–Schönlein purpura
    abdominal symptoms 295
    alveolar hemorrhage 236–7
    American College of Rheumatology classification 32
    angiitis of the central nervous system 255
    arthralgia 404
    C3d levels 297
    childhood 376, 403–10
    clinical manifestations 403–4
    clinicopathologic correlations 406
    complement cascade activation 40
    diffuse proliferative endocapillary glomerulonephritis 406
    end-stage renal failure 409
    endo-/extra-capillary glomerulonephritis 406
    familial Mediterranean fever 413
    fibrinogen deposits 405
    focal/segmental glomerulonephritis 406
    Goodpasture's syndrome simulation 404
    hypertension 408
    idiopathic IgA nephropathy relationship 297–8
    IgA
        deposits 405
        rheumatoid factors 297
        system activation 295, 296
    mesangiopathic glomerulonephritis 406
    methylprednisolone 409, *410*
    mucosal infection 298–9
    nephritis 403–10
    pathogenesis 297
    pathology *30–1*, 405–6
    plasma exchange 410
    prednisone 409
    pregnancy 408
    prognosis 237, 296, 406–8

Henoch–Schönlein purpura *contd*
  properdin levels 297
  protein-losing enteropathy 296
  proteinuria 408
  pulmonary vasculitis 237
  rash 295, 404
  renal disease 296
  renal manifestations 404–5
  renal transplant 299
  serum complement abnormalities
    297
  small-vessel necrotizing vasculitis
    295–9
  steroid therapy 237, 299
  testicular torsion 296
  treatment 299, 408–10
  *vs.* microscopic polyarteritis 139
Hepatitis, cryoglobulinemic 284
Hepatitis A virus 285
Hepatitis B virus 10, 24
  essential mixed cryoglobulinemia
    286
  polyarteritis nodosa (PAN)
    association 121–2, 124, *125,*
    129–30, 332–3, 337–8, 550–1
  surface antigen 111
  vaccines 129
Hepatitis C virus 10
  essential mixed cryoglobulinemia
    286–8
  polyarteritis nodosa (PAN) 124
Hepatotropic viruses 285–8
Herpes zoster infection 249
High walled endothelium (HEV) 86
HIV infection 10
  p-ANCA pattern 50
  polyarteritis nodosa (PAN) 124
HLA alleles 74
HLA-B5 antigen in Behçet's syndrome
    203, 207, 422
HLA-B51 antigen in Behçet's
    syndrome 202, 417
HLA-DR+ cells in Behçet's disease 208
HLA-DR antigen expression in giant
    cell arteritis 174
HLA-DRB1 alleles in giant cell arteritis
    173
HTLV–1 124

HTLV-III 249
Human herpes virus-6 386
Humanized anti-T cell antibody
    treatment 370–1
Hydroxyl radicals 40
Hypersensitivity, bacterial 10
Hypersensitivity angiitis 24
  angiitis of the central nervous
    system 255
  immune complex 139
Hypersensitivity vasculitis 10, 136
  American College of Rheumatology
    classification 32
  pathological characteristics *30–1*
Hypertension 127, 128
  polyarteritis nodosa (PAN) 131
  renovascular 182
Hypocomplementemia 40
  urticaria association 299
Hypocomplementemic urticarial
    vasculitis, *see* Urticarial vasculitis,
    hypocomplementemic
Hypoxanthine–guanine
    phosphoribosyl transferase
    mutants 324

ICAM-1 74, 85, 215, 216
  disease levels 89–90
ICAM-2 215
Idiopathic crescentic
    glomerulonephritis 225
  necrotizing 50, 52
    anti-MPO 53, 55, 59, **60**
IgA
  deposits in Schönlein–Henoch
    purpura 405
  mesangiopathy 296
  role in Henoch–Schönlein purpura
    298–9
IgA nephropathy
  familial 298
  Henoch–Schönlein purpura
    relationship 297–8
  mucosal infection 298–9
IgG
  antineutrophil cytoplasmic
    antibodies 387–8

B cell secretion in Kawasaki
   syndrome 387
complexes 72
deposits in juvenile
   dermatomyositis 378
Wegener's granulomatosis 158
IgM
   antineutrophil cytoplasmic
      antibodies 387–8
   B cell secretion in Kawasaki
      syndrome 387
   deposits in juvenile
      dermatomyositis 378
ILAM expression 43
Immune complex 24
   Churg–Strauss syndrome 123
   circulating 39, 40
      vasculitis 236
   classical PAN 122
   clearance 41
   complement binding 41
   deposition 24, 40–1
      in pulmonary vasculitides 223
   glomerulonephritis 284
   hypersensitivity angiitis 139
   mixed cryoglobulin 300
   perivascular deposits 40
   plasma exchange clearance 350
   solubilization capacity 289
   vessel wall deposition 40
Immune complex-mediated
      glomerulonephritis, alveolar
      hemorrhage 225
Immune system, classical pathway 41
Immune vascular damage mechanisms
   39
Immunoglobulin
   extracellular domains 85
   intravenous in Wegener's
      granulomatosis 165
   lung deposits in smokers 241
   superfamily 84, 96, 215
      adhesins 89–91
   receptors 85
   see also IgA; IgG; IgM
Immunoprecipitation inhibition 289
Immunosuppressive therapy 3, 25
   cyclophosphamide effects 324

systemic vasculitis 362–3
[111]Indium-labeled autologous
   polymorph scans 366
Infection 9
Inflammation 87
Inflammatory bowel disease
   angiitis of the central nervous
      system 257
   lung vasculitis 241
Inflammatory cell infiltrate 26
Inflammatory disease 96
Inflammatory mediators 311
Inflammatory myopathy 66
Inflammatory response 83
   inappropriate 87
Influenza B virus 357
Inhibition of immunoprecipitation 289
Injury mechanisms 39–45
Integrin-binding 96
Integrins 83–4, 84–5, 96, 215
Intercellular adhesion molecule, *see*
   ICAM-1
Interferon-α
   HBV associated polyarteritis nodosa
      (PAN) 338–9
   juvenile dermatomyositis 378
Interferon-gamma 74, 75
Interleukin-1 (IL-1) 43, 44
   Behçet's disease 217, 218
   Kawasaki syndrome 387
Interleukin-2 (IL-2), gene inhibition
   311
Interleukin-2 receptor (IL-2R)
   systemic vasculitis 365
   T cells in giant cell arteritis 174
   Wegener's granulomatosis 156, 158
Interleukin-6 (IL-6)
   Behçet's disease 217
   Wegener's granulomatosis 158
Interleukin-8 (IL-8) 159
Intravenous drug use 129
IVIg therapy for systemic vasculitis
   369–70

Juvenile dermatomyositis 376, 377–9,
   **380**, 381–2
   complement C3 deposits 378
   corticosteroids 381

Juvenile dermatomyositis *contd*
  cutaneous lesions 379, **381**
  cyclophosphamide 382
  cyclosporin 382
  endothelial cells 378
  factor VIII 379
  fibrinopeptide A 379
  immunoglobulin deposits 378
  intravenous gammaglobulin 381
  livedo reticularis 379
  management 381–2
  methotrexate 382
  muscle pathology 378
  nail bed abnormalities 378–9, **380**
  nail-fold capillary loop
    abnormalities 378, **379**
  serial nail-fold microscopy 381
  vasculitis 377–9
Juvenile rheumatoid arthritis 376

Kaplan–Meier survival curves 322, **323**
Kawasaki syndrome 6, 75–6
  ANCA 385–6, 394
  angiotensin converting enzyme
    levels 394, 395
  antiendothelial antibodies 42
  antiendothelial cell antibodies
    (AECA) 385–6, 394
  aspirin 386, 395
  B cell IgG/IgM secretion 387
  cardiac abnormalities 394
  cardiovascular complications 385
  cathepsin G antibodies 388
  childhood 376, 385–8
    circulating immune complexes
      394
    clinical signs *393*
    Italian study 392–5
    laboratory studies 393–4
  cholesterol levels 394
  clinical features 384–5
  color Doppler echocardiography 394
  complement-dependent endothelial
    cell toxicity 387
  cytokine levels 387
  dipyridamole 386, 395
  endothelial activation antigens 387
  endothelial injury 387–8

  marker 395
  etiopathogenesis 386–8
  incidence 385
  infection association 386–7
  intravenous gammaglobulin 386,
    395
  laboratory investigations 385–6
  neutrophils 378
  prognosis for children 386
  T cell abnormalities 387, 394
  treatment 386
  urokinase 386
*Klebsiella aerogenes* 357
*Klebsiella pneumoniae* 350

La/SSB antibodies 278
Labeled leukocyte scan 11
Lactoferrin autoantibodies 54
Laminin 70
Large-vessel vasculitis
  angiography 112–14, **115**, 116
  Behçet's disease 114, 116
  cutaneous polyarteritis nodosa 303–
    4
  cyclosporin A 319
  livedo reticularis disorders 304–5
  methotrexate 319
  temporal arteritis 112–13
  treatment 319
Leprosy 5–6
Leu8, *see* L-selectin
Leukocyte
  adhesion molecules 43, 87
    VCAM/VLA–4 role 91
  cell adherence 87
  circulation 86, **87**
  elastase antibodies 52, 54
  extravasation to sites of
    inflammation 96
  integrins 85
  margination 112
  migration to infection source 86
  radiolabeled scanning 117
Leukocyte-selectin, *see* L-selectin
Leukocytoclastic vasculitis 24, 26, 278,
  279
  complement cascade activation 40
  complement system 41

erythema elevatum diutinum (EED)
302
histopathology *29*
immunoglobulin deposits 40
LFA–3 expression 216
Livedo reticularis 126
cholesterol emboli 304
disorders 304–5
juvenile dermatomyositis 379
polyarteritis nodosa (PAN) 399
skin biopsy 304–5
with ulceration 304
Livedo vasculitis 304
Lung
disease 8
granuloma 8
infection 240
infiltrate 240
obstructive syndrome 240
Wegener's granulomatosis 103, *104*,
**104–5**, 106, 107, 108
Lung vasculitides
alveolar hemorrhage 224–5
Behçet's syndrome 237
Churg–Strauss syndrome 230–3
connective tissue disease 240
cryoglobulinemia 236
diagnosis 224–5
giant cell arteritis 233–6
Henoch–Schönlein purpura 237
hypocomplementemic urticarial
vasculitis 241
inflammatory bowel disease 241
pulmonary artery aneurysms 237,
**238**, 239 **239**
systemic lupus erythematosus 240
systemic rheumatoid vasculitis 240
Wegener's syndrome 225–7, **228–9**
Lupus anticoagulant 72, 259
Lupus nephritis 69–71
laminin 70
methylprednisolone i.v. 311
pulse cyclophosphamide 320
sera reactivity 77
von Willebrand factor 70
Lymph node homing receptor, *see* L-
selectin
Lymphocytes

adhesion molecules in circulation 87
homing 87
trafficking to target tissues 96
*see also* B cells; T cells
Lymphocytic vasculitis 26, 278
histopathology *29*
Lymphoid granulomatosis 257, 259

Macrophage infiltration in giant cell
arteritis 174, 175
Magnetic resonance angiography 114
Magnetic resonance imaging (MRI),
Wegener's granulomatosis 99,
**100**, 101, **102**, 103, 107–8, 156
Malignant angioendotheliomatosis
**258**, 259
Maxillary sinuses, Wegener's
granulomatosis **100**, *102*, **102**, 103
Membrane attack complex (MAC) 41,
45
Membrane cofactor protein CD59 45
Membrane-associated adhesion
molecules 43
Methotrexate 16, 325
giant cell arteritis 319
juvenile dermatomyositis 382
large-vessel vasculitis 319
Takayasu's arteritis 319
Wegener's granulomatosis 163
Methylprednisolone
intravenous administration 311
microscopic polyarteritis 142
pulse 15
Schönlein–Henoch purpura 409, *410*
MHC antigens 216
Microscopic polyangiitis 121, 122–3
anti-MPO 53
classification *148*
definition 138–9
feature overlap with Wegener's
granulomatosis 149–50
involving respiratory tract 151
pulmonary vasculitides *223*
*vs.* Wegener's granulomatosis 149,
151
Microscopic polyangiitis/arteritis 50,
52, 139
Microscopic polyarteritis 135–6

Microscopic polyarteritis *contd*
  alveolar vessel involvement 141
  ANCA 141
  azathioprine treatment 318
  cyclophosphamide treatment 318
  cyclophosphamide/steroid regime
    142
  definition 138–9
  escalation therapy 142
  fibrin deposits 138
  frequency of presentation 141
  incidence 315
  nodosa, *see* Microscopic polyangiitis
  c-PAN comparison 139–41
  prognosis 141, 143
  pulmonary renal syndrome 140
  renal involvement 142
  renal limited vasculitis and
    polyarteritis 136–8
  renal vasculitis 137–8
  segmental necrotizing
    glomerulonephritis 136, 137, 138
  small vessel vasculitis 138–9
  steroid therapy 318
  systemic small vessel vasculitis 137
  therapy 141–2
  treatment 317–18
  *vs.* classical polyarteritis nodosa
    139–41
  *vs.* cryoglobulinemic vasculitis 139
  *vs.* Henoch–Schönlein purpura 139
  *vs.* Wegener's granulomatosis 141
Monoclonal-antibody therapy,
  Wegener's granulomatosis 165–6
Mononeuritis multiplex 126, 231–2
  rheumatoid vasculitis 267, 268
  Sjögren's syndrome 277
Motor deficiency in polyarteritis
  nodosa (PAN) 126
Moyamoya disease 259
Mucocutaneous lymph node
  syndrome, *see* Kawasaki
  syndrome
Muscle weakness, diagnostic approach
  **56–8**
Myalgia
  Churg–Strauss syndrome 231
  polyarteritis nodosa (PAN) 126

*Mycoplasma gallisepticum* 249
Myeloid granular protein
  autoantibodies 52–4
Myeloperoxidase 44
  antibodies 10, 51, 52, 123, 363–4, 388
    assays 55
    Churg–Strauss syndrome 223
    positivity **56, 58**
  autoantibodies 53–4
  autoantigen therapeutic
    immunoabsorption 368
  deposition on endothelial cells 161

Nails, juvenile dermatomyositis 378–9,
  **380**
Nasal cavity, Wegener's
  granulomatosis **100**, 101, **102**, 103,
  107
NCAM 215
Necrotizing angiitis 24, 26
  overlapping **28**
  systemic 352
Necrotizing glomerulonephritis 136
  crescentic 356
  segmental 136, 137, 138
Necrotizing vasculitis 9, 22, 23
  cyclophosphamide 323–4
  histopathology *29*
  mortality **323**
  peripheral nerve 255
  systemic 22, 24, 225, 256, 320
Neutrophil
  activation
    cytokines 44
    vascular wall damage 43
  ANCA activation 44, 388
  ANCA-associated necrotizing
    vasculitis/glomerulonephritis **60**
  binding to vessel wall 43
  cytoplasmic antigens 364
  granule constituents *51*
  influx in inflammatory response 83
  lysosomal enzyme 43
    release 55
  reactive oxygen species production
    55
  recruitment 40
  TNFα priming 55, 59

Neutrophilic vasculitis 26
Nitric oxide 211, 214
Nitrogen mustard 310
Non-steroidal anti-inflammatory
  drugs 300

Orbit in Wegener's granulomatosis
  **100**, *102*, 103
Orchitis 122, 128
Overlap syndrome 121
Oxygen radical generation 43

Paranasal sinuses in Wegener's
  granulomatosis 101, 108
Parvovirus 10
Parvovirus B19 124
Penicillin 400
Pentoxyphyline 16
  rheumatoid vasculitis 271
Periarteritis nodosa 22, *23*, 24
  *see also* Polyarteritis nodosa
Periglomerulitis granulomatosa 155
Peripheral blood mononuclear cells
  288
Peripheral neuropathy **56–8**
Plasma exchange 15
  Churg–Strauss syndrome 319, 337,
    342, 351
  Goodpasture's syndrome 362
  HBV associated polyarteritis nodosa
    (PAN) 338
  immunosuppressive agent
    combination 325
  polyarteritis nodosa (PAN) 337, 342,
    351
  Schönlein–Henoch purpura 410
  side-effects 347–8
  systemic necrotizing angiitis 353
  Wegener's granulomatosis 316
Plasmapheresis 325–6
Plasminogen activator 210
Plasminogen activator-
    inhibitor:antigen levels 210
Plasminogen activator-inhibitor-1
    (PAI-1) 210, 214
  plasminogen activator balance 210
Plasminogen activator-inhibitor-2
    (PAI-2) 210

Platelet
  adhesion and clotting cascade 72
  alpha granules 86
  IIb/IIIa integrin 85
Platelet-activating factor (PAF) 210
Platelet-selectin, *see* P-selectin
Pleura, Wegener's granulomatosis 103,
  **105**
*Pneumocystis carinii* pneumonia 163,
  227
Polyarteritis 3
  and renal limited vasculitis 136–9
  survival time 310
Polyarteritis nodosa (PAN) 7, 8, 9, 22,
  121
  acute phase response 399
  American College of Rheumatology
    classification 32
  aneurysms 109, 110, 111
  angiitis of the central nervous
    system 252, 255–9
  angiography 130
  anti-MPO 54, 55, 61
  antiviral treatment 333
  arteriography 401
  asthma association 339
  azathioprine treatment 318
  biological analysis 130
  cardiovascular manifestations 127–8
  childhood 398–401
  Churg–Strauss syndrome 123
  classical (c-PAN) 109–12, 121, 122
    ANCA 141
    angiography 109–12
    features at presentation 140
    frequency of presentation 141
    organ involvement 140
    prognosis 141, 143
    *vs.* microscopic polyarteritis
      139–40, 140–41
  classification 121–3
  clinical aspects 124, *125*, 126–30
  clinical features 303–4, 339, *340*
  corticosteroids 400
  cutaneous 303–4, 401
  cyclophosphamide treatment 318,
    332, 336, 337
    duration 352

Polyarteritis nodosa (PAN) *contd*
  disease activity 336
  epidemiology 124
  erythematous nodules 398–9
  etiology 124
  familial Mediterranean fever 413–14
  five factors score *131*
  GI involvement 128
  group A streptococcal infection 398
  hepatic manifestations 130
  hepatitis B virus association 121,
    122, 124, **125**, 129–30, 334, 338–9,
    550–1
  histopathology **29**
  incidence 315
  kidney involvement 127
  laboratory findings 341–2
  livedo reticularis 399
  lung manifestations 128
  markers 32
  microscopic polyangiitis 121, 122–3
  morphological changes 137
  mortality 346–7, *348–9*
  muscle pain 399
  neurological symptoms 126
  ocular manifestations 129
  pain 231–2
  pathological characteristics *30–1*
  pathology 124
  penicillin 400
  plasma exchange 337, 342, 351
  prednisolone long-term therapy 335,
    337, 338
  prednisone treatment 339
  prognosis 131
  recurrent disease 400–1
  rheumatic fever association 401
  rheumatic symptoms 126
  rheumatoid vasculitis overlap 270
  severity criteria 131
  skin manifestations 126–7
  steroid therapy 318
  streptococcal 397
    disease flares 401
  therapy trial results 339–50, *348–9*
  treatment 304, 317–18, 332–4, *340*,
    341–5, *348–9*, 347–53
    side effects 347, 350

  vasculitic illness 401
  viral etiology 10
  *see also* Periarteritis nodosa
Polymorphonuclear cells (PMN)
  ANCA-activated 159, 364
  lysis 161
  Wegener's granulomatosis 159, **160**,
    161
Polymyalgia rheumatica
  azathioprine treatment 324
  clinical signs 176
  corticosteroid therapy 177
  in giant cell arteritis 173, 175, 176
  prednisolone treatment 324
  synovitis 176
Polymyalgia/temporal arteritis 10
Polyvasculitis 9
Post-acute phase management 312
Poststreptococcal reactive arthritis 397
PR3-ANCA marker 150
  Wegener's granulomatosis 156
  *see also* Proteinase 3 (PR3)
Prednisolone
  azathioprine combination 324
  Churg–Strauss syndrome long-term
    therapy 335, 337, 338
  cyclophosphamide combination
    320–3, 326, 352
  giant cell arteritis 177, 178
  polyarteritis nodosa (PAN)
    syndrome long-term therapy 335,
    337, 338
  polymyalgia rheumatica 324
  systemic vasculitis treatment 366–8
  toxic effects *322, 323*
  Wegener's granulomatosis
    treatment 316
Prednisone
  Churg–Strauss syndrome treatment
    339
  familial Mediterranean fever 415
  polyarteritis nodosa (PAN)
    treatment 339
  Schönlein–Henoch purpura 409
  side-effects 347–8
Properdin 297
*Propionibacterium acnes* 386
Prostacyclin (PGI$_2$) 211, 214, 218

inhibition 72
release from endothelial cells 73
rheumatoid vasculitis 271
Protein C 72
Protein S 72
Proteinase 3 (PR3) 44, 51, 388
autoantibodies 52–3
deposition of endothelial cells 161
recognition 77
translocation to cell membrane 161
Proteinase, *Streptococcus pyogenes*
production 398
Proteolytic enzymes 40
Protracted febrile myalgia 414–15
Pulmonary artery aneurysms
Behçet's syndrome association 237
pulmonary vasculitis 237, **238**, 239–
40
rupture 237, 239
Pulmonary hemorrhage 225
Pulmonary hypertension 239
Pulmonary renal syndrome 122, 140
Pulmonary vasculitides 222–3
immune complex deposition 223

Radical scavenging, limited Wegener's
granulomatosis 317
Radiolabeled leukocyte scanning 117
Rapidly progressive
glomerulonephritis 122, 370
Raynaud's phenomenon 277, 300
Reactive oxygen species production 55
Relapse rate 14, 16
Renal allograft rejection 76
ICAM-1 90
VCAM-1 91
Renal failure, polyarteritis nodosa
(PAN) 131
Renal limited vasculitis and
polyarteritis 136–9
Respiratory syncitial virus 357
Reticuloendothelial system, circulating
immune complexes 289
Reticulum cells, interdigitating 174
Retrovirus 386–7
Rheumatic arteritis 24
Rheumatic disease, vascular lesions 22
Rheumatic fever

group A B-hemolytic streptococcus
397
vasculitis 9
Rheumatoid arthritis 26
AECA 66
p-ANCA pattern 50
complement cascade activation 40
complement deposit 40
giant cell arteritis association 173
ICAM-1 90
immunoglobulin deposits 40
lactoferrin autoantibodies 54
late consequences 273
perivascular cell infiltrate 40
Takayasu arteritis association 185
vascular involvement 267
VCAM-1 91
VCAM/VLA-4 interactions 92
Rheumatoid factor 43, 271–2
IgA 297
IgMk 285
monoclonal 279–81
Wegener's granulomatosis 156
Rheumatoid vasculitis 267–8
antiendothelial cell antibodies
(AECA) 272
antinuclear antibodies 272
azathioprine 271
cardiac involvement 268
clinical features 267–8
current status 268–70
cyclophosphamide 271
cytokines 273
etiopathogenesis 271–2
IgG/IgM rheumatoid factors 271,
272
incidence 315
mononeuritis multiplex 267, 268
mortality 269, 270
nail edge/nail fold lesions 272
nodules 268
pentoxyphyline 271
peripheral gangrene 267
polyarteritis nodosa overlap 270
prostacyclin infusion 271
skin lesions 268
small vessel involvement 272–4
subclinical disease 272–4

Rheumatoid vasculitis *contd*
  systemic 7, 240, 269, 270, 318–19
    organic involvement 270
    thalidomide 271
    therapy 270–1
    treatment 318–19
*Rickettsia* 386
Ro/SSA antibodies 278, 279

Sarcoidosis, angiitis of the central
    nervous system 252, 257
Scarlet fever 397
Schönlein–Henoch purpura, *see*
    Henoch–Schönlein purpura
Scleroderma 3
  AECA 66
Selectin 84, 85–6, 96, 215
  cytokine induction 43
  E-selectin 14, **84**, 85, 86, 215
    blood levels *90*
    disease states 88–9
    expression 86
    sepsis levels 89
  L-selectin 85, 86, 215
    blood levels *90*
    isoforms 88
    ligands 86
    lymphocyte homing 87
    release 88
  P-selectin 85, 86, 215
    blood levels *90*
    disease states 88
    expression 86
Sepsis, E-selectin levels 89
Serum sickness 41
  angiitis of the central nervous
    system 255
  vascular lesions 135
Sialyl-Lewis recognition 86
Sister chromatid exchange 324
Sjögren's syndrome, primary 277
  acute necrotizing vasculitis 278
  classification 278
  clinical presentation 277–8
  cryoglobulins 279
  endarteritis obliterans 278
  La/SSB antibodies 278
  leukocytoclastic vasculitis 278, 279

lymphocytic vasculitis 278
  manifestations 277
  monoclonal rheumatoid factor cross
    reactive idiotypes 279–81
  mononeuritis multiplex 277
  purpura 277, 278
  Ro/SSA antibodies 278, 279
  skin ulceration 228, 277
  type II cryoglobulinemia 279–81
  urticaria 277
Skin disease, descriptions 5–6
Skull, Wegener's granulomatosis *102*,
    103
Small-vessel cutaneous angiitis 26
Small-vessel leukocytoclastic vasculitis
    markers 33, **35**
Small-vessel necrosis 33, **35**
Small-vessel necrotizing vasculitis
    294–303
  benign hypergammaglobulinemic
    purpura of Waldenström 302
  causes 294–5
  cryofibrinogenemia 301–2
  cryoglobulinemia 300–1
  erythema elevatum diutinum (EED)
    302–3
  Henoch–Schönlein purpura 295–9
  urticarial vasculitis 299–300
Small-vessel vasculitis 10
  ANCA 363
  angiography 112
Smoking, immunoglobulin lung
    deposits 241
Sneddon's syndrome 304
Spondyloarthropathy 185
*Staphylococcus aureus* 357, 358
  toxic shock syndrome toxin
    producing 387
Steroids
  Churg–Strauss syndrome 319
  epidemic 269
  microscopic polyarteritis treatment
    318
  polyarteritis nodosa treatment 318
  Wegener's granulomatosis
    treatment 316
  *see also* corticosteroids
Streptococcus 397–402

hemolytic 10
molecular mimicry 401–2
pyrogenic exotoxin producing 387
*Streptococcus pyogenes* 397–8
*Streptococcus sanguis* 386
Streptococcus type 5M protein 401
Streptokinase 398
Sulfonamides
erythema elevatum diutinum (EED) 303
hypersensitivity 9
of vascular lesions 135
Synovitis, polymyalgia rheumatica 176
Systemic lupus erythematosus (SLE) 3, 26
AECA 66, 67–74
alveolar hemorrhage 225
p-ANCA pattern 50
aneurysms 111
angiitis of the central nervous system 256
antiendothelial antibodies 42
antiphospholipid syndrome 71
childhood 376
CNS disease 73–4
complement cascade activation 40
endothelial damage 65
genetic of AECA 74
immune complexes 41–2
lung vasculitis 240
lupus nephritis 69–71
serum antigen recognition 77
vasculitis 68–9
VCAM-1 91
Systemic necrotizing angiitis 353
Systemic necrotizing arteritis
angiography 109–12
classical polyarteritis nodosa (PAN) 109–12
with granulomatosis 112
Systemic necrotizing vasculitides 22, 24
alveolar hemorrhage 225
angiitis of the central nervous system 256
pulse cyclophosphamide 320
Systemic rheumatoid vasculitis 7, 269, 270
azathioprine treatment 319

chlorambucil 319
cyclophosphamide 319
pulmonary involvement 240
treatment 318–19
Systemic sclerosis, childhood 376
Systemic vasculitis
adhesion molecule measurement 365
AECA and pathogenic mechanisms 74–7
alternative treatment 362–71
American College of Rheumatology classification 26, 32
ANCA in management 364–5
autoimmunity 363–4
azathioprine 321, 368
B cell function regulation 369–70
C reactive protein measurement 365
conventional treatment 366–8
cyclophosphamide
regimes 321–3
therapy 367–8
cytokine measurement 365
epidemics 6
erythrocyte sedimentation rate 365
granuloma formation 363
humanized anti-T cell antibody treatment 370–1
IL2-R measurement 365
immunotherapy 362–3, 368
[111]Indium-labeled autologous polymorph scans 366
IVIg therapy 369–70
management 364–6
organ function tests 365
relapse prediction 365
specific immunoabsorption 369
steroid therapy 367–8
Wegener's granulomatosis 356

T cells 11, 14, 17
abnormalities in Kawasaki syndrome 387, 394
endothelial cell interaction 40, 75
HLA-DR antigen expression in giant cell arteritis 174
humanized monoclonal antibodies 370–1

T cells *contd*
  immunomodulation function 368
  infiltration in giant cell arteritis 174
  mediated immune mechanisms 44
  polyclonal origin in giant cell
    arteritis 174
  systemic vasculitis 363
  Vβ8 chain receptors 401
  VCAM-1 interaction 92
  Wegener's granulomatosis 158, *159*
Takayasu aortitis, abdominal 182
Takayasu arteritis 181–2
  active phase 186, 189, **190**
  age at onset 185
  American College of Rheumatology
    classification 32
  anatomic types 181
  aneurysm 185, **186**, **187**
  angiitis of the central nervous
    system 255
  angiography 113–14, **115**
  aorta **187**, **188**, **189**
  aortic insufficiency 182
  arterial dissection 185
  biopsy 186, **189**
  classification 234
  clinical features 182–5
  complications 185
  cyclosporin A 319
  diagnostic criteria 182–5
  disease progression 235
  end stage 194, **195**
  erythrocyte sedimentation rate 183–
    4
  fibrosis 234
  geographical differences 113
  giant cell types 186, **190**, 233
  gross changes 185
  histology 234
  late phase 191, 194
  lung vasculitides 233, 234–5
  medial infarction 189, **192**
  methotrexate 319, 325
  musculoelastic tissue destruction
    189, **191**
  pathological characteristics *30–1*
  pathology 185–6, **187–8**, 189, **190**,
    191, **192–3**, 194, **195**

prognosis 185
progressive replacement fibrosis
  189, **191**, **193**
pulmonary artery involvement 115,
  **116**, 234–5
pulmonary vasculitides 222, *223*
racial differences 113
rheumatoid arthritis association 185
sex incidence 181
spondyloarthropathy association
  185
steroid therapy 235
subacute phase 194, **195**
surgical treatment 235
synonyms 181, *182*
topographic classification **183**
transmural sclerosis 191, **195**
treatment 319
*vs.* arteriosclerosis obliterans 194
Temporal arteritis 3, 24, 172
  angiitis of the central nervous
    system 255
  angiography 112–13
  in giant cell arteritis 175
  lung vasculitides 222, *223*, 235–6
  *see also* Giant cell arteritis
Temporal artery biopsy 176
Thalidomide 16
  rheumatoid vasculitis 271
Therapeutic apheresis 325–6
Therapeutic regimes *15*
Therapy 14–15
  aggressive 3, 12, 25
  angiography 117
  maintenance regimes 16
  pulse regimes 14, *15*
  strategies 312–13
  target and VCAM/VLA-4
    interactions 91–6
  of vasculitides 315–16
Thromboangiitis obliterans 26
  histopathology *29*
Thromboembolism 116, 117
Thrombomodulin 211, 213
  Behçet's disease 218
  vascular endothelial damage marker
    89
Thrombophlebitis, Behçet's disease 116

Thrombosis 71–3
  AECA role 72
  Behçet's disease/syndrome 207, 237
  pulmonary vasculitis 239–40
Thromboxane A$_2$ 211
Thromboxane B$_2$ 211
Tissue factor (TF) 209–10
Tissue injury, immune complex-
    mediated 24
Tissue plasminogen activator (tPA) 211
Tissue plasminogen activator-inhibitor
    (tPA-1) 72, 210
*Tolypocladium inflatum Gams* 325
Trimethoprim/sulfamethoxazole 162
  Wegener's granulomatosis 357, 359
    limited 317
Tuberculosis 227
  Takayasu arteritis 233–4
Tumor necrosis factor alpha (TNF-α)
  Behçet's disease 217, 218
  PR3 translocation to cell membrane
    162
  release by corticosteroids 311
  Wegener's granulomatosis 158, 159
Tumor necrosis factor (TNF) 43, 44
  Kawasaki syndrome 387

Ureteral stenosis 127
Urokinase 386
Urticarial vasculitis 26, 299–300
  hypocomplementemic 241, 300
  treatment 300

Vascular cell adhesion molecule, *see*
    VCAM
Vascular cell adhesion molecule-1, *see*
    VCAM-1
Vascular damage, antibodies 42
Vascular endothelium
  clotting cascade 72
  coagulation balance 72
  fibrin deposition 72
  hemostatic function 209–11, **212**,
    213–14
Vascular homeostasis 210
Vascular nephropathy 127
Vascular occlusion, Behçet's disease
  **219**

Vasculitis 3
  adult 6
  assessment 11–14
  blood vessels **27**
  classification 7–9, 22, 23, 24–6, 48, 49,
    315–16
    rational 9–11
  damage index (VDI) 13
  definition 21
  diagnosis 11
  early descriptions 4–5
  endothelial cells 68–9
  histologic types 26
  histopathologic specificity 26, 27–31,
    32–3
  history 4–7
  hypersensitivity 10
  major syndromes 26, 27
  manifestations 5
  primary 21
  rheumatic fever 9
  secondary 21
  SLE 68–9
  syndromes 24
  therapy 14–15
VCAM
  identification 92–3
  soluble 14
VCAM-1 **84**, 85, 215
  adhesion assays 95–6
  cytokine induction 43
  deletion mutants 93–4
  disease levels 90–1
  domain alignment 93
  expression 86
  receptor binding sites 95
  T cell interaction 92
  VLA-4 binding sites 93–6
VCAM/VLA-4 adhesion pathway 91,
    96
VCAM/VLA-4 interactions 91–6
  rheumatoid arthritis 92
Very late antigen-4, *see* VLA-4
Vidarabine 337, 350–1
VLA-4 **84**, 85
  binding sites in VCAM-1 93–6
  blockade 92, 96
  fibronectin binding site 95

von Willebrand factor (vWF) 13–14, 209, 210
  Behçet's disease 210
  lupus nephritis 70
  vascular endothelial damage marker 89
von Willebrand factor-antigen 218

Wegener's arteritis 112
  leukocyte margination 112
Wegener's granulomatosis 10, 22, 24, 25, 145
  alveolar hemorrhage 153, 225, 227
  American College of Rheumatology classification 32
  ANCA 156, 223
    as marker 49–50
    persistence 357–8
    proteinase 3 specificity 363
    role 159, **160**, 161
  ANCA-mediated vascular injury 159
  c-ANCA 147
    marker 225
    pattern 50–1
  aneurysms 111
  angiitis of the central nervous system 255
  anti-MPO 53
    specificity 61
  anti-PR3 55
  antibiotic use 357
  c-reactive protein 156
  classic 145, 147, 153–4
  classification 148–52
    criteria 147, *148*, *149*, 152
  clinical aspects 226–7
  clinical manifestations 152–4
  co-trimoxazole treatment 359
  computed tomography (CT) 99, 103, *104*, **104–5**, 106–7, 108, 227, **228–9**
  corticosteroid treatment 158
  cotrimoxazole treatment 158
  cyclophosphamide therapy 226, 316
  cyclophosphamide/glucocorticoid therapy 162
  cyclosporine A 163, 165
  cytokine levels 158
  cytokine production induction 159

  definitions 148–52
  diagnosis 99
  disease
    activity monitoring 156, **157**
    extent index 13, 147
    progression 155
  ELK triad 147–8
  ENT region 152, 154
  erythrocyte sedimentation rate 156
  feature overlap with microscopic polyangiitis 149–50
  features at presentation *140*, 141
  fibrosis 226
  fulminant 147, 153–4
  giant cells 155
  granuloma formation 157–9, *160*, 161–2, 226
  high-dose intravenous immunoglobulin 165
  histology/histogenesis 154–5
  IL-2R levels 156
  incidence 315
  incomplete 146–7
  indolent lesion 152
  infection association 357–8
  initial phase 157
  kidney pathology 155
  laboratory investigations 155–6
  latent 145–6
  leukocyte counts 155–6
  limited 151, 317
    immunosuppressive therapy 317
    radical scavenging 317
    trimethoprim/sulfamethoxazole 317
  lung
    infiltrates 226
    involvement 152, 153
    pathology 155, 226
    vasculitides 225–7, **228–9**
  macronecrosis 155
  magnetic resonance imaging (MRI) 99, **100**, 101, **102**, 103, 107–8, 156
  markers 32, **33**
  mastoiditis 152
  methotrexate 163, 325
  micro-organisms 357
  micronecrosis 155

monoclonal-antibody therapy 165–6
nasal mucosa ulceration 152
necrotizing crescentic
    glomerulonephritis 356
organ involvement 225
c-PAN comparison 139–41
pathological characteristics *30–1*
pathological process 159, **160**
pathology 226
patient follow-up 52–3
plasma exchange 316
polymorphonuclear cells (PMN)
    159, **160**, 161
PR3 autoantibodies 52
PR3-ANCA marker 150, 156
prednisolone treatment 316
prognosis 153–4, 162
pulmonary nodules 226
pulmonary vasculitides 150, 222, 223
pulse cyclophosphamide treatment
    163, *164*
relapse 52–3, 156, 356–9
renal failure 153
respiratory tract inflammation 356
rheumatoid factor 156

serum antigen recognition 77
signs 225
sinusitis 152, **153**
*Staphylococcus aureus* 357, 358, 387
steroid therapy 316
survival time 310
symptoms 152–4
systemic vasculitis 356
T cell activation 44
therapy 316
tracheobronchial lesions 226
treatment 162–3, *164*, 165–6
triad 145, 146
trimethoprim/sulfamethoxazole
    combination 162, 357, 359
two-phase course 152–4
upper respiratory tract involvement
    152
vasculitic phase 154
vasculitis 157–9, *160*, 161–2
vessel involvement 149–50, 154
viral infection 357
*vs.* microscopic polyangiitis 151
*vs.* microscopic polyarteritis 141
Weibel–Palade bodies 86